Annual Session of Scientific Papers - IMT ORADEA 2020

IOP Conference Series: Materials Science and Engineering
Volume 898

Oradea, Romania
28 - 29 May 2020

Editors:

Gavril Grebenisan
Alexandru-Viorel Pele
Florin Sandu Blaga

ISBN: 978-1-7138-3624-7
ISSN: 1757-8981

Printed from e-media with permission by:

Curran Associates, Inc.
57 Morehouse Lane
Red Hook, NY 12571

CURRAN ASSOCIATES INC. proceedings.com

Some format issues inherent in the e-media version may also appear in this print version.

Institute of Physics
Dirac House, Temple Back
Bristol BS1 6BE UK

Phone: 44 1 17 929 7481
Fax: 44 1 17 920 0979

techtracking@iop.org

Additional copies of this publication are available from:

Curran Associates, Inc.
57 Morehouse Lane
Red Hook, NY 12571 USA
Phone: 845-758-0400
Fax: 845-758-2633
Email: curran@proceedings.com
Web: www.proceedings.com

TABLE OF CONTENTS

IOP Conf. Series: Materials Science and Engineering **898** (2020) 011001 doi:10.1088/1757-899X/898/1/011001

Preface

The ANNUAL INTERNATIONAL SESSION OF SCIENTIFIC PAPERS - IMT Oradea 2020 is an International Conference which held, usually, in Felix Spa, near Oradea, in the North-Western part of Romania. We are delighted, proud, and honored to be your hosts every year at the scientific events of the Faculty of Managerial and Technological Engineering at the University of Oradea. But, this year, our Conference held virtually, due to the mass disruption caused by the coronavirus.

We had to find solutions so as not to affect the continuity in the annual event of the Conference, even if on March 16, 2020, a state of emergency was declared on the Romanian territory. The decision of the Romanian authorities was referred to as travel, transport, and accommodation restrictions, respectively, the closure of restaurants and the prohibition of public meetings, even the formation of groups of more than three people. Thus, we were morally obliged towards our fellow researchers, university professors, scientists, to find feasible solutions for the Conference. Therefore, we found that the best option is to change the format of the Conference to a virtual format, with the presentation of works online, live video-audio.

We prepared the schedule of the works presented by distributing fifteen minutes for the presentation of the one paper each and discussions on the topic of the article. The timing was strictly adhered to, and there were no delays or postponements.

The organizers had at their disposal eight different rooms in the Campus of the University of Oradea from where they coordinated the development of the Conference. Six members of the Scientific Committee of the Conference were permanently online, plus the two co-chairs of the Conference. We had eight office rooms at the disposal of the members of the Scientific Committee because personal contact was not allowed, as the social distance was restricted.

The authors presented their works from their office, from home or other personal locations. From time to time, during presentation sessions, we saved a few screenshots, images from the Virtual Conference, and kept them as pdf files, which we publish in the preliminary section of our volume. Permanently was online about 20-27 participants.

The platform used for the online meeting and presentation of the papers was Cisco Webex Meetings, which worked exemplary, without interruptions or distortions.

The Conference took place over two days: Thursday, May 28, and Friday, May 29, 2020, respectively, from 9:30 to 17:00, every day. The start of the meeting was 30 minutes ahead to have time for possible questions or concerns about how the session was going. The end of the session was extended by one hour compared to the schedule of the works, to include possible delays or postponements.

IOP Conf. Series: Materials Science and Engineering **898** (2020) 011001 doi:10.1088/1757-899X/898/1/011001

All the participants in our Conference expressed their enthusiasm to the organizers for the elegant way in which the Conference took place. All works were presented at the Conference. We had a smaller number of papers (i.e., 55 works), half of the number of presentations in 2019, but this is due to the coronavirus pandemic, which prevented the development of research activities. We cite here a message from a researcher in Hungary, present to the Conference, *Erdei Timotei, timoteierdei@gmail.com: "Dear Assoc. prof. Ph.D. Eng. Gavril Grebenisan!*

Due to the current circumstances, many conferences canceled. However, not only the „Annual Session of Scientific Papers "IMT Oradea 2020" international Conference not canceled, but it held at its usual high standards. Seeing how others solve particular problems serves is always valuable to us.

Thank You for this opportunity, we were glad to partake in this international event on the 28-29th of May 2020. I hope that we will meet again next year, in person!

I wish you and your colleagues good health!

"

After the publication of the papers in the IOP Conference Series: Materials Science and Engineering, we will ensure the printing of physical copies, in paper format, Conference's Proceedings, and we will send them by regular mail to the address provided by the first author.

Editorial Board

IOP Conf. Series: Materials Science and Engineering **898** (2020) 011001 doi:10.1088/1757-899X/898/1/011001

THE CONFERENCE'S EDITORS AND COMMITTEES

IOP Conf. Series: Materials Science and Engineering **898** (2020) 011001 doi:10.1088/1757-899X/898/1/011001

IOP Conf. Series: Materials Science and Engineering **898** (2020) 011002 doi:10.1088/1757-899X/898/1/011002

Presentations-first day of Conference IMT Oradea 2020

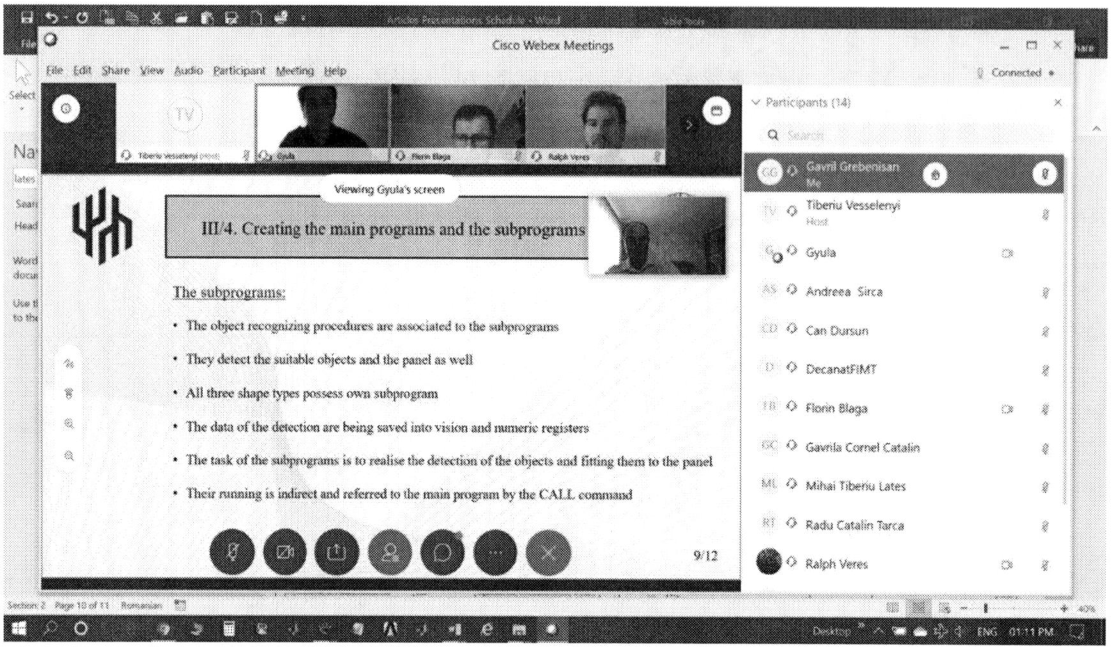

Presentations-second day of Conference IMT Oradea 2020

Ioan Szabo

Dan Moldovanu

IOP Conf. Series: Materials Science and Engineering **898** (2020) 011002 doi:10.1088/1757-899X/898/1/011002

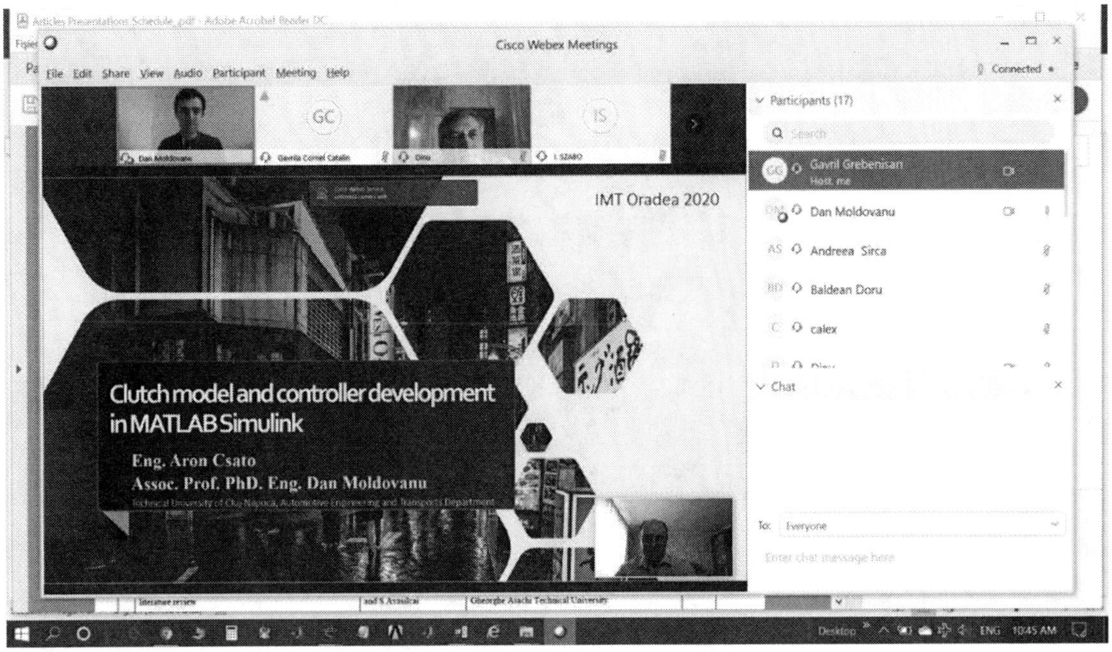

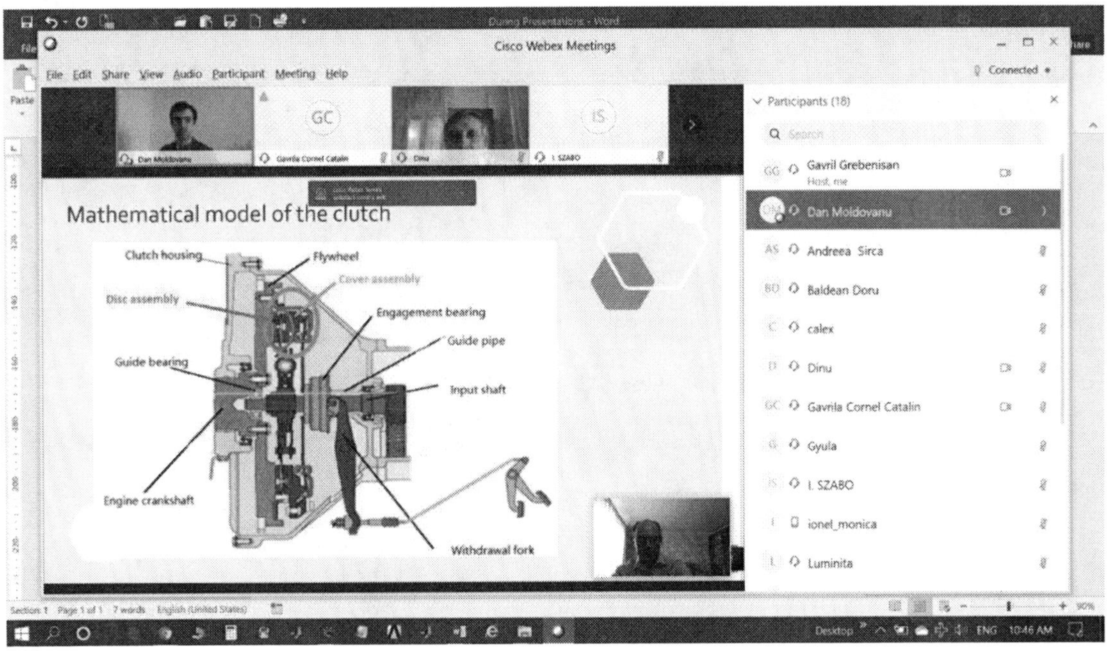

Second paper

IOP Conf. Series: Materials Science and Engineering **898** (2020) 011002 doi:10.1088/1757-899X/898/1/011002

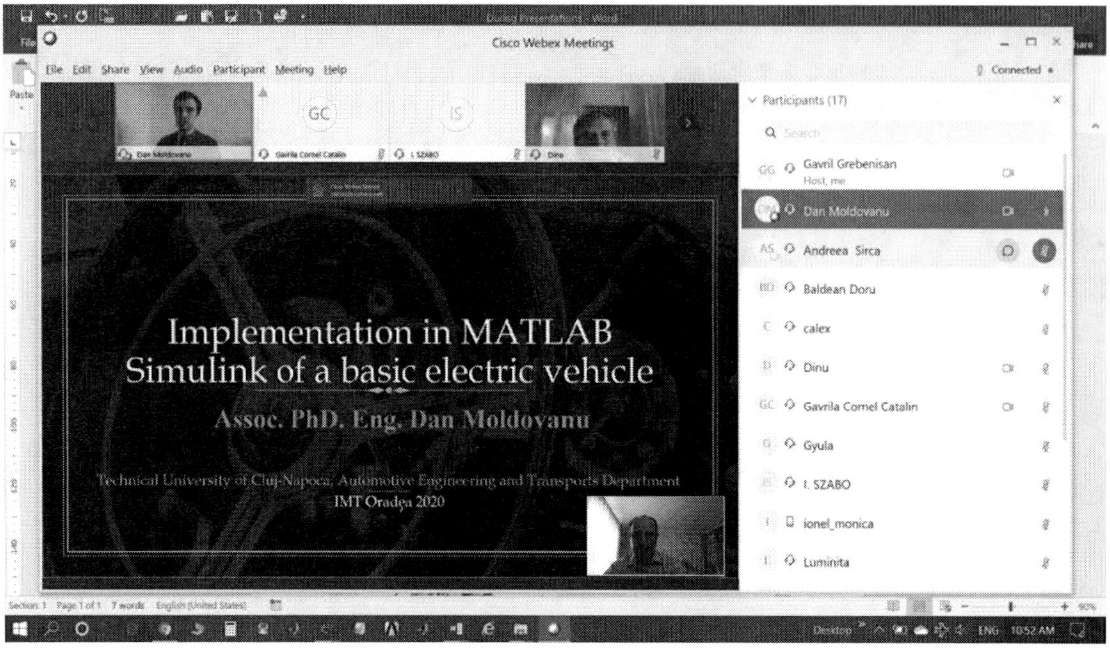

Baldean Doru

First

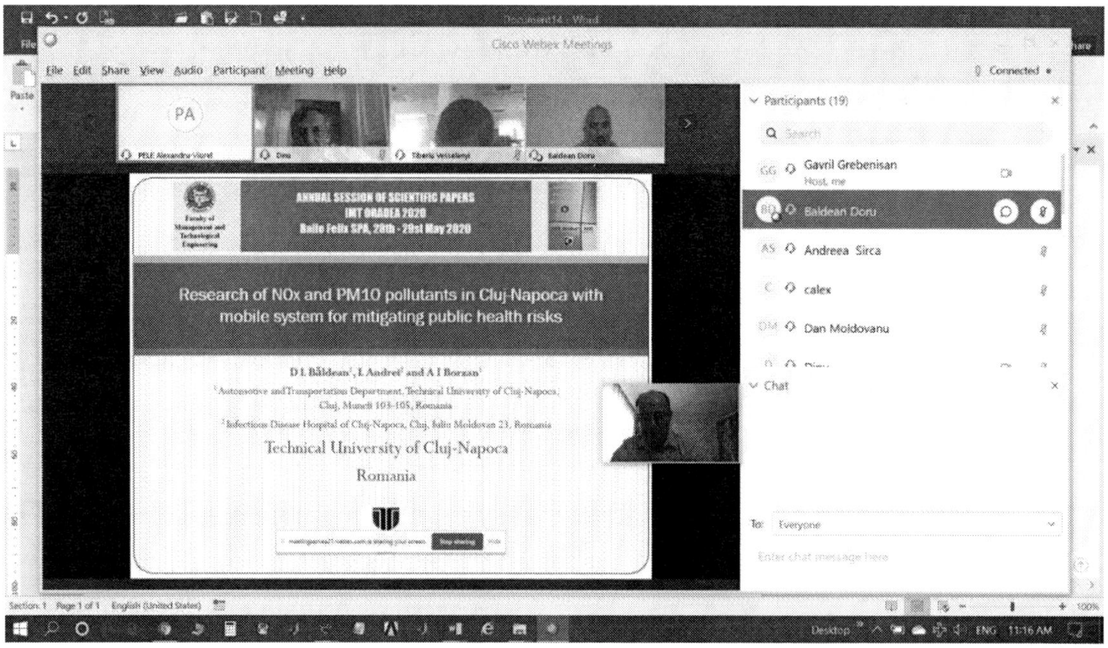

Catalin Alexandru

first

IOP Conf. Series: Materials Science and Engineering **898** (2020) 011002 doi:10.1088/1757-899X/898/1/011002

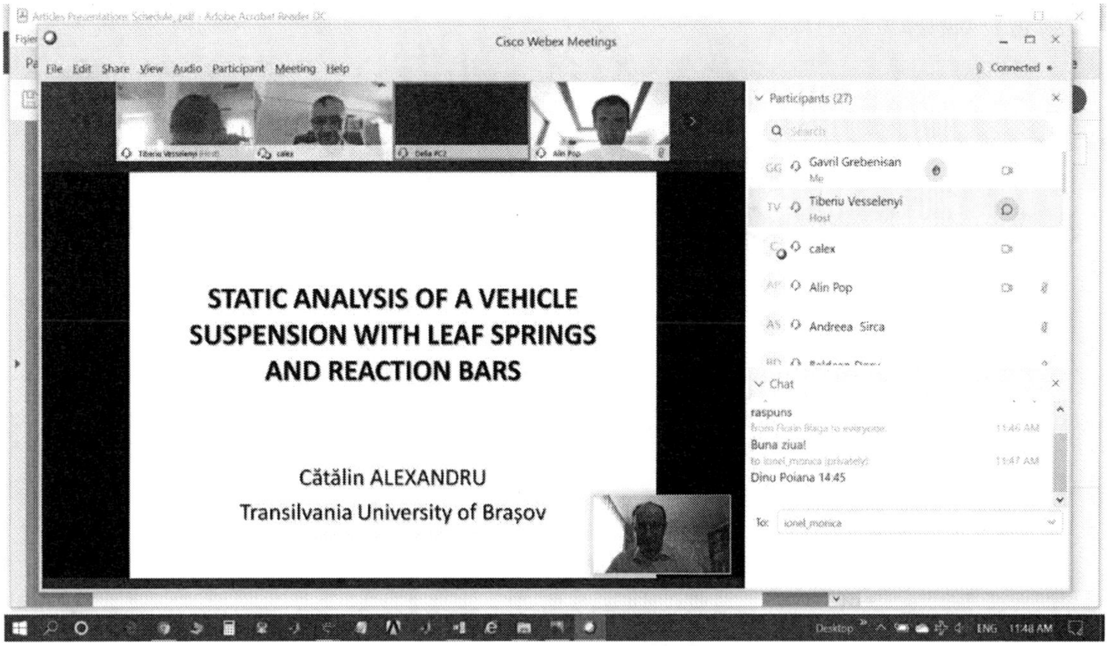

Second

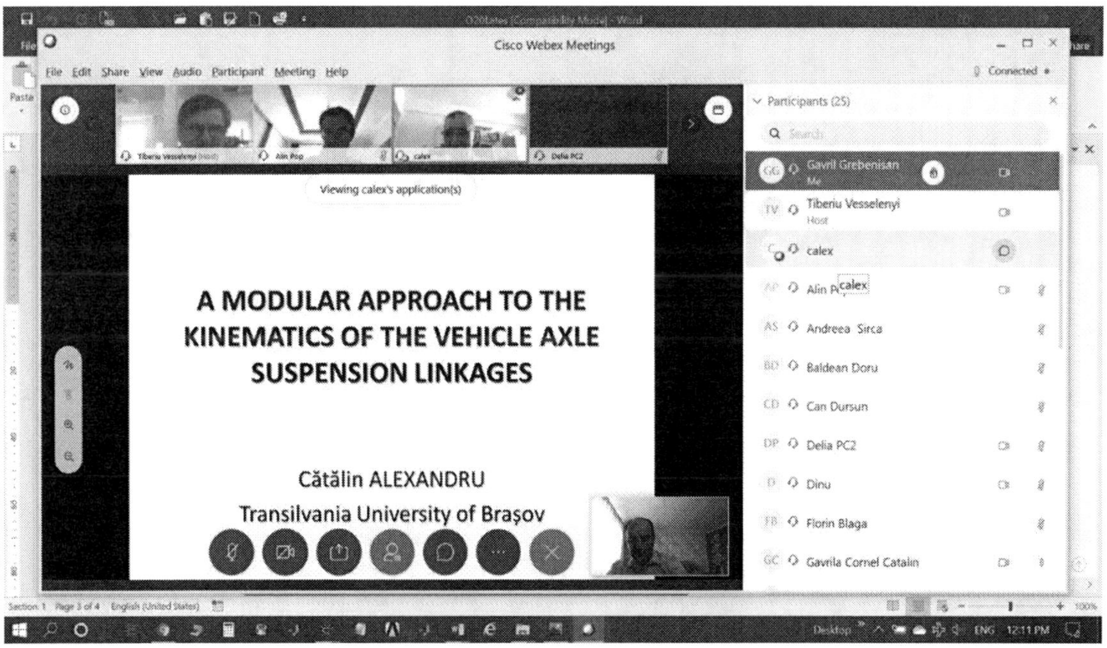

I Pascu, I Popescu and A Didu
First

I Pascu, I Popescu and A Didu
first

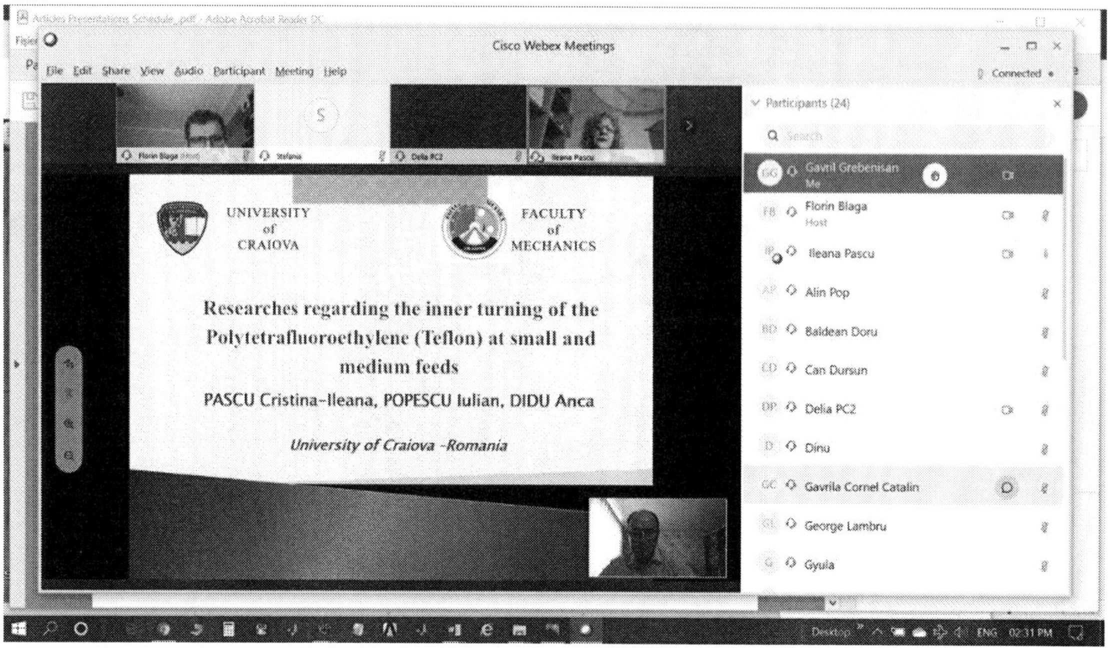

Second

M B Tudose, G Agafitei and S Avasilcai

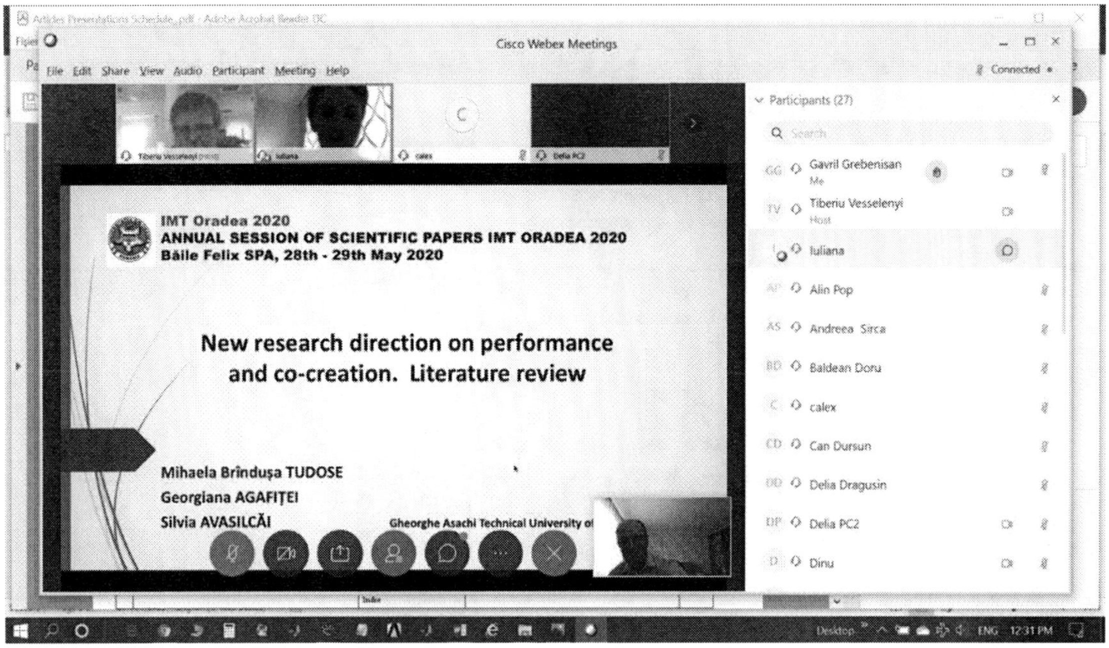

Delia Dragusin

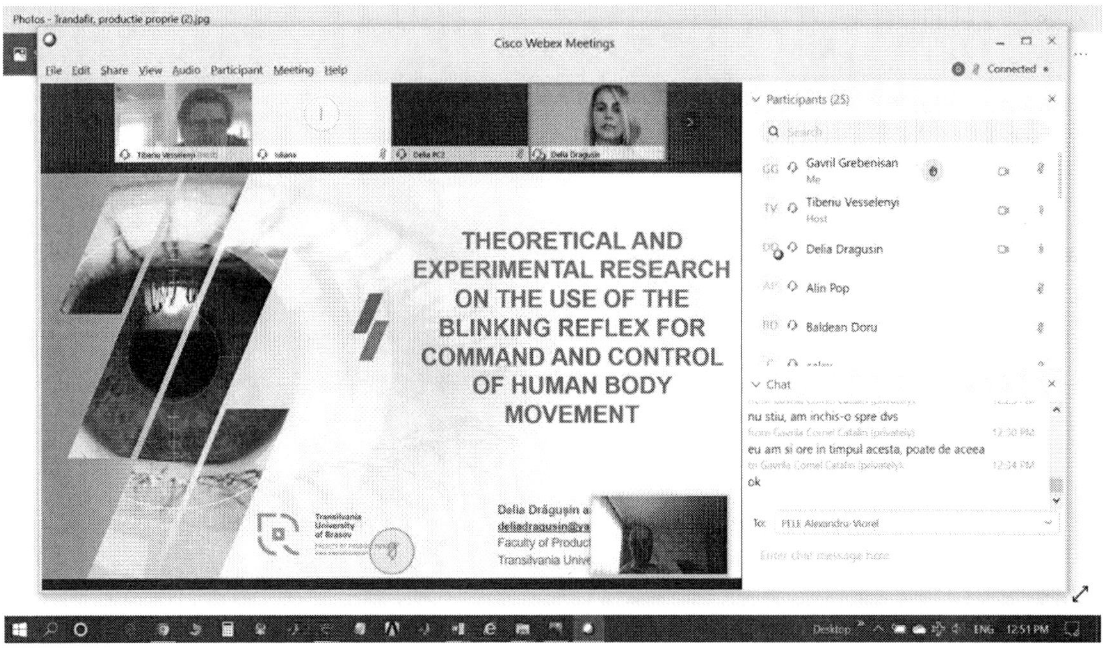

IOP Conf. Series: Materials Science and Engineering **898** (2020) 011002 doi:10.1088/1757-899X/898/1/011002

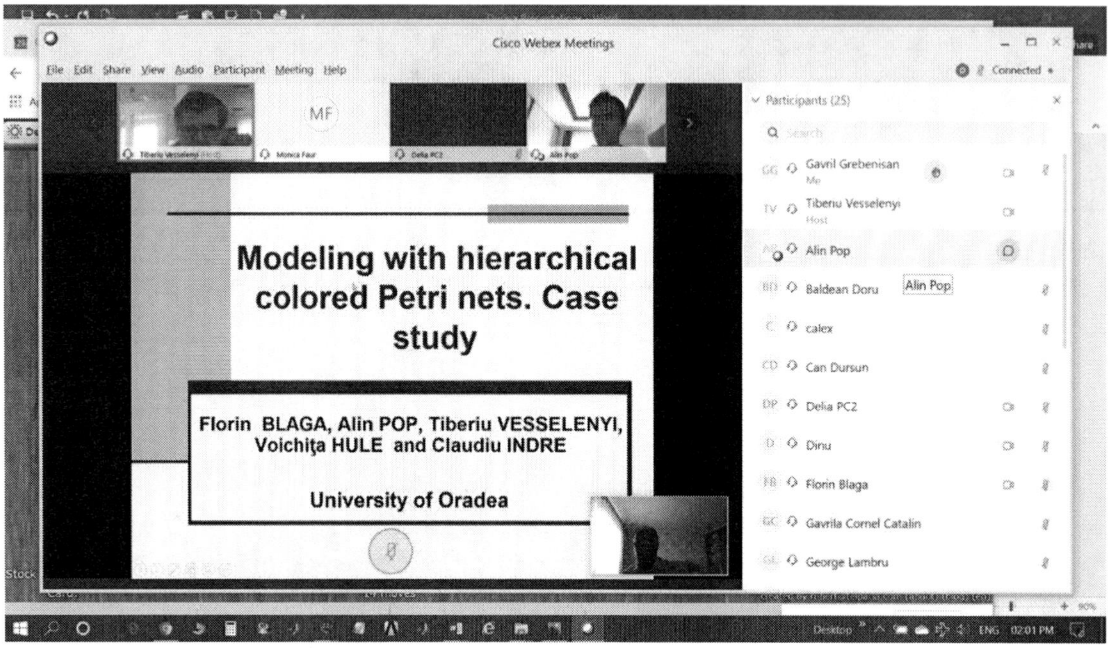

St Borta and M Baritz

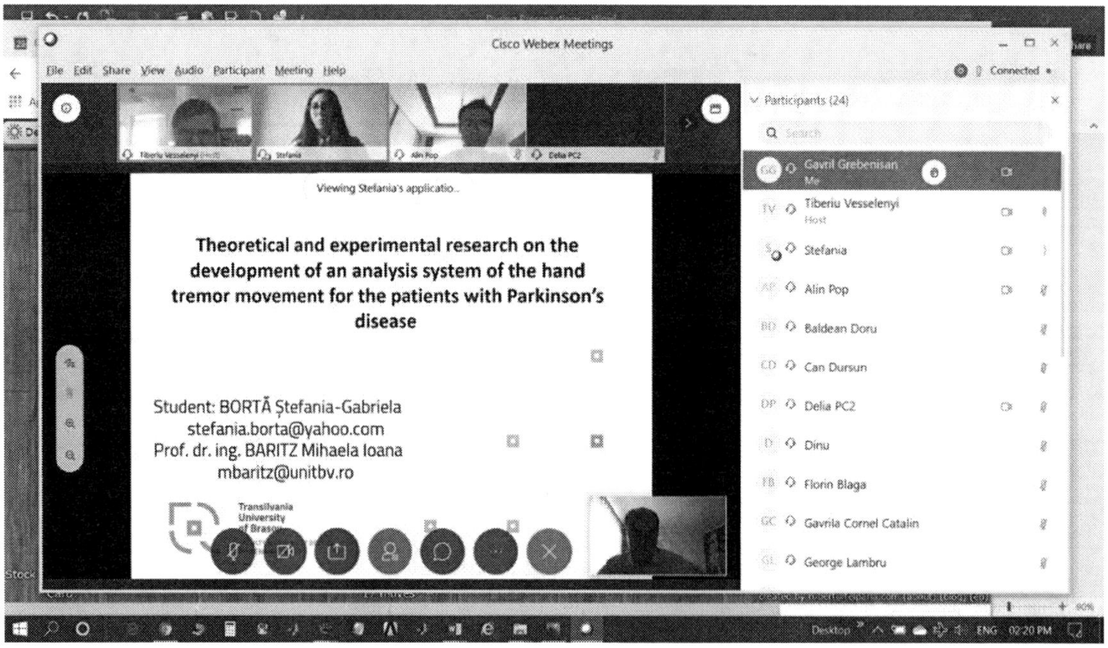

C Vartolomei and S Avasilcăi

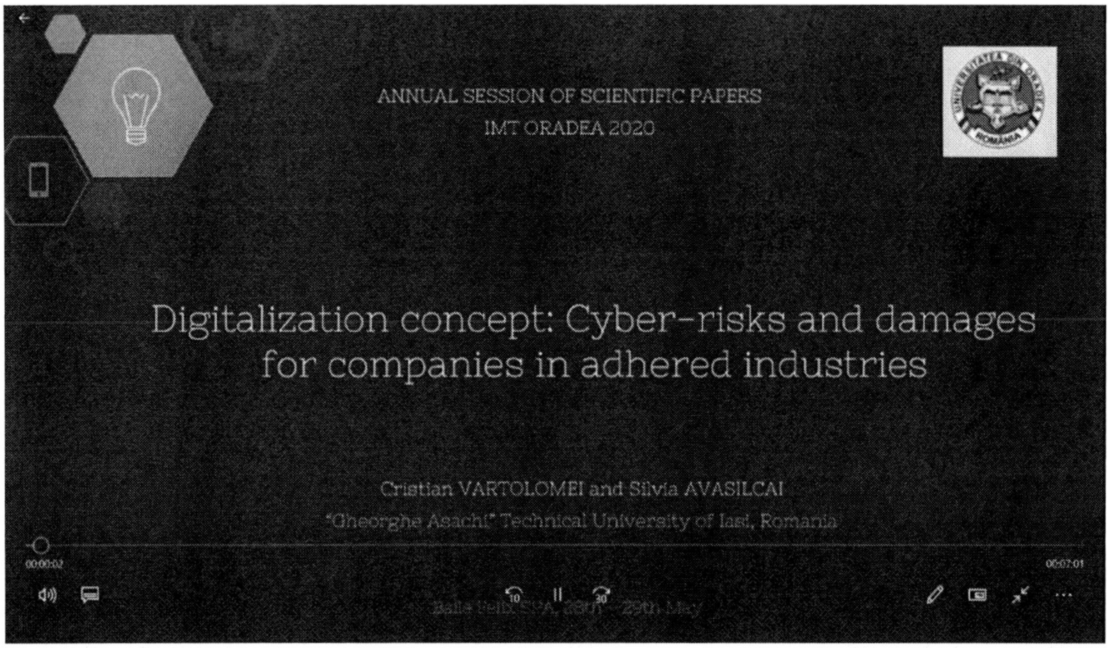

N Bagiu, S Avasilcăi, L Alexa

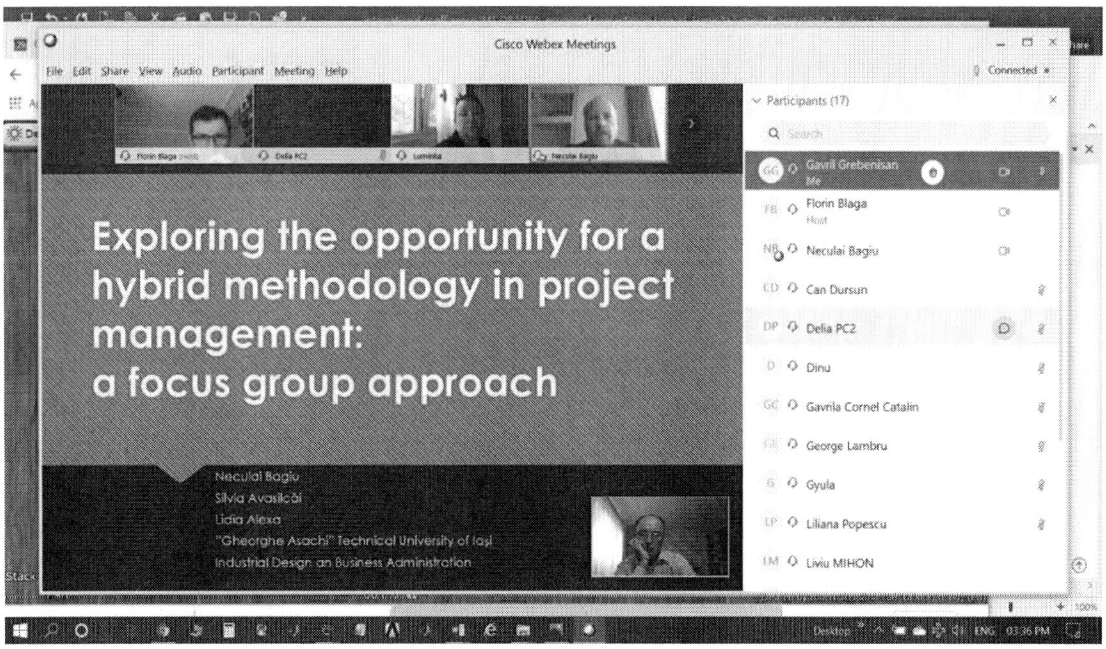

L C Simion and S Avasilcai

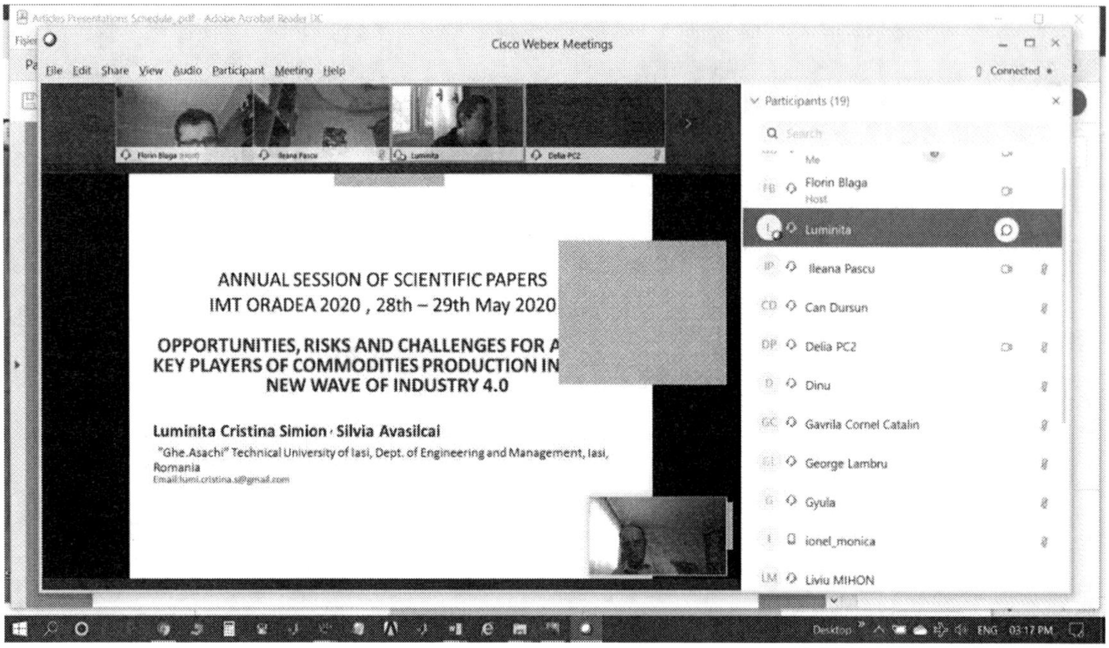

G L G Mitrea -Curpanaru

IOP Conf. Series: Materials Science and Engineering **898** (2020) 011002 doi:10.1088/1757-899X/898/1/011002

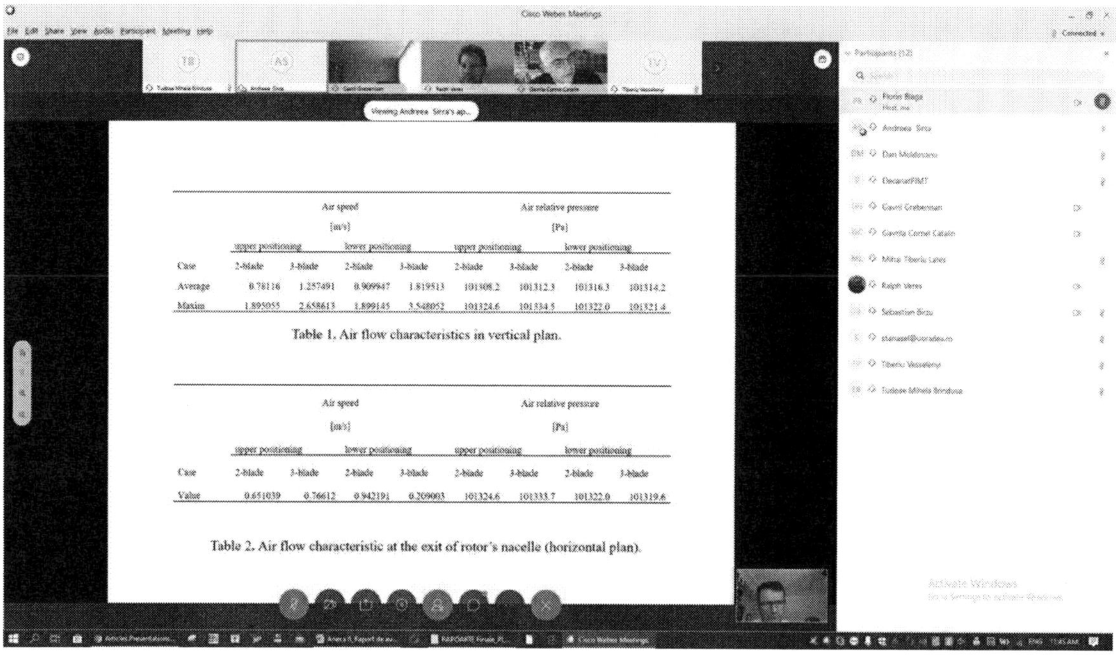

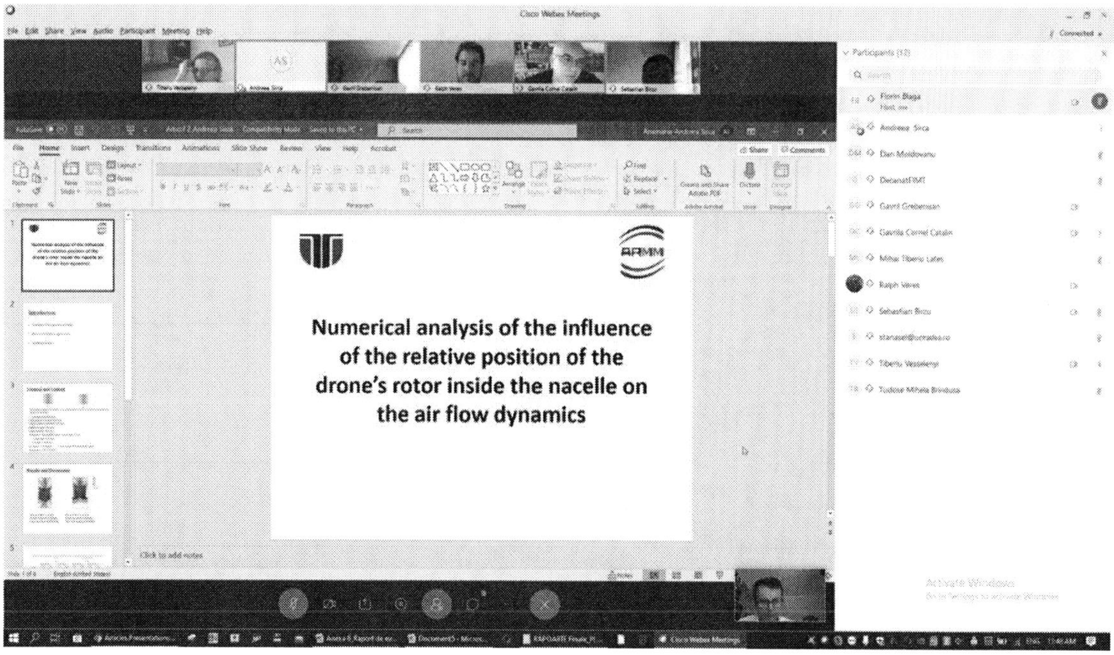

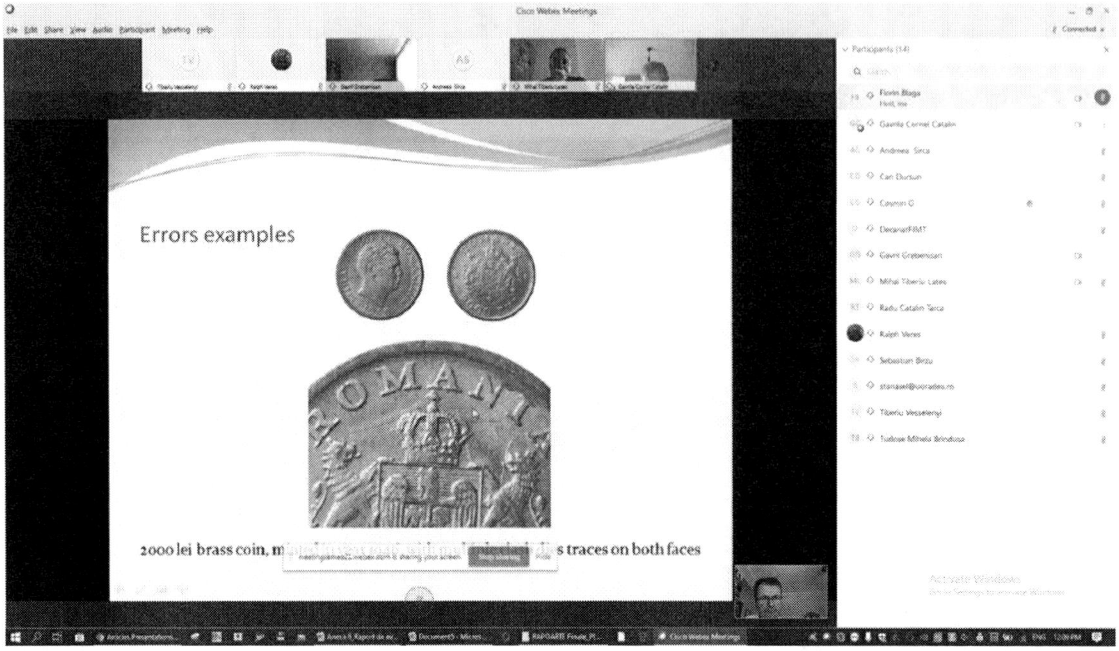
Cisco Webex Meetings
Errors examples
2000 lei brass coin,
s traces on both faces

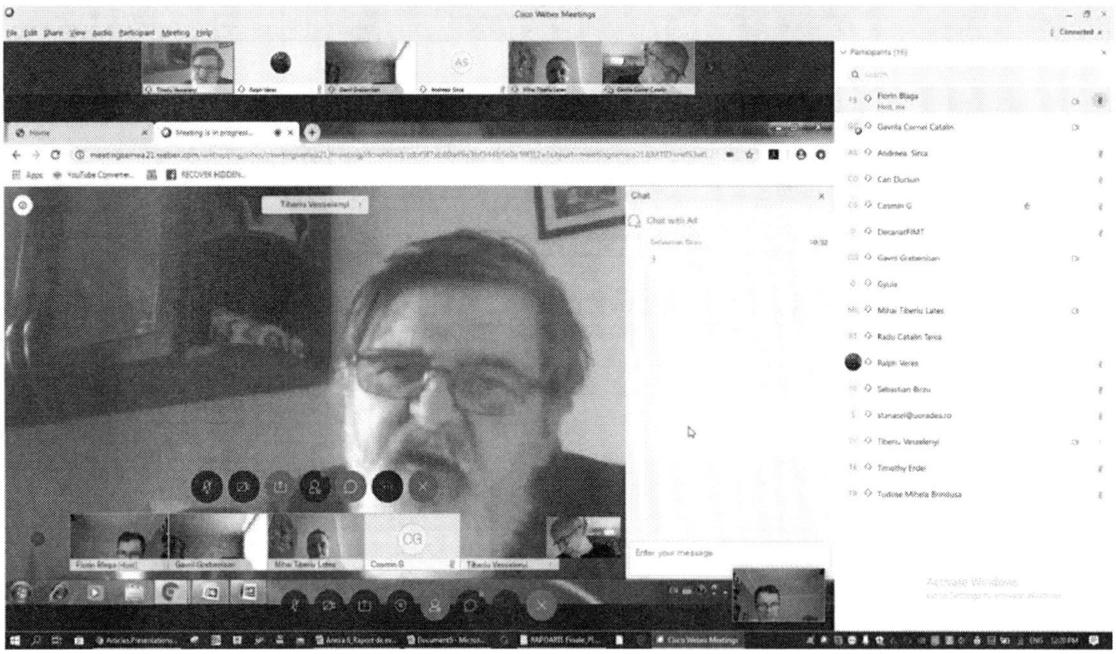
Cisco Webex Meetings

IOP Conf. Series: Materials Science and Engineering **898** (2020) 011002 doi:10.1088/1757-899X/898/1/011002

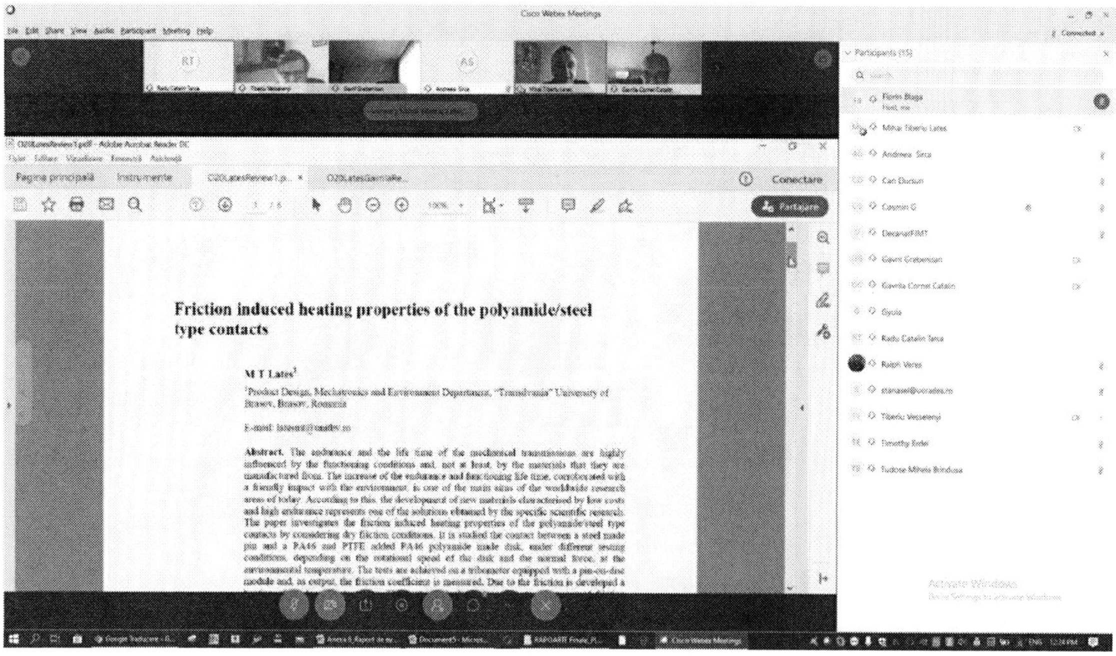

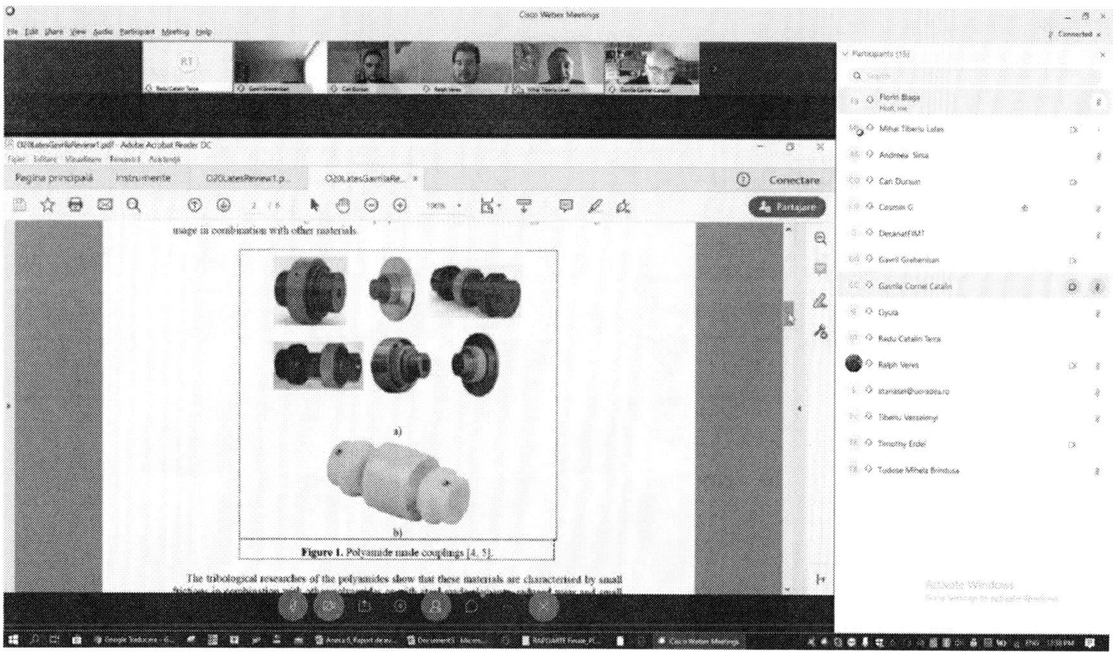

IOP Conf. Series: Materials Science and Engineering **898** (2020) 011002 doi:10.1088/1757-899X/898/1/011002

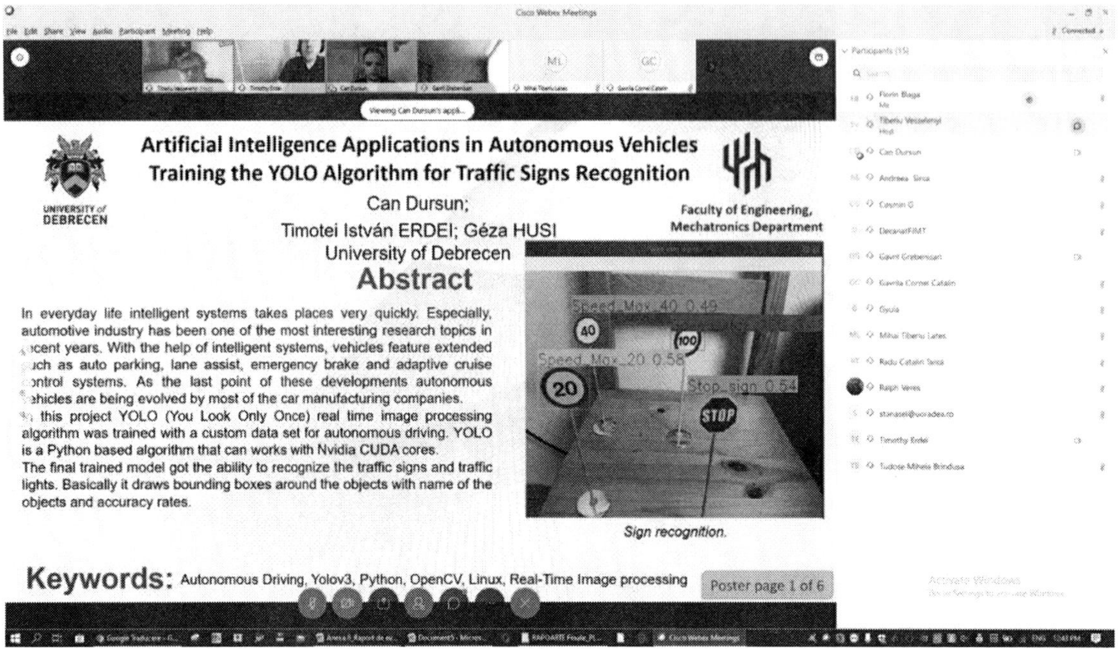

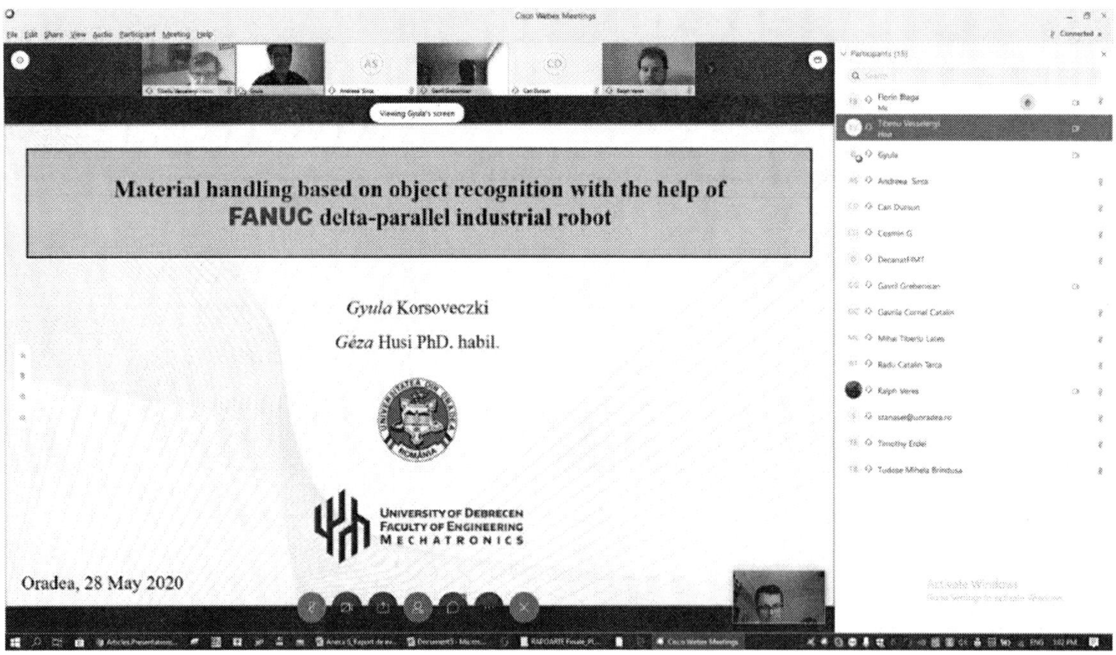

IOP Conf. Series: Materials Science and Engineering **898** (2020) 011002 doi:10.1088/1757-899X/898/1/011002

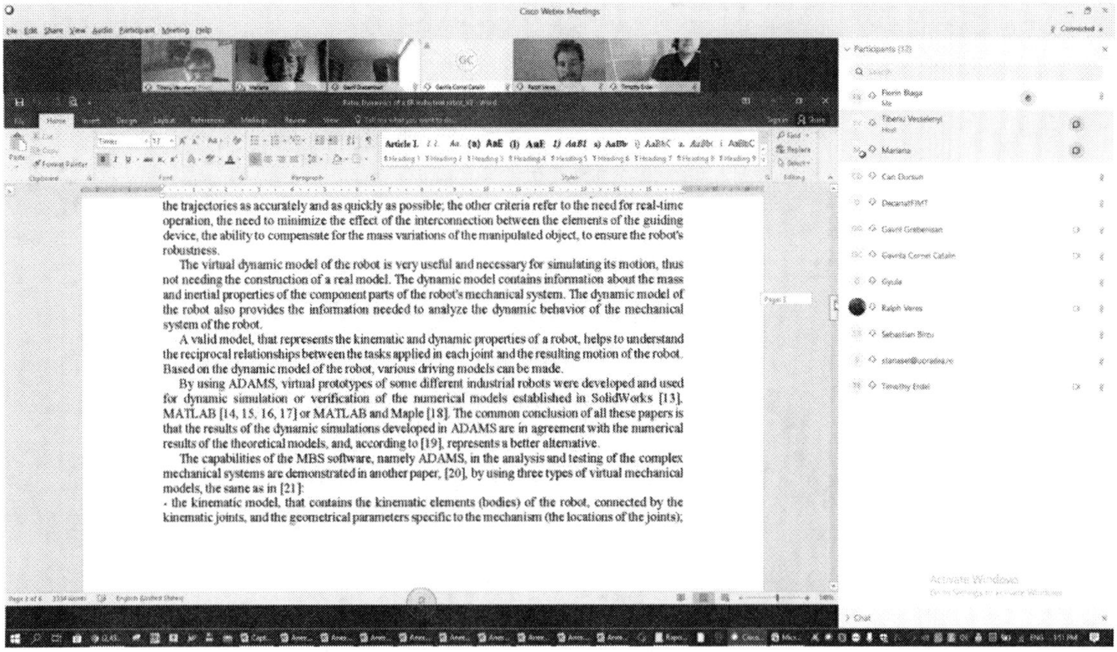

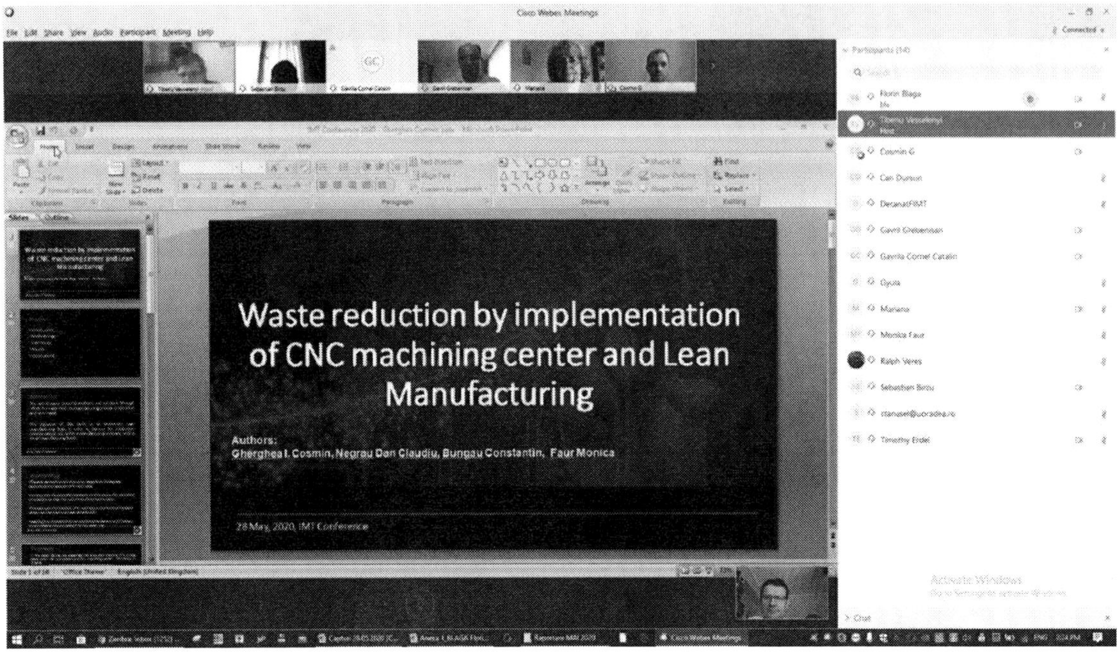
Waste reduction by implementation of CNC machining center and Lean Manufacturing
Authors:
Gherghea I. Cosmin, Negrau Dan Claudiu, Bungau Constantin, Faur Monica
28 May, 2020, IMT Conference

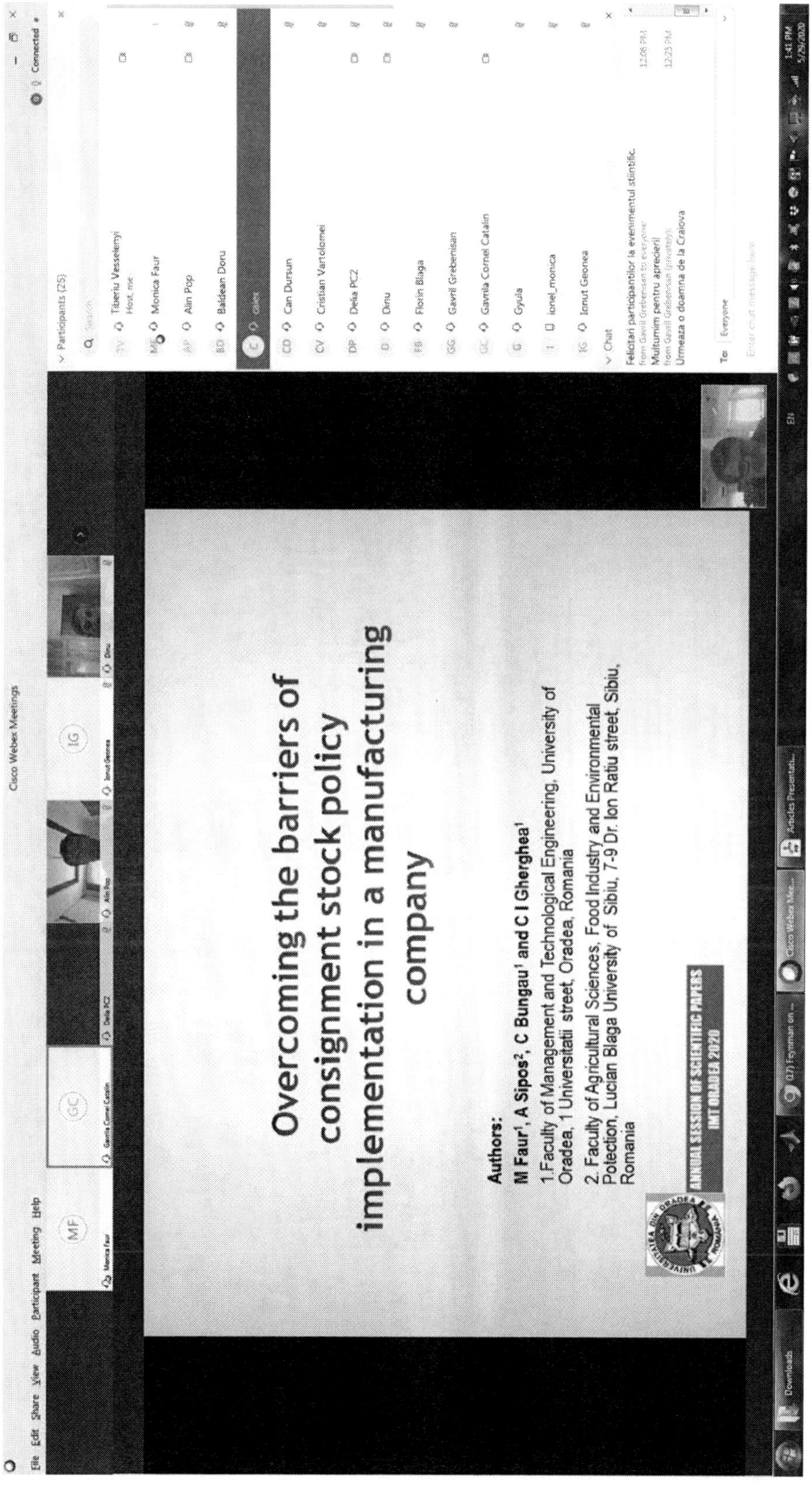
Cisco Webex Meetings
Overcoming the barriers of consignment stock policy implementation in a manufacturing company
Authors:
M Faur1, A Sipos2, C Bungau1 and C I Gherghea1
1.Faculty of Management and Technological Engineering, University of Oradea, 1 Universitatii street, Oradea, Romania
2. Faculty of Agricultural Sciences, Food Industry and Environmental Potection, Lucian Blaga University of Sibiu, 7-9 Dr. Ion Ratiu street, Sibiu, Romania
ANNUAL SESSION OF SCIENTIFIC PAPERS
IMT ORADEA 2020
Participants (25)
Tiberiu Vesselenyi
Monica Faur
Alin Pop
Baldean Doru
Can Dursun
Cristian Vartolomei
Delia PC2
Dinu
Florin Blaga
Gavril Grebenisan
Gavrila Cornel Catalin
Gyula
ionel_monica
Ionut Geonea
Chat
Urmeaza o doamna de la Craiova

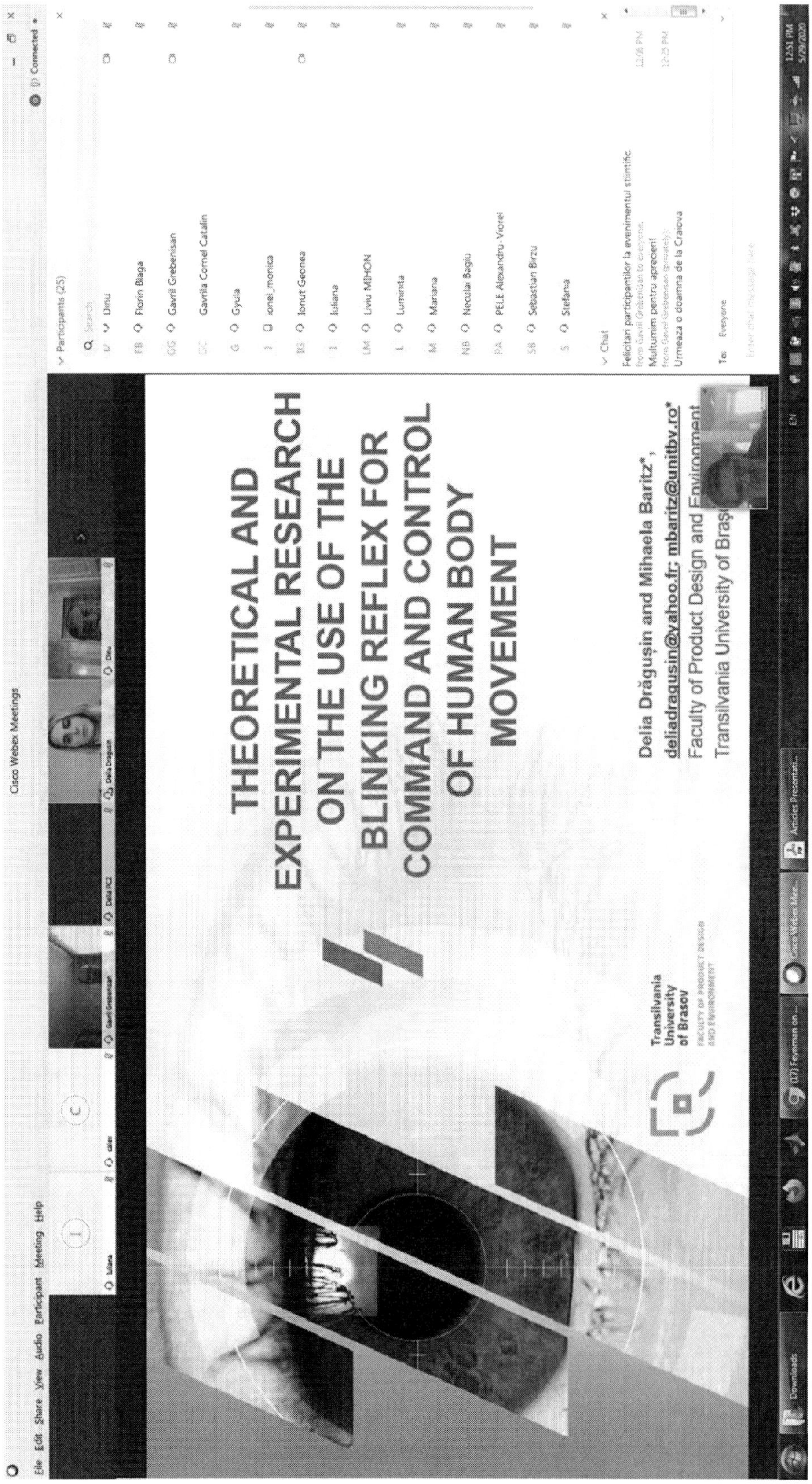
Cisco Webex Meetings
THEORETICAL AND EXPERIMENTAL RESEARCH ON THE USE OF THE BLINKING REFLEX FOR COMMAND AND CONTROL OF HUMAN BODY MOVEMENT
Delia Drăguşin and Mihaela Baritz*,
deliadragusin@yahoo.fr; mbaritz@unitbv.ro*
Faculty of Product Design and Environment
Transilvania University of Braşov
Transilvania University of Brasov
Participants (25)
Chat

IOP Conf. Series: Materials Science and Engineering **898** (2020) 011002 doi:10.1088/1757-899X/898/1/011002

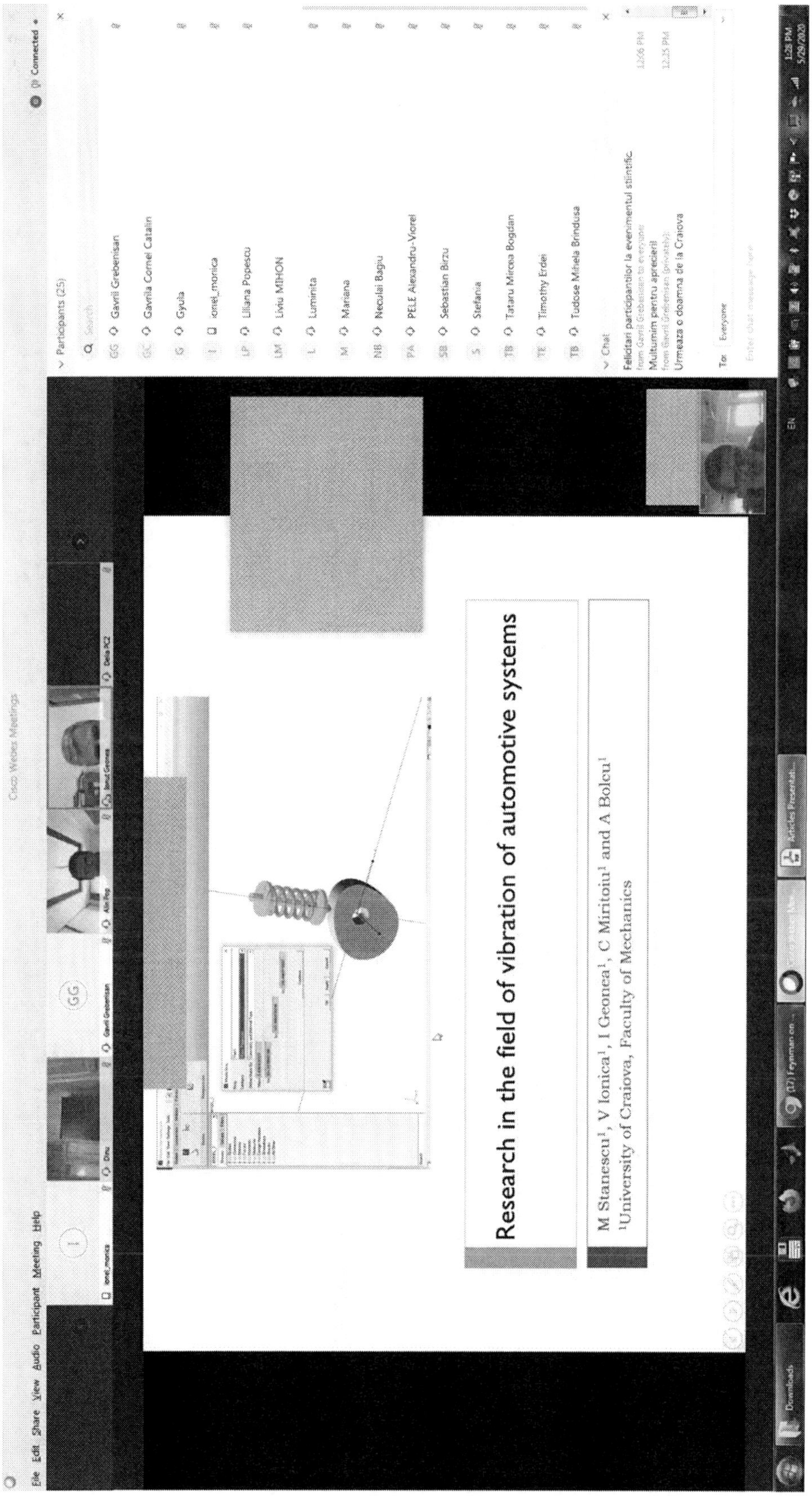

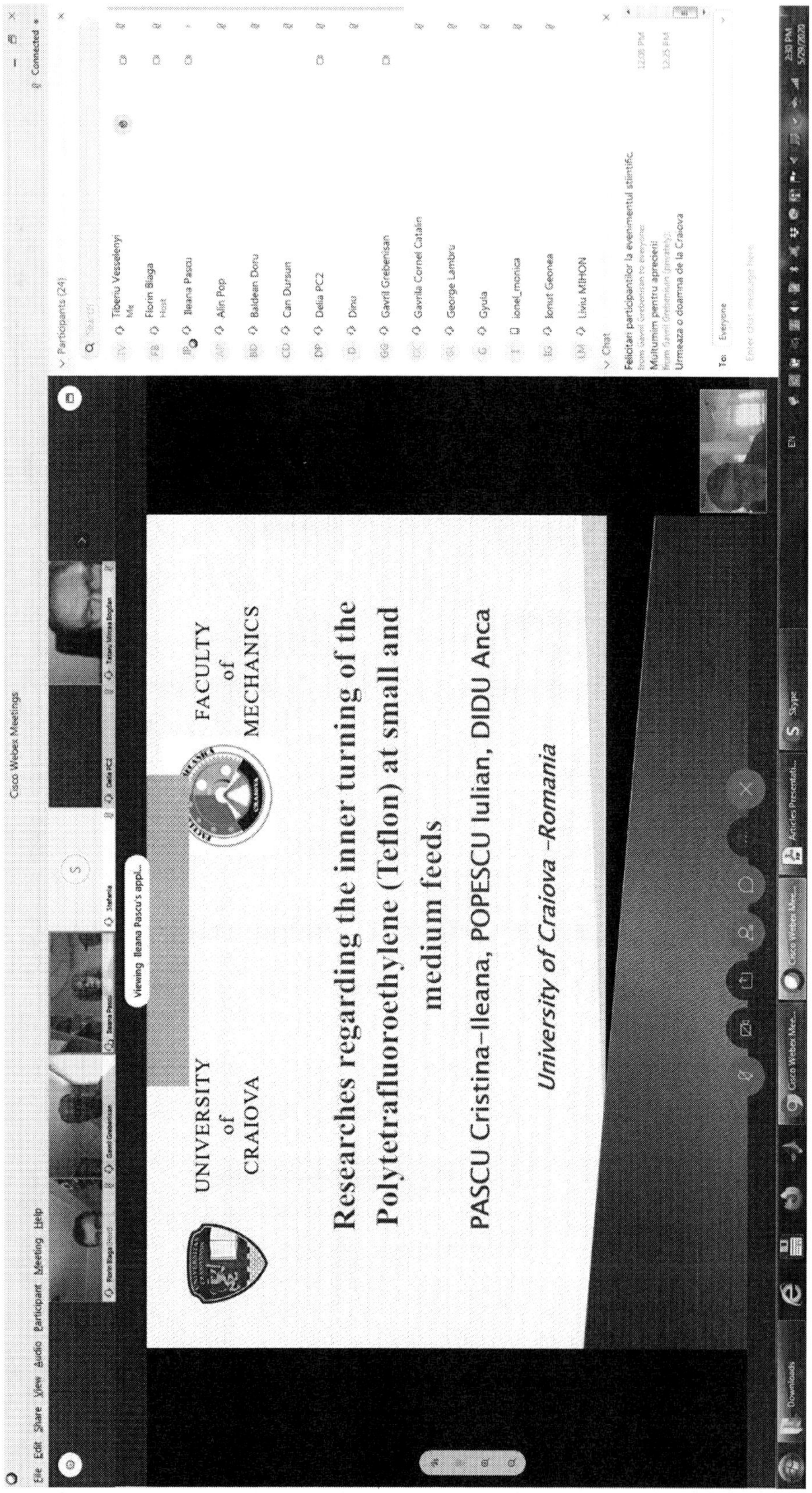
Cisco Webex Meetings
UNIVERSITY of CRAIOVA
FACULTY of MECHANICS
Researches regarding the inner turning of the Polytetrafluoroethylene (Teflon) at small and medium feeds
PASCU Cristina–Ileana, POPESCU Iulian, DIDU Anca
University of Craiova –Romania
Viewing Ileana Pascu's appl...
Participants (24)
Tiberiu Vesselenyi
Florin Blaga
Ileana Pascu
Alin Pop
Baldean Doru
Can Dursun
Delia PC2
Dinu
Gavril Grebenisan
Gavrila Cornel Catalin
George Lambru
Gyula
ionel_monica
Ionut Geonea
Liviu MIHON
Chat
Felicitari participantilor la evenimentul stiintific
Multumim pentru aprecieri!
Urmeaza o doamna de la Craiova

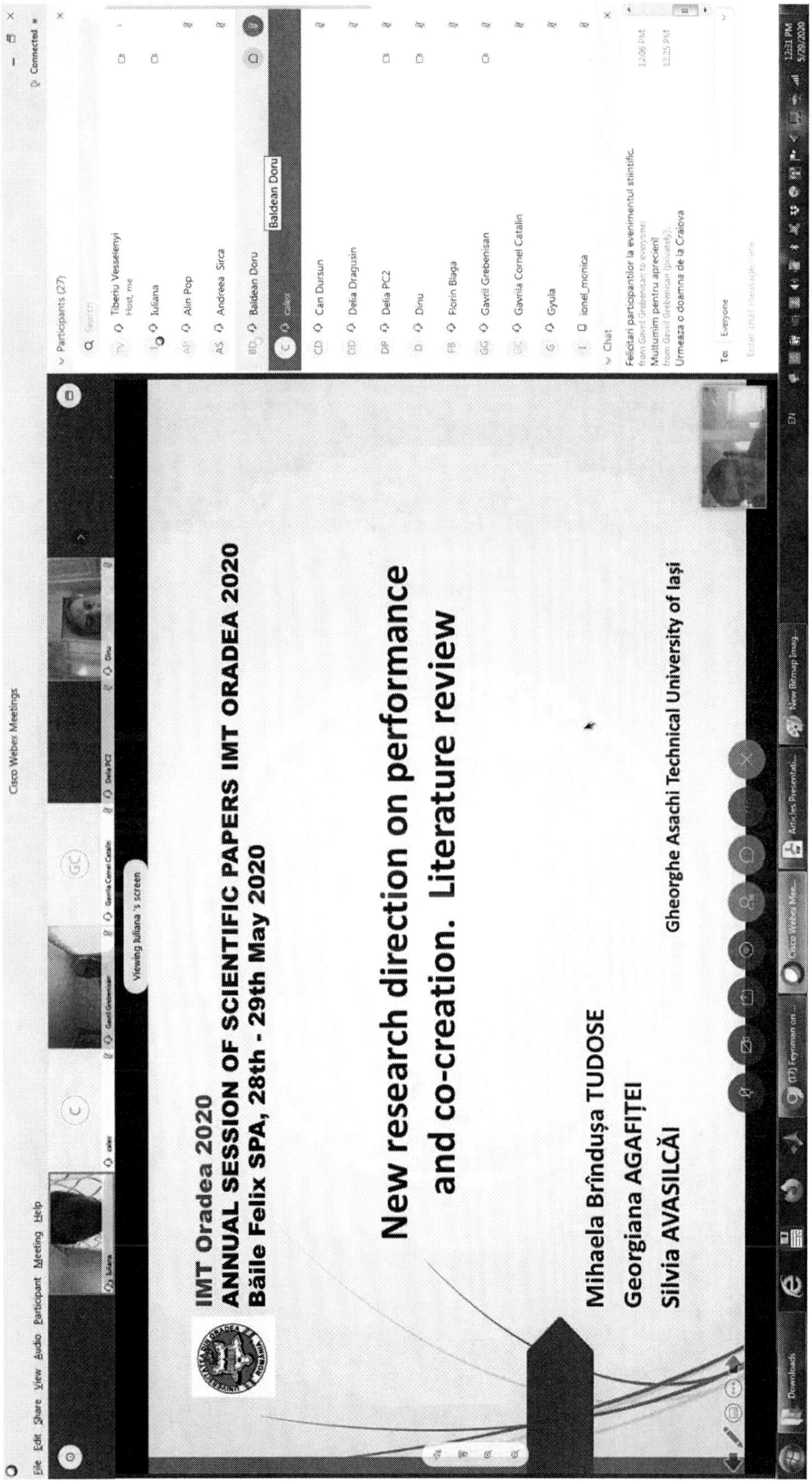
Cisco Webex Meetings
Viewing Iuliana 's screen
IMT Oradea 2020
ANNUAL SESSION OF SCIENTIFIC PAPERS IMT ORADEA 2020
Băile Felix SPA, 28th - 29th May 2020
New research direction on performance and co-creation. Literature review
Mihaela Brîndușa TUDOSE
Georgiana AGAFIȚEI
Silvia AVASILCĂI
Gheorghe Asachi Technical University of Iași
Participants (27)
Chat

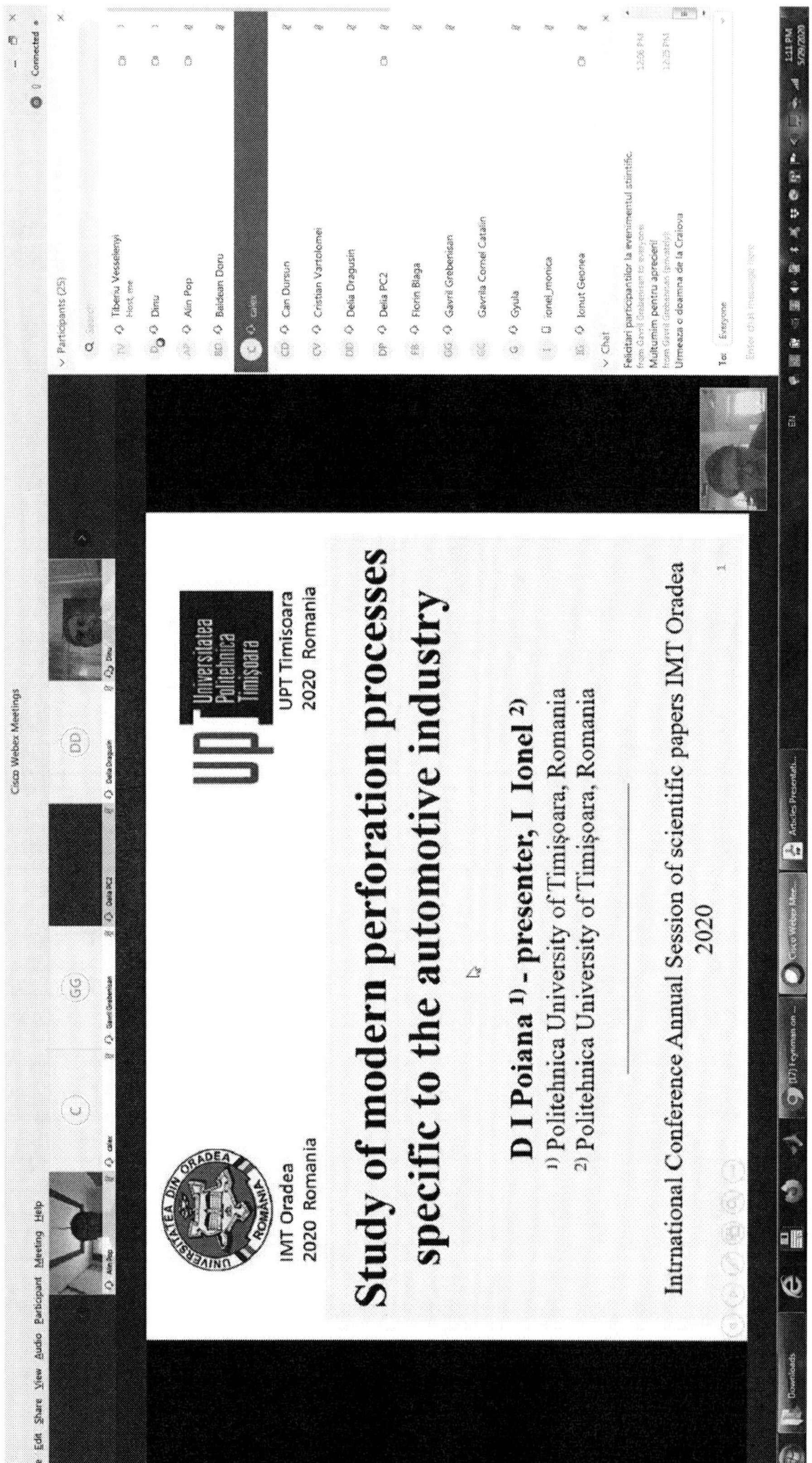
Cisco Webex Meetings
IMT Oradea
2020 Romania
Universitatea Politehnica Timişoara
UPT Timisoara
2020 Romania
Study of modern perforation processes specific to the automotive industry
D I Poiana 1) - presenter, I Ionel 2)
1) Politehnica University of Timişoara, Romania
2) Politehnica University of Timişoara, Romania
Intrnational Conference Annual Session of scientific papers IMT Oradea 2020

IOP Conf. Series: Materials Science and Engineering **898** (2020) 011003 doi:10.1088/1757-899X/898/1/011003

Peer review declaration

All papers published in this volume of IOP Conference Series: Materials Science and Engineering have been peer reviewed through processes administered by the Editors. Reviews were conducted by expert referees to the professional and scientific standards expected of a proceedings journal published by IOP Publishing.

- **Type of peer review: Single-blind**
- **Conference submission management system: Conference's independent platform**
  https://imt.uoradea.ro/conference/conf.detail.php?workID=0
- **Number of submissions received: 80**
- **Number of submissions sent for review: 65**
- **Number of submissions accepted:58**
- **Acceptance Rate (Number of Submissions Accepted / Number of Submissions Received X 100):72.5%**
- **Average number of reviews per paper: 2**
- **Total number of reviewers involved:6**

Călin Băban cbaban@uoradea.ro	Horia Beleş horia.beles@gmail.com	Florin Blaga fblaga@uoradea.ro
Izabela Dembinska izdem@wp.pl	Gavril Grebenişan grebe@uoradea.ro	Alexandru Viorel Pele apele@uoradea.ro

- **Any additional info on review process: -**
- **Contact person for queries: Gavril Grebenisan, grebe@uoradea.ro**

IOP Conf. Series: Materials Science and Engineering **898** (2020) 012001 doi:10.1088/1757-899X/898/1/012001

Numerical analysis of the influence of the number of blades on the dynamic performance of a drone

A A Sirca[1], F Mariasiu[1], S Morariu[2], N Vlad[1]

[1]Technical University of Cluj-Napoca, Department of Automotive Engineering and Transports, B-dul.Muncii 103-105, Cluj-Napoca, ROMANIA
[2]Inovo Finance SRL, Cluj-Napoca

andreea.sirca@auto.utcluj.ro

Abstract. The use of Unmanned Aerial Vehicles and known under the generic name of drones is increasingly present in different economic sectors, due to the versatility they offer in solving specific problems. The specific problems to be solved, however, raise the need for the technical characteristics of the drones to be optimized according to the specific usage. In the case of drones used in technological processes in agriculture, they must offer a high autonomy, that can be increased by optimizing their weight. The paper proposes the study of the dynamics of the air flow according to the configuration of the drone rotor (2 and 3 rotor respectively) by numerical analysis methods for the further development of a drone that will be used in the phytosanitary treatment process of agricultural crops. The parameters evaluated and compared are the specific parameters for characterizing the airflow: velocity and relative pressure (at different nacelle's section, both in horizontal and vertical directions). The obtained results show that the most constant, dense and homogeneous velocity field was obtained from the 3-blade configuration of the rotor.

1. Introduction

Since ancient times, agriculture was the basic occupation of mankind. The main result of the agricultural activities was the production of food, with a huge role in the development of the human civilization. This more important nowadays as the world population is constantly growing. Thus, in the last 100 years there has been a massive population growth, respectively 4 times more due to: advanced medical procedures, a lower mortality rate and a continuous increase of productivity in the agricultural field. Historians estimate that around 1800 the world population had an approximate value of 1 billion, but after 1800, this changed fundamentally: the world population increased 7 times. Generally speaking, it is estimated that 108 billion people have ever lived on our planet [1].

In a study by United Nations Population Division, it was predicted that by the end of the 21st century, the population will reach 11 billion [2]. However, a study conducted by Sanjeev S. bring counterarguments regarding this prediction, arguing that global fertility will fall in 2020 below the replacement rate and the world population will reach a threshold of less than 9 billion by 2050, followed by a decline on long term [3]. Any of these theories, however, show that a huge volume of food will still be needed to meet the needs of the world's population.

The only constant in the agricultural processes that provide food since the beginning of human society is the limited agricultural area, which is why (in order to counteract the increase of worldwide pollution) in order to increase the production and productivity in agriculture modern methods and methodologies must be used. These principles can be found in the concept of Agriculture 4.0, a concept that includes development directions in order to achieve this goal. The concept of Agriculture 4.0 has brought with it an environment in which all the elements that compose it (Intelligent resources, Devices, Products or Machines, Data Transfer and Infrastructure, Data Analysis, People Processes and Systems) are in continuous and direct connection, reaching a very high level of coordination. For example, the use in agriculture of existing technologies that are used in other fields (e.g. Bluetooth, GPS, or RFID) leads to an agricultural supply with self-optimization function, due to the rapid and direct communication created between people and agricultural machinery.

One of the modern technologies used in the application of the Agriculture 4.0 concept is the use of UAVs (Unmanned Aerial Vehicles), with immediate applications in: the prevention and the treatments of agricultural crops, the monitoring of the vegetation status of the crops, the monitoring of environmental factors, transport, search and rescue , surveillance, security, etc.

At present, due to the use of UAV in the field of phytosanitary prevention and treatment of agricultural crops, there are developed many practical applications and carried out research for this purpose. The Food and Agriculture Organization of the United Nation [4] investigated their effect on agricultural production and how UAVs can simplify and facilitate the work of farmers by automating processes, which leads to correct and rapid decisions regarding: irrigation plants, pesticide treatments or establishing the degree of harvest.

Muraru et al. [5] dealt with the ways in which UAV use revolutionizes agriculture. Regarding the phytosanitary treatments a first essential role is the health of the crops and the identification of bacterial or fungal infections. By scanning the culture with visible light close to the infrared the drones can identify plants that reflect different amounts of green light. With this information, multispectral images are created that follow the changes of the plants and indicate their state of health. The faster the disease is discovered, the more crops can be saved and farmers can apply and monitor the remedies more accurately. Furthermore, Wen S. et. of [6] have investigated the possibilities of implementing artificial intelligence in the use of a UAV for plant and crop protection activities, and propose a system based on neural networks that optimize the spray process depending on the environmental conditions.

The conclusions drawn by Gupta et al. [7] regarding the immediate benefits of using drones in agriculture were related to reducing pollution, costs and losses, finding problems much faster, optimizing treatments, etc. Even if these benefits exist and are recognized, there are still some limitations regarding the large-scale applicability, limitations related to the cost and performance of the sensors used or the high price of a complete equipment.

The present paper aims to present the partial results of some researches carried out in order to optimize the spray process, by proposing to use the spray nozzles built inside a drone's rotor nacelle, compared to the classic systems that use a common ramp. spray. For this purpose, using numerical analysis methods, the effect of the number of blades (2 and 3) forming a drone rotor, on the dynamic air flow parameters (air flow velocity and pressure) was studied.

2. Material and Method

For the numerical analysis of the velocity and pressure fields caused by the air flow under the rotor of a drone that is intended to be used in the process of plant protection (agricultural crops), a model was developed using Solid Works software. The general geometrical dimensions of the model are: inlet platform diameter 550 mm, output platform diameter 550 mm, platform length 550 mm, blade diameter 450 mm, thickness of the platform walls: upper part 16 mm, lower part 1.58 mm and are shown in figure 1. As boundary conditions, the fluid flow was considered to be air with a density of 1.205 kg/m3, at an ambient temperature of 293.2 K (20°C). Rotor speed is 250 rad/s (2387 rpm).

The analysis of the velocity and pressure fields of the air flow was performed for the cases in which the rotor has 2 and 3 blades in design, and the measurements of the obtained values were made both in the horizontal plane and in the vertical plane of nacelle.

IOP Conf. Series: Materials Science and Engineering **898** (2020) 012001 doi:10.1088/1757-899X/898/1/012001

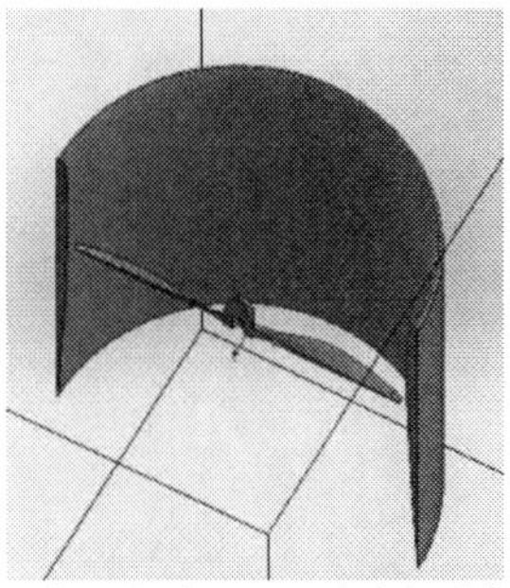

Figure 1. Simulation model (2-blade case).

3. Results and Discussions

The results obtained from computerized numerical analysis (CFD) are presented comparatively in Figures 2-17. It can be observed that differences in the speed and pressure field model appear depending on the number of blades of the rotor.

The graphs of the variation of the air flow velocity are shown in Figures 10-13. From the point of view of the measurements recorded in a vertical plane the maximum speed is 3.45 m/s for the 2-blade rotor case and 3.25 m/s for the 3-blade rotor case. The average speed values are 7.64 m/s for the 2-blade rotor case and 7.41 for the 3-blade rotor case.

The structure of the pressure field is shown in Figures 6-9. Horizontally and vertically for the 2-blade rotor case in figures 6-7 and for the 3-blade rotor case in figures 8-9.

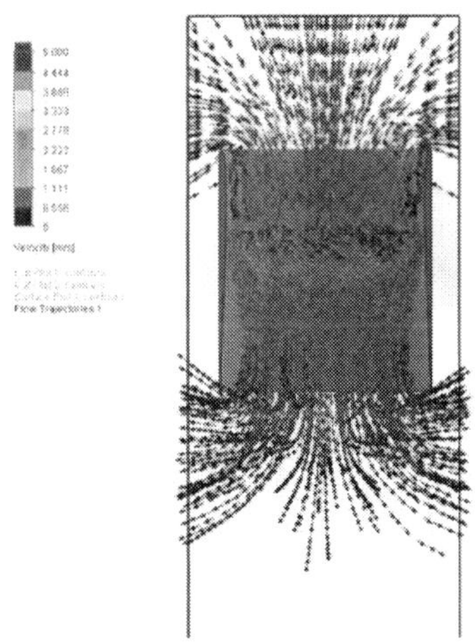

Figure 2. Velocity field in vertical section of nacelle (3-blades rotor case).

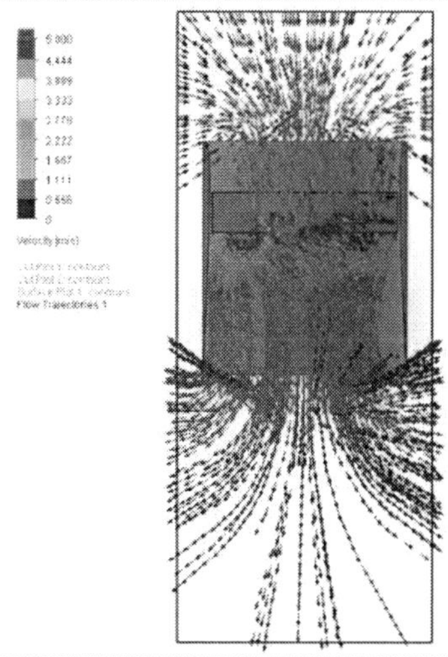

Figure 3. Velocity field in horizontal section of nacelle (2-blades rotor case).

IOP Conf. Series: Materials Science and Engineering **898** (2020) 012001 doi:10.1088/1757-899X/898/1/012001

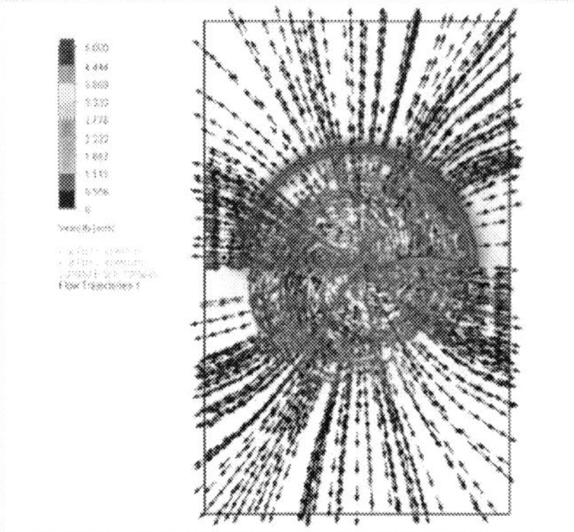

Figure 4. Velocity field in vertical section of nacelle (2-blades rotor case).

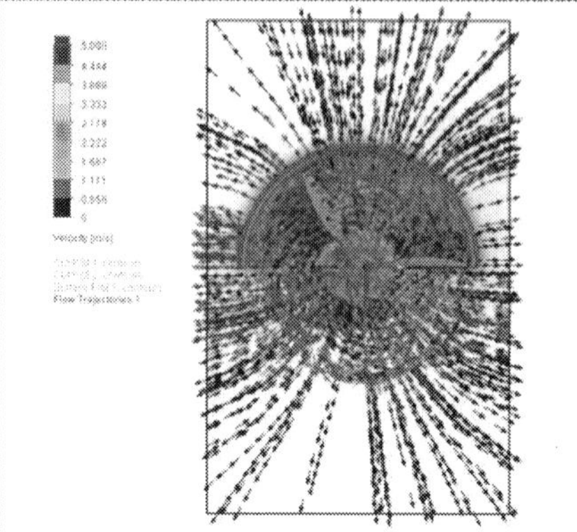

Figure 5. Velocity field in horizontal section of nacelle (3-blades rotor case).

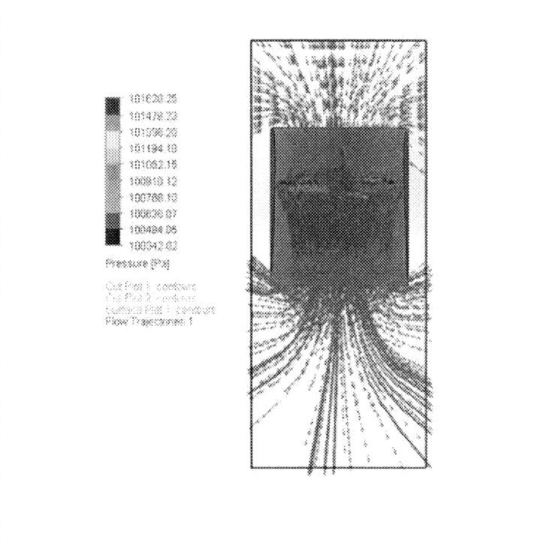

Figure 6. Pressure field vertical section of nacelle (3-blades rotor case)

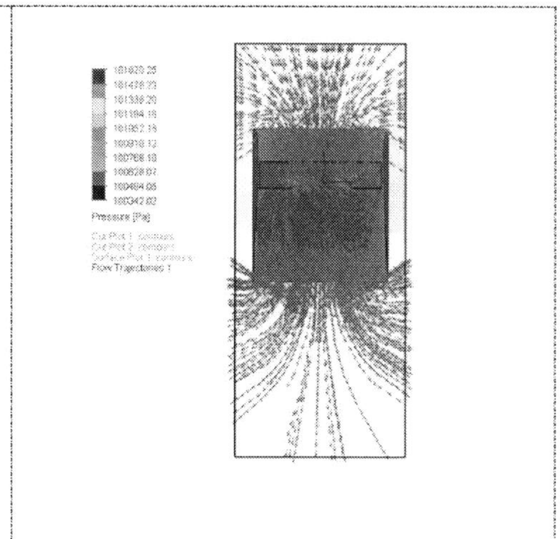

Figure 7. Pressure field in horizontal section of nacelle (2-blades rotor case)

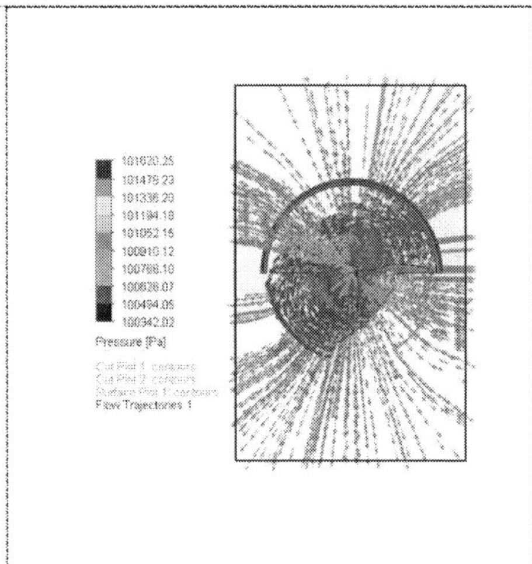

Figure 8. Pressure field vertical section of nacelle (2-blades rotor case).

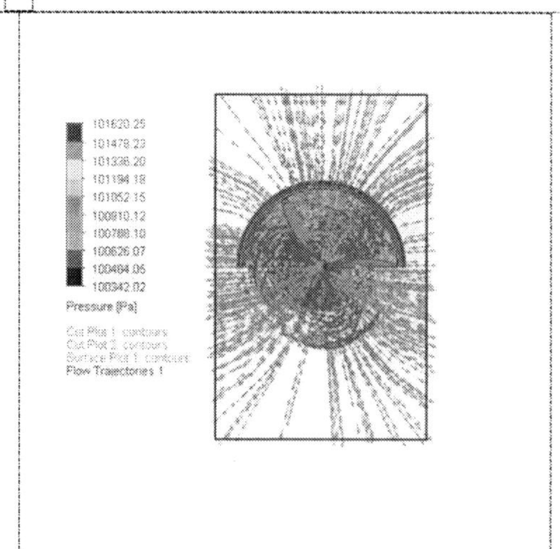

Figure 9. Pressure field in horizontal section of nacelle (3-blades rotor case)

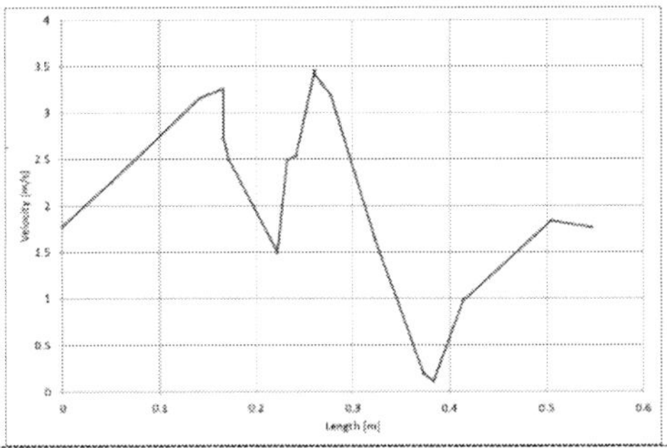

Figure 10. Air speed variation in vertical direction (2-blades rotor case).

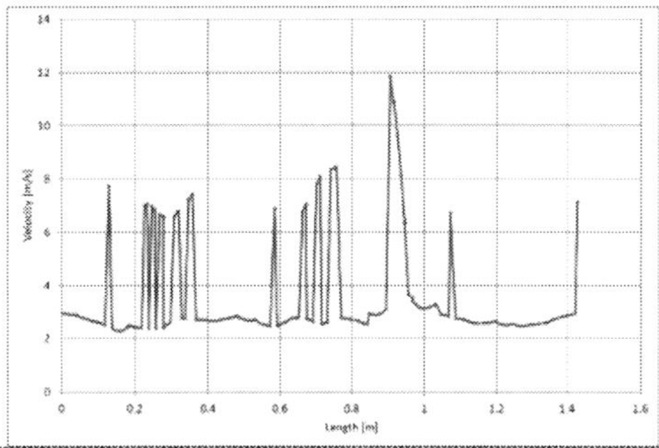

Figure 11. Air speed variation in horizontal direction (2-blades rotor case).

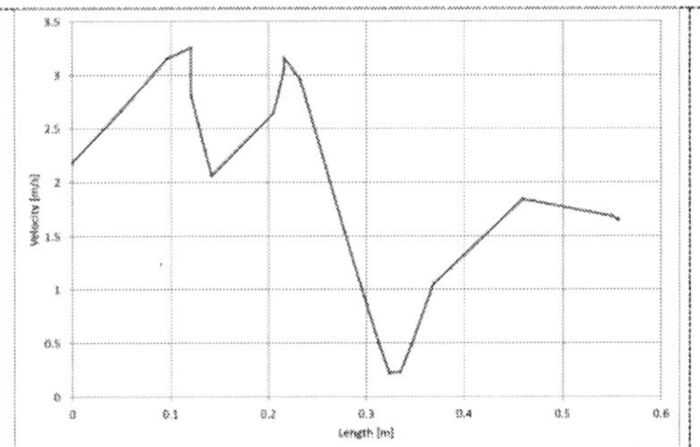

Figure 12. Air speed variation in vertical direction (3-blades rotor case).

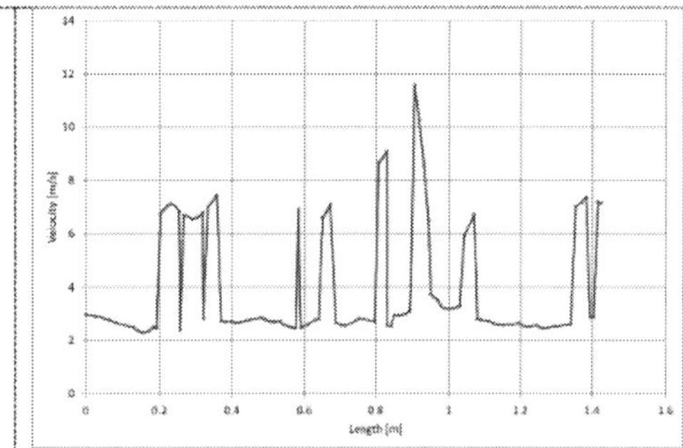

Figure 13. Air speed variation in horizontal direction (3-blades rotor case).

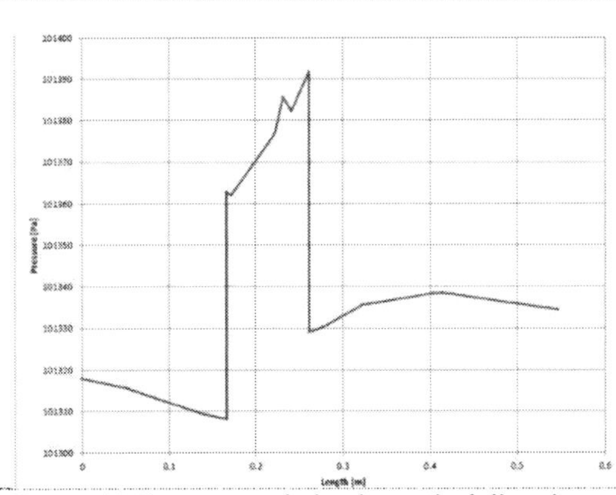

Figure 14. Pressure variation in vertical direction (2-blades rotor case).

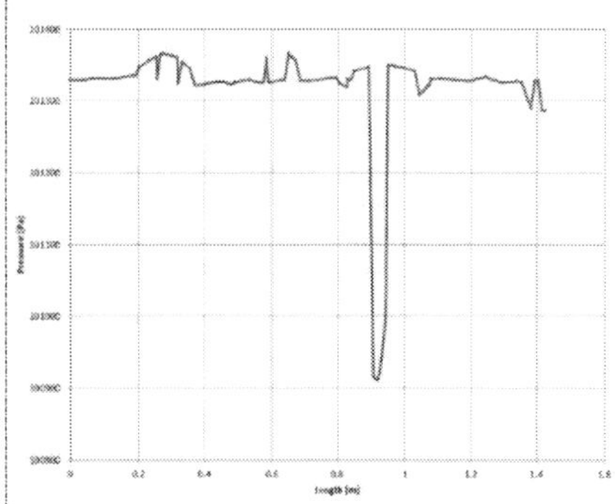

Figure 15. Pressure variation in horizontal direction (2-blades rotor case).

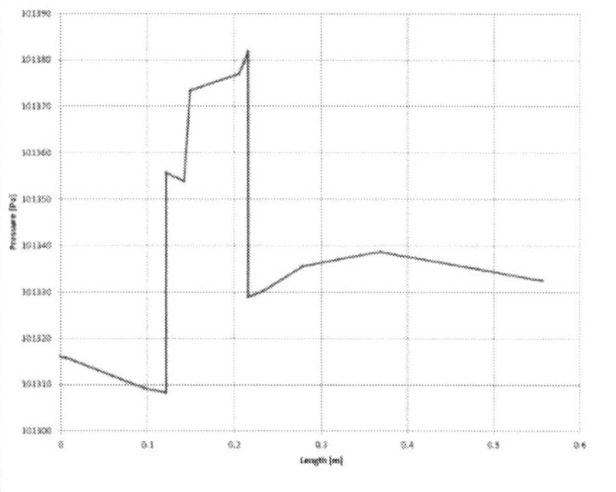

Figure 16. Pressure variation in vertical direction (3-blades rotor case).

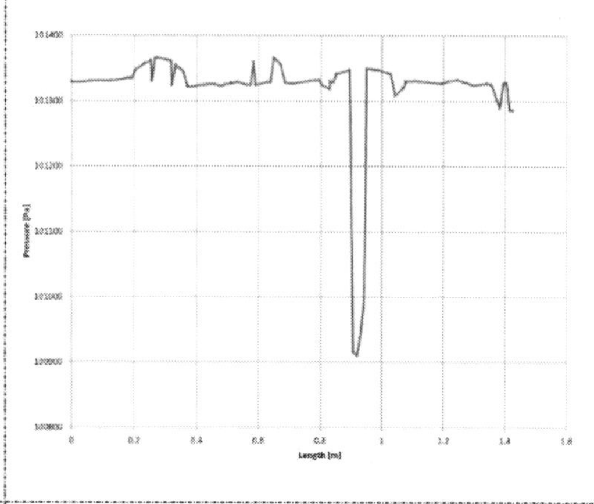

Figure 17. Pressure variation in horizontal direction (3-blades rotor case).

Regarding the pressure variation, the graphs are shown in Figures 14 - 17. The maximum pressure registered horizontally in the case of the 2-blade rotor case is the same as in the case of the 3-blade rotor case (101366 Pa), being recorded in the blade rotation area. In the case of the vertical plane, the following maximum values have been recorded: for the 2-blades rotor case a value of 101391 Pa respectively a value of 101381 Pa for the 3-blades rotor case.

From the point of view of the air-speed recorded at the exit of the rotor nacelle the highest values were obtained for the case of the 2-blade rotor. The average flow through the platform is 2.60 m/s for the 2-blade rotor and 2.45 m/s for the 3-blade rotor case. The relative difference between these values is relatively small by 5.77%.

The pressure fields created by the operation of the rotor at a speed of 2,387 rpm have different values for the studied cases. The highest uniformity of the pressure field is in the case of the 3-blade rotor case but the highest value of the air flow pressure appears in the case of the 2-blade rotor case. The average flow pressure through the nacelle is 101338.05 Pa for the 2-blade rotor case and 101326.89 Pa for the 3-blade rotor case. The relative difference between these values is 11.17%.

Based on the results obtained it can be said that in the case of a 2-blade rotor case can accommodate fewer nozzles spray a common rail of spraying of a drone, but the pressure of the jet is greater which results in the formation of a liquid jet with smaller droplets. In the case of a drone using a rotor with 3 blades in construction, it can accommodate more spray nozzles on the spraying common rail, but the droplet size of the resulting jet (that reach the crop/plant) will be higher.

4. Conclusions

Following the computerized numerical analysis on the influence of the number of blades of a UAV rotor (drones) used in the processes of plant protection in agriculture, the following conclusions can be issued.

UAVs are an emerging technology that offers high potential for use in agriculture with immediate beneficial effects on increasing agricultural productivity. Further studies are needed to improve their performance in order to reduce the costs of application and to obtain material benefits (cost reductions with exploitation).

It is necessary that in order to increase the values of the flow velocity and the pressure, a topological optimization of the shape of the platform will be carried out in the future, which will favor the geometry of the spray in order to make the area covered with the liquid used in the phytosanitary treatments of the crops as uniform.

The obtained results will be used further in the researches to optimize the geometric construction of the platform of a UAV used in the technological processes of protection of agricultural plants, in order to introduce spray nozzles in the construction of the platform and to eliminate the external common ramps currently used. This would eliminate the effect of the influence on the drone's displacement velocity spray and allow a more accurate dosage of the protective liquids used on agricultural crops.

Acknowledgments
The authors thanks to "POC-A1-A1.2.3-G-2-15 Knowledge transfer partnerships, Project title: Technologies for intelligent urban electric vehicles URBIVEL, ID: P_40_333" research program for help and support.

References

[1] Kaneda T, and Haub C 2020 *How many people have ever lived on Earth?,* https://www.prb.org/howmanypeoplehaveeverlivedonearth/ (accesed in 20 February 2020)
[2] United Nations Department of Economic and Social Affairs 2019 *World population Prospects 2019:Highlights,https://population.un.org/wpp/Publications/Files/WPP2019_10KeyFindings.pdf,* (accesed in 20 February 2020)
[3] Sanyal S 2011 *The end of Population Growth,* https://www.straitstimes.com/world/sanjeev-sanyal-the-end-of-population-growth (accesed in 20 February 2020)
[4] Food and Agriculture Organization of United Nation, 2018 *E-Agriculture in Action: Drones for agriculture,* Bangkok, http://www.fao.org/3/I8494EN/i8494en.pdf, (accesed in 20 February 2020)
[5] Muraru L, Cardei P, Muraru V, Sfîru R, Codruz P 2019 *Researches regarding the use of drones in agriculture,* Proceedings of 9th International Multidisciplinary Scientific GeoConference SGEM 2019, 30 June - 6 July, 2019, Sofia, Bulgaria.
[6] Wen S, Zhang Q, Yin X, Lan Y, Zhang J, and Ge Y 2019 *Sensors* **19** 1112
[7] Gupta S K, Kumar S, and Thombare P B 2019 *Drones for future agriculture* 2019 In book: Advances in agricultural extension – volume 4, Chief Editor Dr. Saurabh Sharma, Publisher: Akinik Publication, New Delhi, e-ISBN 978-93-5335-695-8, pages:41-62

IOP Conf. Series: Materials Science and Engineering **898** (2020) 012002 doi:10.1088/1757-899X/898/1/012002

Comparison of passenger vehicle braking distance when travelling on snow and asphalt at different velocities

A I Radu[1] and B A Tolea[2]

[1] Department of Automotive and Transport, University "TRANSILVANIA" of Brasov, Romania
[2] Department of Mechanical Engineering, University of Oradea, Romania

E-mail: alexandru.radu@unitbv.ro

Abstract. The objective of this paper was to determine the braking distance of a passenger vehicle when traveling at the speeds of 30, 50 and 70 km/h, on low friction surfaces (snow, deep snow) and also on asphalt. A number of tests were conducted on different surfaces. The vehicle was accelerated and then braked completely. Based on the braking distance values obtained by GPS, the adhesion coefficient was calculated for each surface for every velocity value tested.

1. Introduction

In winter time, cars travel on different road surfaces influenced by the weather, causing modifications to the dynamic characteristics of the vehicle [1]. Statistic show that in the winter time, on icy roads, the injury risk is believed to be 20 times higher than in the summer where the road is dry [2]. Vehicle parameter will depend on various external parameters that intervene when driving in winter conditions unpredictable to the driver [3].

For the study, by using the GPS data, the vehicle velocity was obtained. Acceleration and distance travelled were determined analytically using the formulas [4]:

$$a = \frac{dV}{dt} \tag{1}$$

$$v = \frac{dS}{dt} \tag{2}$$

Where: V – velocity (m/s), a - acceleration (m/s2), S – distance (m), t – time (s)
Formula (3) defines the velocity in relation with the gravitational acceleration and adhesion coefficient [5]:

$$\left|\frac{dV}{dt}\right|_{\max} = g \cdot \varphi \tag{3}$$

Where: g – gravitational acceleration (m/s2) and φ – adhesion coefficient. To obtain the braking distance, the formula was used [6, 7]:

$$S_f = \frac{V_{\max}^2}{26 \cdot g \cdot \varphi} \tag{4}$$

In this relation, the velocity Vmax is in (km/h). The variation of the adhesion coefficient is given by the type of surface the wheel travels on. In the figure it is illustrated the value of the coefficient in regards to the substances that are between the wheel and the road [8].

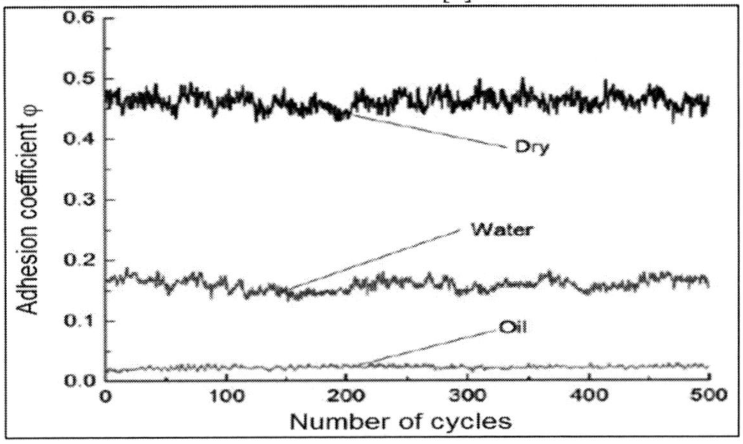

Figure 1. Adhesion coefficient variation.

Tire-road adhesion is a property of rubber that causes it to stick to other materials, as we see with adhesive tape. Adhesion is generally thought to be the result of momentary molecular bonding between the two surfaces [9].

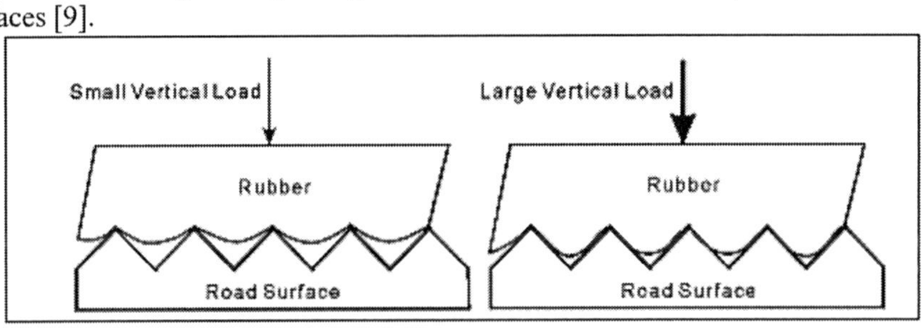

Figure 2. Example of how adhesion works.

When velocity is applied to a rubber slider on a rough surface it results in the deformation of the rubber by high points on the surface called irregularities or asperities. In the figure is shown how the rubber of the tire is dispersed when vertical load is applied along with the forward velocity [10].

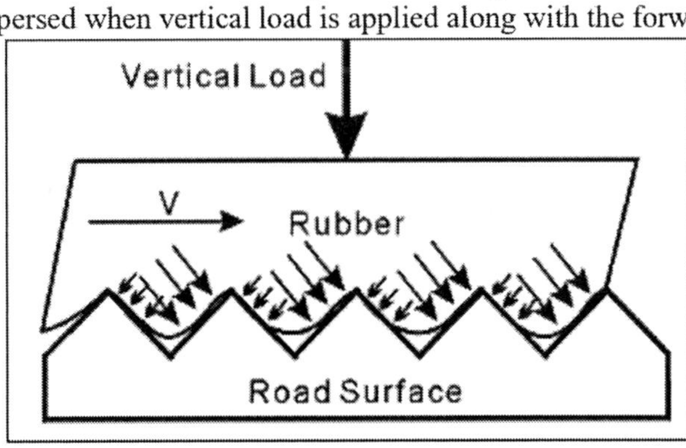

Figure 3. Adhesion influenced by velocity.

2. Methodology

The method used to determine the braking distance was the mean of experimental tests using a passenger vehicle. A controlled environment was established were the road was covered by snow, and another environment were the asphalt road was partially dry. The vehicle was accelerated at the desired velocities (30 km/h, 50 km/h and 70 km/h) then the brake was applied using the full available brake force. To study the key parameters of the vehicle during braking, a GPS system was mounted on the vehicle. By using the data from the GPS, the deceleration, distance, time and velocity parameters were obtained. For the vehicle, a standard sedan vehicle was used equipped with winter tires type Michelin Winter Sport 4D with the dimensions: 225/55 R16 92E M+S.

Figure 4. Tires used for the vehicle and measuring the snow on the road.

In the figure below the test scenario is presented. The vehicle is accelerated at the desired velocity and then the full brake force is applied until the vehicle comes to a complete stop.

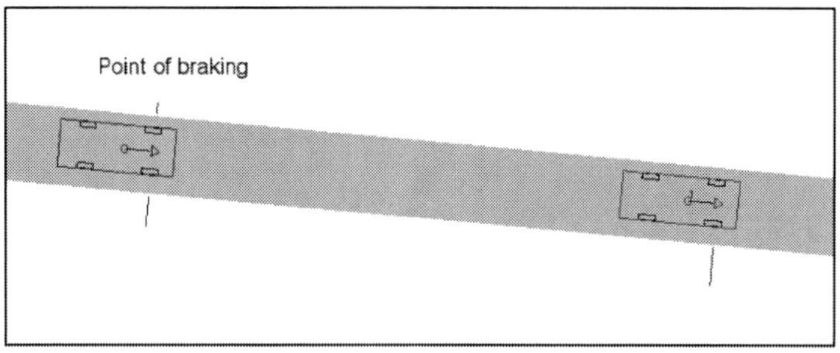

Figure 5. Testing scenario.

There are total of 7 tests conducted with the following velocities: 30 km/h, 50 km/h and 70 km/h on 2 types of surfaces: snow and asphalt. One extra test was conducted at 50 km/h on fresh snow.

3. Results

The primary results were the vehicle distance of braking for the different surfaces at the desired velocities. By studying the data provided by the GPS, the deceleration of the vehicle could be obtained and compared for each surface.

This reflects the braking force of the vehicle, meaning a higher deceleration value resulted in a higher adhesion coefficient between the tire and the road. Also by using the data from the GPS, the adhesion coefficient was accurately calculated for each surface and for each test. In the next two figures the GPS data is presented for the velocity of 30 km/h on snow and asphalt.

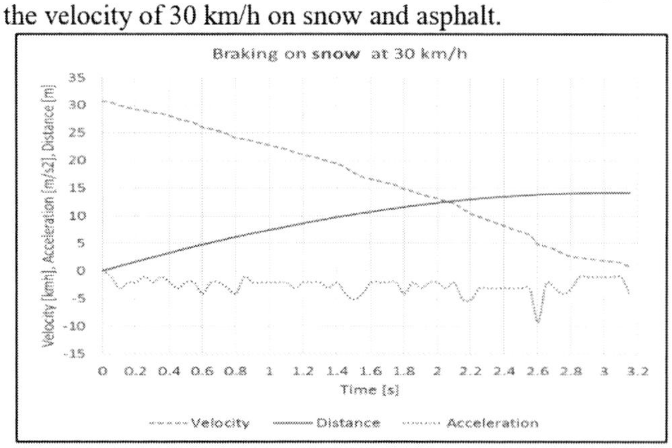

Figure 6. GPS data for 30 km/h on snow.

It can be observed that the distance travelled by the vehicle to completely stop it was 14 m in a period of 3.1 seconds and having a maximum deceleration of 9 m/s2 but with an average of 3-4 m/s2.

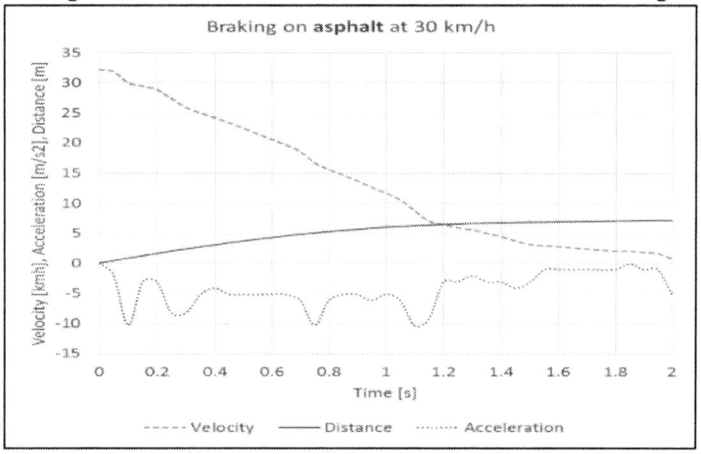

Figure 7. GPS data for 30 km/h on asphalt.

Compared to braking on snow, on asphalt the deceleration of the vehicle is higher, around 10 m/s2 and the duration of the braking phase shorter, 2 seconds compared to 3.1 s. The braking distance is a lot shorter as well, only 6 m. Next are three test at the velocity of 50 km/h on snow, fresh snow and asphalt done in the same conditions.

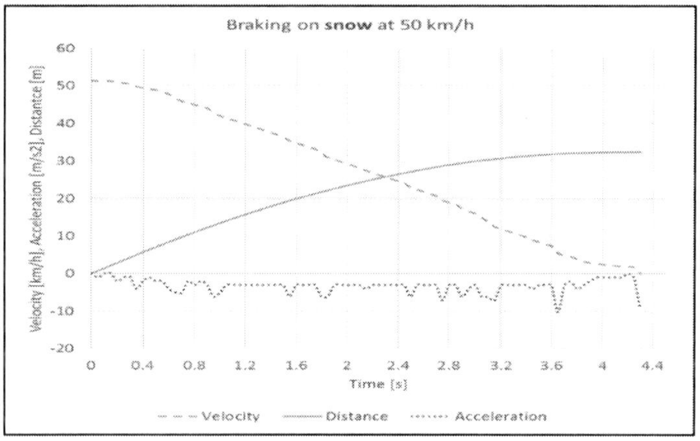

Figure 8. *GPS data for 50 km/h on snow.*

The deceleration in this case is similar to the 30 km/h test, with an average of 5-6 m/s2 and braking distance of 31 m in a period of 4.3 seconds.

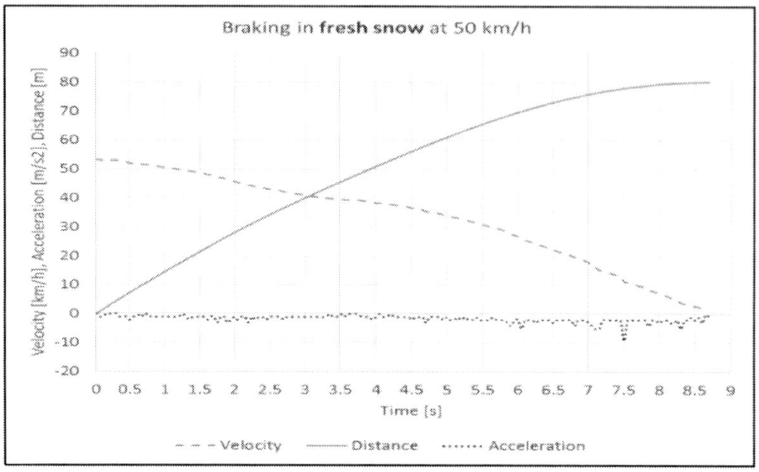

Figure 9. *GPS data for 50 km/h on fresh snow.*

In this case the braking distance increases dramatically compared to regular snow covered road, with a total travel distance of 80 m in a period of 8.5 s and a deceleration of 3 m/s2 on average.

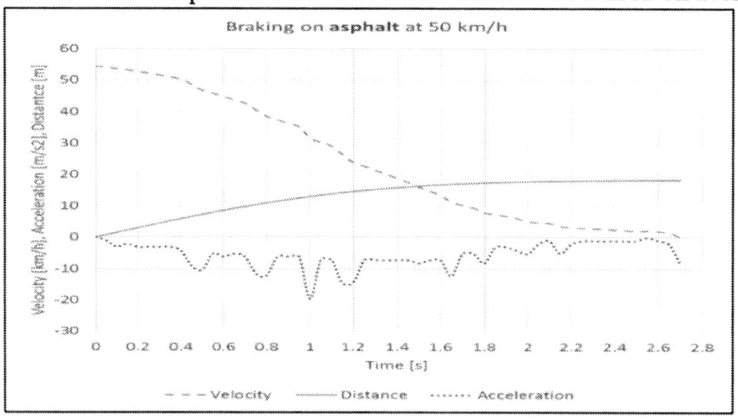

Figure 10. *GPS data for 50 km/h on asphalt.*

On asphalt the braking distance at 50 km/h was about 19 m in a period of 2.7 s and a deceleration of 10 m/s2 on average. In the last 2 tests, the velocity was increased at 70 km/h on 2 surfaces, snow and asphalt.

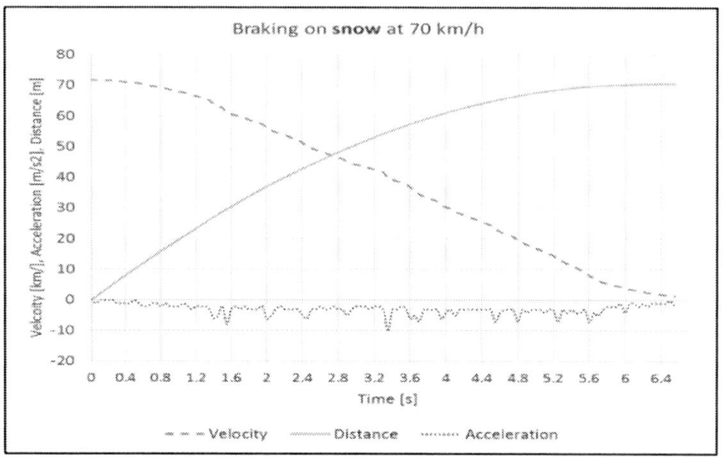

Figure 11. *GPS data for 70 km/h on snow.*

At this velocity, the braking distance on snow was about 70 m in a period of 6.4 s and an average deceleration of 6-7 m/s2.

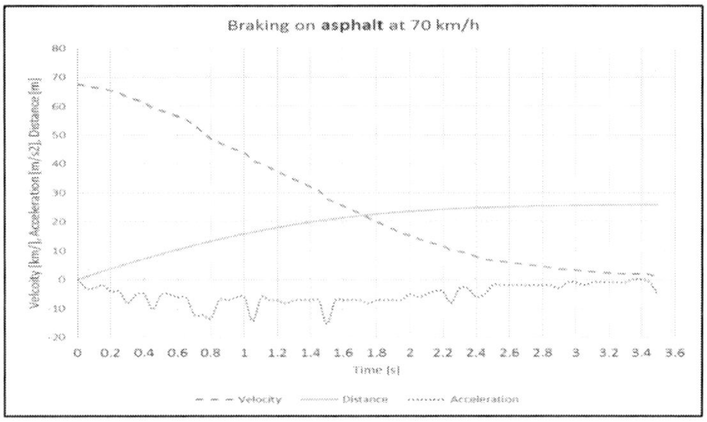

Figure 12. *GPS data for 70 km/h on asphalt.*

On asphalt the braking distance is considerably lower, with 26 m, achieved in a period of 3.5 s and an average deceleration of 11 m/s2. In table 1, a summary of the results are presented and also, using the formulas mention earlier, the adhesion coefficient predicted between the wheel and the road based on GPS data.

Table 1. Summary of the results.

Velocity [km/h]	Surface	Braking distance [m]	Adhesion coefficient
30	Asphalt	5	0.72
30	Snow	14	0.26
50	Asphalt	12	0.70
50	Snow	31	0.33
50	Fresh snow	66	0.16
70	Asphalt	20	0.68
70	Snow	70	0.3

It was found that there was a considerable difference between the regular snow and fresh snow. On fresh snow the adhesion coefficient calculated was 0.16 similar to ice, and in the case of snow, it had a value of 0.33. This difference doubles the braking distance from 31 m to 66 m. In the figure below, there is a graphical representation of the braking distance values and how can different surfaces affect these values.

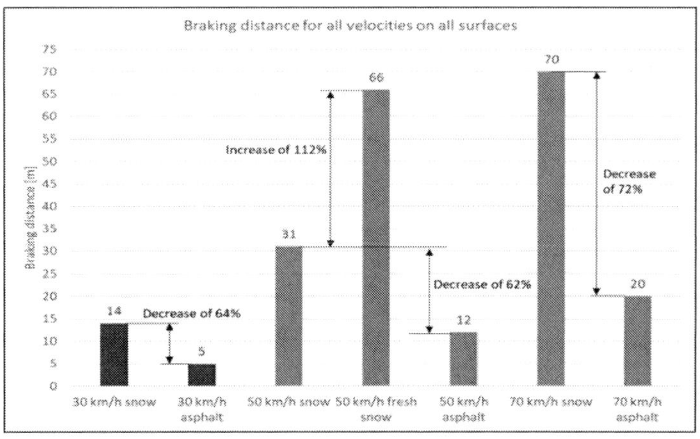

Figure 13. *Braking distances for all the tests.*

In can be observed that the greater the velocity is, the greater the braking distance increases. At lower velocities, at 30 km/h from snow to asphalt, the distance decreases by up to 64 % similar to higher velocity of 50 km/h where the difference is 62%. At 70 km/h from snow to asphalt, the distance decreases by up to 72%.

The huge difference is between snow and fresh snow, on fresh snow the braking distance increases by 112%, doubling it, increasing the chances of accidents.

4. Conclusions

The main conclusion was that the difference in braking distance between asphalt and snow was more than 60-70% increasing the chances of accidents. It was also found that there was a considerable difference between the fresh snow and regular snow. On the fresh snow, the adhesion coefficient was only 0.16, and in the case of shallow snow the coefficient had a value of 0.33. This difference doubles the braking distance of the vehicle.

5. References

[1] Sokolovskij E 2007 Automobile braking and traction characteristics on the different road surfaces *Transport* **22** 275-278

[2] Rämä P & Kulmala R 2000 Effects of variable message signs for slippery road conditions on driving speed and headways *Transportation research part F: traffic psychology and behaviour* **3** 85-94

[3] Gillespie T D 1997 *Vehicle dynamics* (Warren dale)

[4] Minguzzi E 2005 Differential aging from acceleration: An explicit formula *American journal of physics* **73** 876-880

[5] Kato H Momiyama M Yasui Y Muragishi Y Imoto Y & Aizawa H 2008 *U.S. Patent No. 7,398,145* (Washington, DC: U.S. Patent and Trademark Office)

[6] Mannering F Kilareski W & Washburn S 2007 *Principles of highway engineering and traffic analysis* (John Wiley & Sons)

[7] Aashto A 2001 Policy on geometric design of highways and streets *American Association of State Highway and Transportation Officials* **1** 158

[8] Wang W J Zhang H F Liu Q Y Zhu M H. Jin X S 2016 Investigation on adhesion characteristic of wheel/rail under the magnetic field condition *Proceedings of the Institution of Mechanical Engineers, Part J: Journal of Engineering Tribology* **230** 611-617

[9] Haney P 2004 Rubber Friction *Inside Racing Technology* **27**

[10] Rubber Friction 2018 http://insideracingtechnology.com/tirebkexerpt1.htm, last accessed 2018/05/10

IOP Conf. Series: Materials Science and Engineering **898** (2020) 012003 doi:10.1088/1757-899X/898/1/012003

Research of NO_x and PM10 pollutants in Cluj-Napoca with the mobile system for mitigating public health risks

D L Băldean[1], L Andrei[2] and A I Borzan[1]

[1] Automotive and Transportation Department, Technical University of Cluj-Napoca, Cluj, Muncii 103-105, Romania

[2] Infectious Disease Hospital of Cluj-Napoca, Cluj, Iuliu Moldovan 23, Romania

doru.baldean@auto.utcluj.ro

Abstract. The present research is conducted in order to show the possibilities of creating a smart and advanced mobile system capable of investigations regarding the chemical composition of the atmospheric air in the metropolitan areas using flexible technologies and some part of the existing infrastructure. Experimental measurements are made using a mobile station for data recording connected to the internet and with a link to the air quality stations. Taking into consideration the values of different chemical compounds within the atmospheric air, the data processing station may create a chemical map indicating the higher risk of air toxicity. Nitrogen oxides and particulate matter are some of the essential chemical pollutants generated these days by modern industries and transportation activities. By installing the proper technologies and networking capabilities on a current car the research team has made a practical investigation of actual chemical values in the metropolitan area of Cluj-Napoca by accessing the NO, NO_2, NO_x, and PM 10 data retrieving lines. Necessary changes in hourly retrieved data show that road traffic and other human activities do change the air quality level and its chemical composition. An air map is provided for some moments of the day, including the four stations in Cluj-Napoca city. The practical research is intended to present the technological capability and its embedding potential for the new series models of cars which may thus indicate a proper route in order to avoid critical points and to reduce the toxic chemical overload of some areas.

1. Introduction

Today, when health is a global concern investigating air quality is one of the most important activities that may be done. The air intake is more important than water and food ingestion due to the fact of the frequency of this act and its importance for each living being. Atmospheric air is the first, and the most important transfer supports for disease agents, both biologic and physic-chemical origins, making it more important for scientific research and engineering.

There were already developed a series of applied experiments and testing for improving diagnostic devices and systems in polluted metropolitan environments with nitric and carbon oxides [1], but further research of the problem is needed. Some pollutants [2] are quite closely related to the operation of transport vehicles and their internal combustion engines [3, 4]. Temperature, pressure and engine speed have an important influence on the operating regime [5], but in order to optimize the engines, either conventional ones or the alternative types, an experimental approach must be considered [6, 7]. Researching the actual operating values and their influence upon emissions [8], as well as their on-board display [9], could support the analysis of chemical composition with portable devices in order to gain the environmental quality profile [10] and the status of the embedded after-treatment system [11]. Using

IOP Conf. Series: Materials Science and Engineering **898** (2020) 012003 doi:10.1088/1757-899X/898/1/012003

precise sensing methods [12] and equipment [13] gas composition may be established, and specific compounds may be identified or precisely located.

Figure 1 presents the state-of-the-art images regarding the chemical pollution of the atmospheric air (a) and environmental mapping with portable profile-mapper of air quality (b) in order to optimize transportation activity [14].

Figure 1. Capture of air pollution in today industry (a) and environment investigation device (b) [14].

2. Research methodology, materials and results

Today's fast networking and advanced mobile systems allow the researchers and the engineers to develop new smart applications for environmental investigations and atmospheric chemical profile. The reference methodology applied when measuring the nitrogen dioxide and nitrogen oxides has been the one given by the standard SR EN 14211 known as the Ambient air quality. Standard applied method for investigating the actual values of both nitrogen dioxide and nitrogen monoxide is with chemi-luminescence. The applied method for the sampling and determining of PM_{10} and $PM_{2.5}$ actual values is the one specified in the standard EN 12341 Ambient air - Standardized methodology for gravimetric measurement to determine the PM_{10} or $PM_{2.5}$ mass fraction of particulate matter in suspension [13]. NO_x and PM_{10} are generated by road-traffic and industry, both investigated here.

The experimental research car is a series model available at the Technical University of Cluj-Napoca. It was equipped with a mobile advanced system having internet connection capabilities and data transfer via Wi-Fi, as shown in figure 2. It has a wheel train 1, an internal combustion engine 2, a clutch 3, gearbox 4, clutch control 5, engine unit 6, transmission control 7, traction control 8, on-board instrument cluster 9, steering system 10, axle drive 11, exhaust control 12, radar system 13, air quality check point 14 (metropolitan chemical measurement station CJ1), 15 satellite connection. Applied research is showing the hourly levels that may impact the public health and are a risk factor for life.

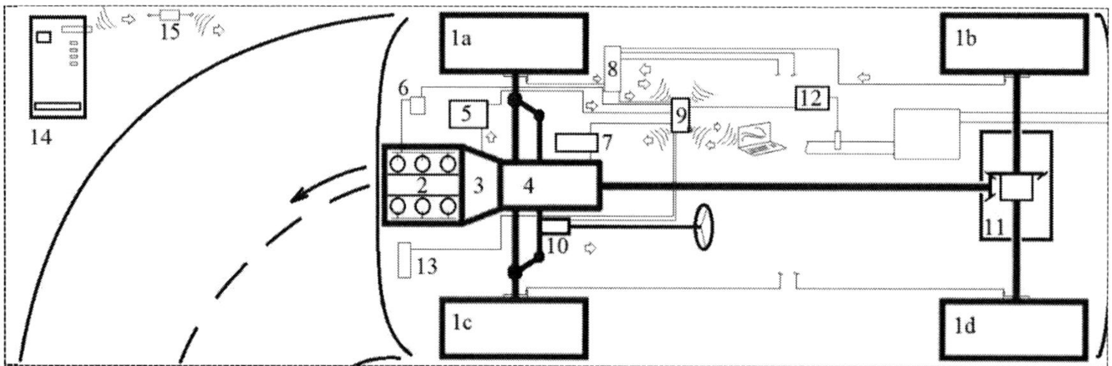

Figure 2. Schematics of the research methodology and equipment placement in the testing structure

In figures 3 to 10 are shown the experimental measurements with the mobile advanced system regarding the air quality check with flexible automotive technology in relation to the infrastructure.

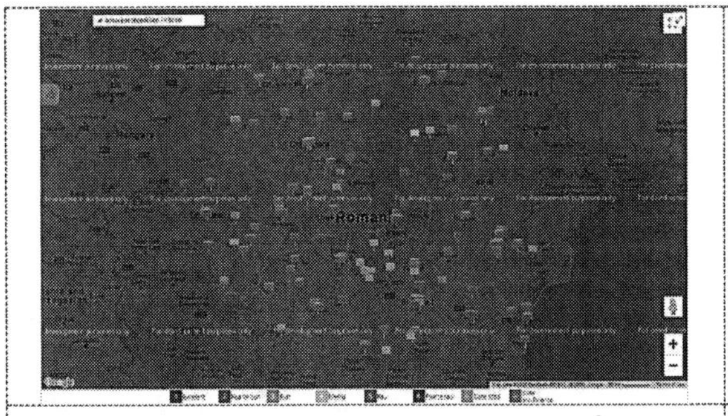

Figure 3. Romania air quality map before noon.

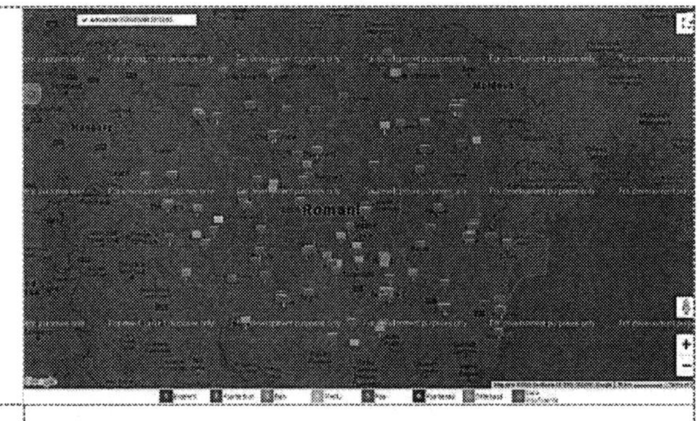

Figure 4. Country's evening air quality map.

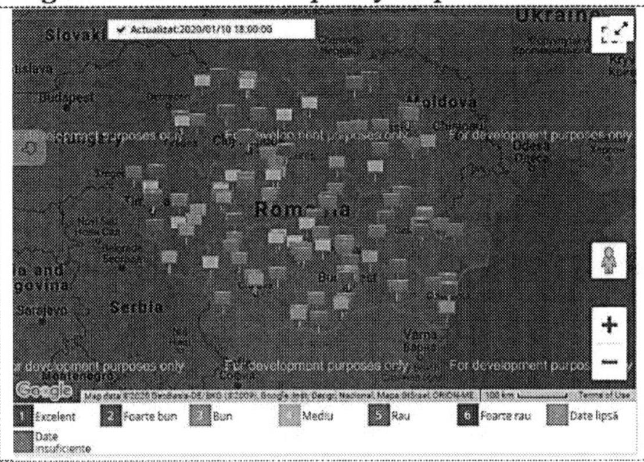

Figure 5. Air quality in the afternoon, at 18:00.

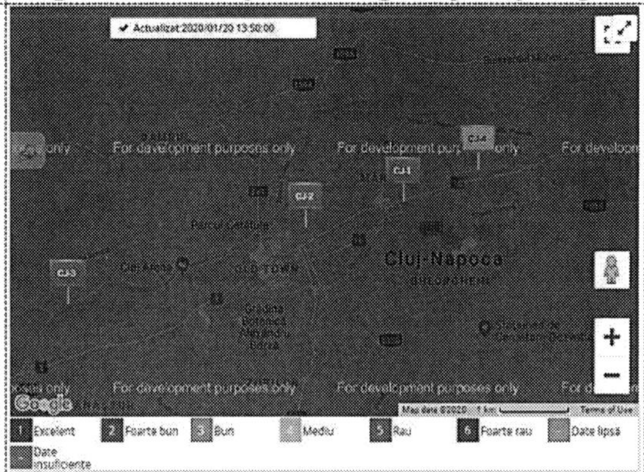

Figure 6. Measuring stations in Cluj-N. city.

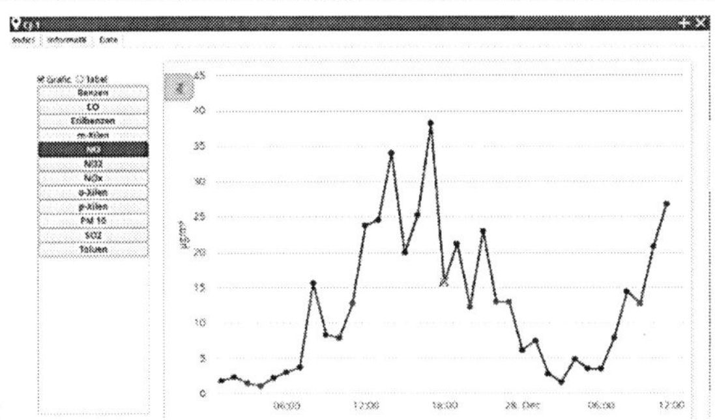

Figure 7. NO variation during 27 and 28 dec.'19

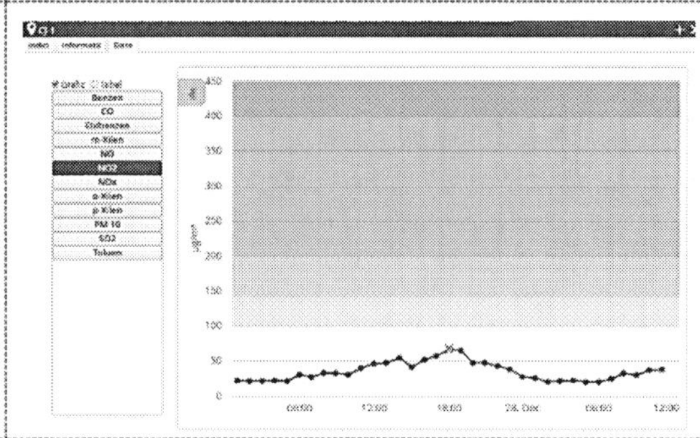

Figure 8. NO_2 variation in 27 and 28 dec.'19

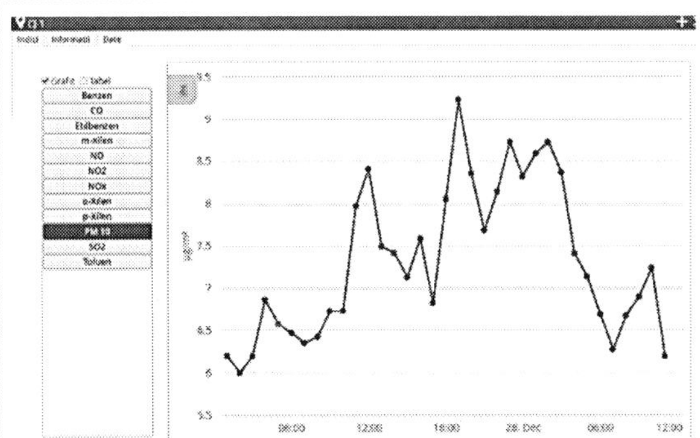

Figure 9. PM 10 variation in 27 and 28 dec.'19

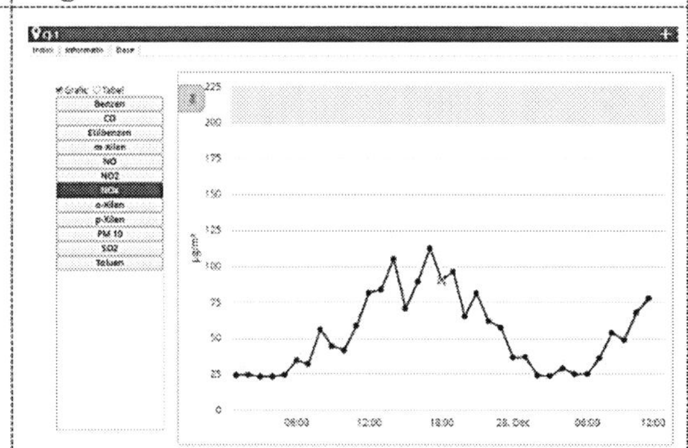

Figure 10. NO_x variation in 27 and 28 dec.'19

The tests were made for a prolonged period in order to take multiple sets of data.

The partial data base concerning NO_2 and PM10 actual values are given in figures 11 and 12. It is observed that before important events, such a Christmas, intense human and road activities are doubling the levels of NO_2 and PM_{10} emissions in comparison to actual holidays and 2020 lockdown.

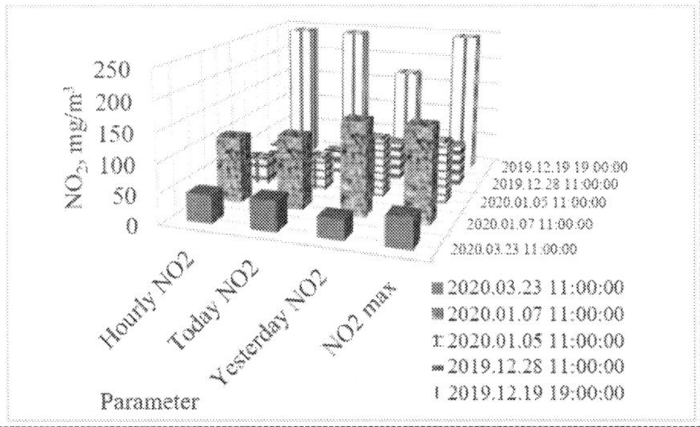

Figure 11. NO_2 values in dec.'19 and mars 2020

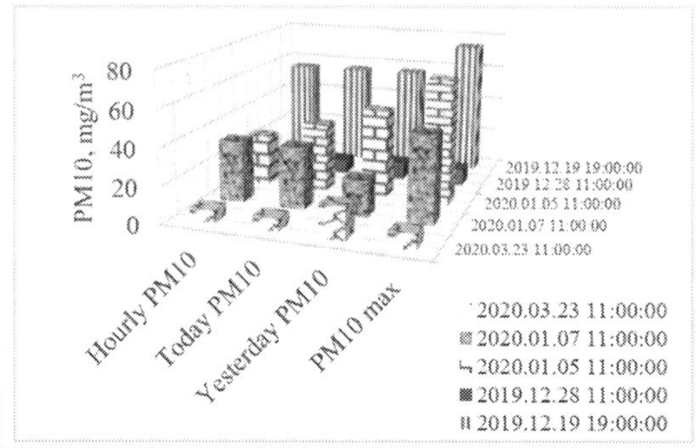

Figure 12. PM 10 values in dec.'19 and spring

The location specifications and investigated pollutants are presented in table 1, presenting the coordinates and addresses of the stations. Data from station CJ1 was represented graphically.

Table 1. Technical specifications for the air quality check points

Station CJ1	A. Vlaicu Street, no. 1; Latitude 46.78; Longitude 23.62; Altitude 336.00
Station CJ2	N. Bălcescu Street, no. 6; Latitude 46.78; Longitude 23.60; Altitude 336.00
Station CJ3	1 Decembrie 1918 Bvd., no. 1; Latitude 46.77; Longitude 23.55; Alt. 346.00
Station CJ4	Dâmboviței Street, no. 80; Latitude 46.78; Longitude 23.63; Altitude 323.00
Investigated Pollutants	NO, NO_2, NO_x, PM 10
Accuracy, [%]	98-100

Measurements taken with the equipment in experimental research allowed us to build a polynomial model on a multi-point determination as follows:

$$NO_2 = 22.779x^2 - 166.66x + 351.95 \quad (1)$$

where x is the timetable (day) measurement variable consisting in the type of working or holiday when taking the measurement, for a polynomial model of NO_2 evolution.

3. Conclusions and observations

The flexible system of measurement works properly and allows the receiver to access data from stationary infrastructure in order to gain insight of air quality in traffic at any given time. For development purpose the infrastructure gives access at this point to the 4 air quality check points in the city of Cluj-Napoca. These checkpoints may be accessed at any time regardless the location of the mobile receiver with the condition of internet connection being available.

For public health concerns hourly NO emissions increase in afternoon, which mean respiratory difficulties, irritations of the respiratory tract and lung's disfunctions such as emphysema. Lower values are recorded during the night fall and in the first hours of the day. The peak point for NO compounds and for PM10 were recorded around afternoon at 18 o'clock by connecting the flexible database with the measuring station from the metropolitan infrastructure. The investigation is required to be developed in order to complete database and expand the applied study.

References

[1] Andrei L et al 2019 *Applied Measurements and Instrumentation for Improving Diagnostic Devices and Systems in Metropolitan Polluted Environments with Nitric and Carbon Oxides* (6th International Conference on Advancements of Medicine and Health Care through Technology; 17–20 October 2018, Cluj-Napoca, Romania, Springer, pp. 45-49) DOI https://doi.org/10.1007/978-981-13-6207-1_8

[2] Andrei L et al 2018 *Research and reporting development of formic acid emissions in environment pollution management and life health* (In Science and Engineering / Știință și Inginerie AGIR Press) eISSN 2359–828X DOI https://http://stiintasiinginerie.ro/34-71

[3] Barabas I et al 2010 *Performance and emission characteristics of an CI engine fueled with diesel–biodiesel–bioethanol blends* (In Elsevier, Fuel, 89 (12) pp. 3827-3832) Online ISSN 0016 – 2361, https://doi.org/10.1016/j.fuel.2010.07.011

[4] Borzan AI et al 2019 *Experimental design for Diesel supply control in order to improve fuel efficiency* (In IOP Conference Series: Materials Science and Engineering, Volume 568, Issue 1, 17 September 2019, Article number 012038, Annual International Session of Scientific Papers - IMT Oradea 2019; Oradea, Felix SPA; Romania; 30 May 2019 through 31 May 2019; Code 152405) ISSN: 17578981, DOI: 10.1088/1757-899X/568/1/012038

[5] Botean AI 2019 *Influence of the temperature gradient on the speed of a Stirling gamma engine* (In Acta Technica Napocensis Series-Applied Mathematics Mechanics and Engineering, Volume 61, Issue 3, pp. 339-346 Published: Sept. 2018) ISSN: 1221-5872, Accession Number: WOS:000468025900008

[6] Botean AI 2019 *Optimization of a gamma-type Stirling engine performance based on an experimental approach* (In Acta Technica Napocensis Series-Applied Mathematics Mechanics and Engineering, Volume 62, Issue 2, Pages: 309-316 Published: JUN 2019) ISSN: 1221-5872, Accession Number: WOS:000483186000012

[7] Ferenti I et al 2019 *Experimental research in mechanic's field of control for further innovation in the hydraulic and electronic operated injection system* (In E3S Web of Conferences Volume 85, 22 February 2019, Article number 08006, 2018 Sustainable Solutions for Energy and Environment, EENVIRO 2018; Cluj-Napoca; Romania; 9 October 2018 through 13 October 2018; Code 145542) ISSN: 22671242, DOI: 10.1051/e3sconf/20198508006

[8] Ferenti I et al 2014 *Research of some operating parameters and the emissions level variation in a spark ignited engine through on-board investigation methods in different loading conditions* (In Central European Journal of Engineering, Volume 4, Issue 2, June 2014, pp. 192-208) ISSN: 18961541, DOI: 10.2478/s13531-013-0167-9

[9] Jovrea S et al 2018 *Researching on-board display of essential information concerning technical conditions in operation and fuel-economy of a motor-vehicle in operation (*In Science and Engineering/Știință și Inginerie AGIR Press, Volume 31, Issues 67) eISSN 2359–828X

[10] Krzysztof S et al 2019 *The Study on Influence of the Method of Handling of Measuring Head on Measurement Results Obtained with the Use of a Portable Profilometer* (In: Tehnicki Vjesnik-Technical Gazette, Volume 26, Issue 3, pp. 592-595) Online Last modified June 2018, DOI 10.17559/TV-20160601124330

[11] Odenie VS et al 2017 *Experimental researches concerning the clogging level of diesel particle filter (DPF) after 45000 km in case of k9k892 e5 1.5 dci engine* (Bucharest AGIR Publishing Press)

[12] Yin XT et al 2019 *A highly sensitivity and selectivity Pt-SnO2 nanoparticles for sensing applications at extremely low level hydrogen gas detection* (In: J of Alloys and Compounds, Publishing date Oct 15 2019, Volume 805, pp. 229-236) DOI: 0.1016/j.jallcom.2019.07.081

[13] *** 2019 *Air quality. Air quality assessment* (In: National Air Quality Monitoring Network) DOI: http://www.calitateaer.ro/public/assessment-page/assessment-page/?locale=en

[14] *** 2019 *Bosch Mobility Solutions* (In: EN _ Bosch - improving air quality and securing individual mobility) DOI: https://www.bosch-mobility-solutions.com/en/highlights/power train-and-electrified-mobility/urban-mobility-and-air-quality/

IOP Conf. Series: Materials Science and Engineering **898** (2020) 012004 doi:10.1088/1757-899X/898/1/012004

Friction induced heating properties of the polyamide/steel type contacts

M T Lates[1]

[1]Product Design, Mechatronics and Environment Department, "Transilvania" University of Brasov, Brasov, Romania

E-mail: latesmt@unitbv.ro

Abstract. The endurance and the life time of the mechanical transmissions are highly influenced by the functioning conditions and, not at least, by the materials that they are manufactured from. The increase of the endurance and functioning life time, corroborated with a friendly impact with the environment, is one of the main aims of the worldwide research areas of today. According to this, the development of new materials characterised by low costs and high endurance represents one of the solutions obtained by the specific scientific research. The paper investigates the friction induced heating properties of the polyamide/steel type contacts by considering dry friction conditions. It is studied the contact between a steel made pin and a PA46 and PTFE added PA46 polyamide made disk, under different testing conditions, depending on the rotational speed of the disk and the normal force, at the environmental temperature. The tests are achieved on a tribometer equipped with a pin-on-disc module and, as output, the friction coefficient is measured. Due to the friction is developed a heating of the local contact area. The heat is calculated according to the measured friction coefficient, the normal load and the relative sliding velocity. Finally there are given conclusions regarding the friction induced heating properties of the tested materials.

1. Introduction

The environmental friendly solutions are widely researched themes today in many areas as: energy production systems, automotive industry, aircraft and railway industry, food production and processing systems, chemical industry etc. These solutions are characterised by low energy consumptions and low CO_2 and other pollution gases emissions, during their manufacturing process and their usage. According to this, one research direction is represented by the development of new materials; a class of materials which are characterised by small frictional loses and good endurance in the contacts with steel made parts is represented by the polyamides. These materials, in combination with steel made parts, have applications in automotive industry, aerospace engineering, chemical industry and electronics due to their good tribological properties, corrosion resistance and simple and economic manufacturing process [1, 2, 3, 4].

In the scientific literature the tribological properties of the polyamide type materials are determined by performing two categories of tests on the tribometer: tests performed on the pin-on-disc module and tests performed on the reciprocating module.

In [5] there are performed test on a pin-on-disc module for a composite PA66 polyamide with different values for the contact pressure and for the velocity; as results, there are highlighted the friction and wear behaviour of the tested material: low friction coefficients in combination with steel materials and stabilised wear.

The paper [6] presents the mechanical properties of a class of PA66 polyamides. The tests are performed on a pin on disk type tribometer under dry and lubricated conditions. In the case of no lubricated conditions, the friction coefficient is increasing with the increasing of the force and has values between 0.15 ... 0.23 for forces between 50 N ... 250 N, the pin diameter d=10 mm and the velocity v=0.025 m/s; the friction coefficient is decreasing with the increasing of the sliding velocity – bellow 0.1 for 0.1 m/s.

In [7] the friction is studied using a pin-on-disc module (the pin is made of two types of polyamides and the disk is made of steel) with normal loads of 5, 10, 20 and 30 N and rotational speeds of 1000, 1500 and 2000 rpm for a pin of 12 mm diameter, dry friction conditions and 1 km sliding distance. The friction coefficient is decreasing with the increasing of the force with values between 0.14 ... 0.42 for the pure PA66 and with 0.32 ... 0.42 for the graphite reinforced PA.

The dry sliding tests on reciprocating type motions modules are performed in [8] and [9]. In [8] the tests were performed on an Universal Micro Tribometer (UMT) [10] on the reciprocating module equipped with a ball on block device, at a humidity of 40% with a stroke of 5 mm. The test period is about 1800 s and the friction coefficient of PA66 is equal with 0.22 ... 0.28 increasing with the increasing of the normal load and with the increasing of the frequency. The paper [9] presents the study of the friction coefficient and of the wear in the case of a GCr15 steel ball in contact with a PA66 plate; the tests were performed on a Universal Micro Tribometer (UMT) [10] by using the reciprocating module of it. The normal loads were 1 N, 2 N, 3 N, 4 N for a diameter of 4 mm of the ball; the sliding velocities were 31.42, 62.83, 94.25, 1245.66 mm/s. Under dry conditions, the friction coefficient decreases with the increasing of the normal load with values between 0.15 ... 0.4 and is increasing with the increasing of the translational speed.

Under no lubricated conditions, due to the friction, in the local area of contact, it is produced heat. The paper [11] studies the heat produced by a dry friction of PA6 nanocomposite fabrics and steel; the local temperature increases with the increasing of the normal load and the increasing of the test period.

The paper investigates the friction induced heating properties of the PA46 and PTFE added PA46 polyamide on steel type contacts by considering dry friction conditions and different testing conditions, depending on the rotational speed of the disk and the normal force, at the environmental temperature. The heat is calculated according to the measured friction coefficient, the normal load and the relative sliding velocity. Finally there are given conclusions regarding the friction induced heating properties of the tested materials.

2. The tests

The experimental rig used to perform the tests is an Universal Micro Tribometer (UMT) [10] connected to a computer, as it can be seen in figure 1. On the test rig may be performed wear and friction tests.

The wear can be measured by using a sensor which allows the vertical stroke for the slider of 150 mm with an accuracy of 50 nm [10]; the lateral stroke is equal with 75 μm and can be adjusted with an accuracy of 2 μm [10].

The friction tests are performed in order to find out the friction coefficient between different materials being in contact. The friction coefficient is calculated automatically by the test rig's software as the ratio between the force measured about a horizontal direction, same with the motion's direction and the vertical normal load measured about the vertical direction; these forces can be measured by the sensors in an interval of 0.1 ... 1000 N with the resolution up to 50 mN [10].

A pin-on-disc module is mounted inside the tribometer. The steel made pin has the diameter equal with 6.3 mm. Inside the rotary module are mounted the PA46 and PTFE added PA46 polyamide disks which will be in contact with the steel made pin (figure 2).

The tests are performed after a running-in period of 2 hours at a rotational speed of the disk equal with 1000 rpm and a normal force of 50 N. After that, the tests are accomplished at the room temperature of 22 °C, with a set of rotational speeds equal with: 500 rpm, 1000 rpm, 2000 rpm and 3000 rpm; the normal forces used for the tests are equal with:10 N ... 50 N. Each test duration is equal with 10 minutes. The sliding radius is equal with 12 mm.

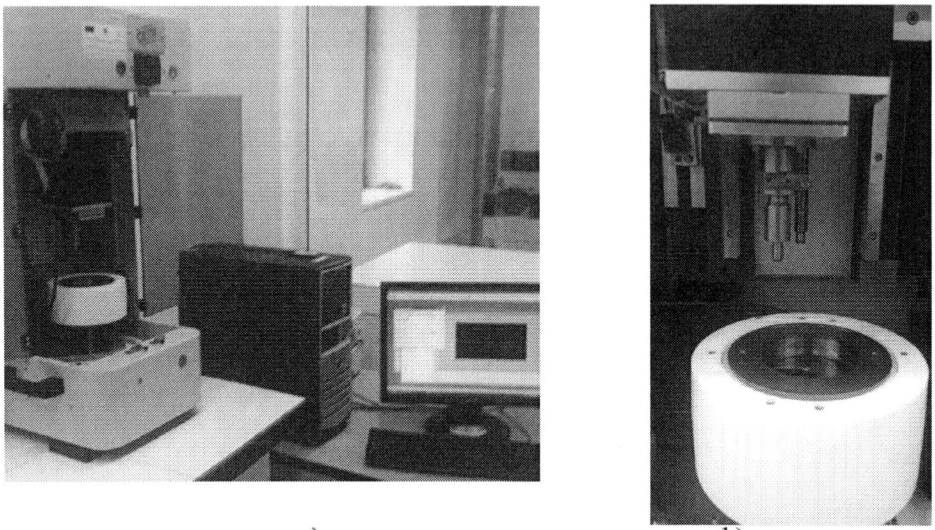

a) b)

Figure 1. The tribometer.

During the tests there are measured the normal force and the horizontal force oriented opposite to the rotation's direction. The software calculates the friction coefficient as the ratio between the two measured forces. The friction induced heat is calculated with [11]

$$Q=\mu F v t \quad (1)$$

where: μ represents the friction coefficient; F – the normal force; v – the speed; t – the test period.

a) b)

Figure 2. The polyamide disks in contact with the steel made pin.

3. Results and conclusions

The results present the evolution of the wear during the running-in test and variation of the friction induced heat with the rotational speed of the disk and with the normal force.

Figure 3 shows the evolution of the wear during the running-in process. The value of the wear, for the PA46 polyamide is stabilised after approximately 60 minute at a value around 0.08 mm. The wear for the PTFE added PA46 polyamide is higher (about 0.11 mm) and it is stabilised after 100 minutes.

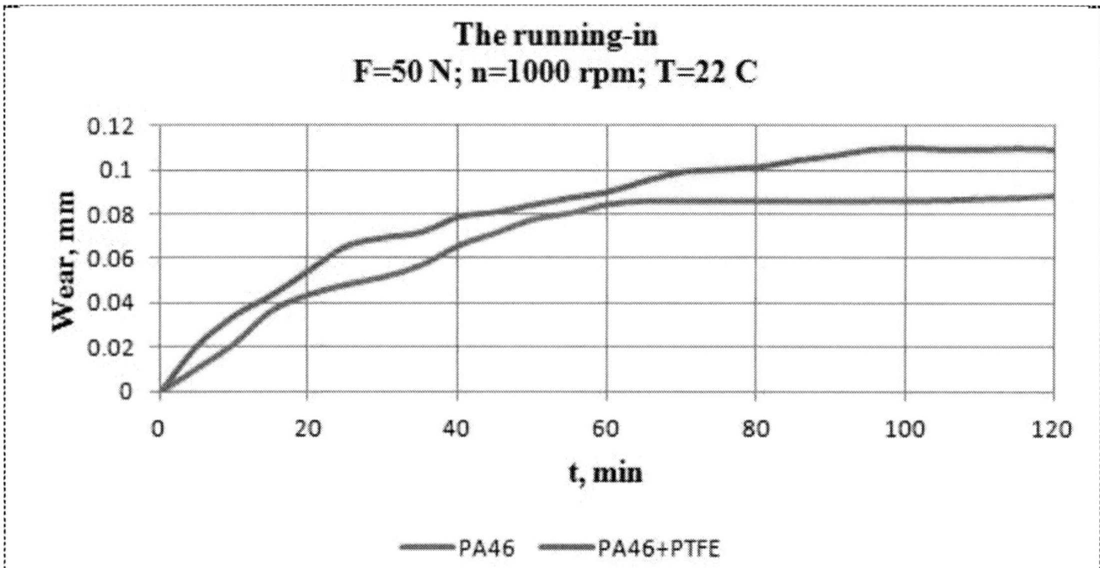

Figure 3. The evolution of the wear during the running-in process.

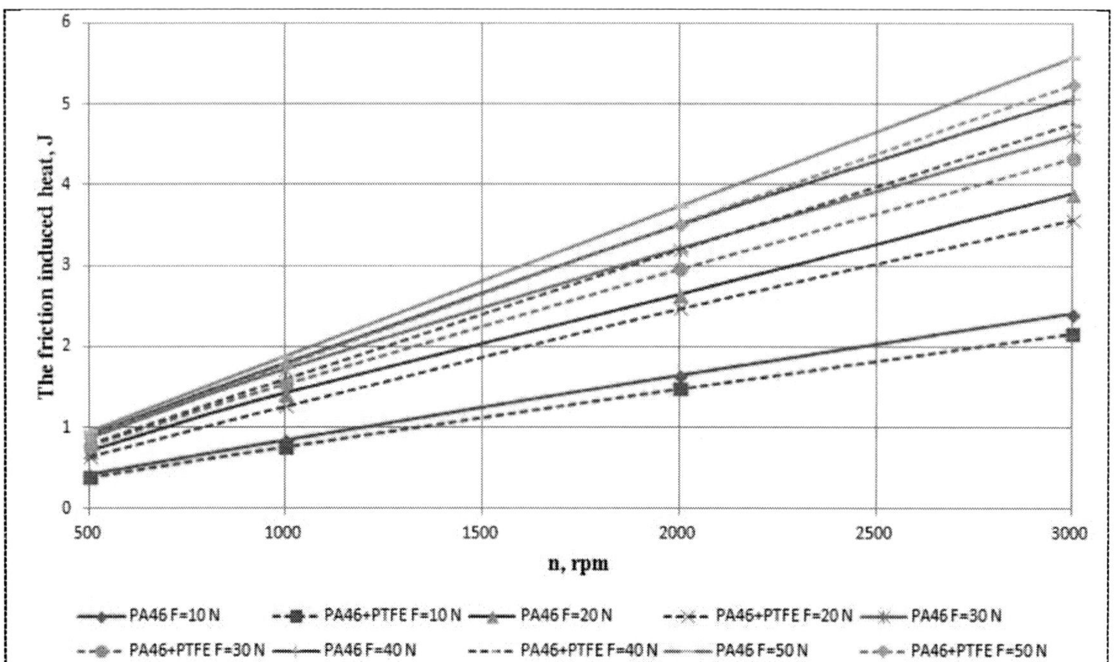

Figure 4. The variation of the friction induced heat with the rotational speed.

The variation of the friction induced heat with the rotational speed and with normal load is presented in figure 4 and in figure 5. Figure 4 highlights that the amount of the friction induced heat increases with the increasing of the normal force and with the increasing of the rotational speed. The PTFE added PA46 polyamide produces smaller amounts of friction induced heat than the PA46 polyamide. According to figure 5, higher differences regarding the evolution of the friction induced heat, between the PA46 and the PTFE added PA46 are noticed at high rotational speeds.

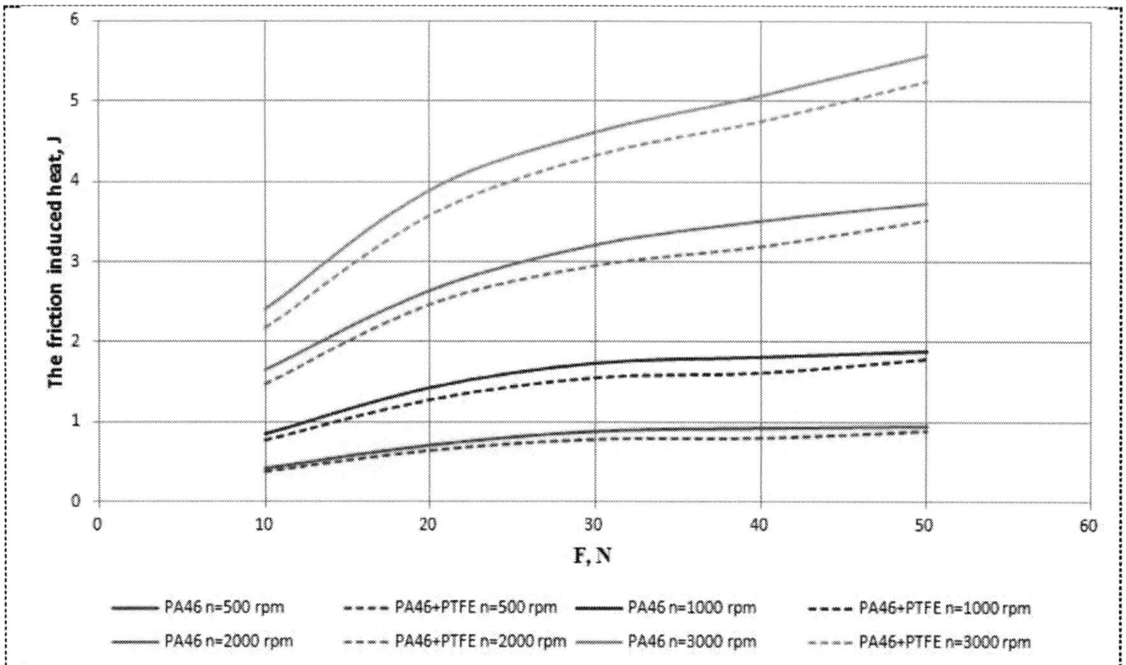

Figure 5. The variation of the friction induced heat with the normal force.

As a general conclusion, the PTFE added PA46 polyamide has a higher wear than the PA46 polyamide and the wear is stabilised at a constant value after a longer period of time, instead of the case of the PA46 polyamide. The friction induced heat increases with the increasing of the speed and of the normal force, for both materials. In all the cases, the PTFE added PA46 has a smaller friction induces heat production. Regarding their applications, the PA46 polyamides are preferable to be used in order to obtain high endurances of the mechanical components. For smaller amounts of produced friction induced heat, the PTFE added polyamide is preferable to be use.

References

[1] Unal H and Mimaroglu A 2012 Friction and wear performance of polyamide 6 and graphite and wax polyamide 6 composites under dry sliding conditions *Wear* **289** pp 132-137

[2] Zhang X R, Pei X Q, Wang Q H, Wang T M and Chen S B 2015 The friction and wear properties of carbon nanotubes/graphite/carbon fabric reinforced phenolic polymer composites *Advanced Composite Materials* **24** pp 147-159

[3] Hoskins T J, Dearn K D, Chen Y K and Kukureka S N 2014 The wear of PEEK in rolling – sliding contact – Simulation of polymer gear applications *Wear* **309** pp 35-42

[4] Zhaobin C, Tongsheng L, Yuliang Y, Xujun L and Renguo L 2004 Mechanical and tribological properties of PA/PPS blends *Wear* **257** pp 696-707

[5] Shibata K, Yamaguchi T, Kishi M and Hokkirigawa K 2015 Friction and wear behavior of polyamide 66 composite filled with rice bran ceramics under a wide range of Pv values *Tribology Online* **10 (2)** pp 213-219

[6] Kozma M 2005 Effect of incorporated lubrication on the tribological properties of polyamides *The Annals of University Dunarea de Jos of Galati* **VIII** pp 65-69

[7] Kumar S S and Kanagaraj G 2016 Investigation on mechanical and tribological behaviors of PA6 and graphite-reinforced PA6 polymer composites *Arab journal of science engineering* **41** pp 4347-4357

[8] Li J and Xia Y C 2010 The friction and wear properties of thermoplastic PA6 composites filled with carbon fiber *Journal of thermoplastic composite materials* **23** pp 337-349

[9] Haixia H, Sirong Y, Mingyu W and Kaixin L 2009 Tribological behavior of polyamide 66-based binary and ternary composites *Polymer engineering and science* **49** pp 2454-2458

[10] *** 2009 UMT Multi-Specimen Test System *Hardware Installation & Application Manual* Center for Tribology, Dell Ave, Campbell, USA

[11] Fasahat F, Dastjerdi R, Mojtahedi M R M and Hoseini P 2015 Wear properties of high speed spun multi-component PA6 nanocompositefabrics; abrasion resistance mechanism of nanocomposites *Wear* **322** pp 17-25

IOP Conf. Series: Materials Science and Engineering **898** (2020) 012005 doi:10.1088/1757-899X/898/1/012005

Waste reduction by implementation of CNC machining center and Lean Manufacturing

I C Gherghea[1], D C Negrau[1], C Bungau[1] and M Faur[1]

[1]Faculty of Management and Technological Engineering, University of Oradea, 1 Universitatii street, Oradea, Romania

cosmin.gherghea@csud.oradea.ro

Abstract. In the context of increasing competitiveness, the companies from manufacturing industry are constrained to provide quality products at a lower price and a shorter delivery time. This brings new challenges during the manufacturing process. One of the main challenges is to reduce or eliminate waste during the production stage. The actual paper presents methods and solutions through which the waste from the manufacturing process is identified and eliminated. As a consequence, increased productive efficiency and higher profit have been gained. The objective of this study is to implement Lean manufacturing tools, in order to improve the production process carried out within a manufacturing company, with a small manufacturing batch. Based on real data from the manufacturing process, the current value stream mapping (VSM) was built, identifying both operational and non-operational times. After waste identification, an optimal solution was chosen for the manufactured parts required by the customer. Implementing the best solution resulted in a productive efficiency increase by 90.93%, which significantly reduces the manufacturing time.

1. Introduction

With the digitization of companies, new challenges have emerged in the operational activities of manufacturing companies, aiming to increase performance by waste reduction. In this context, an important role is played by the tools and the ways that companies use to increase their performance, reducing the waste (different type of waste) within the manufacturing processes. [1-3]

Lean Manufacturing has proven to be one of the best practices that manufacturing companies adopt within organizations to increase their performance by reducing, respectively eliminating waste [4-5].

According to Arunagiri and Gnanavelbabu (2014), a study was conducted on the main waste from automotive industry, ranking the 7 waste defined by Taichi Ohno (overproduction, waiting, transport, over-processing, stocks, unnecessary movements, defects). The most common losses that affect the production process are: transportation, waiting times and unnecessary movements [6-7].

Subsequently, two other types of waste were added (in addition to the 7 waste defined by Taichi Ohno), namely the improper use of human talents and insecurity or ergonomic working conditions [8].

According to Gupta and Kundra (2012), in all manufacturing processes, there are two types of activities, namely value adding activities and non-value-adding activities. The non-value-adding activities are distributed in activities that can be eliminated, referring to those activities for which the client is not willing to pay, and the second category of the non-value-adding activities, which are necessary (these cannot be completely eliminated, but can be reduced by making major changes within the company) [9].

Lean manufacturing is a production system, which aim is continuous improvement of productivity, by identifying and reducing waste [10-11].

According to Azizi and Manoharan (2015), the implementation of the Lean Manufacturing production system involves five principles, as follows: specifying the value, identifying the value flow, creating the value flow, configuring the pull system by the customer and pursuing perfection through continuous improvement [12].

The implementation of the Lean manufacturing concept within a company contributes significantly to increasing productivity and sustainability, improving lead-time and improving quality of product [13-14].

Lead time reduction implicitly reduces the time to make the payment to the raw material suppliers and the time to collect the payment for the product manufactured and delivered to the customer, because the lead time represents the time between the order from the customer is received and the product is delivered to him [15-16].

Even if the transformation of a company based on traditional production principles into a company based on the Lean Manufacturing principles contributes to the increase of the productive efficiency and thereby generates higher profit, according to the study realized by Almanei et al, there are many challenges that the company must face [10].

Regarding the correlation between profitability and waste, Hamed and Soliman (2017) point out in their study that those companies that identify and eliminate waste, automatically increase their profit because the costs of waste are eliminated. Moreover, the savings created by eliminating that percentage of waste can be used to cover other operational expenses (employees' salaries, direct expenses, maintenance costs, etc.) [17].

The operational process analysis is performed using Value stream mapping, which is a visual management tool used in Lean production [18]. VSM is a method that helps to visualize the entire technological process, quickly identifying the existing waste and eliminate it, realizing a streamline flow of raw materials and information, removing non-operational times, respectively waiting times from the technological flow [19-20].

This tool can be automatically generated, in real time, by computer-assisted programming, using a system with wireless RFID monitoring, thus reducing the time to make the value stream mapping and at the same time significantly reducing the errors [21].

Within Lean production, there are certain techniques that once implemented can significantly contribute to the increase of the productive efficiency [22], for example the implementation of the SMED (Single minute exchange of die). This technique improves the production flexibility and the time required to change the manufacturing batch by 50% [15]. In some cases, by implementing the SMED and Kaizen, these times have been reduced by over 60% [12].

The replacement of the traditional production system with the Lean Manufacturing production system significantly contributes to increasing the quality and costs reduction [19].

The implementation of Lean techniques not only has an impact on increasing profitability, but it also has a major impact on the environment. For example the use of VSM helps to identify the production process impact on the environment, 5S helps to maintain cleanliness (oil leaks and a variety of toxic solutions) and a clean working space, Total Productive Maintenance helps prevent oil leaks, dust emissions, chemical vapors in the atmosphere [23].

The term of "leanness" represents the impact of a company to become lean, reaching its objectives, respectively elimination of waste [24].

Lean design in manufacturing on CNC machining centers, or the design process by lean machine tools, according to Gupta and Kundra (2012), have many advantages, such as increased operating time of the machine, less materials, reduced design cycle time, easier manufacture and assembly of machine tools, improves product quality and reliability, easy cleaning and maintenance of machine tools, improves loading and unloading system, fast loading of machined workpiece [9].

The occurrence of errors in the production process involves different challenges, which if not resolved in time cause significant waste, thus reducing the company's profit. Any errors in the production process can be prevented by Poka Yoke mechanisms [25]. Here are some of the causes of the occurrence of possible errors and their impact on productivity:

IOP Conf. Series: Materials Science and Engineering **898** (2020) 012005 doi:10.1088/1757-899X/898/1/012005

- human errors can cause quality problems in the operation of the automated equipment (for loading, unloading and configuration) [26];
- errors of the processed part because of tool [27];
- errors at the workplace can be caused by inappropriate planning, ineffective communication and lack of proper training [7];
- during the mechanical processing performed on CNC machining centers or conventional machine tools, thermal errors can occur, due to the increased temperature of the machine tool elements that cause a relative movement between the machined part and the tool, during the machining process [28].

2. Research methodology

The study started from a human error, respectively inadequate assumptions and measurements were made. Consequently, the price offer was also incorrectly issued, the calculated price being lower than the cost of manufacturing the piece. The assumption for the price offer was made on the fact that the piece will be manufactured on a classic lathe SN 320.

As a result of this human error, the technological process was re-analyzed, reaching the conclusion that manufacturing the piece on a CNC lathe is more profitable than processing it on the classic one, provided in the tender.

The adoption of the production on CNC machining center was considered more cost-effective due to the reduced manufacturing time (considerably shorter than using the lathe SN 320 machine tool), which led to the shortening of the overall operational time, a significant increase of profit and higher product quality.

3. Case study

The present study analyzes the technological flow for the execution of a piece – knurled axle.

The number of pieces executed is 200. Figures 1 and 2 present the 2D, respectively the 3D drawings of the manufactured piece, produced on a CNC machining center - CNC Fanuc Oi Tc lathe. The final piece is shown in figure 3.

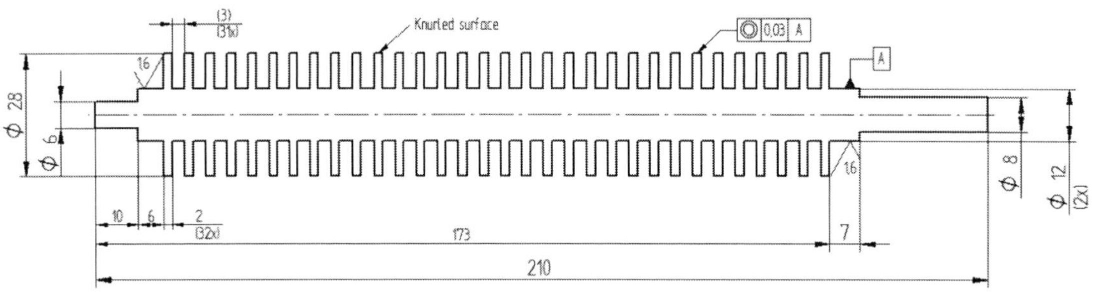

Figure 1. Knurled axle – 2D drawing.

Figure 2. Knurled axle - 3D drawing.

Figure 3. Knurled axle - manufactured on CNC machining center Fanuc Oi Tc lathe.

Name of the piece: Knurled axle.
Material: 16MnCr11.
Manufacturing technology
The processing of this part requires 2 clamps:
Clamping 1.
Exterior turning Φ28 and Φ8x30 respectively Φ12x7;
Knurl Φ28x157, step 0.6, deepness 0.2 mm;
Turning 3 mm;
Cutting Φ28x210 mm.

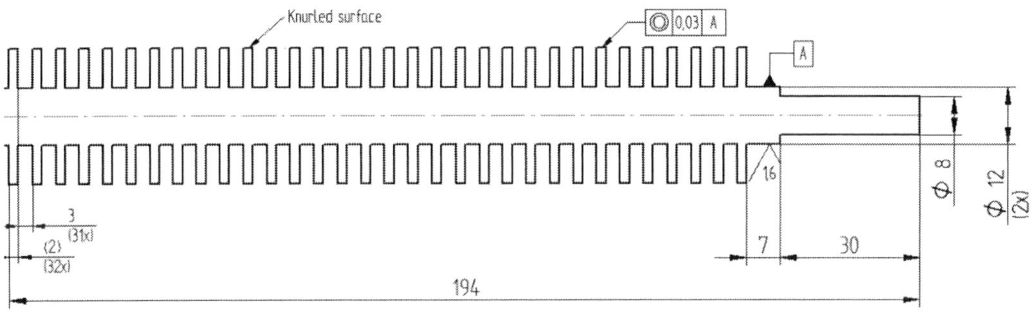

Figure 4. Knurled axle - clamping 1.

Clamping 2.
Exterior turning Φ12X16 and Φ6x10.
The cutting parameters:
speed: 2100 rpm;
feed rate: 400 mm/min (0.2mm/revolution).

The table 1 presents the cutting tools for each operation.

Table 1. Operations within the production process - CNC machining center.

No. crt.	Operation	Tool	Parameters

1	Exterior turning Φ28 and Φ8x30 respectively Φ12x7		Speed: 2000 rpm Feed: 0.1 mm/revolution Ap= 3 mm
2	Knurl Φ28x157, step 0.6, deepness 0.2 mm		Speed: 1000 rpm Feed: 0.5 mm/revolution Ap= 0.2 mm
3	Turning 3 mm	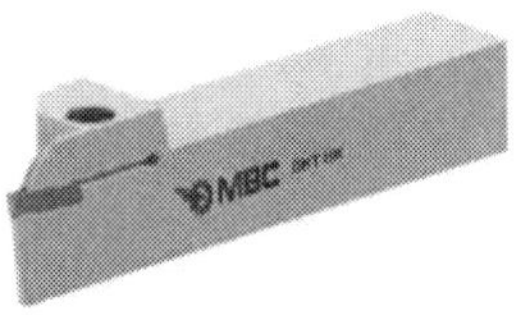	Speed: 2280 rpm Feed: 0.2 mm/revolution Ap= 8 mm
4	Cutting Φ28x210 mm	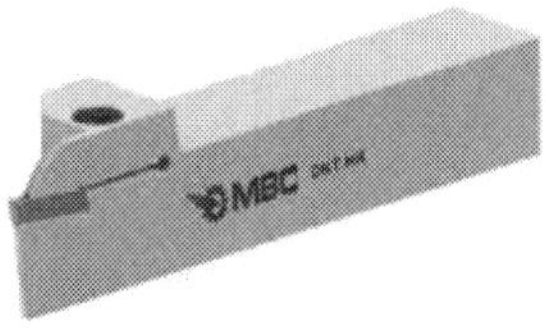	Speed: 2280 rpm Feed: 0.2 mm/revolution Ap= 14.5 mm
5	Exterior turning Φ12X16 and Φ6x10		Speed: 2000 rpm Feed: 0.1 mm/revolution Ap= 3 mm

In order to manufacture the knurled axle on CNC machining center (Fanuc oi-Tc), the below table presents the operations performed within the manufacturing process, with processing times related to each phase.

Table 2. Operations within the production process - CNC machining center.

No. crt	Operation	Machine tools	Measuring device	Operations time
		Clamping 1		
1	Exterior turning Φ28 and Φ8x30 respectively Φ12x7	FANUC 0I TC Lathe	Caliper	10 s - clamping aluminum bar 18 s - processing 5 s - measuring
2	Knurl Φ28x157, step 0.6, deepness 0.2 mm	FANUC 0I TC Lathe	Caliber	25 s - processing 5 s - measuring 7 s - time for cutting tool positioning
3	Turning 3 mm	FANUC 0I TC Lathe	Caliper	38 s - processing 5 s - measuring 10 s - time for cutting tool positioning
4	Cutting Φ28x210	FANUC 0I TC	Caliper	4 s - processing

	mm	Lathe		5 s - measuring 15 s - handling parts and time for cutting tool positioning
		Clamping 2		
1	Exterior turning Φ12X16 and Φ6x10	FANUC 0I TC Lathe	Caliper	10 s - clamping aluminum bar 14 s - processing 5 s - measuring 15 s - handling parts and time for cutting tool positioning
		Total time		**191 s** - Production lead time **99 s** - VAT - Value added time

With the help of Value Stream Mapping, the entire technological process and the operational flow was graphically exemplified. The operational time (CT - cycle time) from each operation separately and the nonoperational time (summing all the nonoperational activities from each operation separately - C / O - changeover time) is included in the VSM, shown in figure 5.

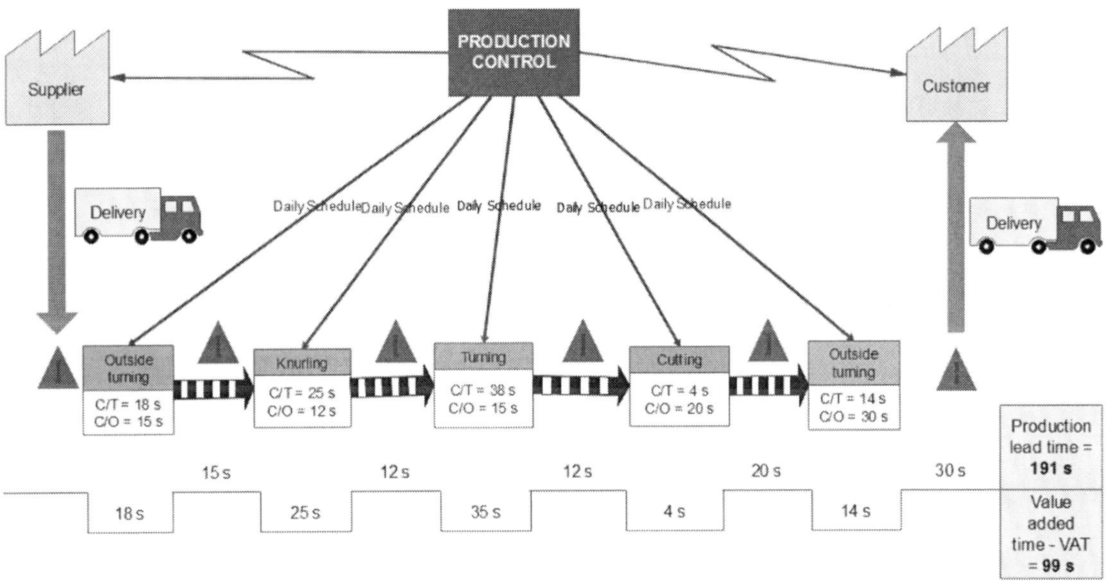

Figure 5. The VSM - manufacturing on CNC machining center

4. Results and discussions

The price offered for the manufacturing process of 200 pieces was Eur 27 per hour. Following the piece manufacturing on a SN 320 lathe, it took 35 minutes to manufacture a single piece, which means a production capacity of 1.71 pieces / hour.

Regarding the necessary manufacturing time for the first piece it was found that the time required to manufacture a single piece is much longer compared to the estimated time on which the price offer was made. Therefore, after analyzing the technological process, new solutions were sought to reduce this time significantly, which would solve human error.

A solution chosen to quickly fix the human error was the implementation of a manufacturing technology that shortens both the operational and non-operational times. As a solution in this regard was the proposal to manufacture these parts on a CNC machining center (Strung Fanuc OiTc). The manufacturing time of one piece in this case was 3.18 minutes, which means a production capacity of 18.86 pieces per hour.

Therefore, starting from the premise that for every working hour Eur 27 are charged, the production cost for a single piece in the case of CNC machining center is Eur 1.43, and in the case of manufacturing on the SN 320 lathe is Eur 15.78 for a single piece.

Regarding the difference between the manufacturing time and cost, respectively the revenues for each manufactured piece, there is a very big difference, reflected in figure 6.

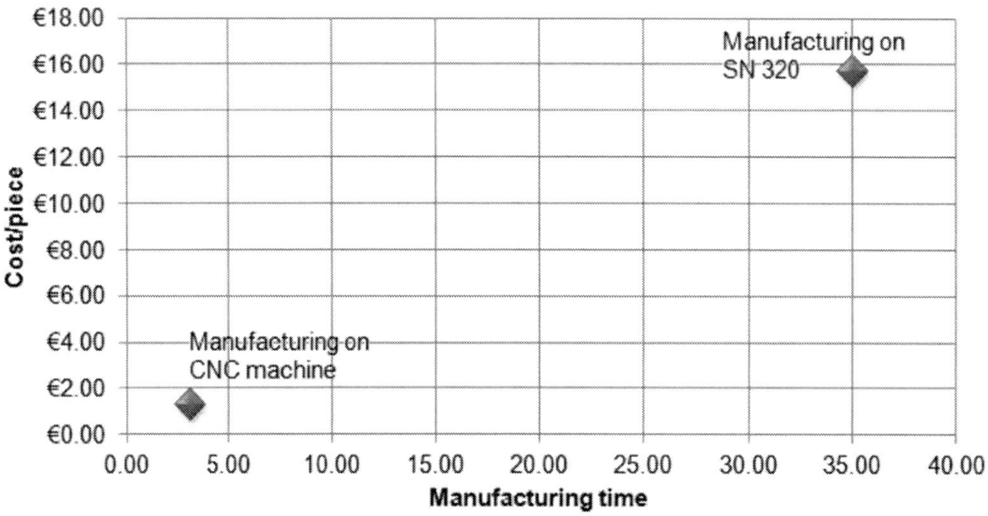

Figure 6. The impact of the manufacturing time on the production cost.

5. Conclusions

The adoption of the manufacturing technology on a CNC machining center led to the increase of the productive efficiency by 90.93%.

It took 15.60 days for the manufacturing of 200 pieces on the lathe SN 320, while in the case of producing the same quantity on the lathe Fanuc Oi Tc 1.41 days were required (working hour per day - 7 hours and 30 minutes - 30 minutes break).

In the time difference between the two presented cases, respectively in the remaining 14.19 days, continuing the production after the completion of the 200 pieces on CNC, an extra quantity of 2008 pieces can still be produced, which is equivalent of Eur 2871.4. This could be valued in case of contracting orders with larger amounts than the 200 parts.

By adopting the manufacturing technology using a CNC machining center, not only was the human error remedied, but it also significantly contributed to the elimination of non-operational times, which led to an increase in the percentage of value-creating activities.

In conclusion, the shorter the required time for a single piece manufacturing, the more it reduces the cost of production, in other words, more will be produced with less resources, which is one of the characteristics of Lean Manufacturing.

Acknowledgments
Published with the support of the University of Oradea because it provided me with scientific resources and with the support of the ID 123008 - POCU/380/6/13 project, - SMARTDOCT – Programe de înaltă calitate pentru doctoranzii şi cercetătorii postdoctorat ai Universităţii din Oradea pentru creşterea relevanţei cercetării şi inovarii în contextul economiei regionale.

References

[1] Bungau C, Blaga F, and Gherghea C 2014 Kaizen implementation for cost reduction in manufacturing process product " driver control board ", *Int. Conf. Prod. Res. – Africa, Eur. Middle East 3rd Int. Conf. Qual. Innov. Eng. Manag* 55–58.

[2] Csokmai L S, Bungău C, Ţarcă R C and Husi G 2015 A comprehensive approach to off-line advanced error troubleshooting in intelligent manufacturing systems, *Int. J. Comput. Commun. Control,* **10**, no. 1 30–37.

[3] Bungau C, Blaga F, and Gherghea C 2011 Method of analysis and audit, used to implement 5s in operational management, *Proceedings of the 2th RMEE Int. Mng. Conf.* (Cluj- Napoca) 36–47.

[4] Gherghea C, Bungau C, and Negrau D C 2019 Best Practices to Increase Manufacturing Productivity - Comparative study *Int. Manufact. Science and Ed. Conf.* MATEC Web of Conferences (Sibiu) **290** 1–9.

[5] Faur M and Bungau C 2019 Exploring the insights of a consignment stock program implementation in a leagile supply chain *Annual Sess. Scientific Papers- IMT Oradea* 291-295.

[6] Womack J P., Jones D T and Roos D 1990 *That Machine That changed the world*, Free Press.

[7] Arunagiri P and Gnanavelbabu A 2014 Identification of major lean production waste in automobile industries using weighted average method *Procedia Eng.*, **97** 2167–2175.

[8] Wyrwicka M K and Mrugalska B 2017 Mirages of Lean Manufacturing in Practice *Procedia Eng.,* **182** 780–785.

[9] Gupta A and Kundra T K 2012 A review of designing machine tool for leanness *Sadhana - Acad. Proc. Eng. Sci.*, **37**, no. 2 241–259.

[10] Almanei M, Salonitis K and Xu Y 2017 Lean Implementation Frameworks: The Challenges for SMEs *Procedia CIRP,* **63** 750–755.

[11] Bungau C, Blaga F and Gherghea C 2012 Continuous improvement of processes in cutting operations *Proceedings of the 3th RMEE Int. Mng. Conf.* (Cluj- Napoca) 208–216.

[12] Azizi A and Manoharan T 2015 Designing a Future Value Stream Mapping to Reduce Lead Time Using SMED-A Case Study *Procedia Manuf.* **2**, no. February 153–158.

[13] Soliman M and Saurin T A 2017 Lean production in complex socio-technical systems: A systematic literature review *J. Manuf. Syst.*, **45** 135–148.

[14] Badulescu D, Bungau C and Badulescu A 2015 Sustainable development through sustainable businesses. An empirical research among master students *J. Environ. Prot. Ecol.*, **16**, no. 3 1101–1108.

[15] Kumar A and Rajenthirakumar D 2016 Reducing the Manufacturing Lead Time of Export Pump Components in an Indian Pump Industry *Natl. J. Technol.*, **12**, no. 2 July 23–31.

[16] Gherghea I C, Bungau C and Negrau D C 2019 Lead time reduction and increasing productivity by implementing lean manufacturing methods in cnc processing *center IOP Conf. Ser. Mater. Sci. Eng.*, **568**, no. 1.

[17] Hamed M and Soliman A 2017 A comprehensive review of manufacturing wastes: Toyota a comprehensive review of manufacturing wastes: Toyota Production System lean principles *Emirates J. Eng. Res.*, **22**, no. 2 April 1–10.

[18] Forno J D, Pereira F A, Forcellini F A and Kipper L M 2014 Value stream mapping: A study about the problems and challenges found in the literature from the past 15 years about application of Lean tools, *Int. J. Adv. Manuf. Technol.*, **72** no. 5–8 779–790.

[19] Manjunath M, Shivaprasad H C, Keerthesh K K S and Deepa P 2014 Value Stream Mapping as a Tool for Lean Manufacturing Implementation-A case study *Int. J. Innov. Res. Dev.*, **3**, no. 5 477–481.

[20] Bungau C, Gherghea I C and Prichici M 2010 Value Stream Mapping Analysis, Efficiency Methods of Operational Management *Proceedings of the 1th RMEE Int. Mng. Conf.* (Cluj-Napoca) 188-198.

[21] Chen J C and Chen K M 2014 Application of ORFPM system for lean implementation: An industrial case study *Int. J. Adv. Manuf. Technol,* **72** no. 5–8 839–852.

[22] Faur M and Bungau C 2018 Supply chain 'leagility' through adopting consignment stock strategy in manufacturing companies *Proceedings of the 6th RMEE Int. Mng. Conf.* (Cluj-Napoca) 623-632.
[23] Chiarini A 2014 Sustainable manufacturing-greening processes using specific Lean Production tools: an empirical observation from European motorcycle component manufacturers *J. Clean. Prod.,* **85** 226–233.
[24] Tekez E K and Taşdeviren G 2016 A Model to Assess Leanness Capability of Enterprises *Procedia Comput. Sci.,* **100** 776–781.
[25] Gherghea C and Bungau C 2018 Poka yoke application synthesis in manufacturing engineering *Proc. 6th Rev. Manag. Econ. Eng. Int. Manag. Conf.* **6** 564–571.
[26] Sundar R, Balaji A N and Satheesh K R M 2014 A review on lean manufacturing implementation techniques *Procedia Eng.,* **97** 1875–1885.
[27] Li X 2001 Real-Time Prediction of Workpiece Errors for a CNC Turning Centre, Part 4. Cutting-Force-Induced Errors Int. *J. Adv. Manuf. Technol.* **17** 665–669.
[28] Ruijun L, Wenhua Y and Zhang H H 2012 The thermal error optimization models for CNC machine tools *Int J Adv Manuf Technol* **63** 1167–1176.

IOP Conf. Series: Materials Science and Engineering **898** (2020) 012006 doi:10.1088/1757-899X/898/1/012006

Investigation of NOx emissions for mitigating public health risk with Mercedes E Coupe

D L Băldean[1], L Andrei[2] and A I Borzan[1]
[1] Automotive and Transportation Dep., Technical University of Cluj-N., Cluj, Muncii 103, Ro.
[2] Infectious Disease Hospital of Cluj-Napoca, Cluj, Iuliu Moldovan 23, Romania

doru.baldean@auto.utcluj.ro

Abstract. Investigation of NO_x emissions is a worldwide concern in order to mitigate or to reduce the health risks that the pollutant poses for the public. It is known that the NO_x are a combination of highly reactive odourless gases containing nitrogen and oxygen in different proportions. The main objective of the present paper is to show the applied investigation and its following results regarding NO_x values and trendlines for the warm-up phase of a Mercedes E 350 CDI Cabrio-Coupe A/C/W207. The secondary objective of the investigation consists in dividing and comparing the results obtained before and after the threshold limit of 40°C in coolant temperature. The warm-up sequence is not the highest temperature stress regarding pollution and NO_x formation. It is although very important due to at least two reasons. In the first place, many of the usual road traffic trips are not exceeding the warm-up duration, meaning that the travel is below 40 minutes long. Secondly, the operation below 60°C is the most stressful both physically and chemically for the engine (its systems) and the environment being defined by high wear and pollution levels. Temperature is a key factor.

1. Introduction

In the early 2020 the Corona virus pandemic brought both significant restrictions and research opportunities. Due to the risk of transmitting the deadly virus authorities worldwide have drastically limited the road traffic and other forms of transportation impacting the air quality and pollution levels. Some researchers took this opportunity to further the science and to gain valuable result by looking into the given problem as it was a great chance to understand the phenomena and to size the benefits [1, 15].

NO_x is produced during the compression, combustion and evacuation strokes when air is compressed at high temperatures and fuel is injected in order to generate mechanical work. NO_x leads to smog formation, acid rain, deterioration of water quality, greenhouse effect, and reduced visibility in urban areas [2, 3, 4]. Controlling the temperature is the key factor to impact engine performance [5]. Smart systems and comprehensive approaches allow the researchers to tackle both the operation and emission problems [6, 7, 8, 9, 10]. Climate emergency mandates the researchers to investigate furthermore the NO_x control technologies and emission model implementation for different vehicles [11, 12, 13, 14, 15].

Because there are two ways of determining the NO_x emissions in relation with the operating regime, the most common one is by taking into consideration the performance, as the power output – which coincides with the high engine speed and high temperatures. At high power output the hourly consumption is significantly high. In this first case the formation of NO_x compounds is sustained by the high temperatures and fuel consumption. Increasing the demand upon the accelerator leads to higher fuel injected quantity, thus increasing the temperature inside the cylinder, increasing the engine speed, thus increasing the power output and NO_x. In this stage, high engine loads lead to high NO_x emissions.

This observation is valid only when we consider the warming up phase and acceleration process, with the precise consequence of increasing the power, the hourly fuel consumption, the engine speed and mixture temperature. Another way of looking into the problem is more practical and considers the fact of deceleration and engine brake when the temperature is already high. Depending on the environmental temperature level, the usual warm-up phase takes between 10 and 30 minutes. In the present case the first 5 to 7 minutes from engine start allowed the coolant to reach the 40°C level and the following 5 minutes led to another increase in temperature from 40 to almost 63°C.

Figure 1 shows the map with the recorded differences between the NO_2 emissions before and after the road traffic restrictions in New York. It highlights the fact that the lower road traffic is the higher air quality will get. On global scale, after animal farming industry, the transportation is an important pollutant [14].

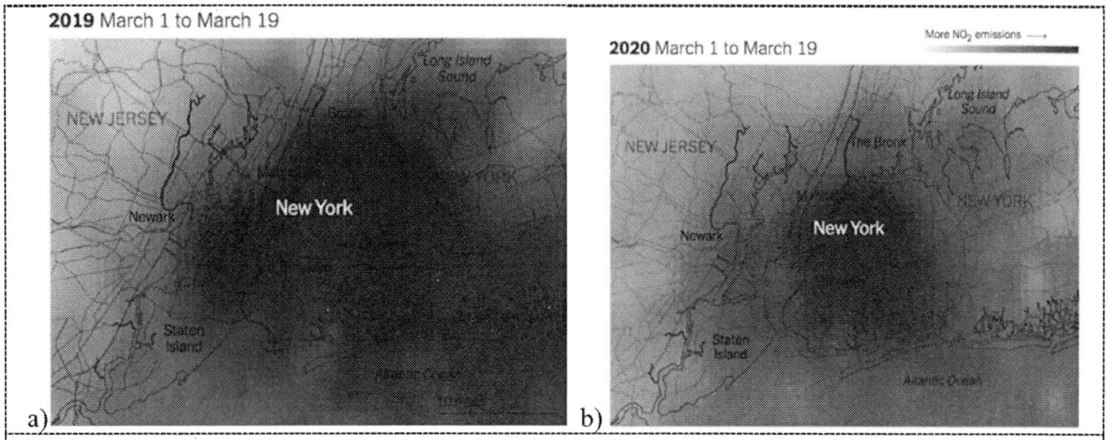

Figure 1. Differences in N.Y. NO_x emissions in March 2019 (a) and 2020 (b) after restrictions [14].

2. Applied research

Applied research is based on advanced computer aided investigation with models and methods used for taking measurements, calculating, predicting or estimating NO_x levels in exhaust gas, as it is allowed by the Law 104/2011 [14], and upon a vehicle which has diesel engine with technical data given in table 1.

Figure 2 presents the vehicle 1, a Mercedes E Cabrio-Coupe A207, having a common platform with C207 and W207 models, used in the current research for NO_x assessment. It has engine, clutch, gearbox and steering system placed on the front axle 2. The on-board data management system 3 records the vehicle performance in operation. Fuel tank 4 and traction 8 are disposed on the rear axle. Tail pipes 7 on the rear end allow the exhaust compounds 9 to affect the atmosphere and develop specific reaction 5 with oxygen. Rain 6 and NO_2 gas negatively impacts the environment 10 and public health [2, 3, 12].

Figure 2. Schematic of the NO_x emissions into the environment and some negative effect on life [11]

In figures 3 to 12 are presented the results for NO_x levels determined when operating the vehicle in the warm-up phase, by taking the reference the point of 40˚C which is also indicated on board.

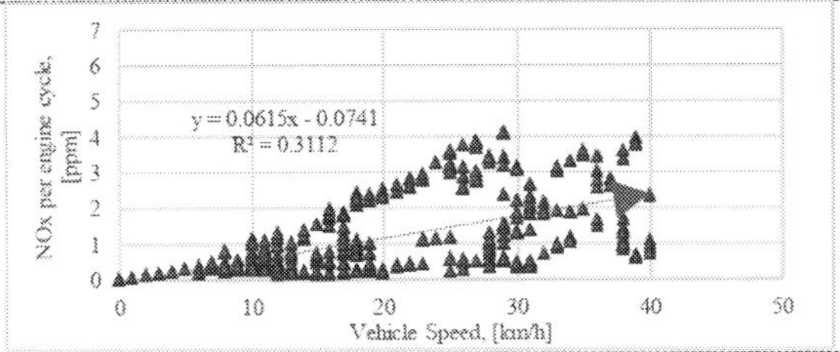

Figure 3. NO_x per stroke vs. speed below 40˚C.

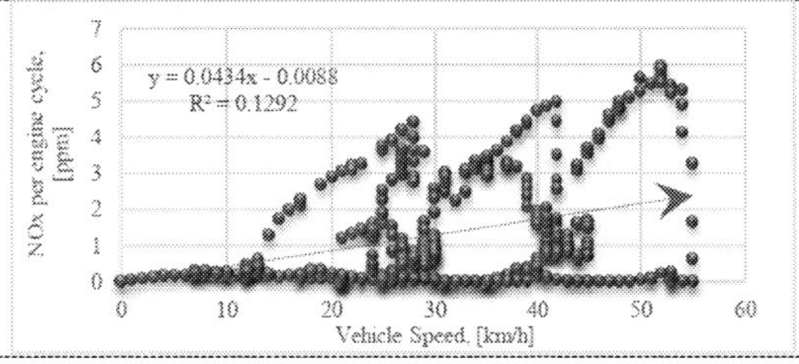

Figure 4. NO_x per stroke vs. speed above 40˚C.

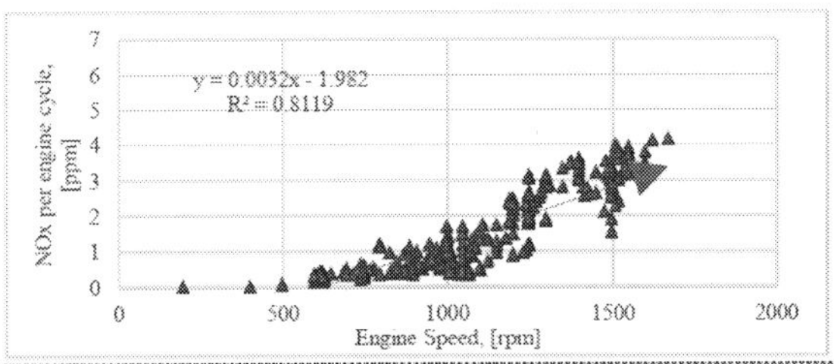

Figure 5. NO_x vs. engine speed, when below 40˚C.

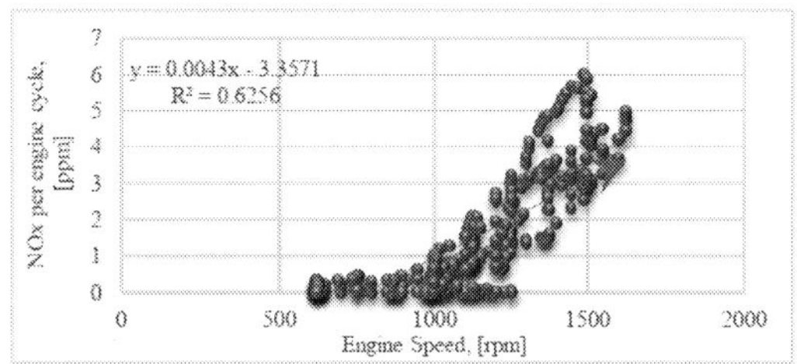

Figure 6. NO_x vs. engine speed above 40˚C.

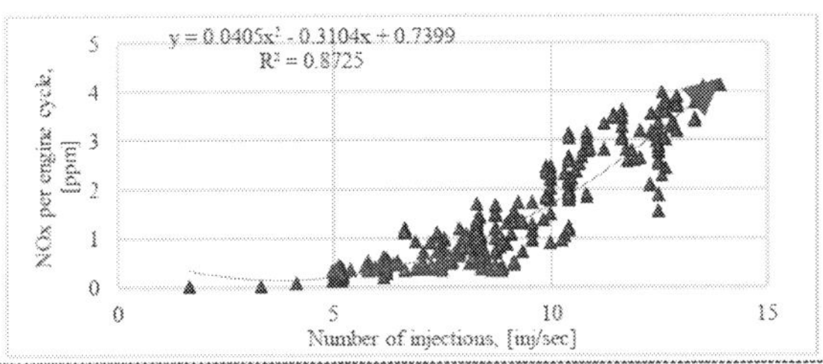

Figure 7. NO_x vs. no. of inj., when below 40˚C.

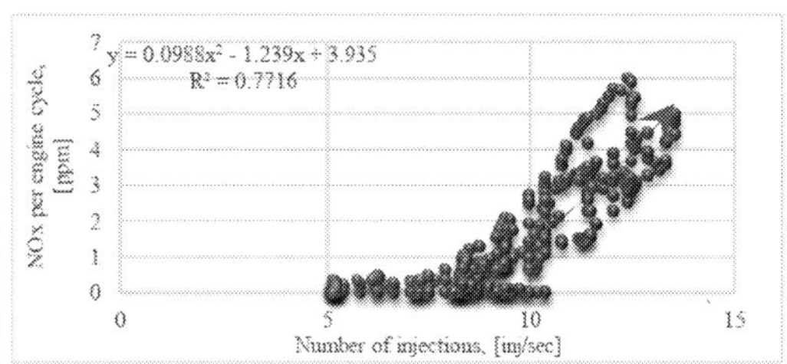

Figure 8. NO_x vs. no. of inj., when above 40˚C.

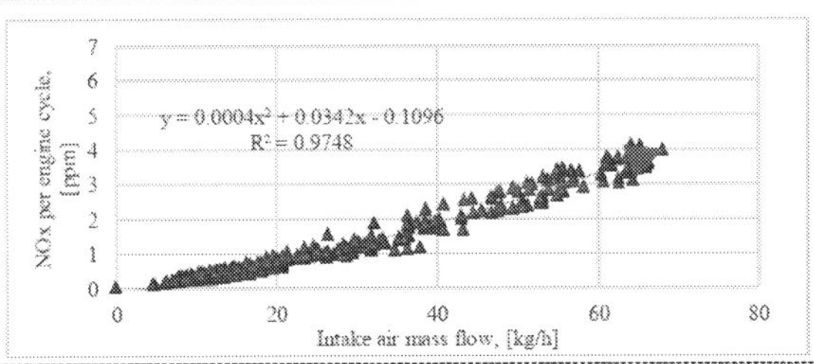

Figure 9. NO_x vs. air mass flow when below 40˚C.

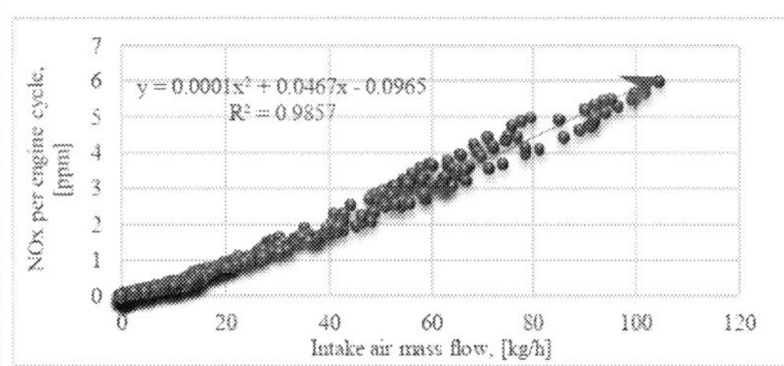

Figure 10. NO_x vs. air mass flow above 40˚C.

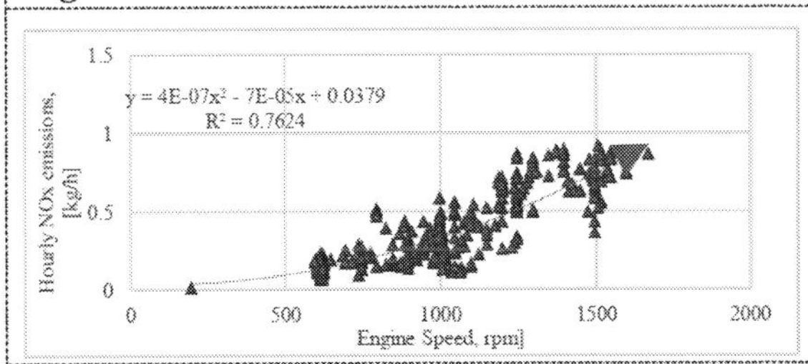

Figure 11. Hourly NO_x vs. rpm, when below 40˚C.

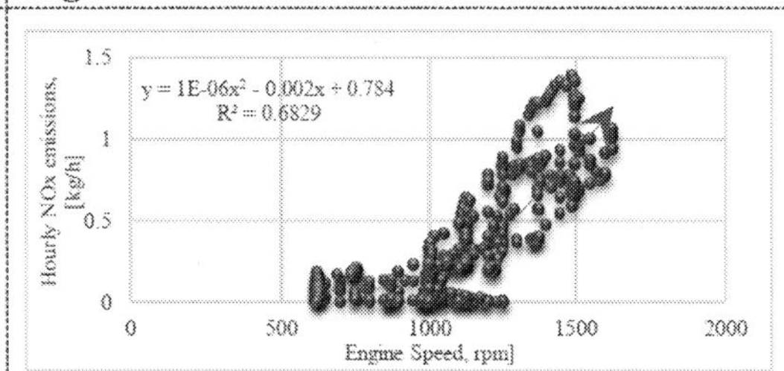

Figure 12. Hourly NO_x vs. rpm, above 40˚C.

The NO_x emissions were determined just for a limited period of time to reach 60˚C in worm up process.

The hourly NO_x emissions versus the selected gear ratio are given in figures 13 and 14.

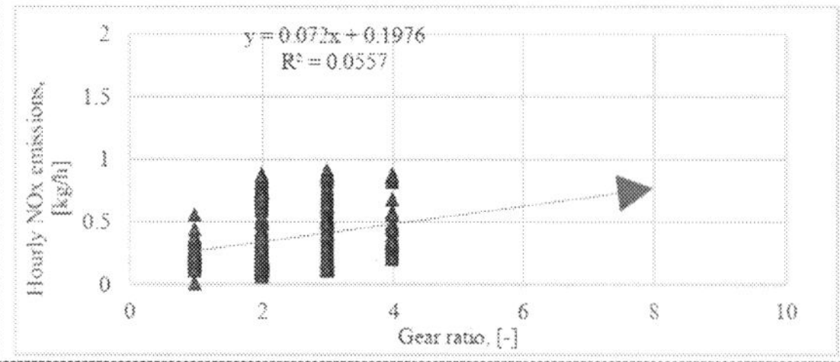

Figure 13. NO_x values vs. gear ratio in first 5 min.

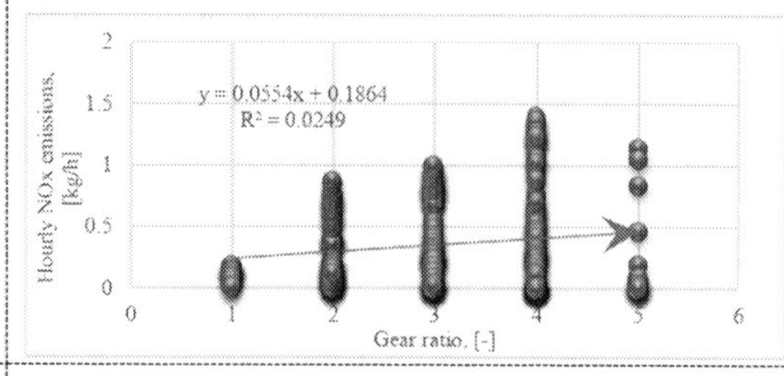

Figure 14. NO_x vs. gear ratio above 40°C

Most significant technical specifications regarding the vehicle's power-train are presented in table 1.

Table 1. Technical data for the vehicle under testing

Parameter	Mercedes-Benz E350 CDI/Bluetec (2013) Model W/A207
Engine	Diesel 350 CDI/Compression ignited/W207/2013 3.0 L OM642 V6 turbo
Fuel supply	Common-Rail/6 injectors
Clutch	Dual
Transmission	7G-TRONIC 7-speed
Axle	Rear axle drive, Front axle steering
Wheelbase	2874 mm

NO_x assessment based on air and fuel consumption makes use of the instant mixture as modeled below:

$$NO_x = 5 * 10^{-2} I_{AFM}/6 = 5 * 10^{-2}[(F_C + A_C) * N_c]/6, \text{ [kg/h]} \quad (1)$$

where I_{AFM} is the instant air-fuel mixture introduced each operational cycle into the engine; F_C – fuel consumption, in kg/h; A_C - air consumption, in kg/h; N_c – number of working cycles.

3. Observations and conclusions

Outside environment temperature during the experimental test was 13°C. The Mercedes E-Class Cabrio under testing in this research has a complex transmission with 7 gear ratios for forward drive and one for reverse, with three optional gear change strategies (M-manual; S-sporty; E- automatic eco-mode). There have been written more than 300 lines of data code and a total of 3920 digital cells were created for the current investigation. Steering excessively leads to higher fuel consumption and less autonomy. In approximately ten to twelve minutes, the engine has been warmed up above 60°C due to traction load and operation in road conditions. Changing in temperature led to a significant improvement in fuel economy due to the reduction of fuel consumption. There are at least 2-3 seconds delay between actuating the accelerator of the car, increasing consequently the fuel injected quantity, and the following process of engine speed increase. After increasing the engine speed there are at least 1-2 seconds delay till the recorded vehicle speed is following up the engine crankshaft.

The average NO_x value, during the first phase of the investigation, when the coolant temperatures increased from 13°C to 40°C, was 1.198 ppm. The same pollutant recorded an average of 1.326 ppm in the second part of investigation, when the engine coolant was between 40°C and 63°C. The NO_x increase trend on average with approximately 10% is also validation proof of the research, showing the effect of a higher temperature upon the formation of nitrogen and oxygen compounds. Estimation allows us to foresee a significant increase in NO_x levels when coolant temperatures are above 80°C and the vehicle operates in engine braking regime. These insights, to be proven and discussed. require further research.

IOP Conf. Series: Materials Science and Engineering **898** (2020) 012006 doi:10.1088/1757-899X/898/1/012006

References

[1] Andrei L et al 2019 *Applied Measurements and Instrumentation for Improving Diagnostic Devices and Systems in Metropolitan Polluted Environments with Nitric and Carbon Oxides* (6th International Conference on Advancements of Medicine and Health Care through Technology; 17–20 October 2018, Cluj-Napoca, Romania, Springer, pp. 45-49) DOI https://doi.org/10.1007/978-981-13-6207-1_8

[2] Andrei L et al 2018 *Contributions to the research of the relation between air pollution with nitric oxides and the impact upon environment and life environment pollution management and life health* (In Science and Engineering / Știință și Inginerie AGIR Press) eISSN 2359–828X DOI http://stiintasiinginerie.ro/34-68

[3] Andrei L et al 2017 *Contributions to the applied research of effects from surrounding environment pollution due to road traffic upon eyes' anatomy and pathology development* (In Science and Engineering / Știință și Inginerie AGIR Press) eISSN 2359–828X DOI http://stiintasiinginerie.ro/32-85

[4] Borzan AI et al 2019 *Experimental design for Diesel supply control in order to improve fuel efficiency* (In IOP Conf. Series: Materials Science and Engineering, Vol. 568, Issue 1, 17 Sept. 2019) ISSN: 17578981, DOI: 10.1088/1757-899X/568/1/012038

[5] Botean AI 2019 *Influence of the temperature gradient on the speed of a Stirling gamma engine* (In Acta Technica Napocensis Series-Applied Mathematics Mechanics and Engineering, Vol. 61, 3, pp. 339-346, 2018) ISSN: 1221-5872, WOS:000468025900008

[6] Ferenti I et al 2019 *Analysis of Intelligent Control and Interface in Pressurized Liquid Injection Systems for Competitive Technical Solutions* (In Hidraulica, Issue 2, 43-48, 1 April 2018, Article number 327395077) ISSN: 1453 – 7303.

[7] Jovrea LD et al 2017 *Study of the semiologic aspects concerning the hidraulics of the injection system in diagnosing 1.8 TDCI engine from Ford Focus* (In Science and Engineering / Știință și Inginerie AGIR Press, ISSN 2067-7138, eISSN 2359–828X) DOI https://http://stiintasiinginerie.ro/31-68

[8] Jovrea S et al 2018 *Researching on-board display of esential informations concerning technical conditions in operation and fuel-economy of a motor-vehicle in operation* (In Science and Engineering/Știință și Inginerie, Vol. 31, Issue 67, AGIR Press) eISSN 2359–828X

[9] Jurchiș BM et al 2019 *Research of Alternative Composite Solutions in Gas After-Treatment System for Diesel Cars* (In SIAR International Congress of Automotive and Transport Engineering: Science and Management of Automotive and Transportation Engineering, First Online 14 Oct. 2019, pp. 160-164) DOI: 10.1007/978-3-030-32564-0_19

[10] Jurchiș BM et al 2019 *Experimental research on Diesel Particle Filter (DPF) in relation to the fuel type (*In IOP Conference Series: Materials Science and Engineering, Volume 568, Issues 1, pp. 12-27) DOI: 10.1088/1757-899X/568/1/012027

[11] Komanoff C et al 2020 *What the coronavirus pandemic can teach us about the climate emergency* (In: The intercept, https://theintercept.com/2020/04/04/coronavirus-pandemic-climate-change-co2-emissions/) Online 4 April 2020

[12] Reşitoglu IA 2018 *NOx Pollutants from Diesel Vehicles and Trends in the Control Technologies* (In Tech Open https://www.intechopen.com/books/diesel-and-gasoline-engines/no-sub-x-sub-pollutants-from-diesel-vehicles-and-trends-in-the-control-technologies)

[13] Wang X et al 2019 *A NOx Emission Model Incorporating Temperature for Heavy-Duty Diesel Vehicles with Urea-SCR Systems Based on Field Operating Modes* (In: Atmosphere 10, 00337, Publishing date 20 June 2019) DOI: doi:10.3390/atmos10060337

[14] *** 2020 *Air quality. Air quality assessment* (In: National Air Quality Monitoring Network) http://www.calitateaer.ro/public/assessment-page/assessment-page/?locale=en

[15] *** 2020 *What the Coronavirus Teaches Us About the Climate Crisis* (In: *LIVEKINDLY* individual mobility https://www.youtube.com/watch?v=LNvCRUWjYyU)

IOP Conf. Series: Materials Science and Engineering **898** (2020) 012007 doi:10.1088/1757-899X/898/1/012007

A brief overview of parallel robots and parallel kinematic machines

M Ratiu[1] and D M Anton[2]

[1] Department of Mechanical Engineering and Automotive, University of Oradea, 410087 Oradea, Romania
[2] Department of Mechatronics, University of Oradea, 410087 Oradea, Romania

E-mail: mratiu@uoradea.ro

Abstract. In this paper, a brief overview is presented, resulting from a recent literature review of some representative books or papers regarding various research on different types of parallel robots, parallel kinematic machines, or hybrid parallel mechanisms. Starting from a short introduction regarding the motivation and necessity of the emergence of such parallel mechanisms, some basic terms are presented, in the context of founding the existence of several terminologies for certain specific expressions of this subject, as well as different definitions for them. Then, after a short presentation of the advantages and disadvantages of the parallel robots, compared to serial ones, and of the parallel kinematic mechanisms compared to the conventional machine tools, some publications related to kinematics, dynamics, modeling and simulation of these types of mechanisms are highlighted. This research is realized in order to identify a solution that will be the basis for the development of a prototype, both virtual and experimental, for such mechatronic system.

1. Introduction

In the last few decades, alongside the development of serial industrial robots, parallel robots attracted the attention of many industries and researchers. This because of their higher precisions, rigidity, dynamic performance and loading capacity, compared with the serial robots.

Also, in order to reduce or eliminate the disadvantages of the conventional machine tools, the modern machine tools, based on the parallel kinematic mechanisms were developed, that are faster, more accurate and less costly and offer higher dynamic performance and operational flexibility. Nowadays, parallel kinematic machines, also called parallel robotic machine tools or parallel robot-based machine tools, are developed and produced by many researchers and companies and used in a lot off applications.

The development in this field continued, and a new concept appeared, followed by the development of different variants of hybrid parallel kinematic machines, that are able to work both in robotics and machining applications.

After a substantial analysis, based on an extensive bibliography, consisting of 123 references, Merlet and Gosselin [1] stated in 2008 that, compared to the serial robots, the analysis of the parallel mechanisms and robots is far from being complete.

Their statement proved to be true, so in the years that followed, the interest of researchers and companies in this field has increased, so that at present, there are a large number of books, articles or websites presenting papers, researches or information on this topic. For example, searching on Google "parallel robot", about 237,000 results were identified [2], and on Web of Science Core Collection [3], 2,314 results. Also, on the Robotic Industries Association (RIA) website [4], for the same search, a total of 5,679 results were identified, out of which 1085 referred to products, 308 were companies, 205 were services, and 384 were industry insights.

According to this general interest, and in the desire to develop such a mechatronic system, we conducted a literature review on the topic of parallel mechanisms, presented, briefly, in this paper.

2. Basic terminology

Consulting numerous books and scientific articles in this field, we found the existence of several terminologies for certain specific expressions of this subject, as well as different definitions for them. Of all these, we chose to take over and use the terminology provided by the ISO 8373 (1994), Manipulating industrial robots-Vocabulary, and the Parallel Mechanisms Information Center. According with these:

- Parallel Robot is "a robot whose arms (primary axes) have three concurrent prismatic joints".
- Parallel Kinematic Machine (PKM) is "a machine tool based on a parallel mechanism".
- Fully-parallel mechanism is "a parallel mechanism with an n-DOF (degree of freedom) end-effector connected to the base by n independent kinematic chains, each having a single actuated joint".
- Hybrid parallel mechanism is "a parallel mechanism with an n-DOF end-effector connected to the base by m ($m < n$) independent kinematic chains, each having one or more actuated joints".

Other definitions for the parallel robots, of some prestigious researchers in this field:

- "Parallel robots are closed-loop mechanisms presenting very good performances in terms of accuracy, rigidity, and ability to manipulate large loads" [5]. After few years, in the second edition of the book, [6], Merlet maintains his definition but adds that parallel robots are sometimes called hexapods or PKM, and notes that this field is in a quickly moving, becoming increasingly popular in the machine tool industry, and being addressed in an extensive bibliography.
- For Gogu [7], a parallel robot is "a robot in which the end-effector is connected in parallel to the reference link by $k \geq 2$ kinematic chains, called limbs or legs" For a fully-parallel robot, the number of limbs is equal with the robot mobility, $k=M \geq 2$, and in each limb exist just one actuator. For an hybrid serial-parallel robot, the end-effector is connected to reference link by just one complex kinematic chain, called complex limb or complex leg.
- Recently, Staicu, in [8], defines the parallel robots as "closed-loop mechanical structures whose mobile platforms are linked to the base by some independent kinematical chains, presenting very good potential in terms of accuracy, rigidity and ability to manipulate large loads with minimal positioning errors".

3. Advantages and disadvantages

Reiterating the advantages of the parallel robots, compared to serial ones, Staicu presents the increased interest of the researchers in this field, as well as the wide spectrum of their use, going up to the non-traditional and innovative applications, such the surgical and rehabilitation robots. Regarding the hybrid industrial robots, they present the advantages of both serial and parallel robots, namely high dexterity, high stiffness, large workspace, and high force-to-weight ratio and minimized disadvantages.

Compared to the conventional machine tools, the parallel kinematic mechanisms offer higher stiffness, higher acceleration, potential higher accuracy, lower moving mass, mechanical simplicity, and reduced installation requirements, and so, they have the potential to be high precision machines, highly modular and reconfigurable, with high dexterity and multi-mode manufacturing capability, and also with a small footprint, thus changing the classical manufacturing technologies. [9]. By citing on Giddings and Lewis, Zhang shows that a Hexapod machine tool is a substantially improved machine tool in terms of speed (about 4 times), rigidity (about 5 times) and precision (about 7 times).

The Hexapod machine tool, which is a parallel kinematic machine with six legs, combines the speed and dexterity of a conventional robot with the stiffness and accuracy of a machine tool.

The Deltapod represents a combination between Delta robot and Hexapod machine tool, which is a less costly and very compact machine, powerful, and that can carry payloads exceeding four times its own mass [10].

An overview on the evolution of the parallel kinematics machine tools is conducted in [11], a survey of their development, performances and practical application compared with the serial machine tools.

In a comparison between Tricept hybrid PKM versus serial robots and traditional machine tools, in [12] is shown that, in terms of speed and flexibility, Tricept hybrid PKM is comparable to serial robots, and in terms of accuracy and stiffness, its performance is more close to the machine tools.

4. Kinematics

Direct kinematics, also known as direct kinematic problem or forward kinematics, consists in finding the generalized coordinates (the variables that describe the pose of the mobile platform) from the articular coordinates (the variables that describe the actuated joints), or, in other words, in determining the position and orientation (the pose) of the mobile platform for a given set of the articular coordinates. For the parallel robots, usually, the direct kinematics is much more complex than the inverse kinematics, but it is necessary for motion planning, control purposes and calibration of the robot.

Inverse kinematics, also called inverse kinematic problem, or reverse kinematics, consists in finding the articular coordinates from the generalized coordinates, or, in other words, in determining the vector of leg lengths for a given pose of the mobile platform, defined by the position vector and the rotation matrix. For the parallel robots, the inverse kinematics is usually uncomplicated and easy to do or understand.

A comparative study on the optimum kinematic designs of the parallel, serial and hybrid machine tools and robotics structures is presented in [13] and few strategies for kinematic designs of some parallel and hybrid mechanism are developed, discussed and analyzed.

For an experimental prototype of a new type of a 2RPU-2SPR parallel manipulator, presented in [14], the direct kinematics solution is calculated by applying an artificial intelligence method, namely the back propagation neural network (where the notation used for the kinematic pairs means: P - prismatic; R - revolute; U - universal or Hooke's or Cardan joint; S – spherical).

To solve the inverse kinematics for any kind of parallel robot used, generally, for drilling or milling, that require three translational and two rotational degrees of freedom, 3T2R, a general kinematics model is introduced in [15].

5. Dynamics

Dynamics of the parallel robots, usually more complicated because of their complexity, is based on the direct kinematics and requires the determination of the dynamic parameters, in order to obtain the dynamic model, which is important for its automatic control.

The most used approaches in the dynamics of the parallel robots are the classical methods that use the Lagrange formalism, Newton-Euler equations, Hamilton's principle, Kane method, and the screw theory and the principle of virtual work. An analogy between the principle of virtual work, and the formalism of Lagrange's equations is presented by Staicu, in [8].

The inverse and direct dynamics of a 3SPS-1S parallel robot is solved in [16] and two different control methods are presented, with some control examples for each method.

A new method for the dynamic modeling of the Stewart parallel mechanism, but which can be used also for other linear multibody systems is proposed in [17] and validated through a numerical study. It is based on the transfer matrix method and considers the mechanism as a flexible multibody system, with the platform rigid and the legs flexible.

By using the matrix method and the Lagrange equation, in [18] is obtained the solution of the inverse dynamics problem for a combined relative manipulation mechanism.

The dynamics formulation for a 5-DOF 2RPU-2SPR parallel manipulator is realized in [19] by using the principle of virtual work, and linear parameterization with the regression matrix form.

The general principle of virtual work is used, also, in [20], for solving the inverse dynamics problem for a 5-DOF modular parallel robot and the achieved results are compared with the Siemens NX dynamic simulations.

For a new and innovative structure of a 6-DOF 6-PGK spatial parallel robot, an inverse dynamic model based on an artificial neural network is performed in [21] and a dynamic numerical simulation is presented in [22].

6. Modelling and simulation

A finite and instantaneous screw based method is used in [23] for dynamic modeling and analysis of a Helix robot, a 2-DOF parallel robot with changeable rotational axes. After the validation of the dynamic model in ADAMS and MATLAB software, the method is declared suitable for the integrated design of different parallel robots.

In the paper [24] is presented a virtual prototyping environment, developed in ADAMS, for the analysis and design of a 3-DOF PKM, but which can be applied to any PKM modeled by rigid bodies.

After the study on the kinematics of a 3-DOF parallel robot, mathematical modeling, and simulation is developed by using the MATLAB software and Optimization Toolbox package in [25] and a prototype of the robot is developed.

In the paper [26], on the research object, which is a 5-DOF 4-SPRR-SPR parallel mechanism, an inverse kinematic position analysis is carried out and solved by using MATLAB and then verified in a virtual simulation, in ADAMS.

Through the paper [27] a novel 5-DOF 4-UPS-RPS spatial parallel mechanism is introduced and its rigid dynamic model is established and analyzed, model that is then verified by numerical calculation in MATLAB and virtual simulation in ADAMS.

A modeling and simulation process in SolidWorks for a 6-DOF 6-SPS parallel robot is used in [28] for the analysis the variations of different motion parameters.

In [29], on the base of a detailed study on the mechanical structure, the kinematics, dynamics, and the control system of a new type of 5-DOF 3T 2R hybrid robot manipulator (composes by a 3-DOF 3T parallel module and a 2-DOF 2R serial module) and a virtual kinematics co-simulation conducted both in ADAMS and MATLAB, a physical prototype and a specially designed control system are developed, followed by few tests for repeatability and accuracy.

Another dynamic modeling of a new type of a 5-DOF gantry hybrid machine tool, in fact, a 2-RPU+2-UPS parallel mechanism connected to the linear guides by two sliding pairs P5, is developed in [30], by using the virtual modeling and simulation, also, in MATLAB and ADAMS.

In a recent paper, [31], a type of 5-DOF 2UPR-RPS-RR hybrid robot is analyzed and is proposed a multi-objective optimization model taking into consideration that objectives which better reflect the motion range (the higher priority), the stiffness and the kinematics performance of the robot.

For a 6-DOF 3-PRUS parallel kinematic machine, a comparative study of the dynamic analytical model, realized through the numerical and ADAMS-based simulation, and, also, a dual experimental prototype are presented in [32].

7. Conclusion

The purpose of this article is to present a brief overview, resulting from a recent literature review on different types of parallel mechanisms, which is realized in order to identify a solution that will be the basis for the development of a prototype, both virtual and experimental, for such a mechatronic system.

References

[1] Merlet J P and Gosselin C 2008 Parallel mechanisms and robots *Springer Handbook of Robotics*
[2] https://www.google.com Accesed in 02.04.2020
[3] https://clarivate.com/webofsciencegroup Web of Science Core Collection Accesed in 02.04.2020
[4] https://www.robotics.org Robotic Industries Association Accesed in 02.04.2020
[5] Merlet J P 2000 Parallel robots, *Solid mechanics and its applications*, Kluwer Academic

[6] Merlet J P 2006 Parallel robots (Second edition), *Solid mechanics and its applications*, Springer
[7] Gogu G 2010 *Structural Synthesis of parallel robots: part 3: Topologies with planar motion of the moving platform* (Springer)
[8] Staicu S 2019 Dynamics of Parallel Robots *Parallel robots: Theory and applications* (Springer, Cham) Chapter 1 pp 1-12, Chapter 9 pp 191-243 and Chapter 12 pp 309-326
[9] Zhang D 2010 *Parallel robotic machine tools* (Springer US)
[10] Soetebier S, Kock S and Legeleux F 2007 The Deltapod: a new parallel robot for flexible fixturing applications. www.parallemic.org
[11] Pandilov Z and Dukovski V 2012 Parallel kinematics machine tools: overview- from history to the future *Annals of Faculty Engineering Hunedoara – Int. J. of Engineering* **X**(2) pp 111-124
[12] http://www.pkmtricept.com
[13] Harib K H et al. 2012 Parallel, serial and hybrid machine tools and robotics structures: comparative study on optimum kinematic designs http://www.intechopen.com
[14] Zhang H, Fang H, Zhang D, Luo X and Zou Q 2019 Forward kinematics and workspace determination of a novel redundantly actuated parallel manipulator *Complexity* **2019**
[15] Schappler M, Tappe S and Ortmaier T 2019 Modeling parallel robot kinematics for 3T2R and 3T3R tasks using reciprocal sets of Euler angles *Robotics* 2019 **8**(3) 68
[16] Chaparro-Altamirano D, Zavala-Yoe R and Ramirez-Mendoza R 2014 Dynamics and control of a 3SPS-1S parallel robot used in security *21st Int. Symp. MTNS*, Groningen, The Netherlands
[17] Gangli C et al. 2018 A novel method for the dynamic modeling of Stewart parallel mechanism *Mechanism and Machine Theory* **126** pp 397-412
[18] Pashchenko V et al. 2020 Inverse dynamics problem solution for the combined relative manipulation mechanism with five DoF *Proc. Int. Conf. ER(ZR)* (Springer) **154** pp 253-263
[19] Zhang H, Fang H, Zou Q and Zhang D 2020 Dynamic modeling and adaptive robust synchronous control of parallel robotic manipulator for industrial application *Complexity* **2020** (Hindawi)
[20] Plitea N et al. 2016 Inverse dynamics and simulation of a 5-DOF modular parallel robot used in brachytherapy *Proceeding of the Romanian Academy* **17** (1) pp 67–75
[21] Moldovan L, Grif H S and Gligor A 2016 ANN based inverse dynamic model of the 6-PGK parallel robot manipulator *Int. J. of Computers Communication&Control* **11** (1) pp 90-104
[22] Moldovan L, Gligor A, Grif H S and Moldovan F 2019 Dynamic numerical simulation of the 6-PGK parallel robot manipulator *Proceedings of the Romanian Academy* **20** (1) pp 67–75
[23] Shao P, Wang Z, Yang S and Liu Z 2019 Dynamic modeling of a two-DoF rotational parallel robot with changeable rotational axes *Mechanism and Machine Theory* **131** pp 318-335
[24] Bianchi G, Fassi I, Molinari Tosatti L and Catelani D 2000 A virtual prototyping environment for parallel kinematic machine analysis and design *Proc.PKM int. conf.* (Ann Arbor, USA)
[25] Artemov D V, Masyuk V M, Yu Orehov S and Pchelkina I V 2019 3DoF Parallel robot analysis *IOP Conference Series: Materials Science and Engineering* **489** 012052
[26] Yu H, Guo Q and Cui G 2019 Kinematics simulation of 5-DOF parallel mechanisms *Proc. of the 2019 Int. Conf. on Robotics, Intelligent Control and Artificial Intelligence* pp 712-717
[27] Chen X, LiangX, Deng Y and Wang Q 2015 Rigid Dynamic Model and Analysis of 5-DOF Parallel Mechanism *International Journal of Advanced Robotic Systems* **12**:108
[28] Miclosina C O, Cojocaru V and Korka Z I 2020 Analysis of a 6-DOF parallel robot motion simulation *Journal of Physics Conference Series* **1426**
[29] Guo W, Li R, Cao C and Gao Y 2016 Kinematics, dynamics, and control system of a new 5-degree-of-freedom hybrid robot manipulator *Advances in Mechanical Engineering* **8** (11)
[30] Li R, Wang S, Fan D, Du Y and Bai S 2017 Dynamic modeling of a 2-RPU+2-UPS hybrid manipulator for machining application *Modeling, Identication and Control* **38**(4) pp 169-184
[31] Li J, Ye F, Shen N Y, Wang Z R and Geng L 2020 Dimensional synthesis of a 5-DOF hybrid robot *Mechanism and Machine Theory* **150**:103865
[32] Thomas M J, Joy M L and Sudheer A P 2020 Kinematic and dynamic analysis of a 3-PRUS spatial parallel manipulator *Chinese Journal of Mechanical Engineering* (2020) **33:13**

Study of Vulnerabilities in Designing and Using Automated Vehicles based on SWOT method for Chevrolet Camaro

P Bec[1], A I Borzan[2], M Frunză[1], D L Băldean[2] and I Berindei[2]
[1] Hitory and Philosophy Dep., Babeş-Bolyai University, Cluj-N., M. Kogălniceanu 1, Ro.
[2] Automotive and Transportation Dep., Technical University of Cluj-N., Muncii 103-105, Ro.

doru.baldean@auto.utcluj.ro

Abstract. Travel sickness of the passenger, lack of human control over driving, high level of complexity in construction and programming, possibility of technical failure, limited or no driving engagement are just some of the possible vulnerabilities when using an automated or an autonomous vehicle. There are also important benefits and clear advantages such as more time for relaxing, resting, sleeping, reading, or working during the travel. Therefore is a "must", to study the process of designing and using automated vehicles in order to outline vulnerabilities which may provide the mitigating solutions. As majority of the studies regarding vulnerabilities of the automated vehicles consist in fundamental research investigations a SWOT analyse of vulnerabilities in designing and using autonomous cars is thus necessary. The present work investigates methodological and designing aspects for researching the vulnerabilities in the process of developing and implementing automated driving, based on SWOT approach, simulation tools, practical measurement, experimental testing, sampling, graphical modelling, programming, and ethical considerations. Moreover, some aspects of the SWOT method led by multiple engineering and ethical problems are brought up and highlighted. Part of the data from conducted research concerning the designing and using autonomous vehicles made in a dynamic-based driving environment and with a real car are also applicable in sustaining the arguments.

1. Introduction

Individual vehicles and driverless taxicabs are nowadays the most important and necessary transport means due to rapid contagion in pandemic period. Designing and using automated vehicles is now an absolute necessity because of the high risk of sickness transfer from human to human (during pandemic). Having such cars with automated driving capacities may contribute to the containment of the spread and help the authorities. Also, smart driver-less automated cars may support vulnerable individuals to seek and obtain help from others. Nonetheless these autonomous vehicles may generate a multitude of vulnerabilities both in designing and during exploitation in road transport and traffic. Designing an automated vehicle is not an easy task due to the repeated programming and testing sessions in order to create a proper and adequate artificial intelligence to control the commands and cars' behavior [1][2][3].

Using automated vehicles is entirely another problem based on the experience already acquired. All the operational commands on the actuators resulting in smooth and safe driving of the vehicle are based upon intelligent control of the car's powertrain [4] and driving systems (steering and braking) during the transportation process. Developing learning-based programs in automotive engineering is a niche effort [5] which may lead to multiple discoveries and benefits such as fuel economy and carbon footprint reduction [6]. There are foreseen both advantages and disadvantages, benefits, and treats in automation of driving and road transportation processes based on future developments [7]. Travel sickness and lack of human individual control over the personal road traffic experience are few of the concern signals [8]. Interaction of human to vehicle essential information is also vital [9]. Electronic control [10] of the powertrain in coordination with the over-all vehicle management [11] system should provide [12] the user and the service staff with essential data properly displayed and grouped in order to minimize harshness and to increase the safety, pleasure and comfort of travel experience and cars' maintenance.

Investigating vulnerabilities and SWOT (strength, weaknesses, opportunities, and treat) aspects in designing and using automated vehicles takes into consideration the multiple and complex interactions between different life stages of the automotive product, especially when it comes to road traffic impact.

Figure 1 presents the virtual reality environment design in Unity 5 modelled in order to facilitate the simulation of automated car control and artificial intelligence applied to driving and to machine learning process or artificial intelligence (AI) in order to prevent transport and road traffic events [1],[2],[3],[4].

Figure 1. Virtual reality environment (a) and virtual automated car (b) designed in Unity [1],[2],[3].

Figure 2 shows the relations facilitated by the development of robotic automated vehicles with significant impact on road traffic, society, and transportation, implementing artificial intelligence (neural networks and machine learning), automation, robotics, and programming [3-5]. Mathematical modelling and computer science represent the fundamentals for robotics and automated systems (a). An input set of parameters shown in figure 2 (b) are processed and filtered through kernel component in order to generate an output value. There are also two hidden layers that control the output value. The neural networks are based on mathematical calculus at first layer, then a non-linear operation function in order to obtain features, followed by a pooling function which replaces the output value with a statistical processed result [1-5]. Artificial intelligence (AI) may be programmed using Bayes theorem.

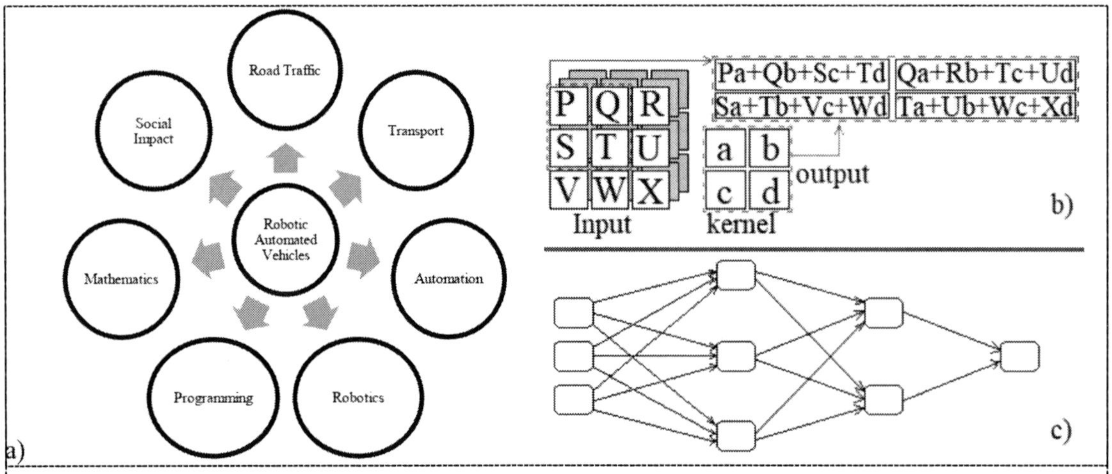

Figure 2. Relations of automated vehicles (a), machine learning (b) and AI neural networks [5]

2. Applied SWOT study and modelling

The applied study of vulnerabilities in designing and using automated vehicles is based on a predefined representation of strengths, weaknesses, opportunities, and treats (SWOT), aspects shown in table 1. In order to address these aspects controlled applied research is necessary. Considering this assumption, may be recorded that little research has been developed around the topic.

Table 1. SWOT descriptive data

Strengths	Weaknesses	Opportunities	Threats
Uniformity of experience	Complexity of protocols	More leisure time	Machine failure
Selfless driver	Lack of empathy	Less road crimes	System hackings
Permanent memory	Digital storage	Rapid data sharing	Data stealing
Remote control	Easy interception	Efficient use	Stealing from distance
High level control	Software failure	More programming	Cyber crimes
Internet connection	Privacy risks	Accessing more info	Personal data leaks
Programmed dynamics	Blind spots	Drive Standardizing	Travel sickness

Considering SWOT method for highlighting the aspects concerning the automated vehicles has been followed a designing of a robotic automated vehicle based on Chevrolet Camaro model which was replicated and tested in Virtual Reality application Unity 5 and afterwards validated with a series car (most similar and closely related to the digital version).

Most important technical data that must be studied in the present case are mechanical, electrical and digital in substance, due to the influence of kinematics (displacement, speed/velocity, acceleration), dynamics (force and torque), electric currents, voltages, and programming on the vehicle movement, travel experience and transportation safety.

Figure 3 highlights the connections and interactions between the shaping factors of the research, testing, and development regarding automated vehicles. This paper shows a study of vulnerabilities and basic considerations regarding materials and methods applied in designing and using Chevrolet Camaro automated vehicles. It will sustain and contribute other applied engineers to choose proper methods, measurements, design features, dynamics, and even ethical rules to be implemented and embedded in automated vehicles which will travel on roads in the future.

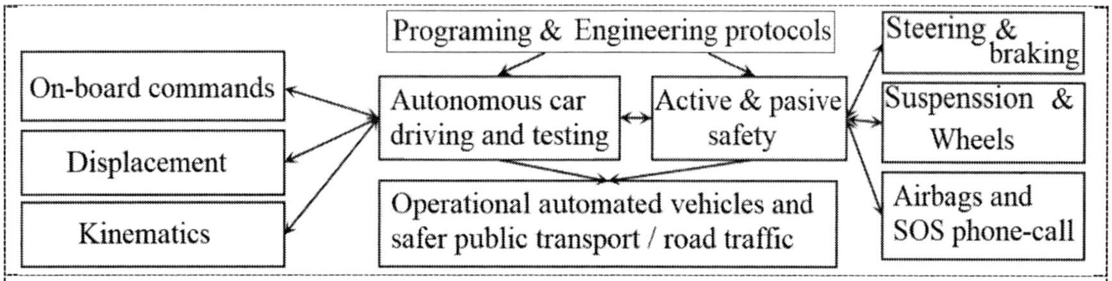

Figure 3. Interactions between factors which may shape automated vehicles future design and use

Programing AI with Bayes theorem where first probability is defined using a training set as shown [5]:

$$P(z|w) = \frac{P(w|z)P(z)}{P(w)} = \frac{P(z)\prod_{i=1}^{N} P(w_i|z)}{P(w)} \quad (1)$$

where P(z) is the prior / first probability; P(w) - last prob.; P(w|z) – class conditioned probability.

IOP Conf. Series: Materials Science and Engineering **898** (2020) 012008 doi:10.1088/1757-899X/898/1/012008

Chevrolet Camaro has been developed in six generations since 1967 till present days of which latest models are suitable for automation. The important technical specifications are given in table 2 [13],[14].

Table 2. Specific data for the Chevrolet Camaro vehicle under the applied study [13]

Parameter	Chevrolet Camaro
Car engine	3.6 L V6 335 HP 250 kW
Fuel system	Gasoline
Gear Box	8L45-automatic
Traction/Steering	Rear drive / Front steering
Tires	245/45R20 (front) 275/35R20 (rear)

In figures 4 to 12 are shown the significant steps in designing and using an automated Chevrolet car.

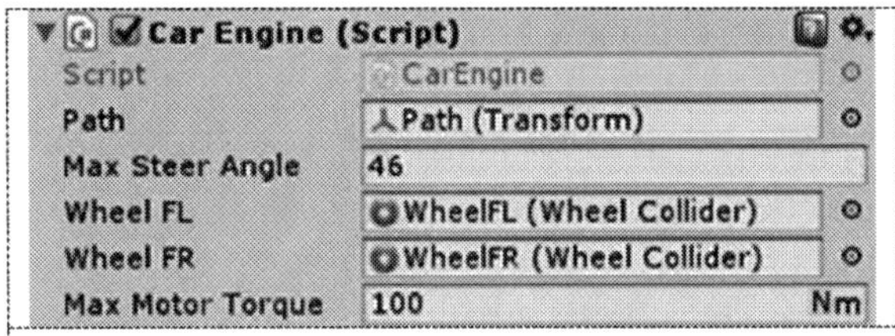

Figure 4. Using Car Engine Script for data set-up.

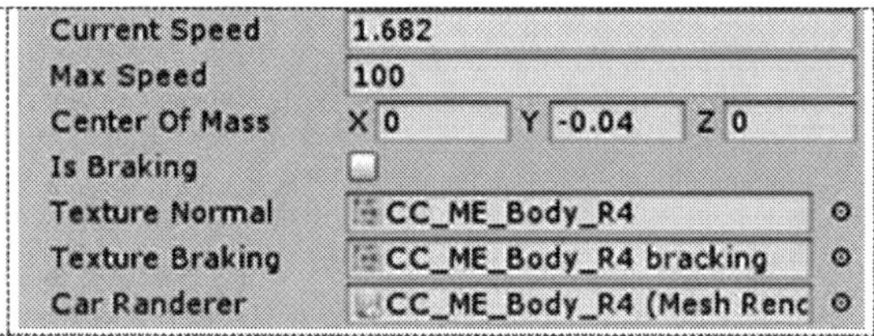

Figure 5. Input the data regarding dynamics.

Figure 6. Real Chevrolet Camaro under the study.

Figure 7. Virtual modelled Chevrolet Camaro.

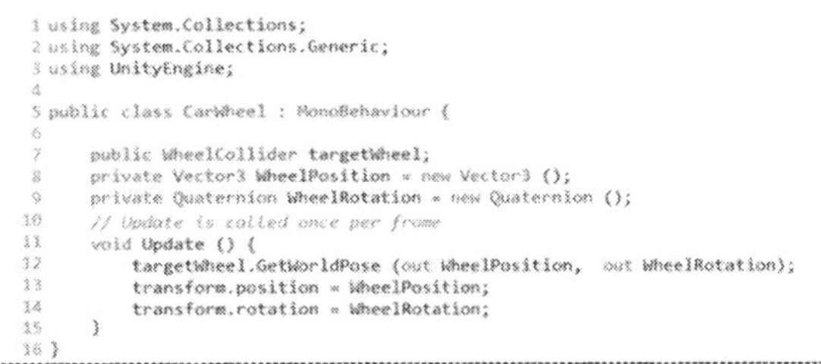

Figure 8. Programming the wheel collider in VR.

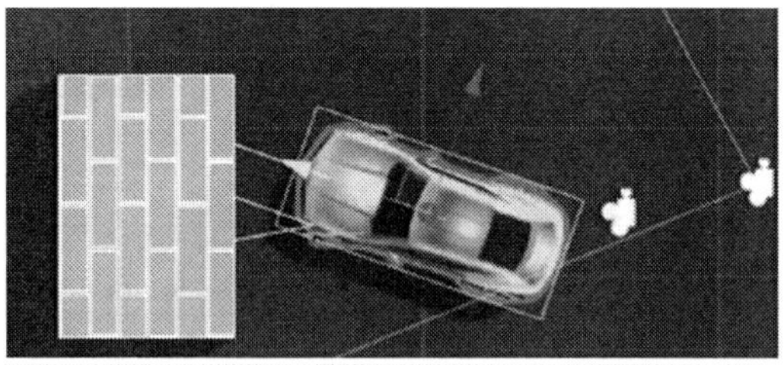

Figure 9. Learning how to avoid the obstacles.

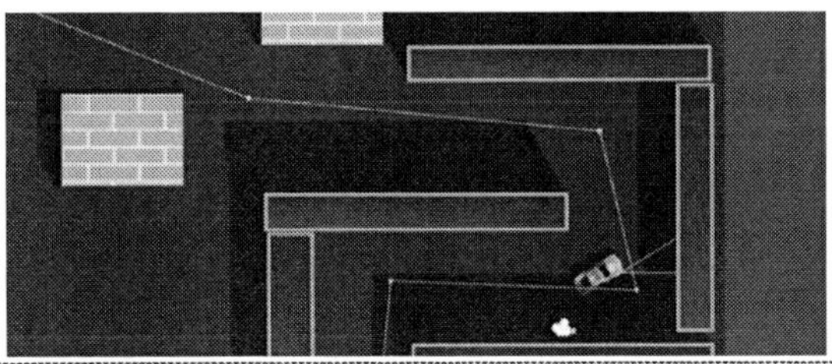

Figure 10. Solving the Maze problem with VR.

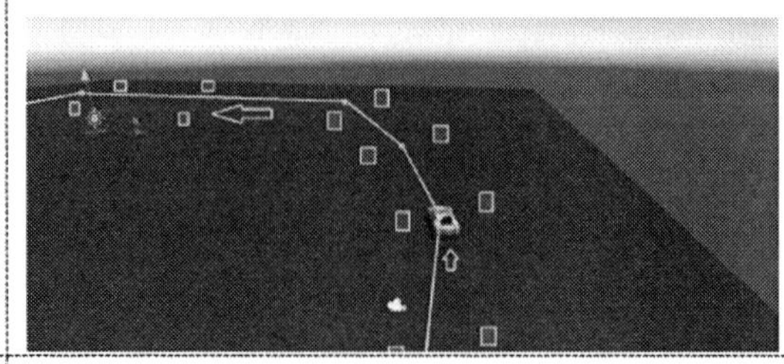

Figure 11. Travel on a specified track in VR.

3. Observations and conclusions

Autonomous or automated cars may contribute to enhance the benefits of the users due to the free time equally available for all the vehicle's occupants to be spent as they see fit in multiple leisure and non-driving tasks. Anyway, this fact leads to some treats and vulnerabilities, such as motion sickness and total lack of control.

The most significant vulnerabilities which were uncovered during the present study are situated both in design phase and in using of the real Chevrolet Camaro vehicle with automated system. For the first step of the study for implementing the automated driven vehicle in virtual environment the weaknesses or vulnerabilities are consisting in the fact that only few of the operational parameters may be replicated and applied at the present stage of the development. The analysed parameters cover mostly kinematic and dynamics of the vehicle and powertrain. Other vulnerabilities are found in the complex machine learning of AI, which is a limited step by step process and has possible negative impacts during travel. The practical testing has although much more vulnerabilities because there is no self-driving vehicle now. Thus, in practical research, now is only possible to drive Chevrolet Camaro with driver assistance.

References

[1] Covaciu F et al 2018 *Design and control of a mobile robot of exploration, using virtual reality as a simulation environment* (In Sci. and Eng. / Şt. şi Ing. AGIR Press) eISSN 2359–828X DOI http://stiintasiinginerie.ro/34-43

[2] Covaciu F et al 2018 *Applied developing of a vehicle solution which drives itself autonomously using Virtual Reality Fuzzy Regulator* (Ing. agr. transp. auto vol 51 pp. 220-4 available in IBN 25 feb 2020) CZU: 629.33.01 DOI: https://ibn.idsi.md/ro/vizualizare_articol/96336 https://ibn.idsi.md/sites/default/files/imag_file/220-224_12.pdf

[3] Covaciu F et al 2020 *Contribution to Research the Applied Engineering Protocol to Implement a Fuzzy Regulator for Autonomous Driving of an Automotive Model Implemented in Virtual Reality* (In: Dumitru I et al (eds) The 30th SIAR International Congress of Automotive and Transport Engineering. SMAT 2019. Springer, Cham) eISBN 978-3-030-32564-0 DOI https://doi.org/10.1007/978-3-030-32564-0_54

[4] Ferenti I et al 2019 *Analysis of Intelligent Control and Interface in Pressurized Liquid Injection Systems for Competitive Technical Solutions* (In Hidraulica, Issue 2, 43-48, 1 April 2018, Article number 327395077) ISSN: 1453 – 7303.

[5] Gomez-Fernandez M et al 2020 Status of research and development of learning-based approaches in nuclear science and engineering: A review (In Nucl Eng and Design 359 110479 Elsevier B. V.) https://doi.org/10.1016/j.nucengdes.2019.110479

[6] Hodor A I et al 2018 *Contributions to fuel economy and carbon footprint study in various operation conditions for a city compact size car with at technical university from Cluj-Napoca* (In Science and Engineering / Ştiinţă şi Inginerie AGIR Press) eISSN 2359–828X DOI http://stiintasiinginerie.ro/34-70

[7] Iordan A E et al 2020 *Improving design of a triangle geometry computer application using a creational pattern* (In Acta Technica Napocensis, 63 (1), 73-8)

[8] Nobel G et al 2012 *Effects of motion sickness on thermoregulatory responses in a thermoneutral air environment.* (In European journal of applied physiology, 112, 1717-1723)

[9] Jovrea S et al 2018 *Researching on-board display of essential information* (In Sci. and Eng. 31-67) eISSN 2359–828X http://stiintasiinginerie.ro/31-67

[10] Marincas C et al 2017 *Contributions to the experimental research of electronic diesel control (EDC) module operation in relation with supply of the n47 engine from BMW 320d (e90) automobile* (In Sci. and Eng. 31-82) eISSN 2359–828X http://stiintasiinginerie.ro/31-82

[11] Moldovan A et al 2017 *Experimental research of the management system from the Peugeot 4007 Sport utility veh.* (In Sci. and Eng. 31-71) eISSN 2359–828X http://stiintasiinginerie.ro/31-71

[12] *** 2020 *Self-driving car* (In: Wikipedia https://en.wikipedia.org/wiki/Self-driving_car)

[13] *** 2020 *Chevrolet Camaro (sixth generation)* (In: Wikipedia accessed 30 April 2020)

[14] *** 2019 *Chevrolet Camaro Transmissions* (In: https://gmauthority.com accessed 30 April 2020)

IOP Conf. Series: Materials Science and Engineering **898** (2020) 012009 doi:10.1088/1757-899X/898/1/012009

Investigation of Failures and Vulnerabilities in Road Traffic Air Quality Management System during 2020 Pandemics

P Bec[1], M Frunză[1], and D L Băldean[2,3]
[1] Hitory and Philosophy Dep., Babeş-Bolyai University, Cluj-N., M. Kogălniceanu 1, Ro.
[2] Automotive and Transportation Dep., Technical University of Cluj-N., Muncii 103-105, Ro.

doru.baldean@auto.utcluj.ro

Abstract. Air deterioration and high toxicity affects life quality and longevity. Most important is pollution and toxicity impact upon health and disease management. Air born sicknesses and health problems are a permanent challenge for air dependent life-forms. Lately have been recorded manifestations and a complex phenomenology of air partially transferred pathogenic agents. The air quality is subject to investigation in order to provide a better management of the pollution control system. Pollution is a major contributor to air quality deterioration and to health problem multiplication. Present day pandemic is thought to be partially transmitted through air and by direct contact. Low air quality may be a serious contributing factor to sickness promotion. Air pollution due to road traffic and livestock is one such factor. The main objective of the paper is to investigate some important failures and vulnerabilities recorded in the road traffic air quality monitoring system during 2020 pandemics in our country and in Cluj. Thus, the evolution of chemical composition has been put under the spot of scientific investigation. Failures and weak points have been acknowledged and considered for improvement. At least twelve significant system failures and multiple vulnerabilities have been recorded due to sensor faults and unit malfunction. Social lockdown reduced all pollutants, but some units failed to indicate it properly.

1. Introduction

Air Quality Management System is implemented and operational by some years now, but its failures and vulnerabilities must be investigated in order to support the proper adjustments for technical and economic optimisations. Directive 2004/107 / EC, 2008/50 / EC, Decision 2011/850 / EC and Directive (EU) 2015/1480 are defining the terms, rules, reference methods, data validation and location points in ambient air quality investigation. Air quality is very important for life development. Many air-born diseases are reported nowadays [1], [2], [3]. High toxicity of the ambient air and the chemical stress are causal factors for many of the current health problems [4]. Benzene compounds (generated by road traffic and by fuel distribution networks), nitric oxides and CO_2 are toxic for human health [5], [6], [7].

Some developments and applied protocols have been realized in order to improve the air quality control and pollutant emissions management in relation with the road traffic and transportation field [8]. These contributions outline the road traffic pollution upon pathologic developments [9]. Some pollutants like particulate matter may be contained by using exhaust after-treatment systems [7], [10]. Fuel and oil combustion, evaporation and spill represent the cause for the environmental pollution and toxicity [10], [11]. Electronics [13] and mathematical modelling of the operational process from internal combustion engines [14], [15] constitutes basic support for pollution management. Improving efficiency regarding fuel consumption and energy transformation depends mostly upon the management system [16]. The electronic engine control [13] or the operational management system [16] facilitate, for both

IOP Conf. Series: Materials Science and Engineering **898** (2020) 012009 doi:10.1088/1757-899X/898/1/012009

compression ignited and for spark ignition engines [15], in the same time improved performances and lower fuel consumption with reduced pollutant emissions in ambient air [16], [17].

Figure 1 highlights the measurements made by the European Space Agency during the first three months of 2020 regarding the pollution with nitrogen dioxide during the COVID pandemics [18], showing the fact that road traffic lockdown in multiple countries have positively impacted the air quality. Reduction of road traffic values generates a reduction in pollutant emissions, as a following reasonable consequence, but there comes a justified question: may be measured correctly the pollutant trend? The present paper investigates the failures and vulnerabilities in road traffic air quality management system.

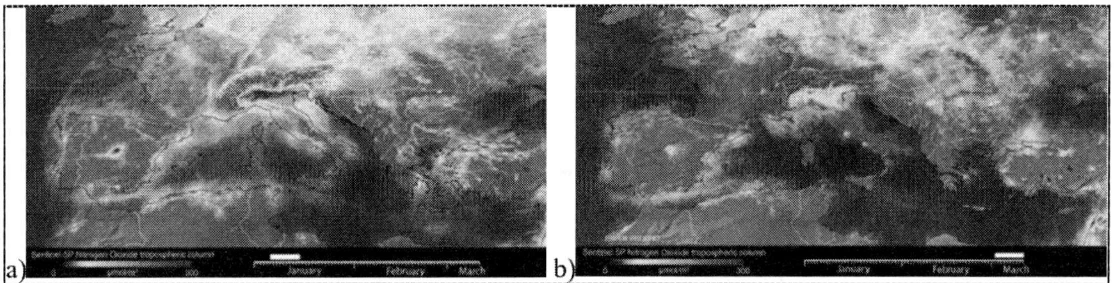

Figure 1. Pollution high level in Italy first part of January 2020 (a) and March (b) same year [18].

2. Applied investigation

Some of the most dangerous pollutants and air toxicity are generated by the road traffic, animal intensive factory farming, some industries and air transport [17], [18]. Main objective of the applied investigation consists in monitoring the actual values of pollutants recorded by four stations in Cluj-Napoca. These units (CJ1, CJ2, CJ3, CJ4) are positioned in central areas of the city, as they measure air quality.

Figure 2 presents the hardware infrastructure available for applied measurements and for pollutant data monitoring and investigation.

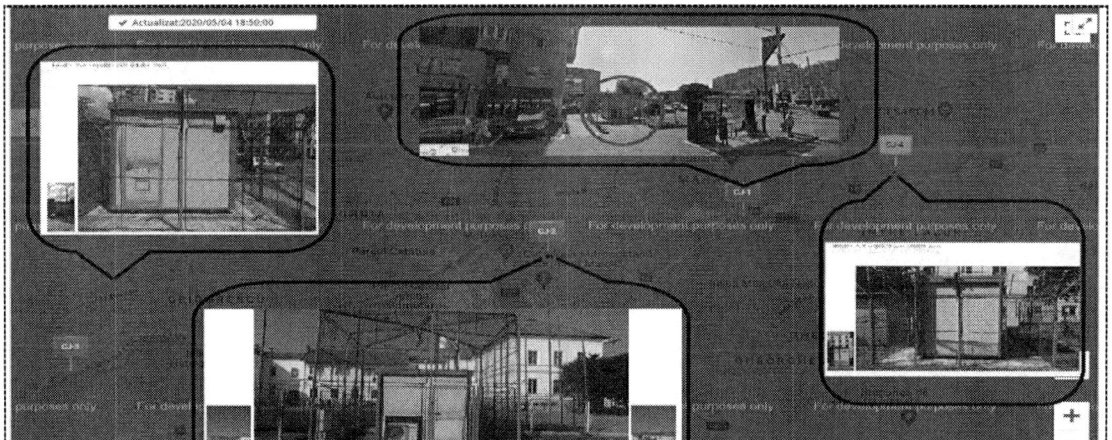

Figure 2. Cluj-Napoca Air Quality Management units distributed in the most populated areas [17]

Definition of PM10 values by empirical polynomial model is presented as follows for CJ1 and CJ4:

$$PM10(CJ1|CJ4) = (1.3 \ldots 2.4) * 10^3 * Mt^2 - (0.2 \ldots 0.4) * Mt + 27, \quad (1)$$

where PM10(CJ1|CJ4) is the PM_{10} estimated average value supported by a polynomial model in the investigated time period of 100 hours for the CJ1 & CJ4 stations; Mt – monitoring time.

Material and method used in applied research of the present investigation is outlined in table 1. Most significant failures were recorded by station CJ2 in monitoring nitric oxide compounds and derivates.

Table 1. Specific technical data regarding Air Quality Management System used in applied study

Pollutant	Method	Station Unit	Chemical formula
Benzene	SR EN 14662, part 1-3	CJ1	C6H6
Ethylbenzene	SR EN 14662	CJ1	C8H10
Nitric oxides	SR EN 14211	CJ1, CJ2, CJ3, CJ4	NO / NO_2
Carbon monoxide	SR EN 14626	CJ1, CJ2, CJ3	CO
Ozone	SR EN 14625	CJ2, CJ3, CJ4	O_3
Sulphur dioxide	SR EN 14212	CJ1, CJ2, CJ3, CJ4	SO_2
Polycyclic aromatic HC	SR EN 15549	CJ1, CJ2	$C_{14}H_{10}$; ... $C_{32}H_{14}$;
Particulate matter	EN 12341	CJ1, CJ4	PM_{10} and $PM_{2,5}$

Most recorded failures during the investigation are based on individual specific posts or system failures. One of the investigated hypotheses is measuring sensors failures due to maintenance faults.

During the investigation presented in the present paper were recorded 12 significant failures of the Air Quality management system which support our conclusion on the importance of remote access and service implementation. An important vulnerability of the system is generated by the un-even type of the measurements. Some pollutants are not monitored on some stations due to the specific differences.

The individual failures shown in figure 3 and the multiple fault values in figure 4 were taken from air quality road traffic pollution management system. In the station unit CJ1 most failures were recorded with m, o, p-xilen emissions, but there were also considerable failures with benzene and some with NO_x.

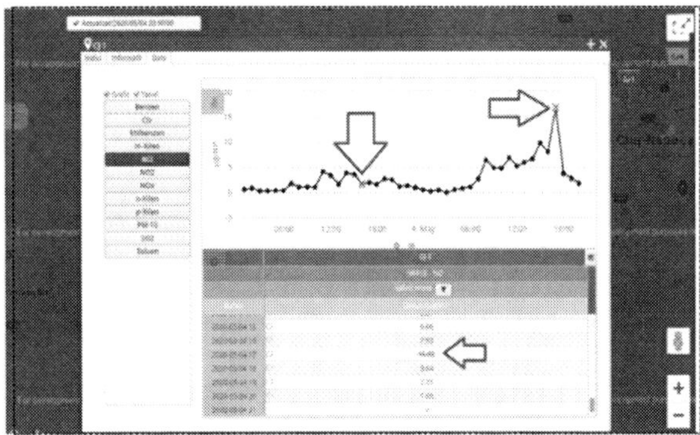

Figure 3. Seldom recorded failures for NO [17].

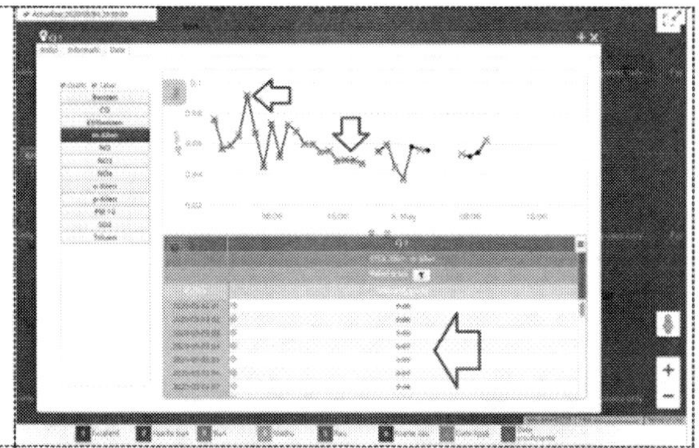

Figure 4. General measuring failure [17].

Definition of NO_x average estimated values by an empirical polynomial model is represented as eq. (2):

$$NO_x(CJ1|100h \ldots 28April \ldots 3May) = 6.1 * 10^3 * Mt - 1.53 * Mt + 105.1, \quad (2)$$

where NO_x (CJ1|10h...28April...3May) is the NO_x estimated average value calculated with a polynomial model on the 100 hours investigated time period; Mt – monitoring time.

In figures 5 to 14 are graphically represented the pollutants hourly recorded values on a 100 hours investigation period. Nitric oxides were specially studied on all the available stations in Cluj-Napoca city in order to properly investigate the failures and vulnerabilities of air quality monitoring units. The graphical representations allow a better study on the failures and vulnerabilities of the monitoring units.

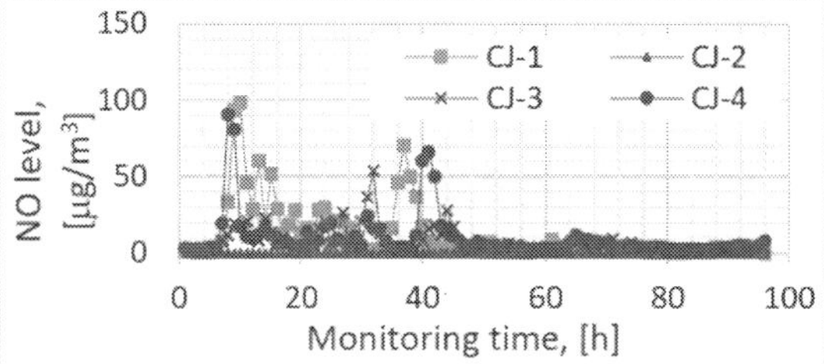

Figure 5. NO values recorded and studied.

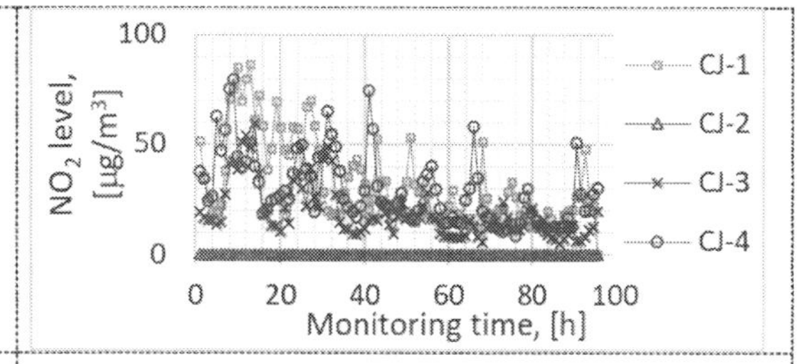

Figure 6. NO_2 actual values for investigation.

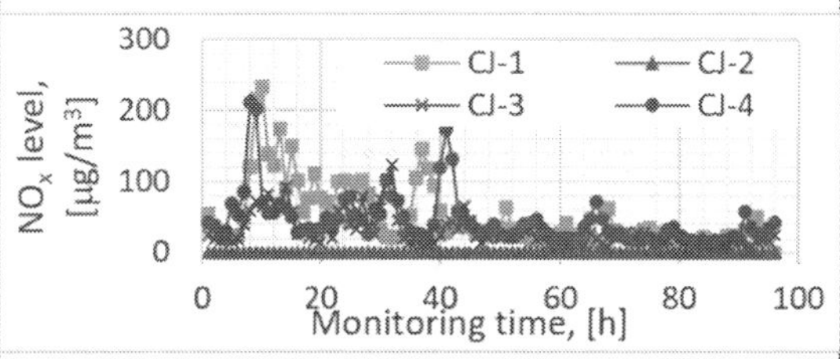

Figure 7. NO_x actual values variation recorded.

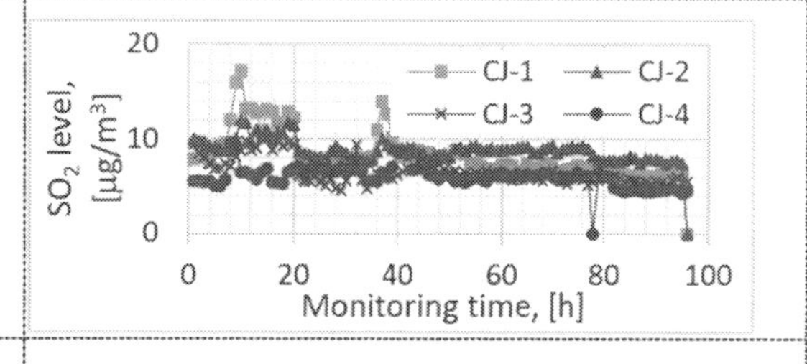

Figure 8. SO_2 actual values for investigation.

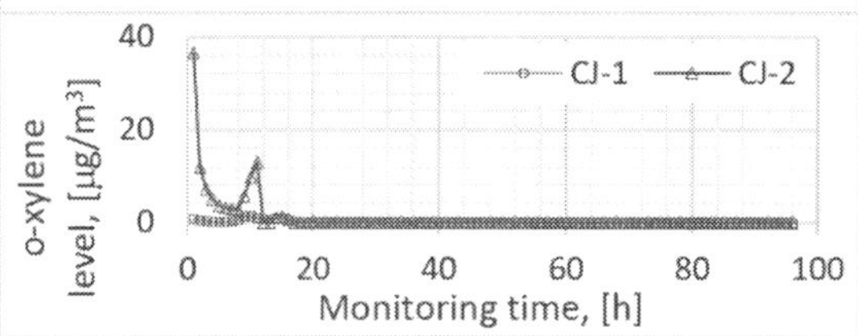

Figure 9. Actual values for o-xilen at CJ1 / CJ2.

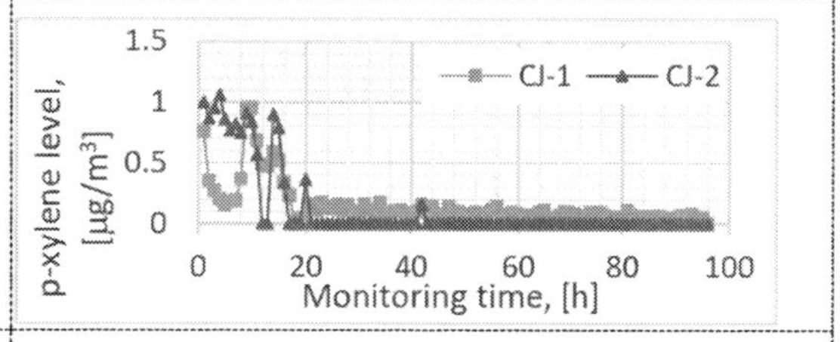

Figure 10. Actual values for p-xilen recorded.

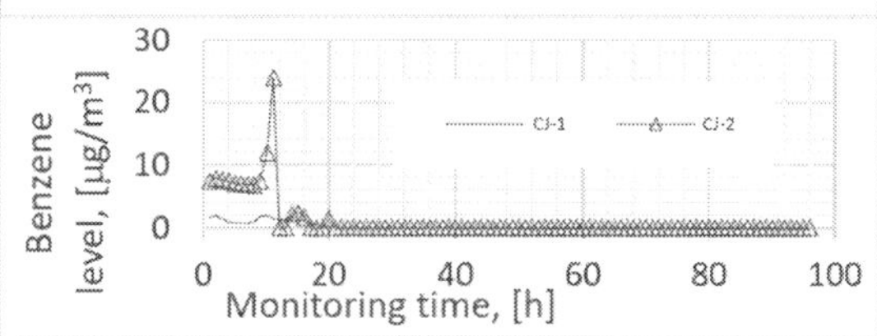

Figure 11. Benzene values at two stations.

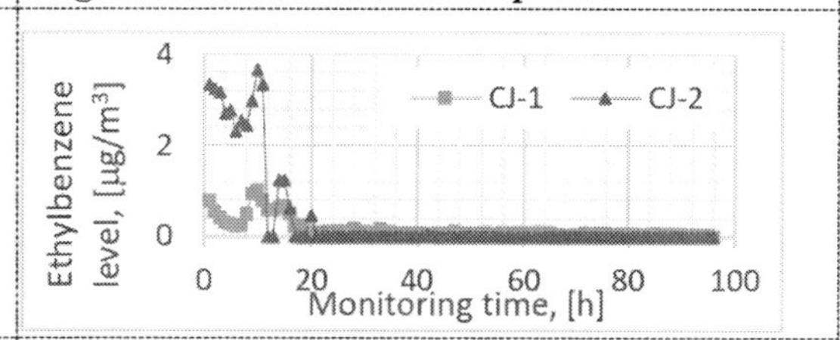

Figure 12. Actual values for ethylbenzene test.

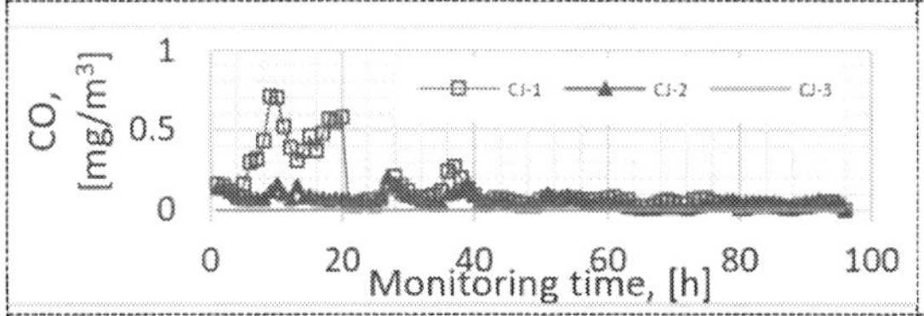

Figure 13. CO values recorded by CJ1, CJ2.

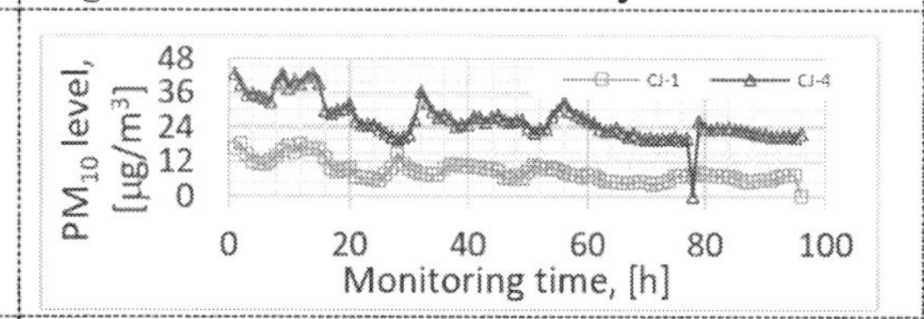

Figure 14. PM_{10} actual values at two stations.

3. Observations and conclusions

Both theoretical and the applied parts of the investigation have been realized successfully as planned. It was possible to show the real values which were recorded and monitored to outline the failures and vulnerabilities of the air quality system.

In the recorded set of values has been highlighted a decrease with an average of 45% of NO level starting from 28 April 2020 till 3rd of May 2020. Also, there was a reduction with 30% of NO_2. The approaching weekend (1-3 May) and the low temperatures (6÷13°C), coupled with the imposed social lockdown measures contributed to overall reduction in pollution. The system failures must be further studied, and faults eliminated.

References

[1] Andrei L et al 2018 *Analytic considerations regarding the link between air pollution with benzene and nasopharyngeal cancer* (In Science and Engineering / Ştiinţă şi Inginerie AGIR Press) DOI http://stiintasiinginerie.ro/34-63
[2] Andrei L et al 2018 *Contributions to research of CO2 emissions in environment pollution management and life* (In Science and Engineering / Ştiinţă şi Inginerie AGIR Press) DOI http://stiintasiinginerie.ro/34-64
[3] Andrei L et al 2018 *Contributions to the research of the relation between air pollution with nitric oxides and the impact upon environment and life* (In Science and Engineering / Ştiinţă şi Inginerie AGIR Press) DOI http://stiintasiinginerie.ro/34-68
[4] Andrei L et al 2018 *Research and reporting development of formic acid emissions in environment pollution management and life health* (In Science and Engineering / Ştiinţă şi Inginerie AGIR Press) eISSN 2359–828X DOI http://stiintasiinginerie.ro/34-71
[5] Andrei L et al 2019 *Applied Measurements and Instrumentation for Improving Diagnostic Devices and Systems in Metropolitan Polluted Environments with Nitric and Carbon Oxides* (In Meditech, IFMBE Proceedings 71, Springer, Singapore) ISBN 978-981-13-6206-4 DOI https://doi.org/10.1007/978-981-13-6207-1_8
[6] Borza E V et al 2018 *Research Concerning Fuel Economy Coefficient and Carbon Foot Print in Various Conditions for a City Compact Size Vehicle with Digital Control for a Green Solution and Method* (In Burnete N., Varga B. (eds) Pr. of the 4th Inter. Congress of Automotive and Transport Engineering (AMMA 2018). Proceedings in Automotive Engineering. Springer, Cham) eISSN 978-3-319-94409-8 https://doi.org/10.1007/978-3-319-94409-8_22
[7] Borzan A I et al 2018 *Research contributions through virtual modeling of an innovative particle filter for alternative application in automotive field using advanced engineering methods* (In the 4th Inter. Congress of Automotive and Transport Engineering AMMA 2018) http://www.amma2018.ro/index.php/amma/2018/paper/view/28
[8] Chereches I A 2018 *Researching the applied engineering protocol to implement a program for monitoring air quality management and carbon-footprint for future green vehicles in urban area* (In the 4th Inter. Congress of Automotive and Transport Engineering AMMA 2018) http://www.amma2018.ro/index.php/amma/2018/paper/view/37
[9] Chereches I A et al 2017 *Contributions to the applied research of effects from surrounding environment pollution due to road traffic upon eyes' anatomy and pathology development* (In Science and Engineering / Ştiinţă şi Ing. AGIR Press) DOI http://stiintasiinginerie.ro/32-85
[10] Jurchiş B M et al 2019 *Experimental research on Diesel Particle Filter (DPF) in relation to the fuel type* (In IOP Conf. Ser.: Mater. Sci. Eng, **568**) 012027
[11] Marc C A et al 2010 *Study of the automotive fuel and oil pollution influences upon some plants in different conditions* (In CONAT, International congress on automotive and transport engineering, 27-29 October, 2010, Braşov, Romania, 3, CONAT20102030) ISSN 2069-0401 http://conat.ro/index.php/conat/2010/paper/view/305
[12] Marc C A et al 2010 *Pollution influence of fuel used for engines on some plant species* (In Acta Mecanica, 2, (3), Tehnical University of Cluj-Napoca, ISSN 2066-9577)
[13] Marincas C et al 2017 *Contributions to the experimental research of electronic diesel control (EDC) module operation in relation with supply of the N47 engine from BMW 320d (E90) automobile* (In Sci. and Eng. 31-82) eISSN 2359–828X http://stiintasiinginerie.ro/31-82
[14] Mitran T et al 2018 *The calculus of the technical-economic parameters of a SI direct inj. engine* (In RoJAE 24 (3) 89-97) http://siar.ro/wp-content/uploads/2018/12/RoJAE-24_3.pdf#page=5
[15] Mitran T et al 2019 *The calculus of the temperatures in the characteristic points of the gasoline direct injection engine cycle* (In IOP Conf. Ser.: Mater. Sci. Eng. **568**) 012073
[16] Moldovan A et al 2017 *Experimental research of the management system from the Peugeot 4007 Sport utility veh.* (In Sci. and Eng. 31-71) eISSN 2359–828X http://stiintasiinginerie.ro/31-71
[17] *** 2020 *AirQuality N.A.Q.M.N.* (In: http://www.calitateaer.ro/ accessed 01 May 2020)
[18] *** 2020 *How nature is thriving during COVID-19* (https://www.livekindly.co/ accessed 04 May)

IOP Conf. Series: Materials Science and Engineering **898** (2020) 012010 doi:10.1088/1757-899X/898/1/012010

The use of finite element analysis in studying deformations of parts clamped in machine tool devices

I Stănăşel[1] , P D Tocuţ[1], R Veres[1]

Faculty of Managerial and Technological Engineering, University of Oradea, Romania

stanasel@uoradea.ro

Abstract. Increasing the productivity and quality of products while reducing production costs are the main objectives of manufacturing companies in the market economy. For companies that produce parts using cutting processes, these objectives are strongly influenced, among other factors,by the devices used during the manufacturing process. Devices must allow the clamping of workpieces to be done in a short period of time, to have simple construction, to be cheap while positioning and installing errors to be minimal to ensure that the parts correspond to the manufacturing documentation. The calculation of clamping errors of workpieces and devices involves complex analysis, therefore in order to accurately determine these errors, data obtained from experiments, sometimes costly and difficult to be done, are used. This paper presents the clamping force calculation of a prismatic part, as well as the study of elastic deformations that appear by clamping it in the device, using finite element analysis.

1. Introduction

Clamping devices are an important component of the technological system that enable complex parts manufacturing, in a short time, at low costs, and ensuring high surface finish [1]. With the increase of computer capabilities and the development of computer-aided design applications, the study of different structures' behaviour during operation can be simulated with acceptable accuracy [2].

The literature [3], [4] in the domain of devices for machine tool processing offers theoretical approaches necessary for the design. There are also researchers who address different device design and operation aspects. In article [5] the authors present a way to reduce workpiece deformations, by minimizing the value of clamping forces, the position and number of pressing points, the mathematical model found being validated by finite element analysis. In article [6], a multi-objective optimization model that allows the reduction of deformation degree and improves the uniformity of deformations' distribution was developed. The researchers [7] study the deformations which appear when clamping parts with thin walls and curved surfaces, using finite element analysis. In article [8] they focus on identifying influencing factors and highlighting the deformations of parts clamped in the device by using finite element analysis. A theoretical approach to improving eccentric clamping device is presented in [9]. The article proposes a mathematical model for calculating an eccentric replacement cam in order to increase the clamping force. Paper [10] presents the finite element analysis stages for establishing optimal construction and loading variant of a modular clamping device for prismatic and circular parts. It shows analysis methodology and the results of its application, under defined loading conditions, in order to identify the most loaded area from the device structure studied.

The paper presents the calculation of the clamping force of a prismatic part, as well as the study of elastic deformations that appear at its installation in the device using finite element analysis.

2. Clamping force calculation

The part to be processed and clamped in the device is shown in figure 1. It is a prismatic piece containing two tolerated holes Ø15 H7 and Ø20 H7. The part material is an aluminium alloy, Al 6061.

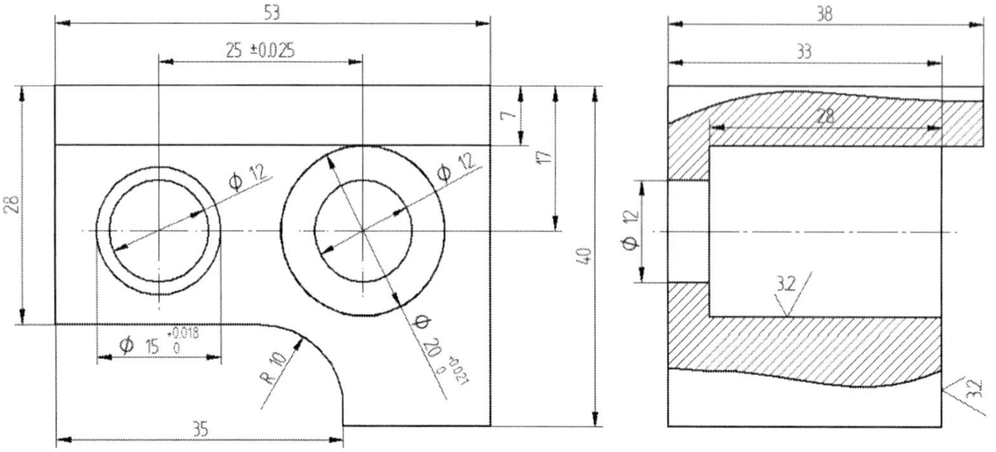

Figure 1. The part to be processed

The machining is done on a milling center. Operations are indicated in table 1, and the cutting parameters in table 2. The outer surfaces had been processed in a previous operation.

Table 1. Machining operations and the tools used.

Machining operation	Tool Description	Tool material	Diameter d [mm]	No. of teeth, z
Face milling la $33^{+0.15}$	End mill	HSS	40	6
Spot drilling	Spot drill	HSS	10	2
Drilling Ø12;	Drill	TiAlN	12	2
Hole milling Ø14.7	End mill	HSS	12	3
Hole milling Ø19.7	End mill	HSS	16	3
Reaming Ø$15^{+0.018}$(H7)	Reamer	Carbide	15	5
Reaming Ø$20^{+0.021}$(H7)	Reamer	Carbide	20	6

Table 2. Cutting parameters.

	DOC a_p/cut [mm]	WOC a_e/cut [mm]	Speed v_c [m/min]	Rotation RPM [1/min]	Feed v_f [mm/min]	Torque Mt [Nm]	Power P [kW]	Cutting force Fc [N]
Face milling $33^{+0.15}$	5	37	163	1297	892	16.20	2.20	828.7
Spot drilling	1.5		30	955	134	0.80	0.08	160.1
Drilling Ø12;			225	5968	2268	1.92	1.20	320.0
Hole milling Ø14.7	3	5	338	6724	2440	2.13	1.50	114.9
Hole milling Ø19.7	3	4	348	6923	2742	2.62	1.90	224.1
Reaming Ø$15^{+0.018}$			60	735	406	2.60	0.20	50.0
Reaming Ø$20^{+0.21}$			60	955	449	1.00	0.10	100.1

Considering that it is a part with thin walls, the problem of designing an orientation and clamping device appears so that the elastic deformations of parts produced by the clamping force do not affect the precision of the Ø15H7 and Ø20H7 after the reaming operation is performed. In order to determine the clamping force, the scheme shown in figure 2 is used, where it was considered that the predominant stress is produced by flat surface milling operation (as shown in table 2).

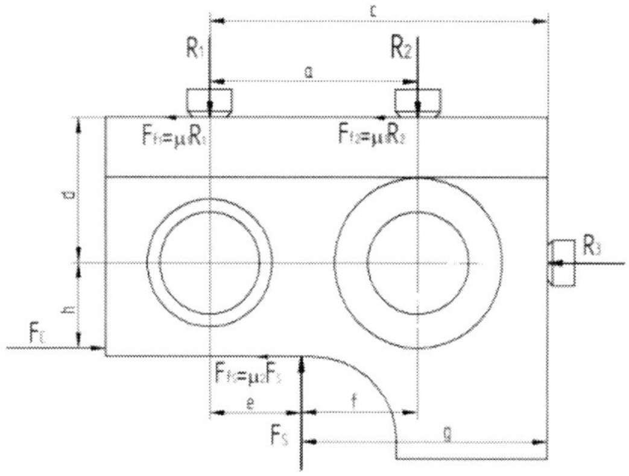

Figure 2. Scheme for retention force and reaction forces calculation.

From force and moment equilibrium expressions one can determine the retention and reaction forces(R_1, R_2):

$$F_S = \frac{k \cdot F_C}{\mu_1 + \mu_2} \quad (1)$$

$$R_1 = \frac{F_C}{c-b}\left(k\frac{g-b}{\mu_1 + \mu_2} - h\right) \quad (2)$$

$$R_2 = \frac{F_C}{c-b}\left(k\frac{c-g}{\mu_1 + \mu_2} + h\right) \quad (3)$$

Where

μ_1 and μ_2 represent the friction coefficient between the workpiece and the support, and between the workpiece and the clamping element, for steel and aluminum $\mu_1=\mu_2=0.45$

F_C – cutting force; $F_C = 828.7$ N

F_S – retention force

a,b,c,d,e,f,g,h – distance between forces' points of application and different characteristical points from the workpiece to be machined

k - safety coefficient, $k=2$ [1]

Replacing the letters in the relations, (1), (2), (3) with the corresponding values in figure 1 and the value of the cutting force for the milling operation in table 2, the values of the retention force and the reaction forces from the supports are obtained (table 3).

Table 3. Calculated values for retention force and reaction forces

Parameter	Symbol	Value
Retention force	F_S [N]	1850.0
Reaction force , support 1	R_1, [N]	704.5
Reaction force , support 2	R_2, [N]	1145.5

3. FEM analysis

The assembly that is studied is composed by the part, the support's elements and by the clamping element.NX Nastran is used to solve the equations. For each component, the material, the type of the element (solid) is defined. The discretization of the model with Ctetra 10 3D elements is shown in figure 3, and the mesh properties in table 4. After the calculations, the results regarding the deformations and stresses that appear in the part when clamped in the device are obtained (figure 4 and figure 5). From the results of the analysis it is observed that although the Von Mises stresses are within limits allowed for the AI 6061 material (tensile yield stress 276 MPa), the deformations are approximately 0.006 mm, which in the case of holes with the diameter at the upper limit can be rejected. It can also be observed that the clamping element creates an evident deformation on the contact surface with the workpiece.

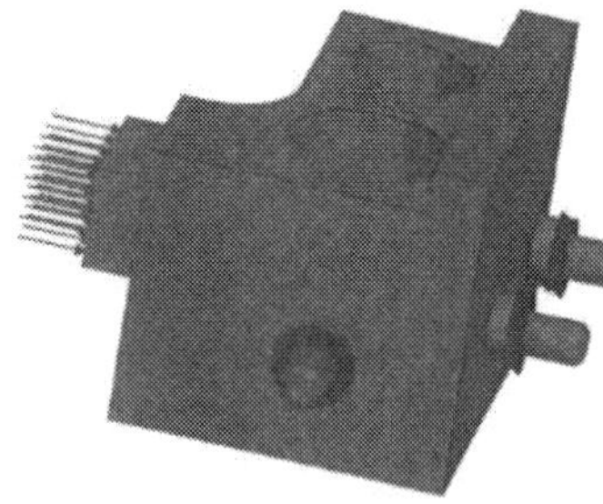

Figure 3. Discretization of 3D model. Applied forces and constraints

Table 4 Mesh information

Type of mesh	3D
Element Type	CTETRA(10)
Element Size	0.5 mm
Number of elements în the mesh	501929
Number of nodes în the mesh	828200

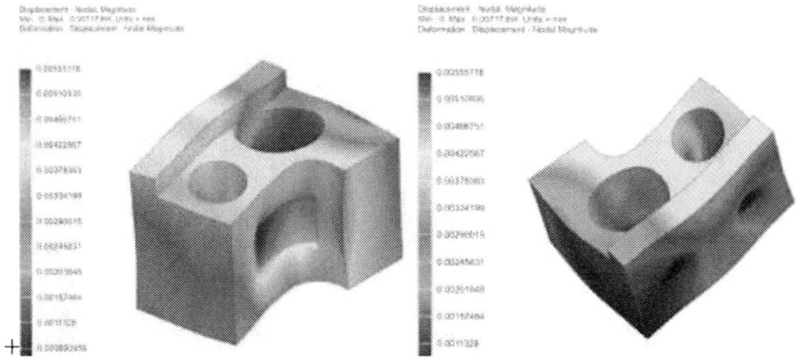

Figure 4. Displacement.

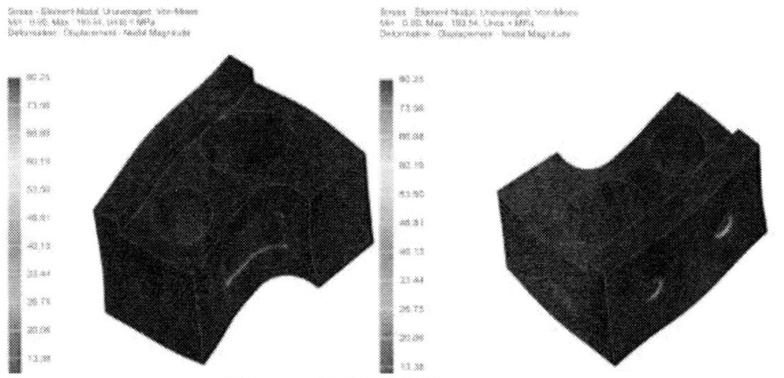

Figure 5. Von Mises stress.

In order to diminish the elastic deformations of the part, it is proposed to make a clamping element with a higher height and at the same time to reduce the clamping force. In order to reduce the retention force, the cutting force for the flat milling operation is reduced by reducing the cutting depth by half, i.e. t = 2.5 mm. Table 5 presents the recalculated values of the forces acting on the workpiece.

Table 5. Recalculated forces' values

Parameter	Symbol	Value
Cutting force	F_C [N]	331.3
Retention force	F_S [N]	740.0
Reaction force , support 1	R_1, [N]	282.0
Reaction force , support 2	R_2, [N]	458.0

The finite element analysis is resumed, and the deformations produced by the retention force are analyzed. From figure 6, it can be observed that this time the deformations reduce below the value of 0.001 mm, which is considered acceptable.

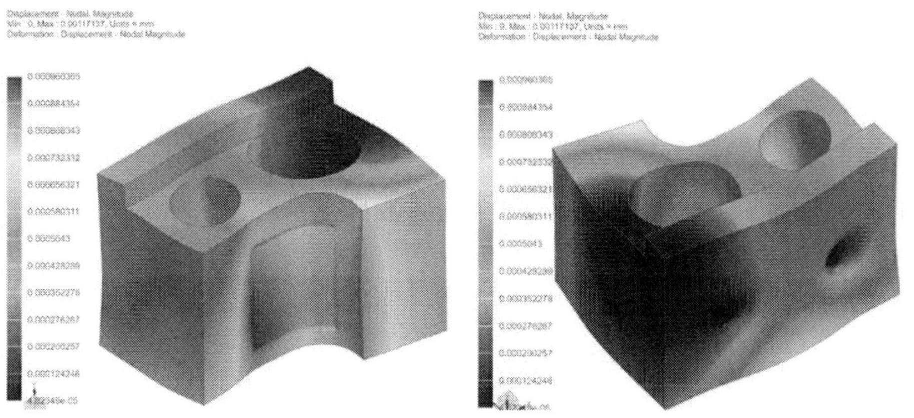

Figure 6. Displacement for recalculated forces.

4. Conclusions

The case study presents the estimation of the deformations that appear when clamping a part with thin walls in a device. Considering the cutting depth 5 mm, the required clamping force and reaction forces from the supports were calculated. The finite element analysis has shown that in this situation, the clamping force produces a deformation of the workpiece by approximately 0.006 mm, which is considered inadequate. In order to reduce this deformation, the contact surface of the clamping element was increased by 100% and the cutting depth to 2.5 mm, which led to a 60% increase of the clamping force. The finite element analysis was resumed, and this time the maximum deformation was below 0.001 mm, which was considered acceptable. Estimating the clamping errors right from the design phase, with the help of finite element analysis, allows the reduction of rejected parts and costs for device modifications.

Acknowledgements

The authors thank to Siemens PLM and ADA Computers for their support with Siemens NX academic software.

References

[1] Tache V., Ungureanu I., Stroe C., 1985, *Outlines of fixture design for machine tools* Editura Tehnica, Bucureşti.
[2] Hoffman E. 2011 *Jig and Fixture Design*, Delmar, USA.
[3] Joshi P.H., - Jigs and Fixtures Design Manual 2003 McGraw-Hill.
[4] Jones E. J. C., 1972 *Jig and Tool Design* Newnes-Butterworths, London 1972.
[5] Selvakumar S., Arulshri K.P., Padmanaban K.P., Sasikumar K.S.K., 2010 Clamping Force Optimization for Minimum Deformation of Workpiece, *World Applied Sciences Journal* 11 (7): 840-846.
[6] Chen W, Ni L, Xue J., Deformation control through fixture layout design and clamping force optimization, *Int J Adv Manuf Technol* (2008) 38:860–867, DOI 10.1007/s00170-007-1153-2
[7] Qu L, and Liu H, Clamping deformation analysis research based on a flexible clamping fixture of thin-walled curved surface parts 2012, *Advanced Materials Research* Vols. 476-478 (2012) pp 2028-2031. Trans Tech Publications, Switzerland.
[8] Shane P. Siebenaler, Shreyes N. Melkote 2006 Prediction of workpiece deformation in a fixture system using the finite element method, *International Journal of Machine Tools & Manufacture* 46 (2006) 51–58p.
[9] Yuanchao Deng, A 2011 Theoretic Study on Cam-Clamping Device, *Advanced Materials Research* Vols 295-297 (2011) pp 1536-1539 Trans Tech Publications.
[10] Niţă R, M, Ghionea A, Rachieru N, 2014 Research regarding performance evaluation using fem of orientation and clamping devices for machine tools , *Annals of the Academy of Romanian Scientists*, Series on Engineering Sciences, ISSN 2066 – 8570 Volume 6, Nr.1/2014.

IOP Conf. Series: Materials Science and Engineering **898** (2020) 012011 doi:10.1088/1757-899X/898/1/012011

Prediction of material removal rate in turning using Response Surface Method

I Stănăşel[1] , F S Blaga[1], T Buidoş[1], F Corb[1]

Faculty of Managerial and Technological Engineering, University of Oradea, Romania

stanasel@uoradea.ro

Abstract. Obtaining quality products at low costs is the main objective of any company. As a result, the efficiency of the production processes is sought by identifying a favorable combination of all the factors that influence the respective process. With regard to metal cutting, the Material Removal Rate (MRR) is an important factor affecting machining time.The paper makes a study to highlight the influence of cutting parameters on material removal rate using Response Surface Method (RSM). Using analysis of variance (ANOVA), you may identify the significant factors of the process and determine a second-order regression model that takes into account both main effects but also and the interactions between factors. It also presents the possibility of optimization of cutting parameters for streamlining the turning process.

1. Introduction

Obtaining a product in economic conditions largely depends on the manufacturing technology. This must ensure that products of appropriate quality are obtained at low cost and high productivity. Within companies, there are continuous changes, which aim is to improve the functional performances of products or manufacturing processes. Therefore, research is needed to establish a relationship that will lead to the optimization of the parameters of the new technological process. In the case of products obtained by cutting processing, production costs are influenced by several factors, including material rate removal.

The optimization of the material removal rate is of great importance in the conditions of large-scale manufacturing. This aspect is noticed by many researchers who propose different methods for adjusting the parameters in order to optimize the cutting process.

The paper [1] investigates by an experimental procedure the optimum process parameters for optimization of material removal rate and tool wear while turning of hardened AISI 52100 steel under dry cutting conditions using Taguchi method. The importance of parameters is studied by using ANOVA. The authors [2] studied the machinability of mild steel in the turning process by using a conventional lathe machine. Two parameters, like tool rake angle and feed, are varied to investigate their effect. The material removal rate is studied, taking into account two variable parameters like tool rake angle and feed. The techniques Taguchi and ANOVA were used to reduce machining time and also the power during processing. Article [3] describes a combined mathematical, graphical method in order to adjust material removal rate in pocket milling operations but with reduction of tooling cost, machining vibration, noise and also to increase surface finish. The authors [4] presents the effect of the spindle speed, feed rate and depth of cut in dry turning of grey cast iron FG 260 in a computer numericaly controled lathe. In order to study the material removal rate the process parameters were varied and were investigated optimum conditions for higher MRR, being used Taguchi method, analysis of variance, multivariable linear regression (MVLR). In [5] is shown an approach of Taguchi method with the aim of optimizing the MRR for an EMCO Concept Turning center, by variation of cutting parameters speed, feed and depth of cut. The study gives a predictive model to determine MRR by combining machining parameters, and it was proposed the optimal solution for increasing efficiency of the machining process.

A method for establishing the relationship between parameters influencing a particular process is the method of factorial experiments. The Response Surface Method is an empirical method that allows the determination of a relationship between different parameters and the response of a process. This method is based on mathematical and statistical calculations. It is used to model and optimize the processes in which the answer depends on several variables.

The paper presents the use of Response Surface Method (RSM) in order to estimate the material removal rate according to the cutting parameters of the turning process and proposes a method of optimizing them over the ranges of values.

2. Material and method

Usually, the influence of different factors is studied one by one at a time, to explore a phenomenon. This means changing the values of the studied factor, and observing the behavior of the process, while all other factors are kept constant. In this way, relations of connection and curves of variation are highlighted, which represent only partially the studied phenomenon, since the interactions between factors are not taken into account. The material for which the study was carried out is Al 7075.

If you know the parameters of cutting at turning: cutting depth (t, mm); feed (sr, mm / rot) and speed (n, rot/min), the relationship that expresses the amount of material removed is:

$$MRR = \pi\left(\frac{D_{med}}{2}\right)t \cdot s_r \cdot n \quad (1)$$

where:

$$D_{med} = \frac{d_i + d_{i+1}}{2} \quad (2)$$

MRR – material removal rate (mm^3/min)
D_{med} - average diameter of the work piece (mm)
d_i – initial diameter (mm)
d_{i+1} – actual diameter (mm)
t – cutting depth (mm)
s_r – cutting feed rate (mm/rot)
n – cutting speed (rot/min).

From the point of view of the working mode, for the study of a process, the method of the factorial experiments involves identifying the influence factors and the response that characterizes the respective process. For each factor, the domain in which it takes values is defined, and the matrix for conducting experiments is also established. The objective of the experiment consists of determining the influence of factors xi on the answer function y and expressing it by the form:

$$y = f(x_1, x_2, ..., x_k) \quad (3)$$

The model obtained from the experiments is an empirical one, in which its statistical estimation replaces the real response function. The form of the empirical model is expressed by mathematical functions that can be polynomial, logarithmic, exponential etc. Usually, the polynomial form is the most used, due to the convenient possibilities for mathematical processing. Higher-order polynomials, most commonly of the second order, expressed by the general relationship, are usually used to describe the optimal:

$$y = b_0 + \sum_{j=1}^{k} b_j x_j + \sum_{j=1}^{k} b_{ij} x_j^2 + \sum_{i,j=1}^{k} b_{ij} x_i x_j + e \quad (4)$$

The aim of the experiments carried out was to obtain the necessary data to determine the coefficients of the presented model. The regression model resulting from the calculations does not cover the entire range of values that process influence factors may take, but it is satisfactory as an approximation for the field of studied values of influence factors.

Using the method of the factorial experiment, it is proposed to estimate the cutting parameters so as to obtain a high MRR, under the conditions of getting an appropriate quality of the surface. The range of factor variation is shown in Table 1.

Table 1. The level of cutting parameters.

Factor	Symbol	Units	Low	High
Curring Depth	t	mm	0.50	5.50
Feed	s_r	mm/rot	0.05	0.55
Speed	n	rot/min	200	2000

3. Result and Discussion

The array of the experiment is presented in Table 2.

Table 2. The experimental array.

Run	A: Curring depth mm	B:Feed mm/rot	C: Speed rot/min	Material removal rate mm^3/min
1	5.5	0.05	2000	1727.88
2	2.75	0.54	1010	4711.92
3	5.4	0.275	1010	4711.92
4	0.625	0.345	1251.67	847.89
5	5.3	0.3	1914.6	9563.68
6	5.5	0.55	2000	19006.6
7	2.75	0.274707	1964	4661.16
8	3.425	0.3425	245	902.89
9	2.75	0.274707	1964	4661.16
10	3.425	0.06	1253	808.93
11	2.75	0.274707	1964	4661.16
12	0.5	0.05	1331.5	104.58
13	5.5	0.55	200	1900.66
14	2.75	0.54	1010	4711.92
15	0.5	0.05	200	15.71
16	3.65	0.05	200	114.67
17	0.5	0.55	2000	1727.88
18	5.4	0.275	1010	4711.92
19	2.75	0.54	1010	4711.92
20	0.5	0.363233	200	114.11

In Table 3 is presented the Analysis of Variance (ANOVA) of the experimental data. Fisher test (F-value) is 209.28, which means that the model is significant. The p value is $0.0001<0.05$ (significance level), which indicates model terms are significant. As you can see in table 3, the significant terms of the model are A, B, C, AB, AC, BC, A^2, B^2, C^2. Also, p-value of the model showed that there is 0.01% chance that an F-value this large could occur due to noise. The coefficient of determination R^2 is 0.9947 indicates a strong correlation between the real model and the regression model. Adjusted R^2 has value 0.99, which means a good correlation of the model, depending on the number of significant variables. Predicted R^2 is 0.9774 indicating how well a regression model anticipates responses to new observations. The difference is less than 0.2, which shows a good correlation between Adjusted R^2 and Predicted R^2. Adeq. Precision measures the signal-to-noise ratio, the ratio value must be greater than 4. The values 57.866 shown a good signal, so the model can be used to navigate the design space.

Table 3. ANOVA for Quadratic model

Source	Sum of Squares	df	Mean Square	F-value	p-value	
Model	20849.91	9	2316.66	209.28	< 0.0001	significant
A-t	4862.07	1	4862.07	439.22	< 0.0001	
B-s	5171.58	1	5171.58	467.18	< 0.0001	
C-n	5062.15	1	5062.15	457.3	< 0.0001	
AB	957.21	1	957.21	86.47	< 0.0001	
AC	947.37	1	947.37	85.58	< 0.0001	
BC	853.11	1	853.11	77.07	< 0.0001	
A^2	372.66	1	372.66	33.67	0.0002	
B^2	364.77	1	364.77	32.95	0.0002	
C^2	310.62	1	310.62	28.06	0.0003	
Residual	110.7	10	11.07			
Lack of Fit	110.7	5	22.14			
Cor Total	20960.6	19				

Final equation in terms of actual factors is:

$$\sqrt{MRR} = -5.04448 + 5.62723 \cdot t + 55.1091 \cdot s + 0.0148941 \cdot n + 21.3868 \cdot t \cdot s + \\ +0.00575621 \cdot t \cdot n + 0.0567796 \cdot s \cdot n - 1.54503 \cdot t^2 - 149.93 \cdot s^2 - 1.10669e^{-05} \cdot n^2 \quad (5)$$

In figure 1 is shown the normal plot of residuals. Residues are distributed relatively evenly on a straight line in both positive and negative direction, which shows that the pattern is appropriate. In figure 2 is presented the Predicted vs Actual graph. As you can see, the estimated values with the regression model are placed very close to the line at 45 degree with the observed points, which shows a good correlation of the model with the observed data.

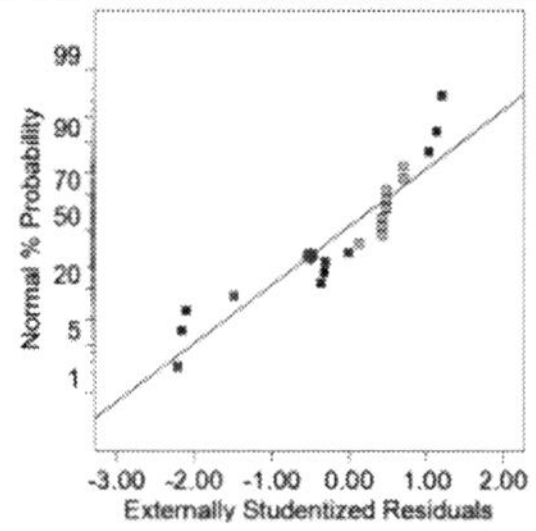

Figure 1. The normal plot of residuals

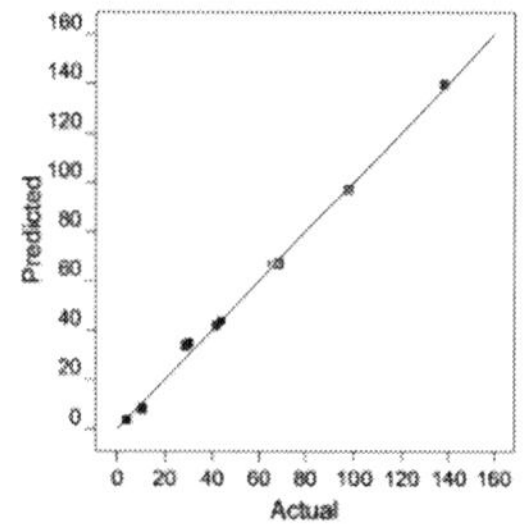

Figure 2. The Predicted vs Actual.

The amplitude of the effects of the cutting parameters on MRR over the studied range is shown in figure 3. It can be seen that the speed and the cutting depth have the most significant influence on the amount of removed material.

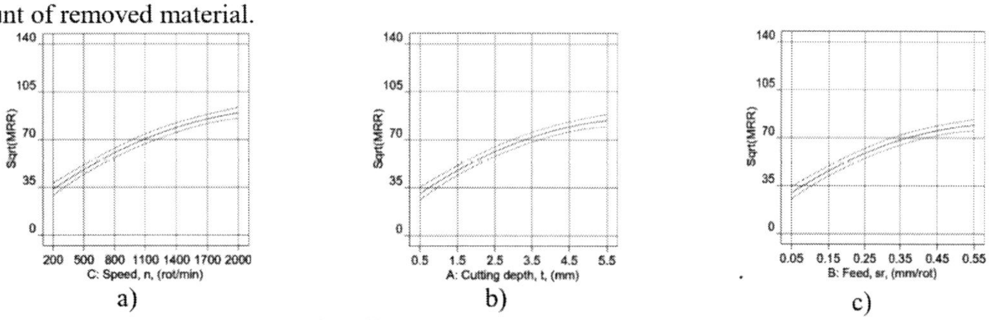

Figure 3. The effects of the cutting parameters on MRR.

a) – $\sqrt{\text{MRR}}$ vs. speed ; b) - $\sqrt{\text{MRR}}$ vs. cutting depth ; c) - $\sqrt{\text{MRR}}$ vs. feed .

In Figure 4 is shown the response surface for material removal rate according to the cutting parameters. The regression model obtained by the factorial experiments offers the possibility of optimizing the process on the field of variation of the influence factors. In the case of finishing operations, it is possible to act in order to increase the quantity of material removed, in conditions of minimizing the working advance in order to obtain an adequate roughness on the processed surface. For this situation, in figure 5 the optimal solution t = 5.5 mm is presented; n = 2000rot / min; sr = 0.24mm/rot; $\sqrt{MRR}$ =0.8964 (MRR = 8035.3296 mm^3/min).

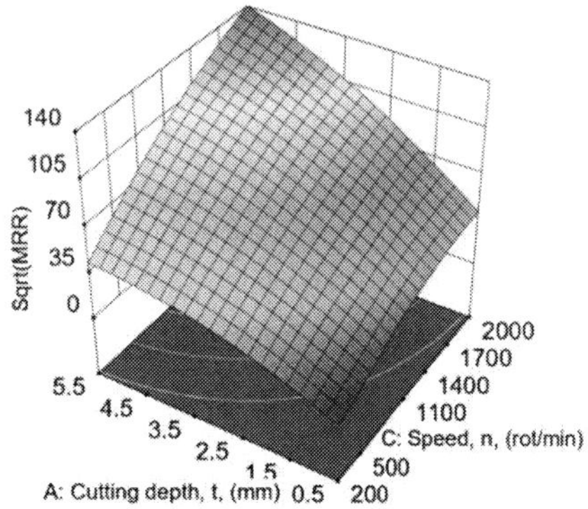

Figure 4. Response surface of the material removal rate vs cutting parameters

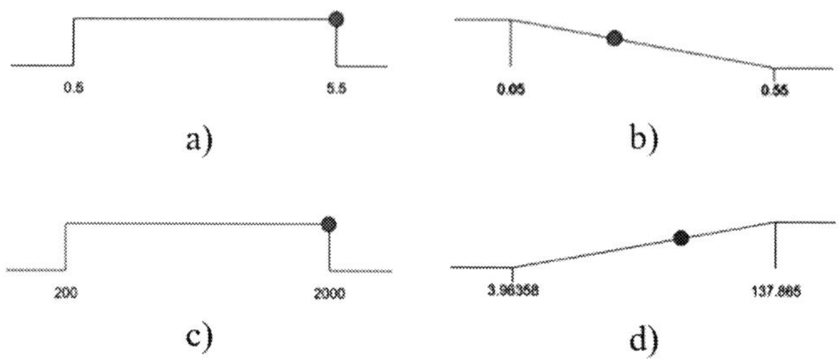

Figure 5. The optimal solution for minimum cutting feed and maximum material removal rate for studied interval. a). Depth=5.5 [mm]; b). Feed=0.247 [mm/rot]; c). Speed=2000 [rot/min]; d). $\sqrt{MRR}$ =0.8964

4. Conclusions

The use of factorial experiments allows the modelling of processes which depend on several variables. In this paper it was analyzed the influence of the parameters of the process of cutting at turning on the MRR. For experiments, an orthogonal matrix L20 was used. Using a second-order polynomial, the regression model was obtained, which was analyzed and the influence of the process parameters on the MRR, as well as the interactions between them were highlighted. From the analysis, it is observed that the feed and the cutting depth have the most significant influence on the material removed rate for the studied range. Based on the obtained model, the cutting parameters can be adjusted so that certain objectives corresponding to a specific purpose are met. For example, it is possible to optimize the maximum quantity of material removed, or the maximum quantity for an advance imposed in order to ensure an adequate roughness of the processed surface.

References

[1] Vijaykumar H.K, Ahamed Kabir and Ganesh Raj, 2014, Experimental investigation of Material Removal Rate and Tool wear în turning of hardened AISI52100 steel using Taguchi Technique *IOSR Journal of Engineering*, Vol. 04, Issue 05. PP 13-18.

[2] Saurabh Singhvi, M.S.Khidiya, S.Jindal, M.A.Saloda, 2016, Investigation of Material Removal Rate în Turning Operation, *International Journal of Innovative Research în Science, Engineering and Technology*, Vol. 5, Issue 3. PP 2890-2895.

[3] Majid Tolouei-Rad, 2011, Efficient CNC Milling by Adjusting Material Removal Rate, *International Journal of Mechanical and Mechatronics Engineering* Vol:5, No:10. PP. 1988-1992.

[4] Mohammed Nooruddin, U. V. Hambire, 2015, Analysis of Process Parameters for Material Removal Rate During Dry Turning of FG 260 Grey Cast Iron, *International Journal of Engineering Research & Technology* , Vol. 4 Issue 04, PP 605-610.

[5] Sayak Mukherjeea, Anurag Kamala, Kaushik Kumarb, 2014, Optimization of Material Removal Rate During Turning of SAE 1020 Material în CNC Lathe using Taguchi Technique, *Procedia Engineering* 97, PP 29 – 35.

[6] Montgomery, D.C.,2001, *Design and Analysis of Experiments*. John Wiley & Sons, New York.

IOP Conf. Series: Materials Science and Engineering **898** (2020) 012012 doi:10.1088/1757-899X/898/1/012012

Cam – clamping device with variable eccentricity

P D Tocuț[1], I Stănășel[1], C Feric[1]

Faculty of Managerial and Technological Engineering, University of Oradea, Romania

tocutpd@yahoo.com

Abstract. Productivity and quality growth, all together with cost reductions, are the main objectives of modern manufacturing companies. Fixture devices shall facilitate that the workpiece can be changed fast, to have simple construction, to be cheap and affordable, and they have minimal positioning and installation errors. Clamping forces are applied to the workpiece with the help of fixture devices, after being correctly oriented by the application of adjusting forces. In the stage of design, many parameters are taken into account, like strokes and forces necessary, operation time, build simplicity, all which determines the cost of the device. In the field of device design, the main requirements are related to rapid fixing of the workpiece, fixing of more workpieces simultaneously and reducing the effort of labour for the operators. In this paper is presented the research and construction of a circular cam with variable eccentricity used in one this kind of fixture devices.

1. Introduction

The main objective of the companies is maximizing the productivity and quality of their products, minimizing production costs. For manufacturing companies that use metal chipping processes, these objectives are strongly influenced, besides many factors, by fixtures used in the manufacturing processes.

The device is a component of the machine-device-tool-workpiece system, their primary role being the orientation of the workpiece and sometimes of the tools in the meantime of manufacturing. By using fixture devices are being watched the reduction of auxiliary times for fixing the workpiece, thus being used rapid fastening mechanisms and also multiple workpiece fastening [1].

The forces must retain the workpiece oriented all the time in the manufacturing process. The direction the fixing force must be perpendicular on the orientation base of the workpiece that cancels the maximal number of grades of freedom.

In the field of manufacturing processes, there are many studies and different approaches related to fixture device design.

In article [2], it's been analyzed the cams geometrical calculations, calculations related to fixing forces and torques, auto-blocking conditions and contact resistance conditions. In this paper, the discussed solution is changing the eccentric with a cam that has a variable curve that can be designed by the needs of auto-blocking. In the patent [3], it's presented a different model with an eccentric drive cam with variable stroke. Authors [4] are evaluating the efficiency of the manipulation systems, by examining recent development of products, and are elaborating new requirements for upcoming systems. The researchers [5] are focusing on aspects related to the development of integrated computerized manufacturing, the automation of devices design, optimizing the fixing forces. Thus, in the paper [6] they are approaching researches related to the performance of fixtures that can be adapted to advanced production. There are presented design methodologies, studies related finite element method, optimization, models for simulation, and fixing forces analyzing, and also numerical research and experiments related to the performance of the device.

In this paper is presented the design of a fixing device for a prismatic workpiece, and also the construction of an eccentric cam with variable eccentricity.

2. The fixture design

In the process of design parameters that were taken into account was the ease of use, the length of the stroke, acting time of the fixture, simplicity of operation and the necessary energy for the operation. In this case, it was used a fixture mechanism with an eccentric cam with variable eccentricity which is part of the rapid fixing systems category, having an operation time between 0.6-1.7 seconds.

The device (figure 1) has two workstations for fixing two workpieces simultaneously. It's made from a base plate (1) on which are fixed elements (2) and (3), in which can be found the support (4). The workpiece (5) is placed on the plates (6) and fixed with the lever (7) that is operated by the eccentric cam (8) being moved by the handle (14). For the release, the elements (15) are withdraw manually, and for the undoing of the eccentric cam, the spring (16) is operating the rod (17) which remove the levers.

For adjusting the eccentricity, we must remove the assembly formed from the sleeve (8), disc (9), handle (12), and the toothed coupling(10) which can be repositioned by rotating over a number of teeth around the axle (11), thereby obtaining the desired eccentricity and operating stroke. The adjustment of the handle can be done in a large domain (100-300 mm) by unscrewing the sleeve (14) and by blocking it by a nut (13).

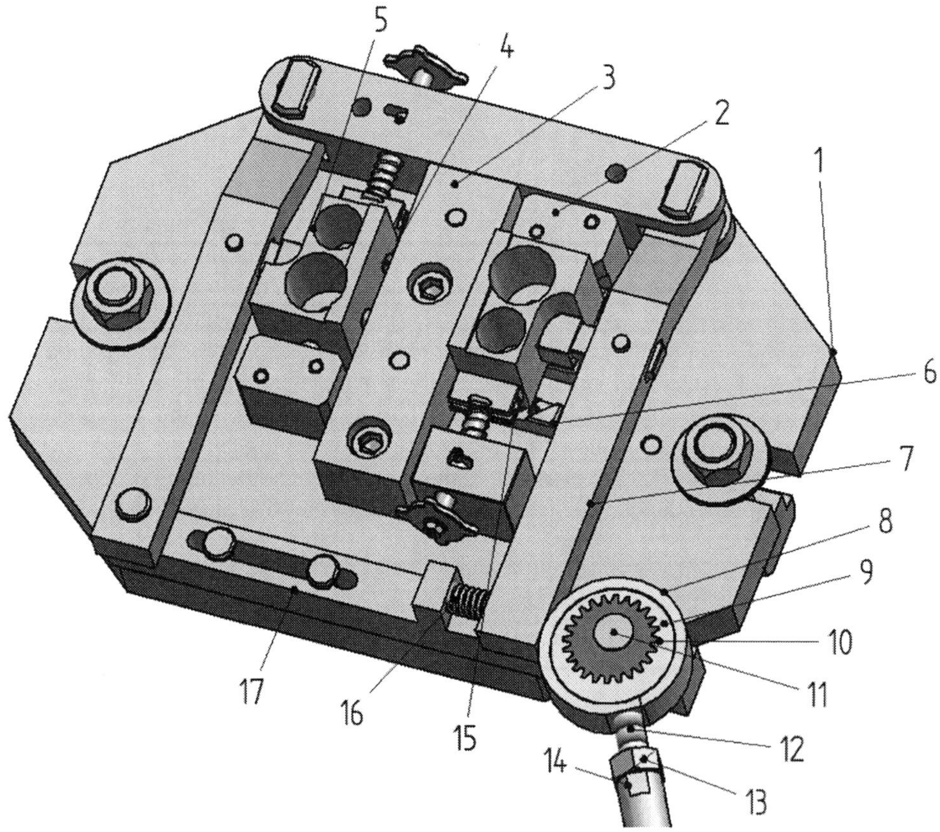

Figure 1. The 3D model of the designed duplex fixture.

The components of the eccentric cam with variable eccentricity is presented in figure 2, and the parameters in table 1. In figure 3 is presented a physical model of an eccentric cam with variable eccentricity.

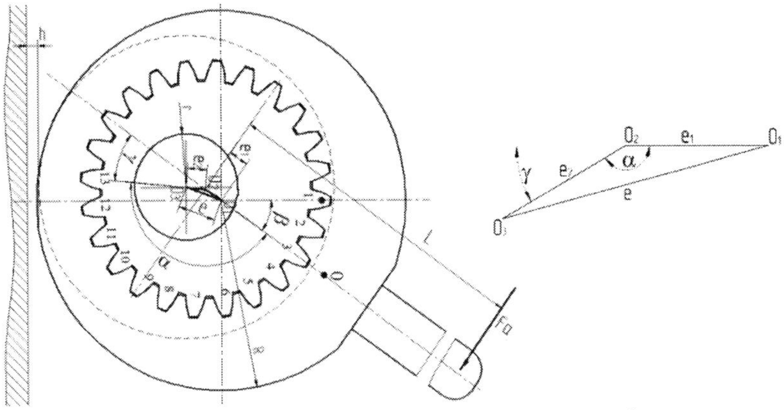

a. b.

Figure 2. Eccentric cam with variable eccentricity.
a. Functional and constructive elements;
b. calculation of the eccentricity

Figure 3. Physical model of an eccentric cam with variable eccentricity.

Table 1. Constructive parameters of the designed eccentric cam.

Parameters	Symbol
Disk radius, mm	R
Rotation axis radius, mm	r
Stroke, mm	h
Toothed coupling eccentricity compared to disk, mm	e_1
Axis eccentricity compared to toothed coupling, mm	e_2
Total eccentricity of the disk compared to axis (adjustable), mm	e
Handle length (adjustable), mm	L

If we know the number of teeth z of the toothed coupling, the angle δ, between 2 teeth can be determined:

$$\delta = \frac{2\pi}{z} \tag{1}$$

The adjustment of total eccentricity *e* can be made by rotating the eccentric disk over the toothed coupling by a number of teeth z_i The corresponding angle in noted γ (figure. 2, b).

$$\gamma = \frac{2\pi}{z} z_i \tag{2}$$

Knowing the eccentricity e_1, relative to the eccentric disk and the eccentricity e_2 of the axis compared to the toothed coupling, using the scheme from figure 2, b can be found the mathematic relation for the total eccentricity *e*:

$$e = \sqrt{e_1^2 + e_2^2 - 2e_1e_2\cos\left(\frac{2\pi}{z}(z - z_i)\right)} \tag{3}$$

The stroke „h" of the eccentric by rotating an angle β can be determined:

$$h = e(1 - \cos\beta) \tag{4}$$

The property of auto-braking in any point selected in the eccentric must verify the following relationship:

$$e \le \mu R + \mu r \tag{5}$$

μ – coefficient of friction between the disk and lever and between the coupling and bolt, μ=0.16.

Knowing the clamping force *Q* made by the eccentric through the length *L* of the handle, we can find the human acting force.

IOP Conf. Series: Materials Science and Engineering **898** (2020) 012012 doi:10.1088/1757-899X/898/1/012012

$$F_a = \frac{Q(e\sin\beta - \varphi) + \mu R + \mu r}{L} \quad (6)$$

$$50 < F_a < 150 \text{ [N]}$$

φ – friction angle tgφ=μ.

3. Numerical study

Numerical studies are simulating the behavior of the eccentric cam for different practical situations: *case I*:e_1=e_2=3 mm; *case II*: e_1=e_2=4 mm; *case III*: e_1=e_2=5 mm;

For the numerical examples were taken in consideration the following values:

- Number of teeth of the toothed coupling: z=24 teeth;
- Eccentric disk radius: R = 30 mm;
- Eccentric axis radius: r = 8 mm;
- Angle between two teeth: γ = 15°.
- Handle length L = 100…300 mm

In figure. 4 is presented the variation of total eccentricity related to the number of teeth over which the disk rotates compared to toothed coupling in three different situations. It can be seen that total eccentricity *e* grows as e_1 şi e_2 are higher.

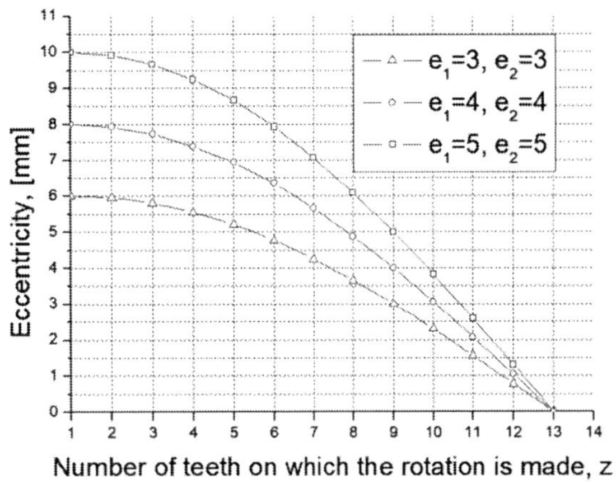

Figure 4. Eccentricity vs. Number of teeth on which the rotation is made, z_i.

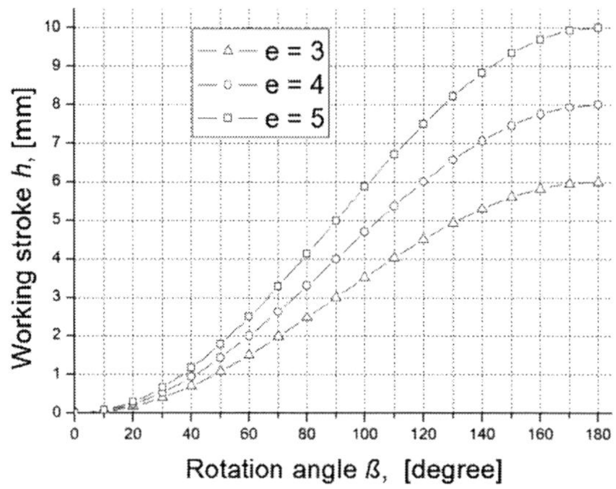

Figure 5. Working stroke vs. rotation angle.

In figure 5 is presented the stroke variation related to rotation angle for three different situations related to eccentricity e_1 şi e_2. We can see that for the same rotation angle, the stroke becomes bigger with the growth of the eccentricities e_1 şi e_2. The variation of the stroke length related to angle of rotation of the eccentric and number of teeth over which the disk is turned are presented in figure 6. The working stroke size increases with increasing the angle of rotation of the eccentric β, and by decreasing the number of teeth over which the disc is rotated relative to the toothed coupling.

Clamping force Q increases with the increase of the number of teeth over which the disk z rotates and towards the ends of the variation range of angle β (figure 7).

Acting human force *Fa* of the eccentric cam compared to the rotation angle β and number of teeth of the rotation is presented in figure 8. This force decreases with the increasing number of teeth on which the rotation is made, and towards the limits of the rotation angles β of the eccentric.

The human acting force *Fa* drops by growing the length of the handle L (figure 9).

Related to auto-braking properties in case I in which e_1=3 şi e_2=3 the condition is respected through the entire variation interval of the angle β ($0<\beta<180$).

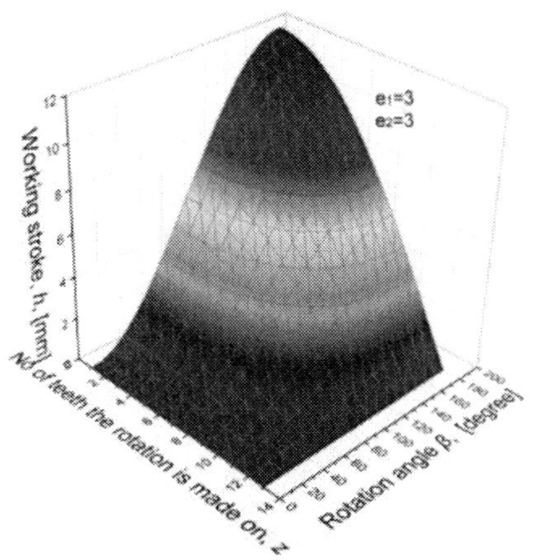

Figure 6. Working stroke *(h)* vs. rotation angle *(β)* and number of teeth of the rotation*(z)*.

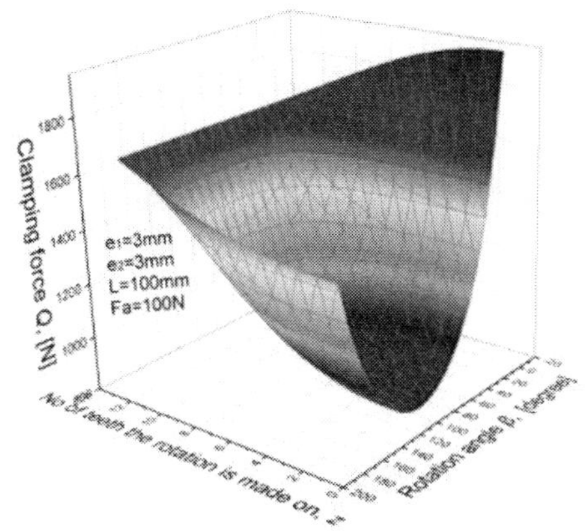

Figure 7. Clamping force *(Q)* vs. rotation angle *(β)* and number of teeth of the rotation*(z)*.

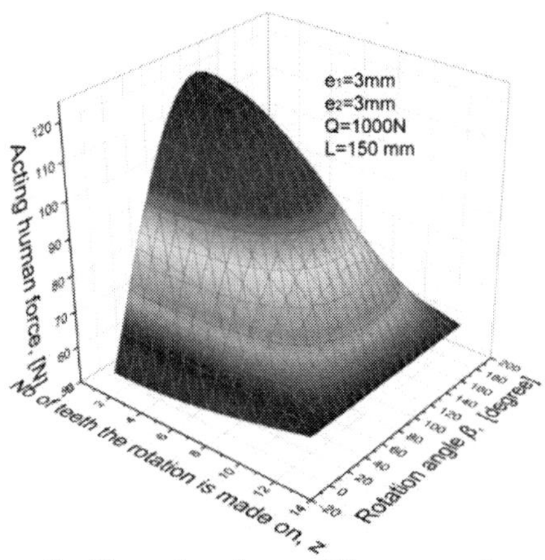

Figure 8. Clamping force *(Q)* vs. rotation angle *(β)* and number of teeth of the rotation*(z)*.

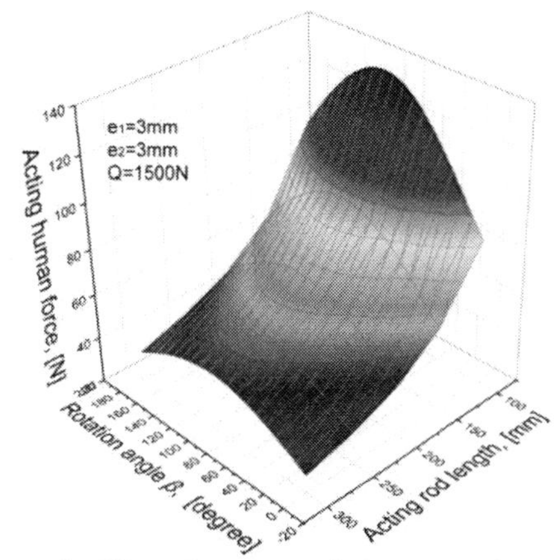

Figure 9. Clamping force *(Q)* vs. rotation angle *(β)* and number of teeth of the rotation*(z)*.

4. Conclusions

It has been designed a clamping device that has a mechanism with eccentric cam with variable eccentricity and makes part of the quick clamping devices. For growing the field of use of the device, a circular eccentric has been designed with the possibility to adjust the stroke and eccentricity. The mathematical relations have been determined and numerical studies were done related to the eccentricity variation, stroke length, clamping force, and rotation angle of the disk compared to the toothed coupling.

References

[1] Hoffman E. 2011 *Jig and Fixture Design*, Delmar, USA.

[2] Yuanchao Deng, A 2011 Theoretic Study on Cam-Clamping Device, *Advanced Materials Research* Vols 295-297 (2011) pp 1536-1539 Trans Tech Publications.

[3] Wehr, Hubert 1999 *Eccentric cam drive with variable stroke* EP 0 974 772 B1, European Patent Office.

[4] Fleischer J, Denkena,B, Winfough, B 2006 Workpiece and Tool Handling in Metal Cutting Machines, *Annals of the CIRP* Vol. 55/2/2006:817-839

[5] Andrew Y. C. Nee, Z. J. Tao & A. Senthil Kumar 2004 *An Advanced Treatise on Fixture Design and Planning*, World Scientific Publishing Co. Pte. Ltd.

[6] Niţă R, M, Ghionea A, Rachieru N, 2014 Research regarding performance evaluation using fem of orientation and clamping devices for machine tools, *Annals of the Academy of Romanian Scientists*, Series on Engineering Sciences, ISSN 2066 – 8570 Volume 6, Nr.1/2014.

IOP Conf. Series: Materials Science and Engineering **898** (2020) 012013 doi:10.1088/1757-899X/898/1/012013

Dynamics of a 6R industrial robot

M Ratiu[1], A Rus[1] and M L Balas[2]

[1] Department of Mechanical Engineering and Automotive, University of Oradea, 410087 Oradea, Romania
[2] Department of Product Design, Mechatronics and Environment, Transilvania University of Brasov, 500068 Brasov, Romania

E-mail: mratiu@uoradea.ro

Abstract. In the context in which, the dynamic behavior and performance of industrial robots are very important for their general performance, we present, in this article, a short dynamic analysis of an industrial robot. First, on the base of a short review of some papers on this topic, a general presentation of some main concepts regarding the dynamics of the industrial robots, and of the actual importance of the virtual prototyping in studying the dynamic behavior and performance of such complex mechanical systems are highlighted. How the modern virtual prototyping tools allow better, more rapid, and less costly dynamic designing of the complex mechanical systems, comparing to the traditional designing and prototyping, and ADAMS MBS of MSC software enables the evaluation of the dynamic behavior of the virtual prototype of a robot during its designing stage, by starting from the kinematic model previously developed, we conducted in this research the process of dynamic modeling and simulation of a 6R articulated robot developed in ADAMS and a short dynamical analysis of it.

1. Introduction

In the last decades, in the context of using, on an increasing scale, and in more and more numerous and diverse fields, of the robots, as well as of the development, in an accelerated rhythm, of the computer-based designing techniques, the virtual prototyping has become much more used compared to the physical prototypes. Having advantages such as reducing the execution time or lowering costs, the possibility of simulation and testing in different stages or alternatives, there are, today, several complex software packages and a lot of studies on the modeling and virtual prototyping of the industrial robots, as presented in [1].

In robots' dynamics, virtual prototyping is an important and ordinary tool used nowadays by several simulating software, ADAMS being declared, in [2], the most known one. Also, in a survey based on the user feedback and presented in [3], the participants indicated more tools currently used by them for the dynamic simulation of the robots, the most known tool being ADAMS (45%). According to these findings, and after a short literature review on this topic, we found that ADAMS software is considered one of the best solutions for analyzing the dynamic behavior of complex mechanical systems, and an efficient alternative to the numerical simulation of the dynamic behavior of industrial robots.

In our research, by using the MBS (Multi-Body System) software ADAMS (Automatic Dynamic Analysis of Mechanical Systems), we developed, in [4], a virtual kinematic model of a 6R articulated robot, with six revolute joints, and conducted, in [5] a short kinematic analysis for it, through direct and inverse kinematics.

In this paper, after a general presentation of some main concepts regarding the dynamics of the robots, by starting from the kinematic model of the robot, developed before, we present the process of dynamic modeling and simulation of the robot, in ADAMS/View, according to [6], [7] and [8].

2. Dynamics of the industrial robots

The dynamic analysis of the multibody systems describes, through the dynamic equations of motion, the relationship between the forces acting on the system and the motion produced by them. The dynamics of robots are important for their design, simulation, and control [9, 10]. There are several parallel algorithms used for dynamic calculation of industrial robots and several approaches, most using either Lagrange or Newton-Euler formalism.

The dynamical analysis is, generally, both direct and inverse. In the direct or forward dynamics, the torques or forces applied to the actuators are specified and the accelerations of the joints are determined, this kind of analysis being required for simulation. In the inverse dynamics, starting from the specification of the robot's trajectory (position, speed, and acceleration), the torques or forces required for the actuators are determined, this type of analysis being used for data control and trajectory planning. There is, in addition, the third type of computation, namely hybrid dynamics, in which some of the accelerations and forces are given and it is necessary to determine the rest of them [11].

In the dynamic analysis are used, also, other two types of computations, namely the joint-space inertia matrix, which leads to the accelerations and the torques or forces in the joints and it is, as a rule, an integral part of many direct dynamics formulations and the operational-space inertia matrix, which leads to the accelerations and forces developed to perform the task, in the operational or Cartesian space, and it is used to control the end effector or the execution level of the robot load [12].

3. The dynamic model of the robot

The dynamic model of a robot expresses the relationship between the torques and/or forces applied to the actuators and the positions, speeds and articular accelerations. In developing the dynamic model of a robot, several criteria have to be considered, cumulatively, the most important being the need to travel the trajectories as accurately and as quickly as possible; the other criteria refer to the need for real-time operation, the need to minimize the effect of the interconnection between the elements of the guiding device, the ability to compensate for the mass variations of the manipulated object, to ensure the robot's robustness.

The virtual dynamic model of the robot is very useful and necessary for simulating its motion, thus not needing the construction of a real model. The dynamic model contains information about the mass and inertial properties of the component parts of the robot's mechanical system. The dynamic model of the robot also provides the information needed to analyze the dynamic behavior of the mechanical system of the robot.

A valid model, that represents the kinematic and dynamic properties of a robot, helps to understand the reciprocal relationships between the tasks applied in each joint and the resulting motion of the robot. Based on the dynamic model of the robot, various driving models can be made.

By using ADAMS, virtual prototypes of some different industrial robots were developed and used for dynamic simulation or verification of the numerical models established in SolidWorks [13], MATLAB [14, 15, 16, 17] or MATLAB and Maple [18]. The common conclusion of all these papers is that the results of the dynamic simulations developed in ADAMS are in agreement with the numerical results of the theoretical models, and, according to [19], represents a better alternative.

The capabilities of the MBS software, namely ADAMS, in the analysis and testing of the complex mechanical systems are demonstrated in another paper, [20], by using three types of virtual mechanical models, the same as in [21]:

- the kinematic model, that contains the kinematic elements (bodies) of the robot, connected by the kinematic joints, and the geometrical parameters specific to the mechanism (the locations of the joints);

IOP Conf. Series: Materials Science and Engineering **898** (2020) 012013 doi:10.1088/1757-899X/898/1/012013

the entry is made by using the kinematic restrictions (the motion generators), through which the position or speed of the driving elements is controlled;
- the inverse dynamic model, that includes the kinematic model and the external and internal forces acting on the system, including the massic-inertial characteristics, this model being used to determine the motor torque / motor force that generate the kinematically prescribed motion of the mechanism;
- the dynamic model, which includes the inverse dynamic model, the input being made by the motor torque / motor force, its purpose being to evaluate the behavior of the mechanism under the action of the forces.

4. Dynamic simulation in ADAMS

By following a so-called master-slave approach, for the kinematic and dynamic analysis of the robot, as proposed in [22], and developed in [5], was obtained the dynamic model of the robot, presented in figure 1, necessary for the investigation of the robot' dynamic behavior.

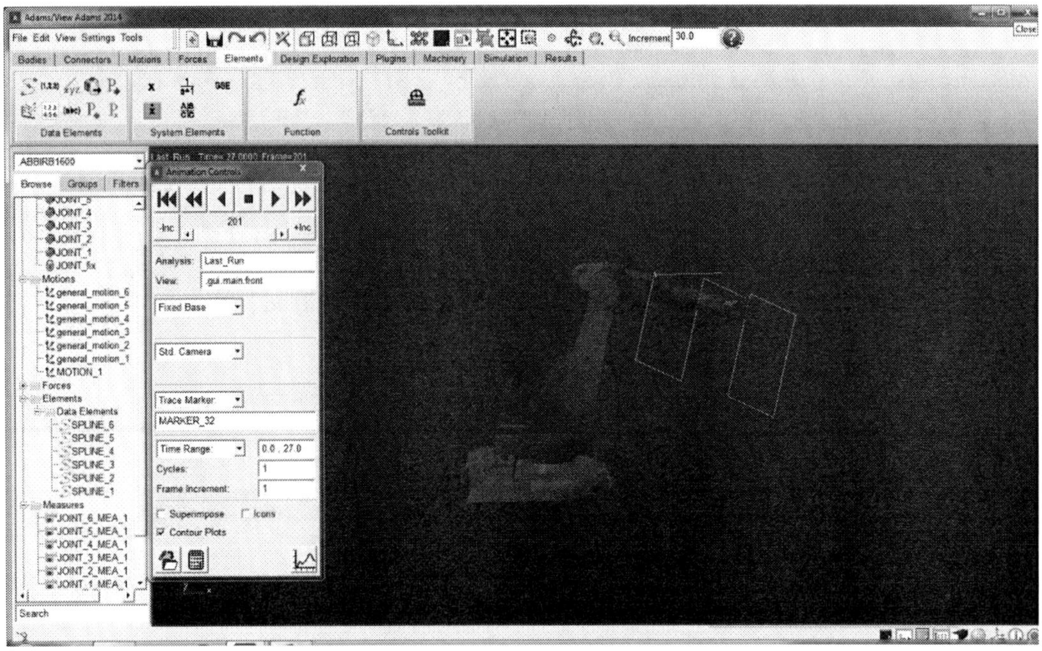

Figure 1. Dynamic model of the robot in ADAMS.

By choosing, in the process of virtual simulation, the aluminum as the material from which the robot is built, with density 2740 kg / m^3 and Young module 71705 newton / mm^2, were obtained the mass and inertia properties of the robot bodies, presented in Table 1.

Table 1. Mass and inertia of the robot bodies.

Link	Mass, m [Kg]	Principal moments of inertia [Kg · m^2]		
		I_{XX}	I_{YY}	I_{ZZ}
Body 1 - Base	79.03	$3.43 \cdot 10^6$	$2.30 \cdot 10^6$	$1.33 \cdot 10^6$
Body 2	109.42	$3.80 \cdot 10^6$	$3.21 \cdot 10^6$	$1.67 \cdot 10^6$
Body 3	21.67	$7.49 \cdot 10^5$	$7.30 \cdot 10^5$	$5.22 \cdot 10^4$
Body 4	44.82	$7.40 \cdot 10^5$	$6.93 \cdot 10^5$	$3.14 \cdot 10^5$
Body 5	7.91	$7.50 \cdot 10^4$	$7.25 \cdot 10^4$	$1.30 \cdot 10^4$
Body 6	0.49	411.71	324.35	264.41
Body 7 - End effector	0.088	39.42	39.10	17.03

For an animation of the robot motion, for 27 s and 200 steps, with the end effector following a spatial trajectory, as presented in figure 1, the variation of the kinetic energy of the mobile elements of the robot is shown in figure 2 and the variation of the potential energy in figure 3.

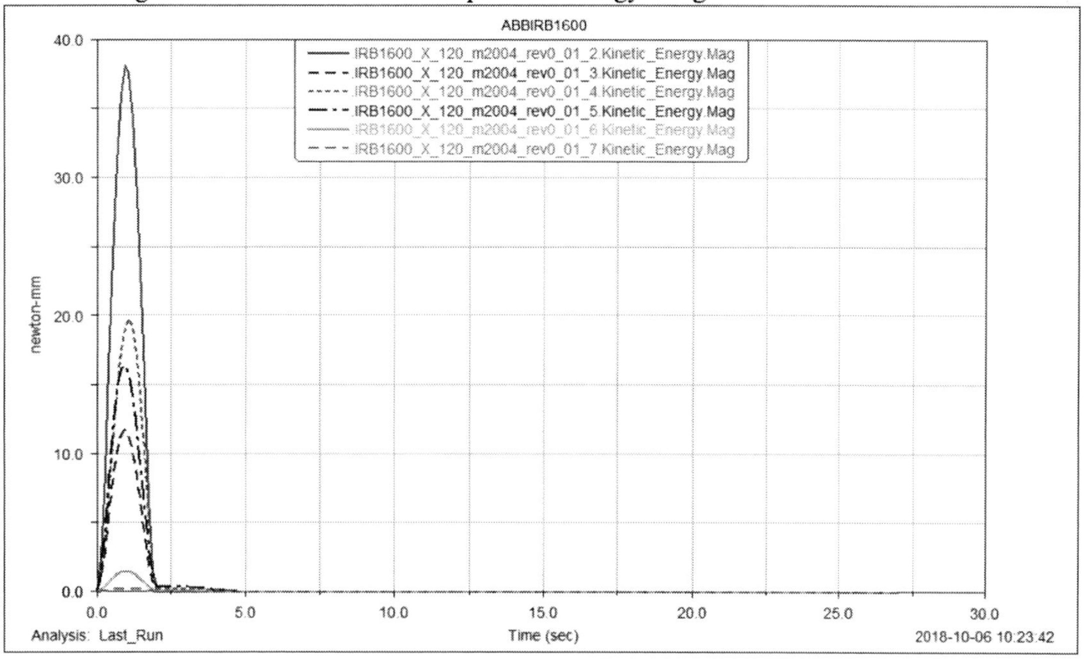

Figure 2. Variation of the kinetic energy of the mobile elements of the robot.

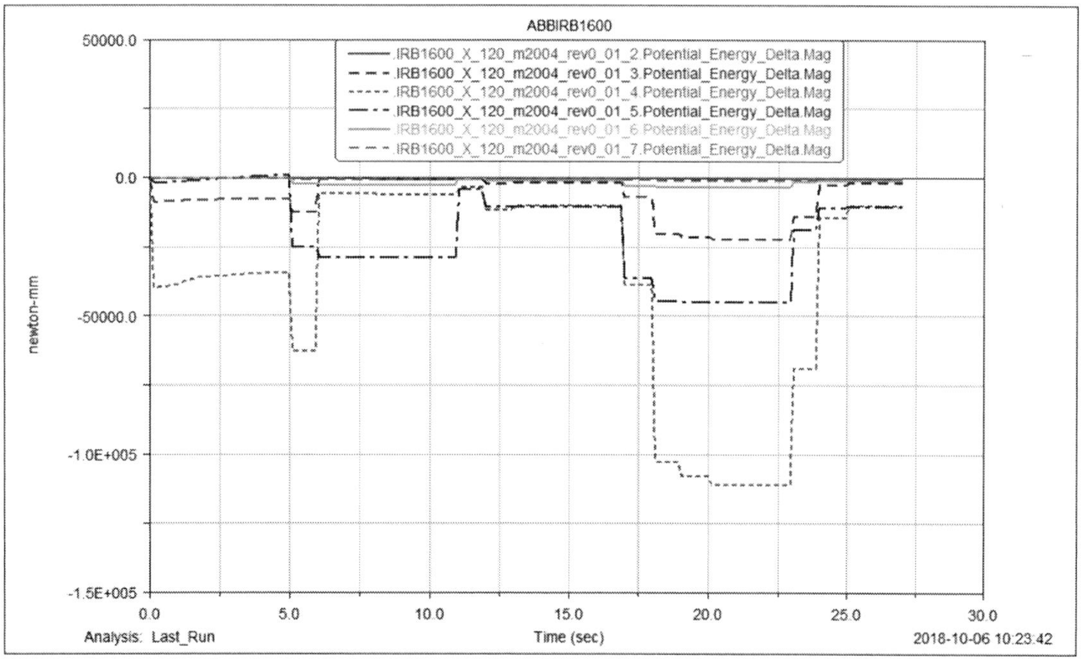

Figure 3. Variation of the potential energy of the mobile elements of the robot.

The variation of forces and torques in the robot joints are presented in figure 4, respectively 5, and they could be used for studying the joints behavior, by using finite element analysis, in future work.

How, normally, the force and torque exerted in a joint due to gravity are in a function to the robot pose, in this case, the bigger values are registered for the first joint, when the robot arm is in the closest pose to the horizontal.

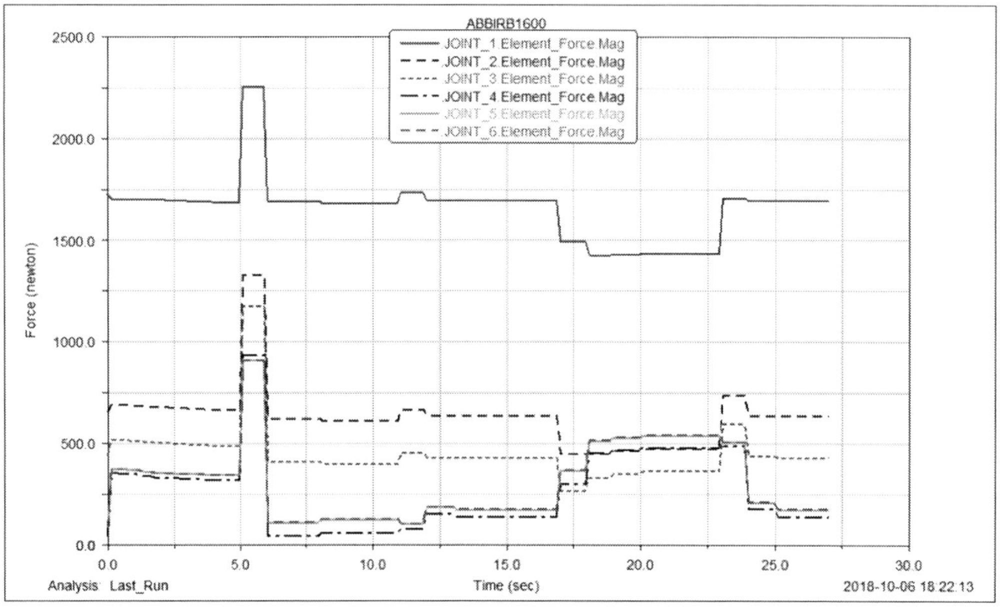

Figure 4. Variation of forces in robot joints.

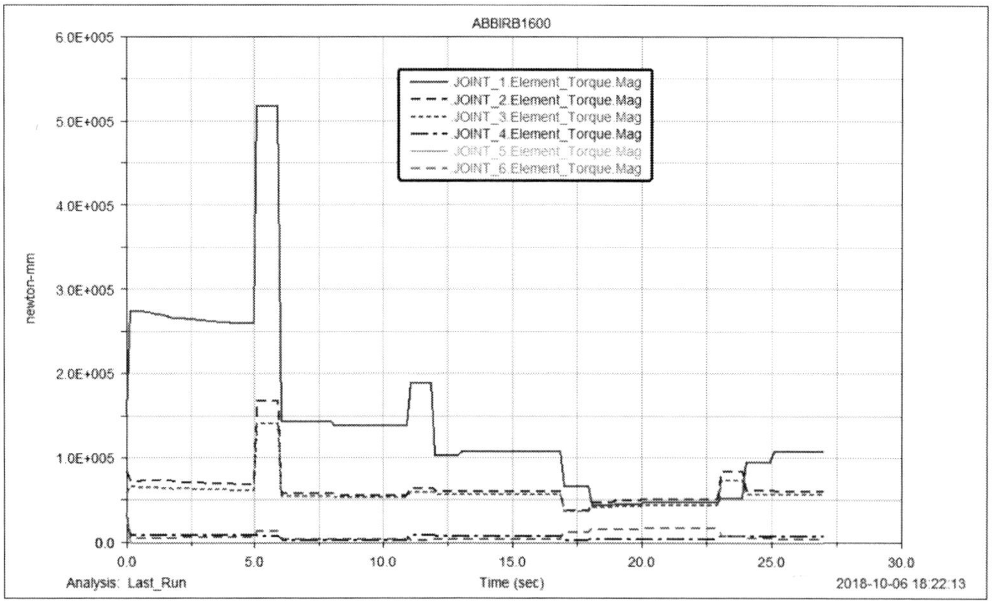

Figure 5. Variation of torques in robot joints.

5. Conclusion

The dynamic behavior and performance of industrial robots are very important for their general performance. The modern virtual prototyping tools allow better, more rapid, and less costly dynamic designing of the complex mechanical systems, comparing to the traditional designing and prototyping.

After a short literature review on this topic, we found that ADAMS enables the evaluation of the dynamic behavior of the virtual prototype of a robot during its designing stage, in much less time and at a lower cost, before realizing the experimental prototype. By adding other opportunities, such as the possibilities to increase the operation speed and maintain the precision of positioning, to avoid the vibration, to validate and optimize the control algorithm, and to evaluate the ability of the virtual model to perform a wide range of applications, Adams offers the possibility to create a better robot in a shorter period.

According to these findings, by using ADAMS in our researches, we developed a virtual kinematic and dynamic model of a 6R articulated robot and conducted a short kinematic and dynamic analysis for it. In this paper, we present the variation of the kinetic and potential energy of the mobile elements of the robot, and, also, the variation of forces and torques in the robot joints. They could be used, in our future work, for studying the joints behavior, by using the finite element analysis method.

References

[1] Ratiu M and Rus A 2017 Modeling of the trajectory-generating equipments *IEEE publication 14th International Conference on Engineering of Modern Electric Systems (EMES)*
[2] Tarabanov E 2006 *The virtual prototyping of robots dynamics*
[3] Ivaldi S, Padois V and Nori F 2014 *Tools for dynamics simulation of robots: a survey based on user feedback* Available on www.codyco.eu EU Project CODYCO
[4] Ratiu M, Rus A and Balas M L 2018 Modeling in ADAMS of a 6R industrial robot, *Annual Session of Scientific Papers IMT ORADEA 2018 (MATEC Web of Conferences)* **184**, 02006
[5] Ratiu M, Rus A and Balas M L 2019 *Kinematic modeling of a 6R industrial robot IOP Conf. Series: Materials Science and Engineering* **568** (2019) 012021
[6] *Getting started using ADAMS/View* 2014 retrieved from mscsoftware.com on September 2015
[7] *Adams Tutorial Kit for Mechanical Engineering Courses (Third Edition)* 2015
[8] *MSC Software: Designing Better Industrial Robots with Adams Multibody Simulation Software* Whitepaper 2016 retrieved from https://files.mscsoftware.com on July 2017
[9] Craig J 2005 *Introduction to robotics. Mechanics and control* (Pearson Prentice Hall: USA)
[10] Spong M W, Hutchinson S and Vidyasagar M 2004 *Robot dynamics and control* (Wiley)
[11] Featherstone R 2014 *Rigid body dynamics algorithms* (Springer)
[12] Featherstone R and Orin D E 2016 Dynamics, *Springer Handbook of Robotics* pp 37-66
[13] Yan Q, Zhang J, Li B and Zhou L 2019 Kinematic analysis and dynamic optimization simulation of a novel unpowered exoskeleton with parallel topology *Journal of Robotics* **2019** (Hindawi)
[14] Kumar v 2019 Kinematics/dynamics analysis with ADAMS/MATLAB co-simulation of a SolidWorks designed spatial robot arm *EasyChair Preprint* **1619**
[15] Parthasarathy T, Srinivasaragavan V and Santhanakrishnan S 2016 ADAMS-MATLAB co-simulation of a serial manipulator *ICMME 2016 (MATEC Web of Conferences)* **95**, 08002
[16] Baglioni S, Cianetti F, Braccesi C and De Micheli D M 2016 Multibody modelling of N DOF robot arm assigned to milling manufacturing *J. of Mech. Sc. and Technol.* **30** (1) pp 405–420
[17] Yang L and Zhang X 2015 Dynamics Modeling and Simulation of Robot Manipulator *Intelligent Robotics and Applications ICIRA* **2015** part III pp 525–535 (Springer)
[18] Li G, Shi B and Liu R 2019 Dynamic modeling and analysis of a novel 6-DOF robotic crusher based on movement characteristics *Mathematical Problems in Engineering* **2019** (Hindawi)
[19] Enescu M 2013 Mechatronic model for the dynamic analysis of a robotic system *Bulletin of the Transilvania University of Braşov Series I: Engineering Sciences* **6** (55) /2
[20] Alexandru C 2006 Modeling in mechatronic concept and virtual testing of orientation mechanisms used in solar panels *Mechanism and Manipulators Magazine* **2/2** pp 49-54
[21] Alexandru C 2019 Multi-body system simulation of the sun trackers used for PV panels *IOP Conf. Series: Materials Science and Engineering* **568** (2019) 012001
[22] Enescu M and Alexandru C 2011 Modeling and simulation of a 6 DOF robot *Proceedings of 2011 International Conference on Optimization of the Robots and Manipulators* (Sinaia, Romania)

IOP Conf. Series: Materials Science and Engineering **898** (2020) 012014 doi:10.1088/1757-899X/898/1/012014

Clutch model and controller development in MATLAB Simulink

D Moldovanu[1*] and A Csato[1]

[1]Technical University of Cluj-Napoca, Automotive Engineering and Transports Department, Bd. Muncii 103-105, Cluj-Napoca, Romania

dan.moldovanu@auto.utcluj.ro

Abstract. With rapid developing of automated gearboxes used in conventional, hybrid and also plug-in hybrid vehicles, clutches are inevitably controlled by transmission control unit. The effectiveness of the clutch control is based on how smoothly are engaged and disengaged in different situations like drive away from standstill, changing gears at high torque demand or changing between different modes in hybrid vehicles. In this paper a clutch model is developed in MATLAB Simulink environment taking in account physical properties of the clutch. For establishing a controller that is robust enough to take in consideration clutch wearing and is smooth enough for ride comfort, multiple control strategies are implemented. Finally the control strategies are compared in different situations.

1. Introduction

The purpose of this paper is to underline the importance of a model in MATLAB for further development use for a clutch and controller model. Especially for hybrid vehicles, this was also studied by Chen et. al. [1], by implementing a torque control during the transition between gears in order to ensure a smooth transition and a continuous torque. This was done by using a reference model, and a controller for the plant. The Lyapunov stability theory was used to make the system have an asymptotic stability. The method was also validated on a SPHEV (series–parallel hybrid electric vehicle) bus. Galvagno et. al. [2] implemented a dynamic and kinematic model of a dual clutch transmission (DCT) on a vehicle with front wheel drive and transversal internal combustion engine and gearbox. For the numerical simulation, several simplifying methods were used. The internal combustion engine model was a steady-state torque map as function of engine speed and throttle position. The implemented maneuvers were a sequence of upshifts and downshifts. By using the transmission model, the performance of specified speed profiles and shift transients were analyzed further and validated. The most problematic issue is the transient state, therefor Kim et. al. [3] implemented several control methods for EV (electric vehicle) / HEV (hybrid electric vehicle) change mode. An open and closed loop analysis was made and as results, methods were evaluated and the method with a slip-less control and ease of calibration was concluded as being the best. The best method was to control the clutch pressure in order to achieve the desired speed difference from each side of the clutch (with motor speed faster than engine idle). Kulkarni et. al. [4] simulated the shift dynamics and control of a DCT. The engine assembly was considered a 2 degree of freedom system (rotation inertia of moving parts and the inertia of the engine and transmission). Engine torque was considered as an interpolation from an engine map with engine speed and throttle position in this case as well. A shift control logic was implemented, with focus on detecting the time of the shift initialization, creating a specific rate of engagement and disengagement of the clutches and determination of the shift completion. The importance of clutch pressure control signals and clutch pressure control were underlined. Letrouve et. al. [5] analyzed the influence of the clutch model in a simulation of a parallel HEV, where simulations were done with and without a clutch and fuel consumption and dynamic performance was underlined. For the NEDC (New European Driving Cycle) the two models gave similar results in terms of fuel consumption, but the models can be further used to implement different strategies, driver behavior and cycles. Liu et. al. [6] also implemented a DCT control model to analyze the dynamic behavior of vehicles for upshifting and downshifting in different situations like launching in first gear, creep and inching or uphill launch with heavy towing load. The implemented controller was a PID with feedback on the difference between a predesigned gear ratio change function and the actual speed ratio. The model was finally validated by measurements on a test vehicle. Smith et. al. [7] implemented a three PID loop control, with a clutch torque control loop, a motor torque control loop and a wheel torque control loop. The control system was implemented on a dSPACE MicroAutoBox on a HIL (hardware in the loop) bench. The experimental study showed that a flying engine start could be performed at 20 kph with little disturbance to vehicle acceleration. Van Der Heijden et. al. [8] simulated and optimized the control of a dry clutch for hybrid vehicles. After implementing the model, a piecewise linear quadratic (PWQL) control was implemented and compared to a PI controller. The PWQL controller was proven to be better since the clutch engages faster and smoother than the PI controller.

The dry clutch has been widely used in manual transmission vehicles and for automated manual transmissions, controlling automated manual transmissions in the most comfortable way for the passengers are one the most important control problem in this field. Various successful products such as the diaphragm spring clutch, self-adjusting clutch, travel adjusted clutch and pre-damped clutch damper have been developed by Valeo, LUK, SACHS and others. Although the development of a highly responsive system with the engine in low speed ranges makes the control task more difficult, the dry clutch is still widely used because of its efficiency, robustness and low manufacturing cost [9].

In order to simulate the clutch behavior all the powertrain must be modelled for more accurate results. The powertrain consists of the engine, clutch, gearbox, differential and vehicle. For a simple model the powertrain is developed based on components inertia and rotational speed.

2. Objectives

In this research a clutch system was developed in MATLAB Simulink environment and controlled by using different control systems. Controlling the engaging process of the powertrain with the engine by using the clutch is the focus of this paper. To achieve an acceptable control strategy two condition must be fulfilled: the engine speed must never get lower than the idle speed and have a smooth torque and rotation transfer from the engine to the powertrain. Different control types were used and compared in the following sections.

3. Mathematical model of the clutch

Therefore, in the clutch engagement and disengagement phase, there is a slip between the drive and the driven part of the clutch, making the clutch a two degrees of freedom mechanical system. When the clutch pedal is released, the pressure between the driving and driven disk increased and transferred torque and rotation is also increased, to the point where the driving disk and the clutch plate achieve synchronous speed, the clutch is locked as a whole and become a single degree of freedom mechanical system [11].

The maximum torque supported by the clutch can be expressed as:

$$T_{max} = \frac{1}{\pi(R_2^2-R_1^2)}\int_0^{2\pi}\int_{R_1}^{R_2} r^2\mu dr d\theta = \frac{2}{3}\frac{R_2^3-R_1^3}{R_2^2-R_1^2}\mu F \tag{1}$$

where, R_1 and R_2 are the inner and outer radius of the contact surface between the driving and driven disk; μ is the coefficient of friction between the driving and driven disk; F is the force to compressing the driven disk. When the rotation speed difference between the driven disk and the drive disk becomes close to zero, the slipping disappears.

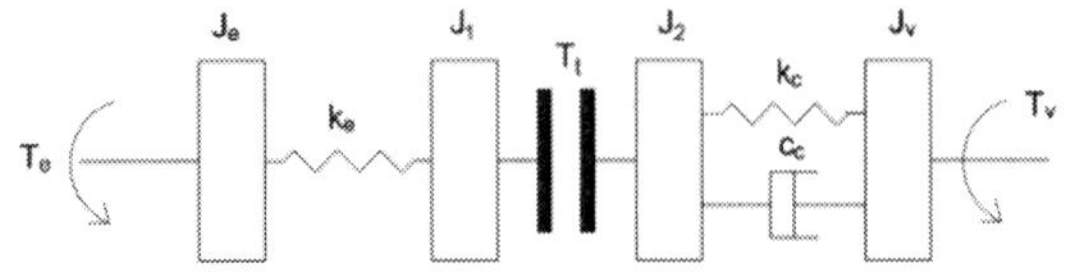

Figure 1. Clutch slipping state

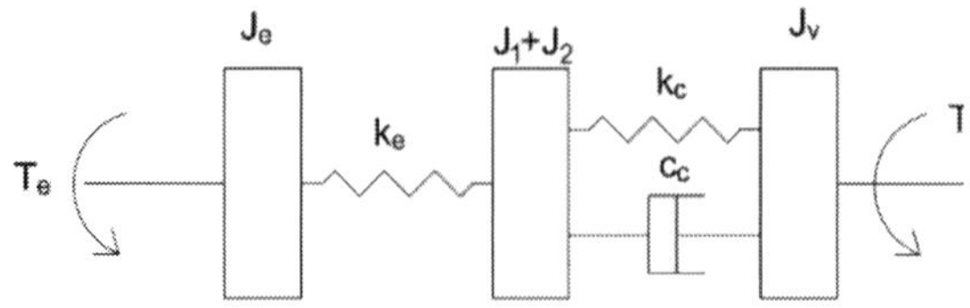

Figure 2. Locked state of clutch

The equations for slipping state of the clutch:

$$J_e\ddot{\theta}_e + k_e(\theta_e - \theta_1) - T_e = 0 \tag{2}$$

$$J_1\ddot{\theta}_1 + k_e(\theta_1 - \theta_e) + T_t = 0 \tag{3}$$

$$J_2\ddot{\theta}_2 + k_c(\theta_2 - \theta_v) + c_c(\dot{\theta}_2 - \dot{\theta}_v) - T_t = 0 \tag{4}$$

$$J_v\ddot{\theta}_v + k_c(\theta_v - \theta_2) + c_c(\dot{\theta}_v - \dot{\theta}_2) + T_v = 0 \tag{5}$$

The equations for locked state of clutch:

$$J_e\ddot{\theta}_e + k_e(\theta_e - \theta_{12}) - T_e = 0 \tag{6}$$

$$(J_1 + J_2)\ddot{\theta}_{12} + k_e(\theta_{12} - \theta_e) - k_c(\theta_{12} - \theta_v) - c_c(\dot{\theta}_{12} - \dot{\theta}_v) = 0 \tag{7}$$

$$J_v\ddot{\theta}_v + k_c(\theta_v - \theta_{12}) + c_c(\dot{\theta}_v - \dot{\theta}_{12}) + T_v = 0 \tag{8}$$

The parameters of the clutch model that were used are [9], [12]: J_e=0.1 [kgm^2], J_1=0.03 [kgm^2], J_2=0.02[kgm^2], J_v=115[kgm^2], k_e=500[Nm/rad], k_c=800[Nm/rad], c_c=0.5[Nms/rad], R_2=220[mm], R_1=150[mm], M=0.42[-].

4. Simulink model of the clutch

The Simulink model of the clutch system is a state dependent model, switching between the slipping state and the locked state is based on speed difference between the drive part and the driven part of the clutch. Equations (1) – (8) were implemented in Simulink. In figure 3 the lock logic block is responsible for switching between states of the clutch. The lock logic has inputs like gearbox speed and engine speed. After the lock, the output is clutch locked speed.

5. Control strategy

Requirements for controlling the clutch engagement are: no stalling of the engine or maintaining a minimal engine speed during the slipping phase, second condition is to maintain the locked clutch status after the driven and drive part of the clutch has the same speed. A closed loop controller is developed to satisfy to above conditions, main focus is on maintaining the engine speed above the idle speed. The control in first step is realized with a PI controller.

The PI controller integrator part is limited to last 50 time step error calculation, which is very helpful with long simulation times and highly nonlinear models.

The second controller used is a transfer function which functions the same as the PI and also the behavior is the same, but no limited integrator was used.

A fuzzy logic controller was implemented. The characteristics of this controller were established with trapezoidal and Gaussian functions. The trapezoidal function is the input function with the rotation speed difference, and the output controls the clutch pedal position.

To test the controllers, two setups were made, both drive-away from standstill situations, the difference between them being the gradient of accelerator pedal with 1 second from 0 to 1 for the high gradient and 5 seconds for the lower gradient. This situation is the most critical because of high demand of engine torque and the high torque at low engine speeds must be delivered fast and smooth without stalling the engine.

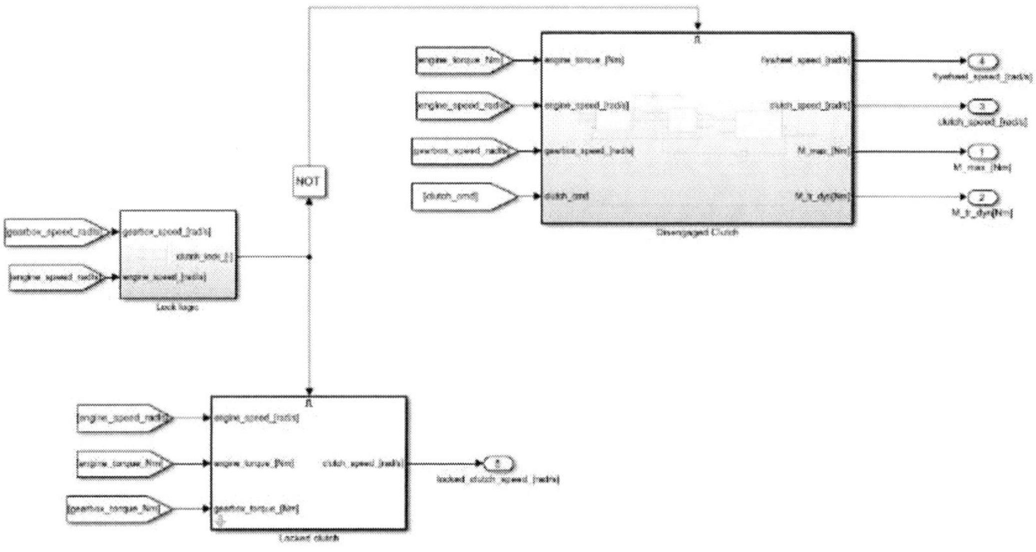

Figure 3. Simulink model of the clutch

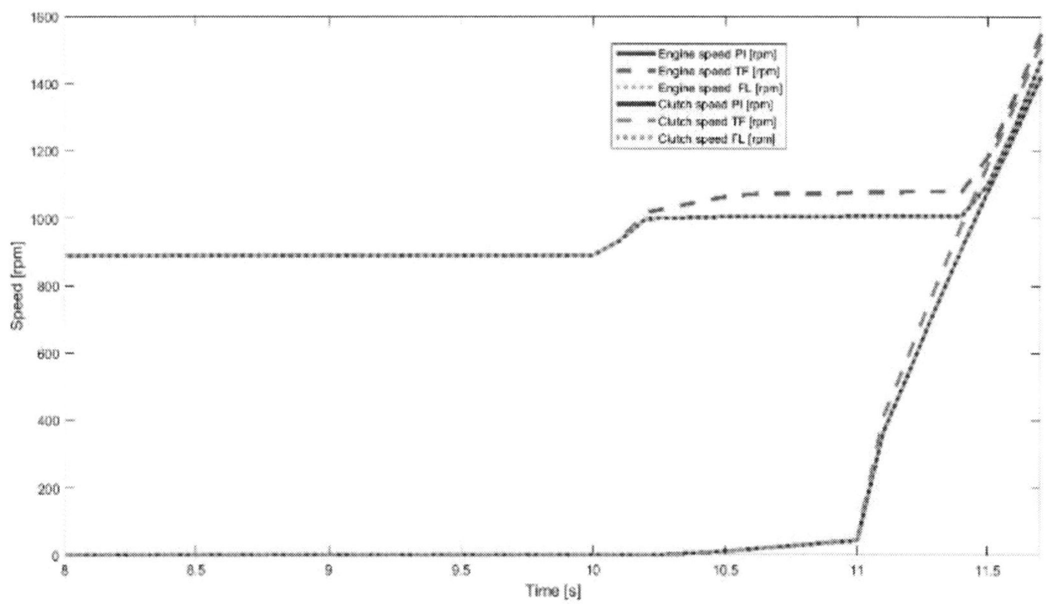

Figure 4. Clutch controller values for high gradient accelerator pedal variation

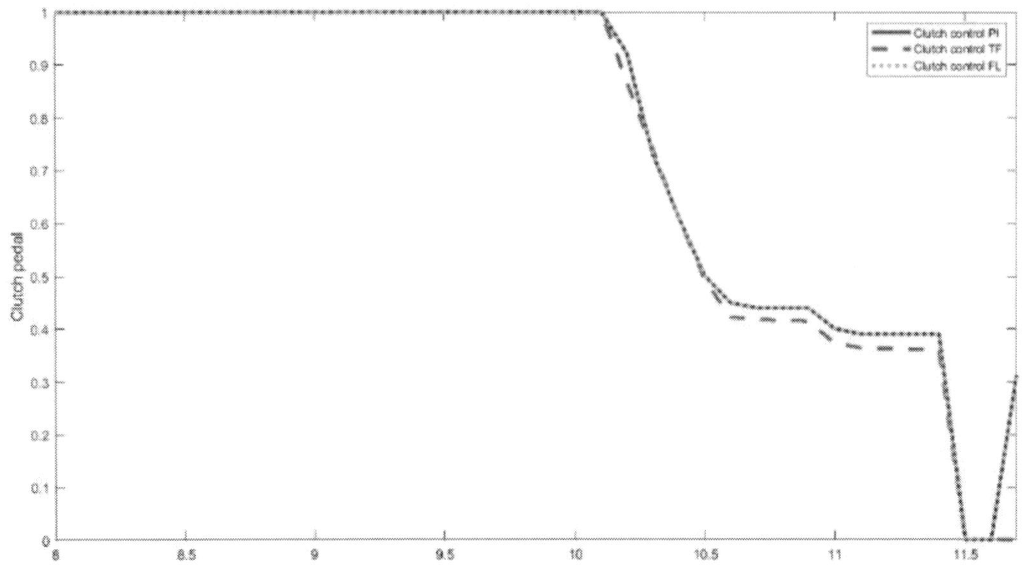

Figure 5. Engine and clutch speed in function of time for high gradient accelerator pedal variation

6. Conclusions

Previous work from Minh and Pumwa [13] has shown that fuzzy logic can be implemented to ensure the successful control of the clutch in HEVs. In this paper, the authors demonstrated that the clutch system model developed for simulation purpose has a similar behavior (figure 4, figure 5). The control strategy that was chosen for this comparison is based on not stalling the engine condition. The three different controllers that were implemented in the model have approximately the same results, engaging times have small differences in all cases. The reason why the three implementations were compared, is because of the computing time necessary to obtain similarly same results, and how that translates into different codes when trying to flash the controller onto a physical control unit.

7. References

[1] Chen L, Xi G, Sun J 2012 IEEE *Trans. Veh. Technol.* **61** 2936-49

[2] Galvagno E, Velardocchia M, Vigliani A 2011 *Mech. Mach. Theory* **46** 6 794-805

[3] Kim S, Park J, Hong J, Lee M, Sim H 2009 *SAE Tech. Pap.*

[4] Kulkarni M, Shim T, Zhang Y 2007 *Mech. Mach. Theory* **42** 2 168-182

[5] Letrouve T, Bouscayrol A, Lhomme W 2009 *IEEE Veh. Power Propuls.* 1330-37

[6] Liu Y, Qin D, Jiang H, Zhang Y 2009 *J. Mech. Des. Trans. ASME* **131** 6 0610121-27

[7] Smith, A., Bucknor, N., Yang, H., He, Y 2011 *SAE 2011 World Congr. Exhib.*

[8] Van Der Heijden A C, Serrarens A F A, Camlibel M K, Nijmeijer H 2007 *Int. J. Control* **80** 11 1717-28

[9] Yan Z, Yan F, Liang J, Duan Y 2019 *IEEE Access* **7** 59100-13

[10] Zanasi, R., Visconti, A., Sandoni, G., Morselli, R 2001 *IEEE Xplore* **1** 416-21

[11] Shi P C, Fang T 2013 *Appl. Mech. Mater.* **313** 1092-95

[12] Vasca F, Iannelli L, Senatore A, Scafati M T 2008 *Amer. Contr. Conf.* 306-11

[13] Minh V T, Pumwa J 2013 *Int. J. Control. Autom. Syst.* **11** 3 526-32

IOP Conf. Series: Materials Science and Engineering **898** (2020) 012015 doi:10.1088/1757-899X/898/1/012015

Implementation in MATLAB Simulink of a basic electric vehicle

D Moldovanu[1*]

[1]Technical University of Cluj-Napoca, Automotive Engineering and Transports Department, Bd. Muncii 103-105, Cluj-Napoca, Romania

dan.moldovanu@auto.utcluj.ro

Abstract. In order to have a better understanding of influencing factors when designing an electric vehicle, the author implemented a model in MATLAB Simulink of a basic electric vehicle taking into consideration several factors, with room for improvements and addons to further studies. The purpose was to have a basic model for teaching. The model takes into consideration the basic aerodynamics of the vehicle, the electric motor (implemented via transfer function), the vehicle velocity calculation from the sum of acting forces and validation of the model. Validation of the model was done by comparing the output data like vehicle velocity with the real measurements done by the manufacturer itself. After the validation, a PI controller was implemented, and two instances were compared. The author considers that students should learn by doing, especially when simulation-based design is easy to understand.

1. Introduction

It is very important to have a model or library of models and systems/subsystems in MATLAB Simulink because they can be used for development or dimensioning or different approximations while in the design phase of the vehicle. Several papers on the issue were studied and analysed in order to have a clear perspective on the steps to take when implementing an electric vehicle.

With many electric vehicles now on the market, with an average range over 250 km on a single charge, the focus is more and more on the development and improvement of the performances for a more sustainable transport. Also because the 2010-2020 time period has been described as a tipping point period for the internal combustion engine to electric propulsion systems.

Butler et. al. [1] modelled an electric and a hybrid vehicle using MATLAB, with the V-Elph package, that offers blocks for the transmission, the internal combustion engine (for the hybrid), the battery, the different drive shafts, the induction motor and the controller, with a clever implementation of the drive cycle from a ".mat" file. The simulation concluded with the fact that all components inserted into the model need a fine tune in order to have accurate results. For all four drive cycles results were analysed for all control strategies and vehicle configuration. Lakshmi et. al. [2] implemented an electric vehicle drive simulation in order to investigate the power flow for both motoring and regeneration instances, with very detailed equations for all blocks like battery model, motor drive and different controllers. Ma [3] implemented a controller for the propulsion system of an electric vehicle to underline the waveforms of phase voltage and current, but also waveform of the power. It also presented a propulsion system design that requires a high-power density. Sri Kaloko et. al. [4] developed a small electric vehicle and modelled it in MATLAB by taking the physical data for batteries and vehicle, in order to know the necessary battery capacity to reach a certain specification. Fan [5] used MATLAB Simulink and ADAMS to model and simulate a hybrid electric vehicle. In ADAMS, the vehicle model was developed with inputs about the chassis, suspension, driveline, tires, braking, steering and terrain info, and in MATLAB, the internal combustion engine (that was implemented using the engine maps for all throttle positions and taking into account the losses when throttle is null), battery, controller, power management and driver input were implemented. The driver controller subsystem is a complex model that considers the desired drive cycle and takes into consideration the actual vehicle speed.

2. Objectives

In this paper, the author presented a model of a basic electric vehicle implemented in MATLAB Simulink, taking into consideration the dynamics of the vehicle (Tesla Model S), a simple generated testing cycle to see the response of two different PI Controller configurations.

3. Method

First, due to the high complexity of a three-phase AC four pole induction motor, it was replaced with a brushed one for similar performances, even though AC motor rotates without contact, therefore more efficient, but an overall efficiency coefficient was taken into account to balance this.
The electric motor is a basic R-L-EMF series circuit, where R is the resistor, L is the inductor and EMF is the electromotive force generated by that motor when rotating. From this circuit, the following equation can be written:

$$V = I(t) \cdot R + L\frac{dI(t)}{dt} + E(t) \tag{1}$$

The generated torque is:

$$T(t) = K_T \cdot I(t) \tag{2}$$

K_T being the motor torque constant defined by the manufacturer.
The generated EMF can be written taking into consideration K_E (EMF coefficient of the motor) and the rotational speed of the motor $\omega(t)$:

$$E(t) = K_E \cdot \omega(t) \tag{3}$$

From the three equations, the current can be obtained with respect to the voltage and the rotational speed of the motor. After using Laplace, the motor can be implemented in MATLAB with a transfer function block with the voltage as an input and the generated torque as an output.
The next subsystem is the vehicle dynamics system, where all forces that act on the vehicle must be taken into consideration, starting with mechanical traction torque of the motor, aerodynamic drag force, roll resistance, and inertia of the vehicle.
The mechanical traction torque of the motor F_{mec} can be calculated taking into consideration the gear ratio G_r, the torque T and the diameter of the wheel, or r the radius of the wheel, but also the efficiency of the transmission eff_{tr}:

$$F_{mec} = \frac{T}{r} \cdot G_r \cdot eff_{tr} \tag{4}$$

The aerodynamic drag force can be written as:

$$D = \frac{1}{2} \cdot \rho \cdot V^2 \cdot A \cdot C_D \tag{5}$$

were ρ is the air density, V velocity of the vehicle, A frontal area of the vehicle, C_D drag coefficient.

The roll resistance is:

$$F_r = C_r \cdot m_T \cdot g \tag{6}$$

IOP Conf. Series: Materials Science and Engineering **898** (2020) 012015 doi:10.1088/1757-899X/898/1/012015

where $C_r$is the roll resistance coefficient, m_T total mass of the vehicle taking into account the mass of the passengers, and g is the gravitational acceleration.

4. Implementation in MATLAB

The constants were introduced and defined in a ".m" file so that it may be adapted to other vehicles as well. After running the m file, the data is introduced to the workspace.

```
clc
clear
close all
% definition of constants
R=5.3*10^-3;          % [ohm]
L=493*10^-9;          % [henry]
Ke=0.12;              % [Vs/rad]
Kt=0.25;              % [Nm/A]

D=0.48;               % wheel diameter [m]
r=D/2;                % wheel radius [m]
Gr=9.73;              % gear ratio [-]
%Ff=T/r*Gr            % Ff=0.45*T

rho=1.225;            % [kg/m3]
A=2.3;                % frontal area [m2]
Cd=0.24;              % drag coefficient [-]
Drag=1/2*rho*A*Cd;    % Drag

Cr=0.02;              % roll resistance coefficient (0.02 for car on dry asphalt)
mveh=2108;            % mass of the vehicle
mpass=200;            % mass of the passenger/passengers
mT=mveh+mpass;        % total mass of the vehicle
g=9.81;               % gravitational acceleration [m/s2]
Fr=Cr*mT*g;           % =413N

efftr=0.7;            % global efficiency of the transmission
Tmax=600;             % maximum torque [Nm]
Vmax=265;             % maximum velocity [km/h]
```

Figure 1. The definition of the constants for the model.

The sum of all acting forces can be written as:

$$\sum F = m_T \cdot a_v \tag{7}$$

Therefore by knowing all the forces and the mass, the acceleration can be calculated. The velocity of the vehicle appears in the 5th equation, but MATLAB allows to calculate it from the acceleration via an integration block with the chosen initial value of zero.

The calculation for the vehicle dynamics were implemented using a MATLAB function block, where the inputs were vehicle velocity and motor torque, and acceleration of the vehicle as output. The final model is presented in figure 2. The main parameters that were followed were: vehicle acceleration, vehicle velocity, motor torque, input velocity, error (calculated by subtracting the actual velocity from the input velocity), PI voltage (voltage given by the PI controller).

In order to have a realistic simulation, there are some limitations that must be inserted: motor torque limitation (by using a saturation block), because the current absorbed by the motor can be limited this way.

The maximum voltage given by the battery must also be limited. This can be done either by inserting a scope or by selecting the output limit of the controller in the output saturation tab.

The validation of the vehicle was done so that the vehicle reaches its maximum velocity (around 250 km/h) and the acceleration matches the acceleration given by the manufacturer (around 6.2 m/s^2).

As an input to the system, the vehicle velocity was chosen. The cycle that was implemented consists of a demand velocity of 50 km/h, follower by a step to 100 km/h (each for 50 seconds), follower by a 0 km/h request and then a maximum velocity demand for 200 seconds follower by motor braking for 300 seconds. When motor braking, the velocity drops gradually since the total mass of the vehicle is 2308 kg and the acting forces are the ones presented before.

For comparison, two PI controllers were compared (using a PID block but coefficients only for P and I, with D=0): PI1: P=100, I=2; and PI2: P1000, I=5; to underline the importance of the coefficients.

The Motor Model was implemented as a transfer function with numerator K_t and denominator $L \cdot s + R$, and the input takes into consideration the rotational speed of the motor $\omega(t)$ calculated from the velocity of he vehicle, knowing the radius of the wheel and the transmission ratio.

IOP Conf. Series: Materials Science and Engineering **898** (2020) 012015 doi:10.1088/1757-899X/898/1/012015

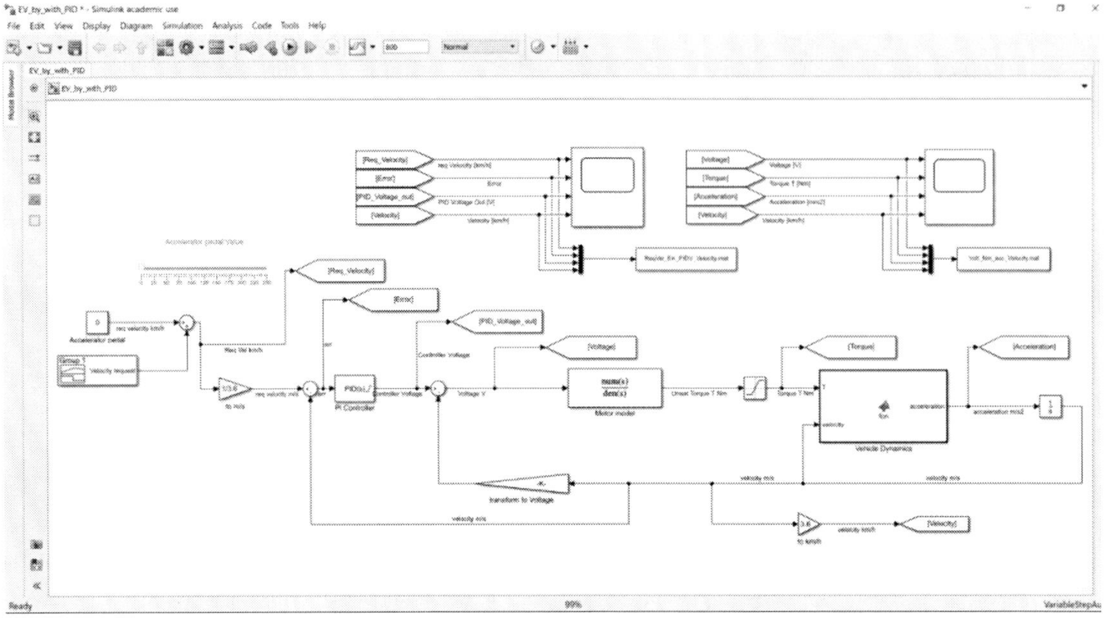

Figure 2. Simulink model implementation.

5. Results

The extracted results are presented in figures 3 and 4. The error and voltage given by the controller are presented in figure 3, and the performance results are depicted in figure 4.

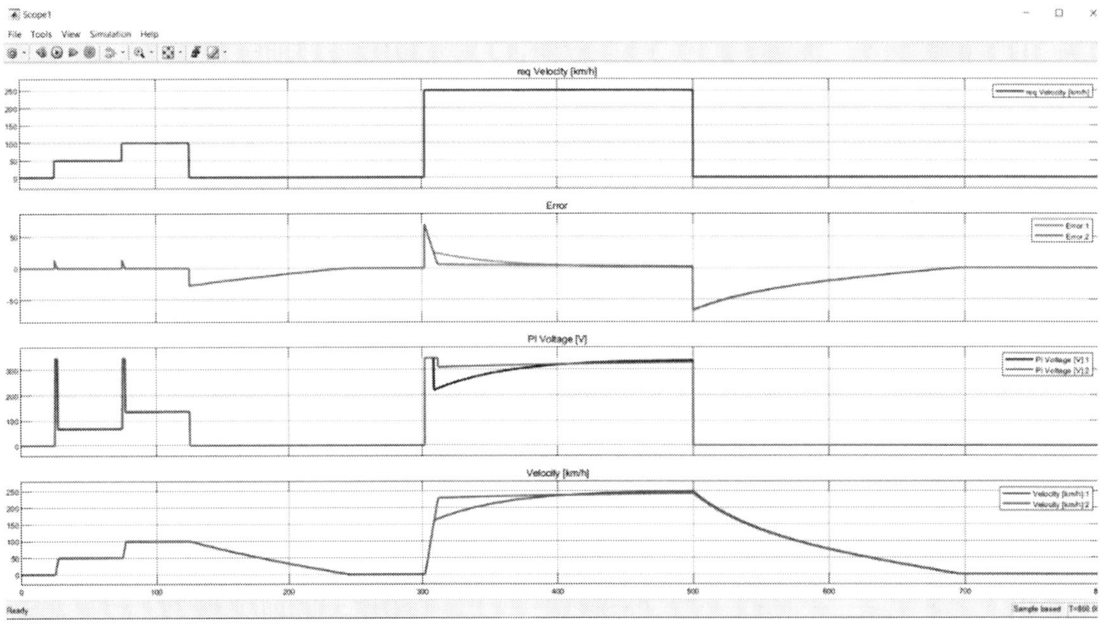

Figure 3. Error and voltage results for the two implemented controllers.

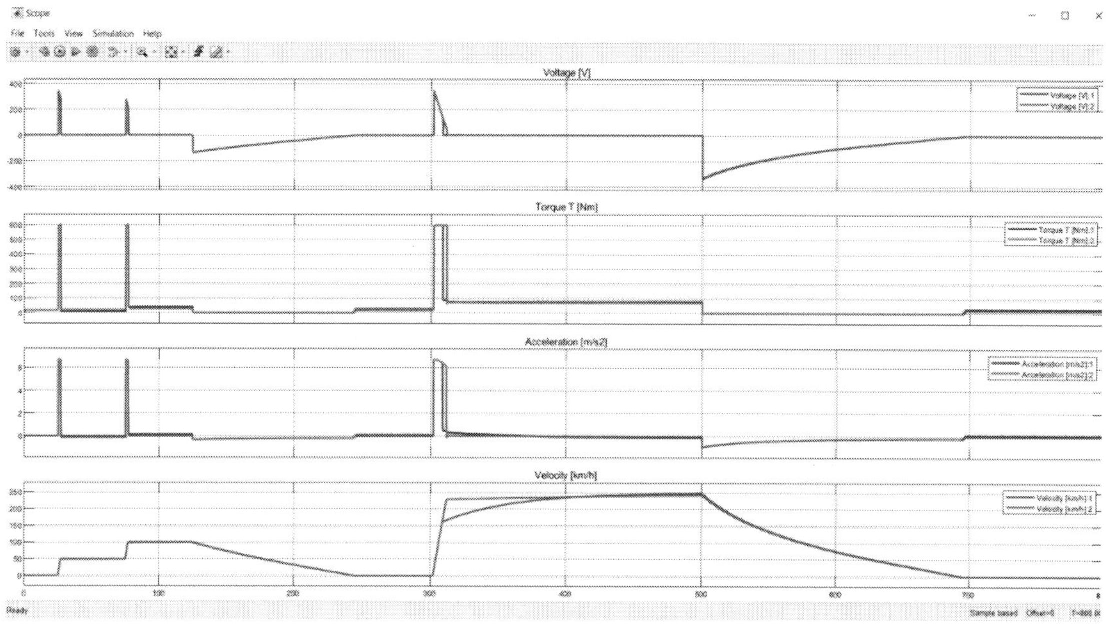

Figure 4. Performance results for the two implemented controllers

6. Conclusion

For PI1, it can be seen from figure 3 that at 300 on the time axis, the demand goes to 250 km/h, and the error starts to reduce, but because of a low proportional gain, when the error is smaller, the rise of velocity is lower, also due to the low integral coefficient.

Figure 3 underlines the fact that both controllers have a good response at low velocity demands (50 km/h errors), but at high velocity demands (step of 250 km/h), PI2 controller responds better because of the higher proportional gain. Figure 4 presents the acceleration and the torque given by the motor for both controllers. Because of the low integral and proportional coefficients, with the PI1 the motor has an early drop of torque and therefore a slower increase of velocity. Since this is a basic model, the following improvements can be considered: introduce wheel slip, modify the transfer function to a time-based model, introduce gradient of the road, filter the acceleration and implement other drive cycles.

References

[1] Butler K L, Ehsani M, Kamath P 1999 *IEEE Trans. Veh. Technol.* ***48*** *6 1770-8*

[2] Lakshmi G S, Fatima K, Madhavi B K 2017 *IEEE Reg. 10 Annu. Int. Conf. Proceedings/TENCON* 147-151

[3] Ma X 2002 *Proc. World Congr. Intell. Control Autom.* **1** 815-18

[4] Sri Kaloko B, Soebagio M, Hery Purnomo M 2011 *Int. J. Comput. Appl.* **24** 6 19-23

[5] Fan B S-M 2007 (thesis) Modeling and Simulation of A Hybrid Electric Vehicle Using MATLAB Simulink and ADAMS

Acknowledgements

This research was possible with the MATLAB license from the campus-wide license of the Technical University of Cluj-Napoca. Portions of this paper are based upon the research of Eliott Wertheimer. The author gratefully acknowledges the importance of his work.

Studies on the improvement of the supply chain in the automotive industry: production of electrical wiring

M Inţă[1], C Purcar[1] and M Bădescu[1]

[1] "Lucian Blaga" University of Sibiu, Engineering Faculty, 4, Emil Cioran Street, Sibiu, 550025, Romania

E-mail: marinela.inta@ulbsibiu.ro

Abstract. Procurement is the beginning of the global supply chain and is driven by customer demand for goods. In the automotive industry and not only, the timely and qualitative delivery of finished products is an important goal, the customer being 100% satisfied. This paper addresses the issue of improving the supply process in case of urgent changes, especially in the automotive industry, in the processing of electrical wiring, for a number of companies producing vehicles. The Romani car industry grew quite a bit after the 1990s. The arrival of the world's major car companies has cemented the domestic sector in adopting best practices in the supply chain. This has led to improved competitiveness and increased quantum exports. However, the Romanian car industry must operate in a unique environment that still presents challenges for the supply chain. It is therefore necessary to continuously study supply chain practices in the automotive sector.

1. Introduction

The concept of SCM has gained over time an increased importance in the world economy due to its impact on the competitive advantages of companies [1-2].

Supply chain management is under pressure from companies that strive to maintain high levels of service for their customers, while being forced to reduce costs and maintain profit margins [3]. Decision support tools, strategic planning and cost reduction are extremely valuable. Each situation is unique and requires a customized model to suit the particular situation and purpose of the company in question [4].

Thus, over time in the automotive industry, standards related to product quality have been developed and implemented, one of which is the ISO/TS 16949 standard, which refers to the quality elements within the supply chain.

Joe Bransky, member of the International Automotive Task Force (IATF) comments: "In the current manufacturing environment, the huge intervention stocks in the inventory have been replaced by logistics and just-in-time delivery. Rooting safety as the ISO TS 16949: 2009 standard ensures major improvements in terms of quality, productivity, delivery is essential."

In essence, supply activity includes the purchase of material resources and inventory management. It is a very important activity, because it has a great financial impact on the company's turnover [4-5].

The organization of the circulation of material values from the supplier to the beneficiary, their reception from a quantitative and qualitative point of view, their proper storage and maintenance and their distribution to the consuming sections and jobs, represent some problems that compose the supply process [6].

2. Procurement and supply - basic logistics operation

The supply of material resources is defined as "the activity that ensures the elements necessary for the consumption of production, in volume and structure to ensure an activity, with a high profit, of economic units." [7]. The good functioning of the enterprises' activity is conditioned, to a large extent, by the timely and complete supply of raw materials that will become part of the finished product manufactured and delivered to the customer.

A competitive and reliable supply means a timely management and adequate quantity of components and raw materials needed in production. This requires a brief analysis / prospecting, a good forecast and optimal management of stocks, a good rotation of them and at the same time a low level of supply disruptions, but also to control, verify, systematically monitor the use of materials, negotiation, contracting.

Figure 1 presents the trends of the automotive industry that have an impact on the supply chain [8]. According to Fisher [9] the supply chain must be adapted to the specific requirements of the manufactured product. This is especially the case for the complex automotive industry, where a car manufacturer has to struggle with the management of a network that includes several supply chains. The macroeconomic cycles of growth, contraction and recovery create extraordinary efforts on the efficiency of the established supply chain, especially in the automotive sector, due to its widespread links with other industries [10].

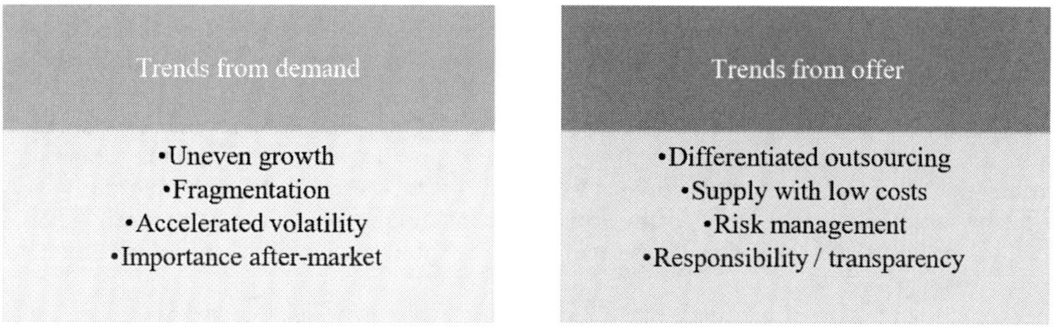

Figure 1. Trends affecting SCM [8]

The share of costs related to raw materials and materials is generally high, sometimes decisive, representing 70-80%, maybe even more. According to some authors, it is an exchange of a commercial nature, through which material values are traded following the elaboration of consumption needs, after the suppliers have been identified; the prices and other conditions of the transfer of ownership between the partners have been negotiated.

3. Evaluation of the efficiency of the supply process - Case study

The case study is carried out on a company producing electrical wiring in Romania that performs cutting operations - automatic and semi-automatic crimping, with and without Seal, tinning operations, printing of strips and automatic banding. The company produces electrical wiring mainly for the automotive industry (lighting systems, safety systems, dashboard, mirrors, handsfree telephony ...). The company has a number of 170 employees, currently having a mainly manual, semi-automatic and very little automatic production. Within the company there are 3 major departments: Automatic/Semi-automatic, Manual/electrical assembly and Final testing. The Automata/Semiautomata Department cuts/crimps the wires, which are then sent to the Manual/Electrical Assembly Department where they perform pressing operations / plug inserts, applying collars, spot bandages, hair dryer operation, etc ... and finally reach the Testing Department where it is checked electrically and visually if the products comply with the customer's requirements.

The main customers of the company are: Harting SCS Romania, Kuhnke Production Romania, Electromagnetica SA Bucharest, and among the final customers are the largest car manufacturers: Mercedes, Audi, Volkswagen, BMW, Aston Martin and Volvo (fig.2).

Figure 2. The company's customers

In the field of engineering and measurement technology, the company has all the necessary facilities for a supplier for the automotive industry, whose products are qualitatively appropriate.

Within the company, the supply department has a very important role, that of maintaining internally a permanent connection with the finance department and with the production department, and externally it relates to the suppliers but also to the company's customers. An essential condition is the synchronization of the production plan with the supply, only in this way it is possible to obtain the reduction of stocks, of the costs related to them, the increase of the quality, of the productivity and of the capacity to adapt to changes [10].

The supply department pays special attention to the way in which the movement of stocks takes place, the evolution of stocks, the conditions and the degree to which they ensured the productive consumption, the critical periods determined by the lack of stock, their speed of movement.

As in any other company, the issue of stock management and the quantity to be supplied is raised by establishing supply rules so that the items necessary for the production process are not missing from the stock.

To optimize the size of the stock, the Wilson-Within model is used, which takes into account two categories of variables:

- C_l - the cost of launching supply orders
- C_s - storage or inventory costs

As stocks of materials are depletable over time, they gradually pass into consumption, storage or inventory costs are calculated as an average of the expenses from the first day of the stock and the expenses from the last day of its existence => any storage costs is weighted (corrected) by 0.5.

Sometimes the company is faced with situations where there is no stock of raw materials, suppliers do not deliver the goods on time or it is delivered with defects.

4. Improving the supply process within the company

In order to be able to start a process of improvement in terms of supply, the main problems were identified. Thus, between 17 - 22.11.2019, the head of the supply department analyzed and identified the problems he faced in the last 6 months. These were grouped according to how often they occurred, importance, degree of occurrence and risk.

The problems identified are:

- Delays in the supply of special conditions (due to changes imposed by customers)
- Impossibility to predict long-term framework orders
- There is no feedback from suppliers to solve the problems identified in the evaluation of suppliers (lack of corrective action plan)
- The supplied products do not correspond to the technical specifications (lack of invoice, documents accompanying the goods)

The Pareto Diagram statistical quality tool was used to solve the problem and apply an improvement [11-12]. The 4 identified problems were given grades, as shown in the table 1.

Table 1. Alternative evaluation matrix

Identifyed Problems	Cronicity 10 %	Importance 40 %	Degree of potential impact 2 %	Expedite 38 %	Risk 10 %	Total
Delays in the supply of special conditions	5	5	5	5	1	21
Impossibility to predict long-term framework orders	2	5	0	5	1	13
There is no feedback from suppliers to solve the problems identified in the evaluation of suppliers	5	5	0	5	1	16
The supplied products do not correspond to the technical specifications	5	2	0	5	2	14

Following the evaluation, the issue with the highest weight and, also the highest priority was considered as "Delays in the supply of special conditions".

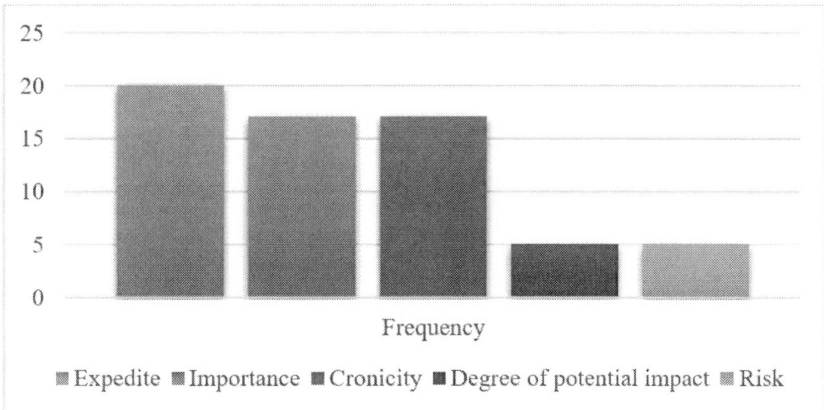

Figure 3. Frequency of the problem

In order to improve the supply process with the raw material necessary to make the car wiring according to the requirements imposed by the company's clients, the DMAIC method was applied, performing the steps [11-13]:

• Defining the problem

The main problem is the supply in special conditions (as a result of changes imposed by customers). Delays in raw materials for special orders are on average 2 weeks. The company's proposal is to reduce delays by 1 week within 6 months. In order to fulfill this mission, a team of project manager and 5 members was created.

• Measuring the main aspects of the current process and collecting the main relevant data

In order to reduce the delays related to the supply of raw materials, it was necessary to identify the main reasons that lead to their occurrence.

To identify the causes, a Fishbone & 5 Why diagram was drawn up, which is presented in figure 4 [12-13].

IOP Conf. Series: Materials Science and Engineering **898** (2020) 012016 doi:10.1088/1757-899X/898/1/012016

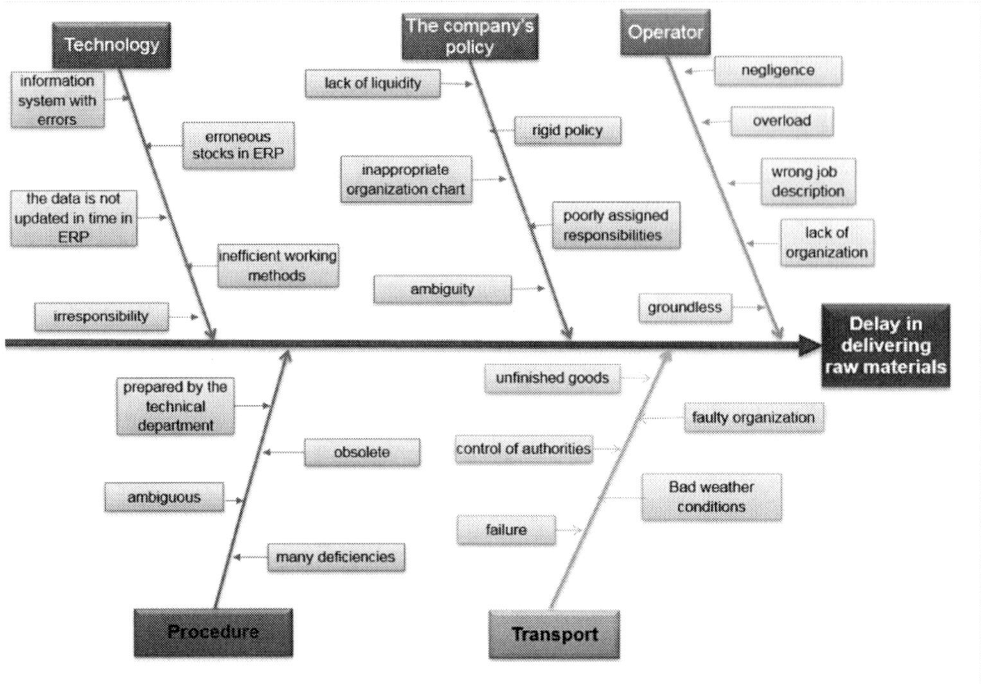

Figure 4. Fishbone diagram combined & 5 Why

• Data analysis and verification of relational causes and effects.

All 130 cases in which there were dysfunctions in the last 6 months were analyzed and the defects that initiated the respective dysfunctions were identified. Based on the frequency of these malfunctions, the Pareto Diagram was drawn up with the help of which the most important defects were identified, which produce 80% of the delays.

Table 2. Alternative evaluation matrix

Defect description	Frequency	Relative frequence	Cumulate relative frequence		TOP defectcs "80%"
Incorrect stocks in the system	**40**	**30.77%**	**30.77%**		
Ambiguous procedure	**40**	**30.77%**	**61.54%**	I	**Incorrect stocks in the system**
Overworked employees	**23**	**17.69%**	**79.23%**	II	**Ambiguous procedure**
Delayed transport	13	10.00%	89.23%	III	**Overworked employees**
Rigid company policy	8	6.15%	95.38%		
Disorganization	4	3.08%	98.46%		
Unclear responsibilities	2	1.54%	100.00%		

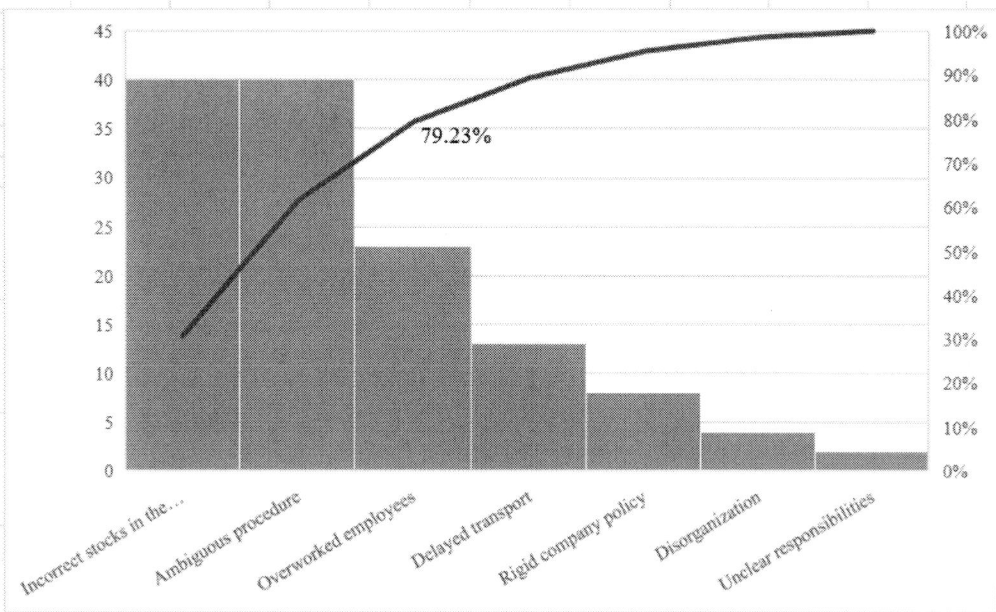

Figure 5. Pareto diagram

So, if the company focuses on eliminating or at least reducing the frequency of these defects, the delays will be reduced by 80%.

• Improving and optimizing the supply process based on the analyzed data

Once the improvement team identified the root causes of the problem, it was ready to identify solutions/improvements. As it emerged from the analysis of the data from the last 6 months, the main non-conformities were:

- Incorect stocks in the system
- Ambiguous procurement procedure
- Overloaded employees

In order to reduce the frequency of non-conformities, several measures have been proposed presented in figure 6.

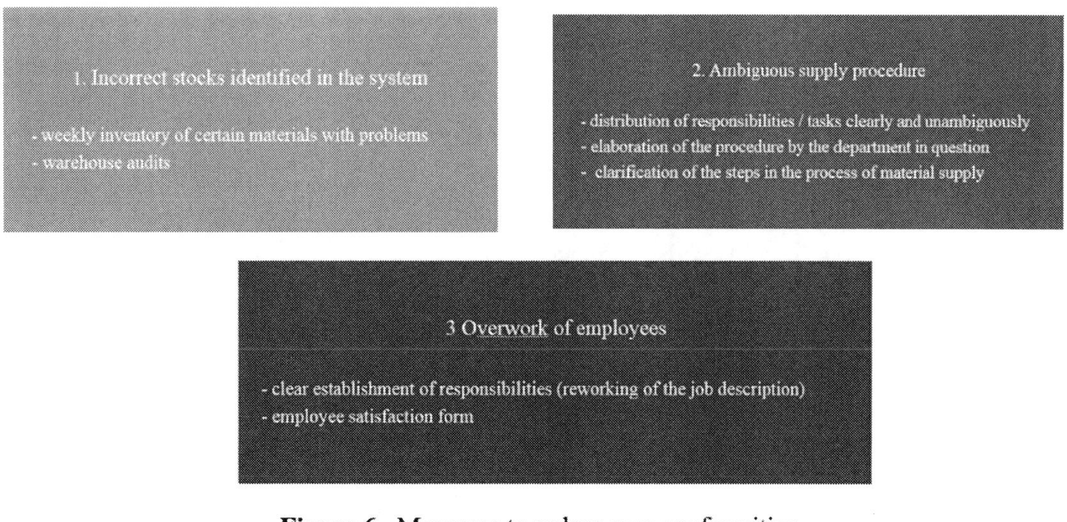

Figure 6. Measures to reduce non-conformities

As not all the improvements found are equally effective, the team had to consider other criteria for evaluating alternatives:

- (3) the impact on the main issue
- (2) implementation time
- (1) the total cost which must not exceed the available resources

The magnitude of the impact of the proposed solution on the current non-conformities was quantified by grades between 1 and 3, where:

- 3 - very favorable impact
- 2 - favorable average impact
- 1 - weak favorable impact

Table 3. Other criteria for evaluating alternatives

Cause	The improvement alternative	Selection criteria 1	2	3	TOTAL score
- Incorrect stocks in the system	**- weekly inventory. to certain materials with problems**	**3**	**3**	**1**	**7**
	- warehouse audits	3	2	1	6
- Ambiguous supply procedure	- redistribution of responsibilities / tasks	2	3	1	6
	- elaboration of the procedure by the department concerned	2	3	1	6
	- clarification of material supply steps	**3**	**3**	**1**	**7**
- overwork of employees	**- clear establishment of responsibilities (reworking of the job description)**	**3**	**3**	**1**	**7**
	- employee satisfaction form	**3**	**2**	**1**	**6**

It is observed that in order to correct the first non-conformity, the incorrect stocks in the system, the biggest impact would have the performance of a weekly inventory for the materials identified as being with problems.

Regarding the ambiguity of the procurement procedure, the measure of clarifying the steps of the process seems to have the greatest impact. With regard to overworked employees, it is recommended to rework the job description in order to clearly establish the responsibilities of each employee.

At the end, we started to plan the improvement process, drawing up the following file.

Table 4. Plan the improvement process

Main causes	Corrective action	Necessary resources	Responsible	Term	Achievement stage%
Incorrect stocks in the system	Weekly inventory. to certain materials with problems	Material resources: scales Human resources: specialist Information resources: ERP	warehouseman	weekly	
Procedure	Clarify material supply steps	Human resources: specialist	Manager supply	2 weeks	
Overworked employees	Stabilirea clară a responsabilităţilor (reelaborarea fisei postului)	Material resources: office Human resources: specialist	Manager Human resources	2 weeks	

Following the implementation of the control elements, the improvement team ensured that the improvements are put into practice and maintained, and at this moment, the delays in the supply of raw materials within the company were reduced by 15%.

5. Conclusions

The importance of supply is materialized by the responsibility assigned to it. It can be said that the procurement process includes both the purchase of material resources and the management of stocks.

The supply department is necessary and very important because it comes to the aid of organizations, so that they maintain internal control, to meet their own requirements and especially those of customers. It has a special importance because it achieves the establishment of the material needs of the production, the continuous reduction of the consumptions of raw materials and materials, the attraction in the economic circuit and in the consumption of the production of new material sources, the increase of the organization's profit.

Good supply of raw materials and materials is the optimal way to accomplish the of production tasks and manufacturing of products. Procurement management can provide the means necessary for improvement activities through a consistent development of its processes. It can generate "good or bad" performance, thus providing the basis for managing these processes.

If the organization does not have a well-defined supply management, it fails to achieve a necessary pace of improvement in order to be able to face the competition in the market, in a very short time. Procurement management that emphasizes continuous improvement can build capacity, lead to significant cost reductions, and create added value.

References

[1] Bottani E and Montanari R 2010 Supply chain design and cost analysis through simulation, *International Journal of Production Research, vol 48 no 10 pp 2859-2886*

[2] Thomas K 2013 *The Automotive Supply Chain in the New Normal: Analysis of the Industry Opportunities* Available: www.scmresources.ca/documents/

[3] Altiparmak F Gen M Lin L and Karaoglan I 2009 A steady-state genetic algorithm for multiproduct supply chain network design *Computers & Industrial Engineering* vol 56 no 2 pp 521-537

[4] D'este G 2001 *Freight and Logistics Modeling* (Pergamon Amsterdam: Handbook of logistics and supply chain management, edited by Brewer A M Button KJ and Hensher DA*)*

[5] Mentzer JT DeWitt W Keebler JS Min S Nix NW Smith CD Zacharia ZG 2001 Defining Supply Chain Management *Journal of Business Logistics* vol 22 pp 1-2

[6] Kearney AT 2013 *The Contribution of the Automobile Industry to Technology and Value Creation* Available www.atkearney.com

[7] Bâşanu G Pricop M 2004 *Supply and sales management* (Managementul aprovizionării şi desfacerii) (Bucharest: Economic Publishing House)

[8] Schwarz M 2008 Trends in the Automotive Industry-Implications on Supply Chain Management *Cisco* Available: www.ictpartner.net/web/about/ ac79/.../ Auto_Trends_WP_FINAL.pdf

[9] Fisher ML 1997 *What is the right supply chain for your product* (Harvard Business Review)

[10] McKinsey 2011 *Building the Supply Chain of the future* Available: www.mckinsey.com/insights/-operations/building_the_supply_chain_of_the_future

[11] Oprean C Kifor C 2012 *Integrated quality management* (Managementul integrat al calităţii) (Bucuresti: Ed. Academiei Romane)

[12] Kamran M Sajid A 2010 Critical analysis of Six Sigma implementation. Total Quality Management

[13] ISO 13053-1:2011. *Quantitative methods in process improvement - Şix Sigma -- Part 1: DMAIC* methodology https://www.iso.org/standard/52901.html

Modeling with hierarchical colored Petri nets. Case study

F S Blaga[1], **A Pop**[1], **T Vesselenyi**[1], **V Hule**[1] **and C I Indre**[1]

[1]University of Oradea, Faculty of Management and Technological Engineering, str.Universității nr.1, Oradea 410087, Romania

E-mail: fblaga@uoradea.ro

Abstract. The paper presents the results of research on the use of hierarchical models with colored Petri nets in the evaluation of automated systems for the preparation and analysis of metallographic specimens. Modeling with hierarchical colored Petri nets was used to evaluate the performance of a preparation and analysis robotic system for metallographic samples. The automated system is composed of three subsystems: the subsystem for the preparation of metallographic samples, the subsystem for the attack with reagent of metallographic samples, the subsystem for the analysis of metallographic samples. Two models were made: a model with colored Petri nets with simple colors and a model with timed colored Petri nets with complex colors. Each model was developed on two hierarchical levels. Level 1 contains the model of the entire flexible system for the preparation and analysis of metallographic samples. Level 2 consists of three sub-models: the sub-model of the system for the preparation of metallographic samples, the sub-model of the system for the reagent attack of metallographic samples, the sub-model of the system for the analysis of metallographic samples. Models with timed colored Petri nets allow the evaluation of system performance in terms of the number of samples analyzed in a time interval. The performance of the modeled system was evaluated by simulation, considering a time interval of eight hours.

1. Introduction

In the field of metallography there had been developed systems which contain a microscope, storage devices and are assisted by a robot which handles samples, although without integrating the sample preparation equipment which it is still made with specialized machines outside the automated systems aided by human operators. The idea of an automatic metallography laboratory is sustained by the ever rising need for an efficient quality control of industrial products. Such a laboratory can be seen as a development in the field of Computer Aided Testing which completes the CAD and CAM concepts. In [1] is presented an automated serial sectioning system with metallographic polisher, robotic arm, ultrasonic cleaner and Inverter Microscope.

The authors analyzed different aspects of this subject in previous work [2].

Modeling and simulation of automated systems are effective tools for evaluating their performance and improving their operation. Among the modeling and simulation tools, one of the most used are Petri nets.

Timed colored Petri nets offer the possibility of making models that describe very accurately the real system and whose structure is simplified compared to models with ordinary Petri nets. Thus, in [3] is present the model with timed colored Petri nets made for a flexible manufacturing cell composed of two CNC machines served by a robot. The use of Petri Nets with complex colors allows the realization of relatively simple models that take into account the fact that in the cells several types of parts are processed. The introduction of the time factor allows the determination of the manufacturing cycle.

The use of models with hierarchical colored Petri nets, allows a more rigorous analysis of the way in which each subsystem within the analyzed system works [4]

The modeling and simulation of a flexible manufacturing cell (FMC) can be done by implementing hierarchical techniques based on Colored Timed Petri Nets. The paper [5] is focused on implementing decisions and strategies in a flexible manufacturing cell with colored and hierarchical techniques. In this context it discusses the decision making for machine, parts and allocated tools in transitions.

In the last years research in the field of automation of specific activities and operations of testing and (biology laboratory, chemical, material sciences). The development of drug industry and biology and medical research needs a huge amount of experimental studies on samples which in turn asks for automation of processes. As examples in the field of biology sciences, we can give Highres Biosolutions [6], Hombrechtikon System Engineering (HSE) [7], or in the field of material science we can give Picoquant [8] or SaxsLab (Xenocs) [9].

In this paper are developed the hierarchical models with colored Petri nets with simple colors and with complex timed colors of the automated system of preparation and analysis of metallographic samples described in [10]. In the paper [10] was presented in detail the model with T-timed Petri nets of the system. The hierarchy of the models highlights the details of the functioning of the subsystems within the analyzed system.

2. Organizing the automated system for preparation and analysis of metallographic samples

The automated system for preparation and analysis of metallographic samples was designed taking into account the existing technical-material resources in the laboratories of Mechatronics and Materials Science of the University of Oradea. The layout of the designed cell is shown in figure 1. Three subsystems can be distinguished:

- S1 subsystem for the preparation of metallographic samples;
- S2 subsystem for reagent attack of metallographic samples;
- S3 subsystem for metallographic analysis.

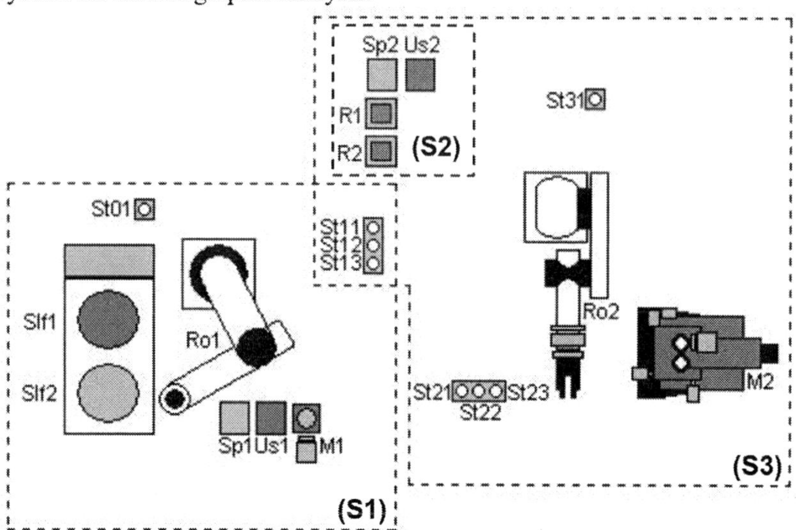

Figure 1. Automated system layout (flexible cell) preparation and analysis of metallographic samples.

Figure 1 can distinguish the following components of the automated system:

- polishing stations for samples: Slf1, Slf2;
- storage units:

- input storage system: St01;
- buffer storage system between subsystem S1 and subsystem S2: St11, St12, St13;
- sample storage devices that are prepared for analysis under the M1 microscope: St21, St22, St23;
- system exhaust storage: St31.

• industrial robots: Ro1, Ro2;
• washing stations for metallographic samples: Sp1, Sp2;
• attack stations with reagents: R1, R2;
• station (microscope) to determine the state of the M1 surface;
• M2 post (microscope) for determining the metallographic structure M2.

In terms of service by robots, those in two subsystems are served as follows:

• the subsystem (S1) which has in its composition the stations: Slf1, Slf2, Sp1, Us1, M1, St01, these being served by the robot Ro1;
• the subsystem (S2) that includes the stations: R1, R2, Sp2, Us2, these being served by the robot Ro2;
• the subsystem (S3) which includes the stations: R1, R2, Sp2, Us2, M2, St21, St22, St23, St31, these being served by the Ro2 robot.

3. Defining hierarchical levels

Two hierarchical levels of the model with colored Petri nets are defined:

- Level 1: is the model of the entire flexible system for preparation and analysis of metallographic samples;
- Level 2 contains three sub-models:
  - the sub-model of the system for the preparation of metallographic samples;
  - the sub-model of the system for reagent attack of metallographic samples;
  - the sub-model of the system for the analysis of metallographic samples.

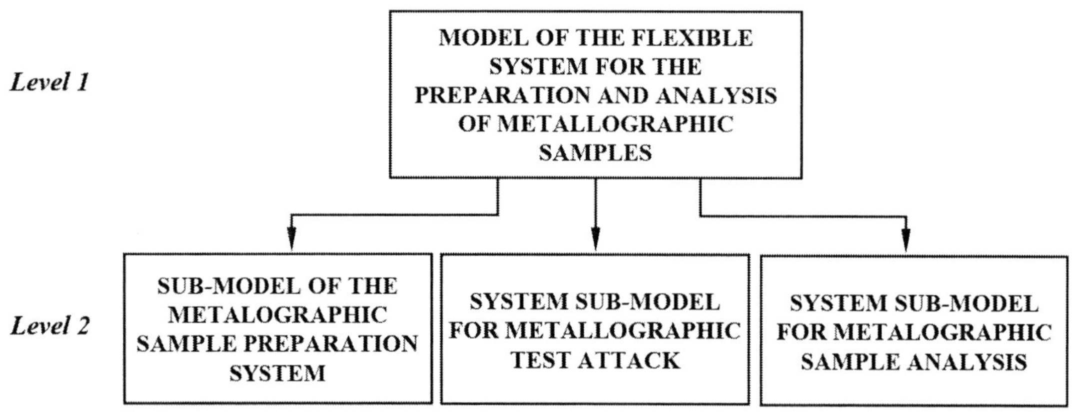

Figure 2. Hierarchical levels of the model with colored Petri nets.

This hierarchical structure corresponds to both the model with colored Petri nets with simple colors and the model with complex colors.

4. The hierarchical model with colored Petri nets with simple colors

4.1. Building the model

A first step in building the model is to define the colors that are associated with the different entities that can be identified in the system.

It is defined the set of colors that will describe the fact that the buffer storage St11, St12, St13 can store, at a given time, a single sample (*colset antiplace=with e)*:

$$antiplace=\{e\} \tag{1}$$

It is also defined the set of colors that will be associated with the samples to be analyzed in the cell. This color will be called sample and will have the following values: A for sample A and B for sample B (*colset sample=with A|B*):

$$sample=\{A,B\} \tag{2}$$

For the two robots, the robot color was defined, with two values: r1 for the Ro1 robot and r2 for the Ro2 robot (*colset robot=with r1|r2*):

$$robot=\{r1, r2\} \tag{3}$$

The color associated with the microscope with which the analysis of metallographic samples is made, is m (*colset microscope=with m*):

$$microscope=\{m\} \tag{4}$$

In the case of the CPN Tools program, each arc is assigned a variable. Thus, the variable r can have the values r1 or r2:

$$r\ \{r1, r2\} \tag{5}$$

The variable *i* can have the values A or B:

$$i\ \{A, B\} \tag{6}$$

The model with colored Petri nets contains two types of nodes: positions and transitions. They are connected by oriented arches. The sequences executed by the industrial robots, the roughing sanding, the finishing sanding, the reagent attack, the washing, the examination under the microscope are modeled by transitions. Execution of a transition (realization of a real sequence) is possible if some conditions are met. These conditions are modeled through positions. Some of the transitions and positions of the model are presented in table 1.

Table 1. Positions and transitions of the model with colored Petri nets with simple colors

Nr. crt.	Symbol	Type	Signification	Characteristics
1.	P1	Pos.	St01 stored one sample A or B	1`A++1`B
2.	T1	Tr.	Ro1 grips one sample A or B	
3.	P2	Pos.	Ro1 has one sample A or B in the grip	
...	...	...	...	...
37.	T18	Tr.	Ro2 transfers sample A or B to R1 and treats it	
38.	P21	Pos.	Ro2 has sample A or B treated by R1	
39.	T19	Tr.	Ro2 transfers sample A or B la R2 and treats it	
...	...	...	...	...
81.	T39	Tr.	Ro2 evacuates sample A or B (St23)	
82.	P44	Pos.	Storage device St31- samples A	
83.	P45	Pos.	Ro1, Ro2 are available	1`r1++1`r2

Compared to the model made with timed Petri nets [10], it is much simpler in terms of the number of positions and transitions.

After the construction of the model, the main model will be defined with colored Petri nets with simple colors and sub-models. Figure 3 shows the main model (*Level 1*) called *System*. Level 2 sub-models are highlighted: Sample *Preparation, Sample Attack and Sample Analysis*. The model and sub-models are displayed on independent pages (in windows), as follows:

- the main model is displayed on the *System* page;
- the sub-model of the system for the preparation of metallographic samples is displayed on the *Preparation* page;
- the sub-model of the system for reagent attack of metallographic samples is displayed on the *Attack* page;
- the sub-model system for the analysis of metallographic samples is displayed on the *Analysis* page.

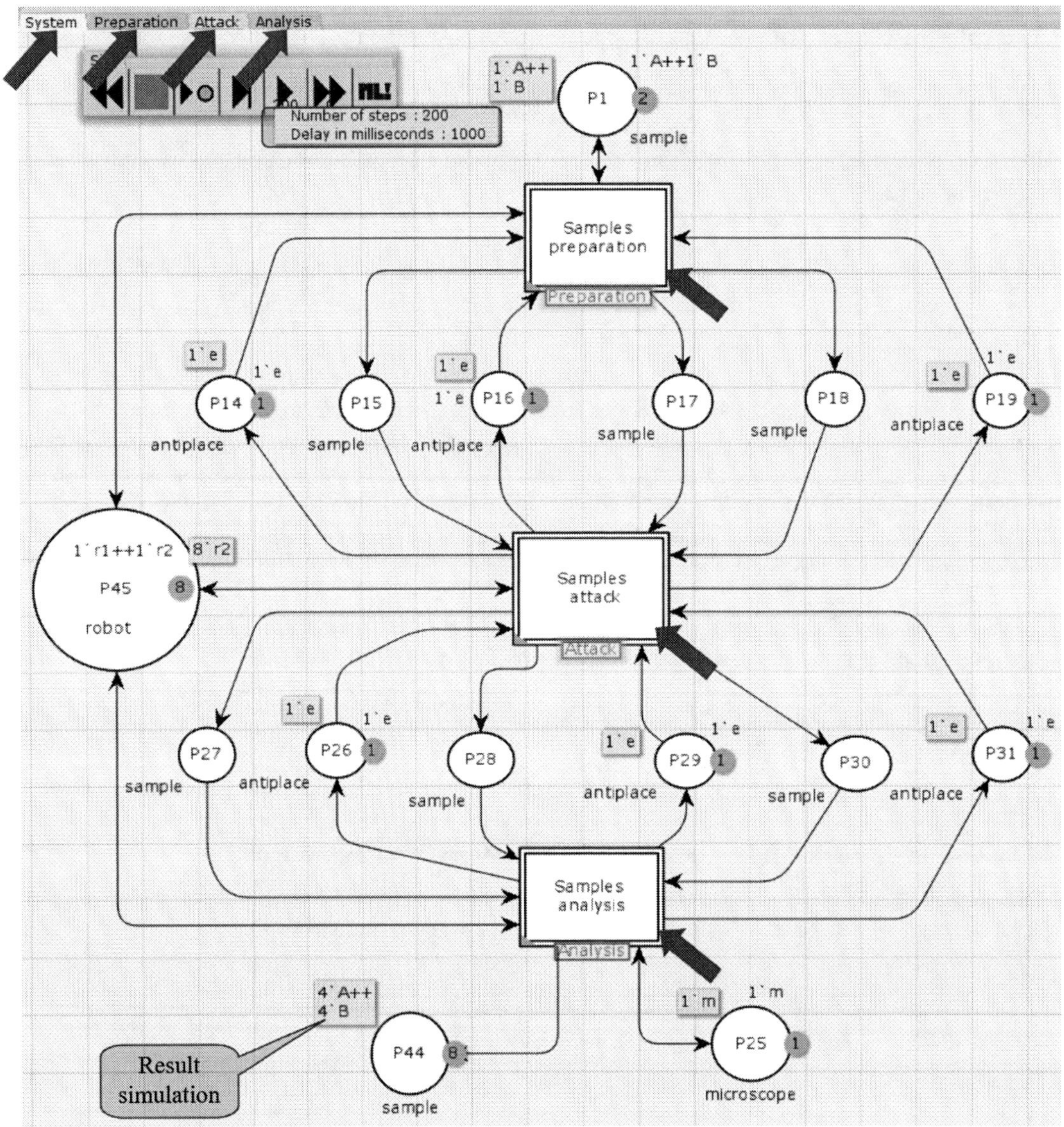

Figure 3. The main model with colored Petri nets with simple colors.

Figure 4 shows the sub-model of the system for preparing metallographic samples. The positions through which the sub-model connects to the main model are highlighted. These positions can be of three types: input positions (P14, P16, P19), output positions (P15, P17, P18) and input / output positions (P1, P45).

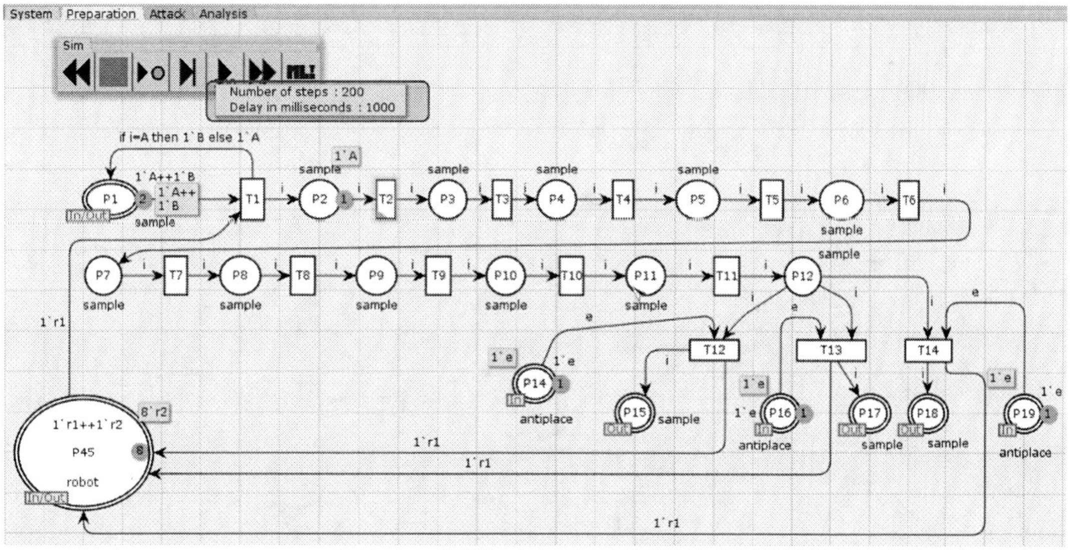

Figure 4. Sub-model of the system for the preparation of metallographic samples.

Figure 5 shows the sub-model of the system for reagent attack of metallographic samples. In this sub-model, the input positions (P15, P17, P18, P26, P29, P31), the output positions (P14, P16, P19, P27, P29, P30) and the input / output position (P45) also appear.

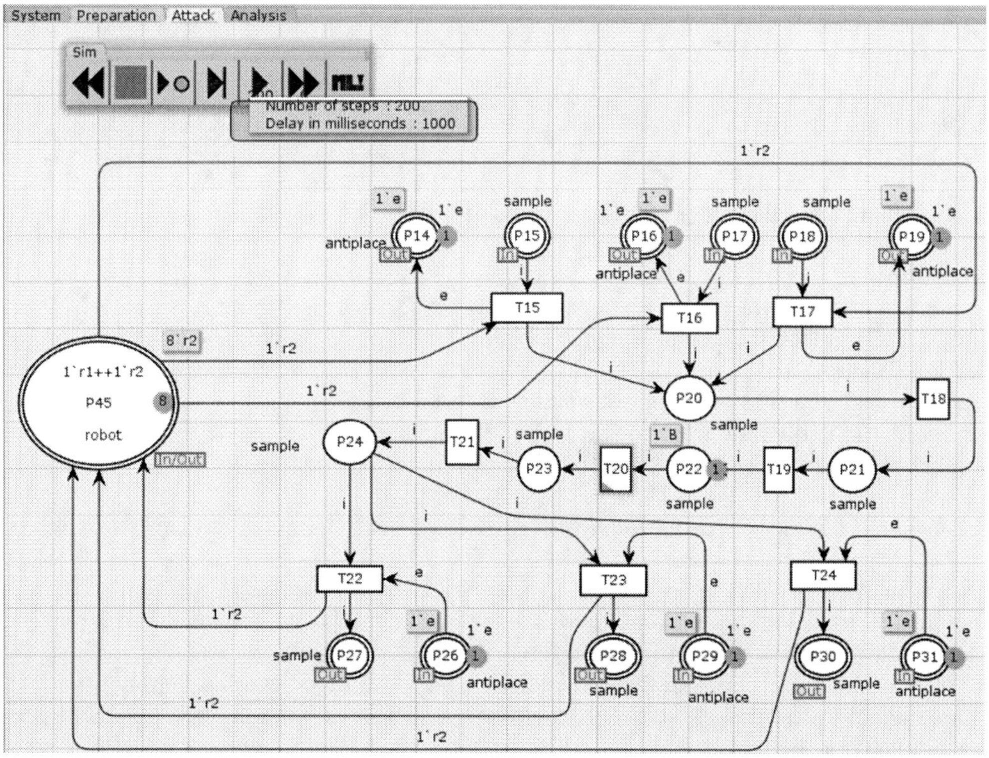

Figure 5. Sub-model of the system for the attack with reagents of metallographic samples.

Figure 6 shows the sub-model of the system for the analysis of metallographic samples. In this sub-model also appear the input positions (P27, P28, P30), the output positions (P26, P29, P31, P44) and the input / output positions (P25, P45).

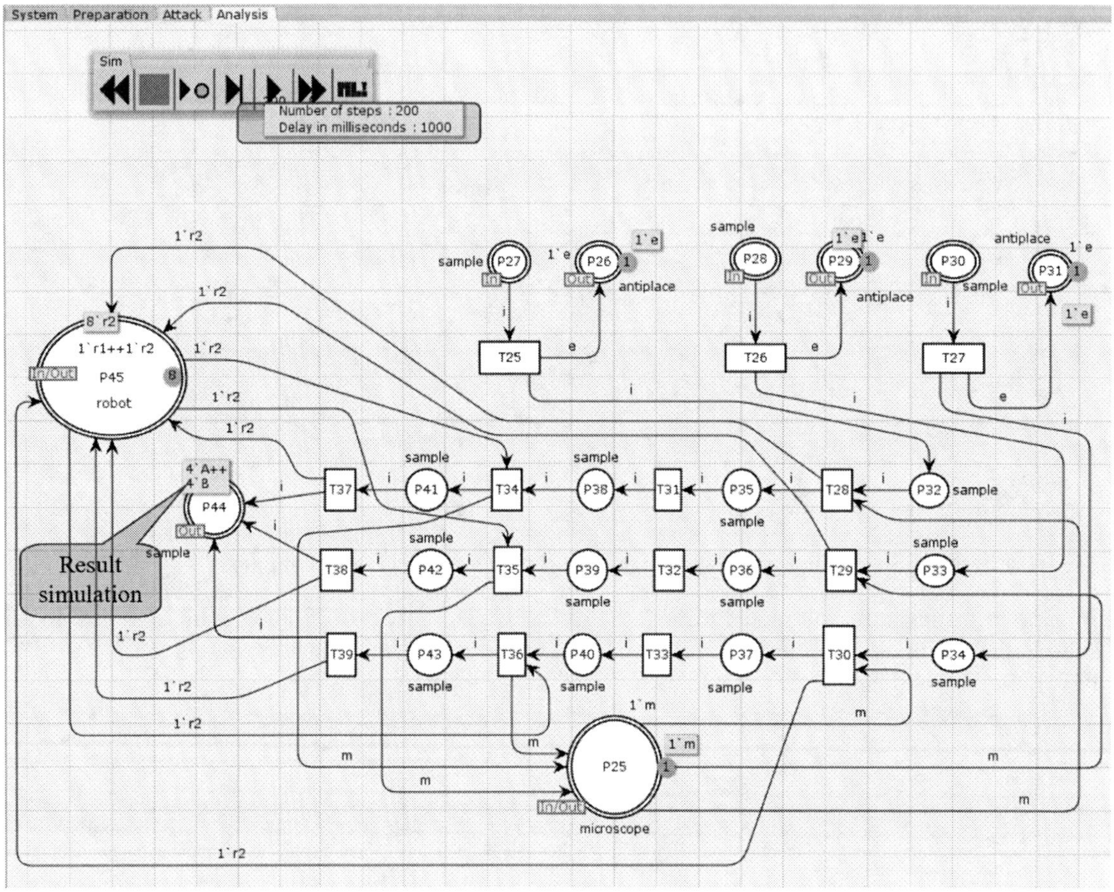

Figure 6. Sub-model of the system for the analysis of metallographic samples.

4.2. Simulation

The system simulation was performed over 200 steps (sequences) (figure 3, figure 4, figure 5 and figure 6). The result of the simulation, the number and type of samples analyzed, is presented in position P44 (figure 3 and figure 6). The simulation showed that four type A samples and four type B samples were analyzed. The two types of samples are analyzed alternately.

During the simulation, the way in which all the components of the model behave are highlighted: the main model and the sub-models. It is found that the model is viable, there are no blockages.

5. Realization of the hierarchical model with timed colored Petri nets with complex colors

5.1. Model building

The hierarchical model with timed colored Petri nets with complex colors has the structure presented in section 3. The main model is presented in figure 7.

The positions and transitions of the hierarchical model with timed colored Petri nets with complex colors are partially presented in table 2. A significant reduction in the number of elements (positions and transitions) can be seen compared to the model with simple color colored Petri nets. More precisely, the number of positions and transitions was reduced from 83 to 50.

Table 2. Positions and transitions of the model with timed colored Petri nets with complex colors

Nr. crt.	Symbol	Type	Signification	Characteristics
1.	P1	Pos.	St01 stored a sample A or B	1\`A++1\`B
2.	T1	Tr.	Ro1 grips a sample A or B	5 sec
3.	P2	Pos	Ro1 has a sample A or B in the gripping device	
4.	T2	Tr.	Ro1 transfers a sample A or B to Sf1	
5.	P3	Pos	Ro1 placed a sample A or B to Sf1	
6.	T3	Tr.	Roughing grinding sequence (Slf1)	
7.	P4	Pos.	Ro1 has the sample A or B ground on Slf1	
8.	T4	Tr.	Ro1 transfers sample A or B to Sf2	2 sec
...	...	...	...	...
23.	T12	Tr.	Ro1 places a sample A or B to St11	
24.	P13	Pos.	A sample A or B is in St12	1\`1++1\`2++1\`3
25.	P14	Pos	Available places left in St11	
26.	T13	Tr.	Ro2 grips a sample A or B from St12	5 sec
27.	P15	Pos	Confirmation- Ro2 has a sample A	
28.	T14	Tr.	Ro 2 transfers sample A or B to reagent 1 and treats it	12 sec
...	...	...	...	...
44.	T21	Tr.	Sample A or B (St21) is examined at M2	60 sec
45.	P25	Pos	Sample A or B (St21) was examined at M2	
46.	T22	Tr.	Ro2 grips sample A or B (St21) from M2	5 sec
47.	P26	Pos	Ro2 are sample A or B (St21) examined	
48.	T23	Tr.	Ro2 evacuates sample A or B (St21)	5 sec
49.	P27	Pos	Storage St1- samples A	
50.	P28	Pos	Ro1, Ro2 is available	1\`r1++1\`r2

In the case of the model with timed colored Petri nets with complex colors (figure 7), the simple colors associated with the samples, robots and the M2 microscope were defined as in the case of the model with simple colors, according to relations (2), (3) and (4).

This variant defines the set of colors that will describe the fact that the buffer storage St11, St12, St13 can store, at a given time, a single sample (*antiplace package = int 1..3*) :

$$antiplace=\{1,2,3\} \tag{7}$$

To highlight the connection between the sample type and the buffer storage on which it is stored before entering the analysis subsystem, the set of complex storage colors (*colset storage = product sample * antiplace)* is defined. This is the Cartesian product of the sample sets (relation (1)) and antiplates (relation (7)), having the following values:

$$storage=\{(A, 1), (A, 2), (A, 3), (B, 1),(B, 2), (B, 3)\} \tag{8}$$

The variables r and i are defined, also in this case, with relations (5) and (6). In the variant with complex colors, another variable is defined, e, which can have three values 1, 2 or 3. Thus,

$$e\ \{1,2,3\} \tag{9}$$

IOP Conf. Series: Materials Science and Engineering **898** (2020) 012017 doi:10.1088/1757-899X/898/1/012017

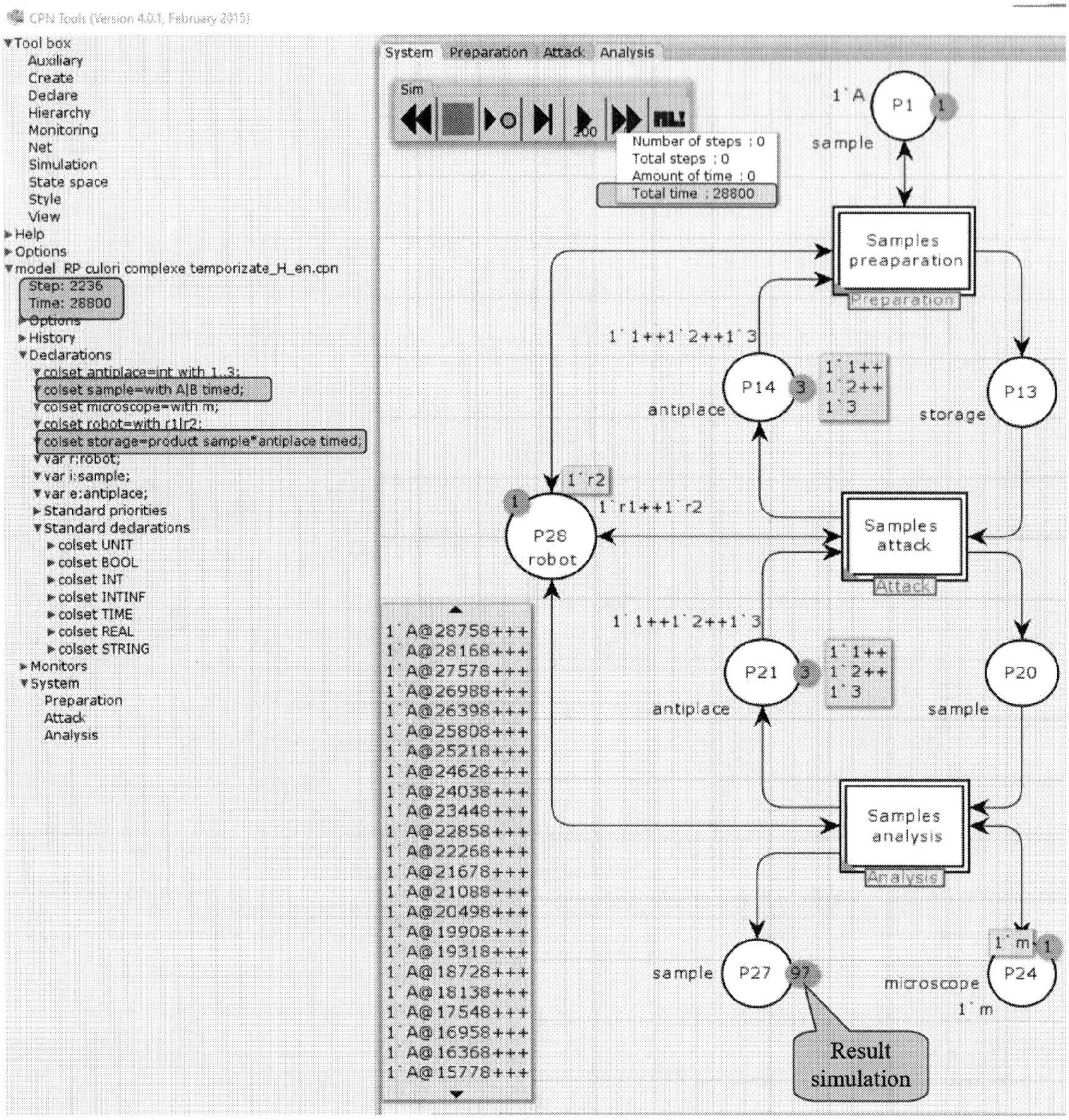

Figure 7. The main model with timed colored Petri nets with complex colors.

Also, there is a complex variable defined directly in the model: (e, i), which can take the following values:

$$(e, i) = \{(A, 1), (A, 2), (A, 3), (B, 1), (B, 2), (B, 3)\} \qquad (10)$$

The variable is associated with the arcs T12 → P13 (figure 8) and P13 → T13 (figure 9).

The introduction of the time factor in the model was done in two ways. Thus, the time factor was associated with the colors that symbolize the samples: *colset sample=with A|B timed* (figure 7) and complex *storage* colors (figure 7). Also, they were associated with timings of the transitions that model sequences (operations) of the processes of preparation and analysis of metallographic samples. For example, the T5 transition that models the finishing grinding operation (sequence) was associated with the 80 sec delay. This is materialized in the model by the expression *@ + 80* (figure 8).

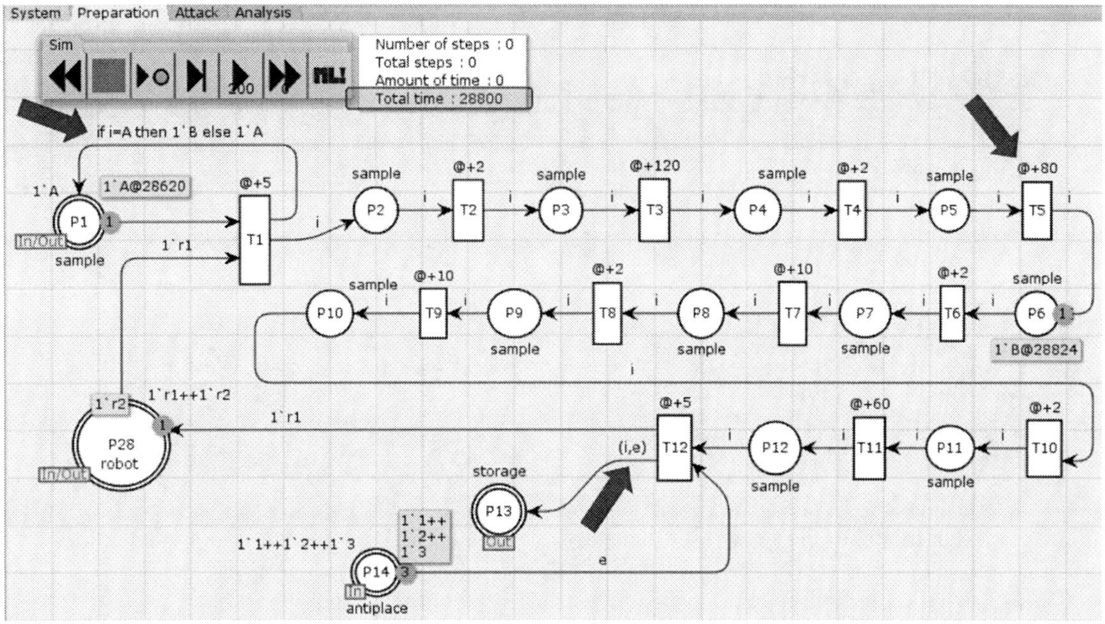

Figure 8. Sub-model of the system for the preparation of metallographic samples - complex colors.

The alternation of the two types of samples is ensured by the decisional function associated with the arc T1 → P1 (figure 8):

$$if = A \text{ then } 1`B \text{ else } 1`A \tag{11}$$

In the case of the hierarchical model with colored Petri nets timed with complex colors, as in the case of the model with simple colors, the interconnection between the main model and the sub-models is made through input positions, output positions and input / output positions.

Figure 8 shows the sub-model of the system for preparing metallographic samples. Figure 9 shows the details of the model of the reagent attack system of the metallographic samples.

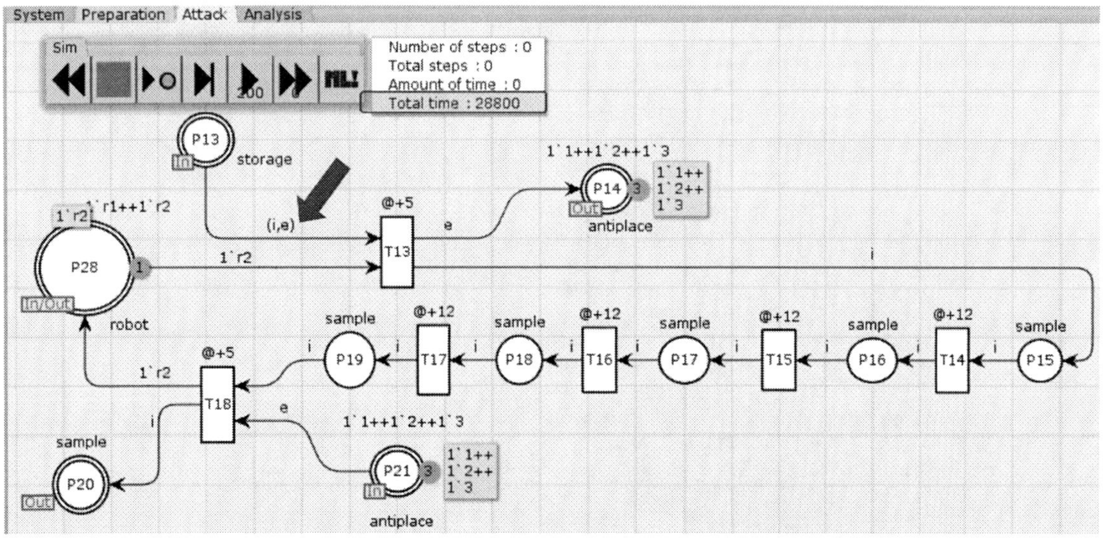

Figure 9. Sub-model of the system for the attack with reagents of metallographic samples - complex colors

IOP Conf. Series: Materials Science and Engineering **898** (2020) 012017 doi:10.1088/1757-899X/898/1/012017

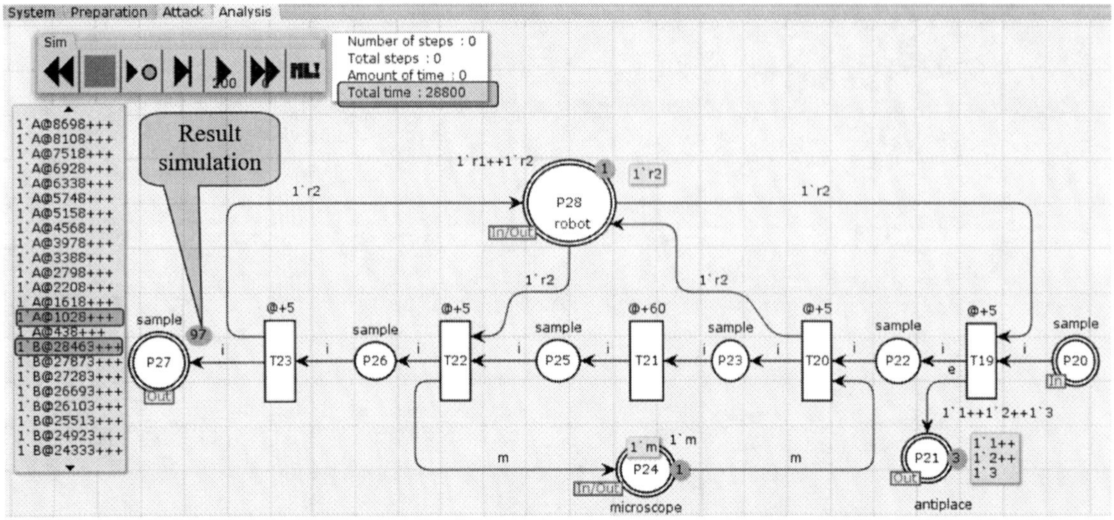

Figure 10. Sub-model of the system for the analysis of metallographic samples - complex colors

Figure 10 shows the sub-model for the analysis of metallographic samples.

5.2. Simulation

For simulation, a time interval of 8 hours (28800 sec) is considered. The simulation duration is set in the *Simulation* window (figure 7, figure 8, figure 9 and figure 10). During the simulation, 2236 steps were performed (figure 7). The program also provides information on when the analysis of each sample is completed. For example, the last type B test was completed at time 28463, and the second type A test at the time (figure 10).

It can also be noted that during the 8 hours a number of 97 samples were analyzed (figures 7 and figures 10).

When introducing the time factor into the model, the CPN Tools software allowed the generation of a report (figure 10) that highlights the times at which each transition is executed. In other words, one can identify the moments at which each sequence (operation) is executed in the real system.

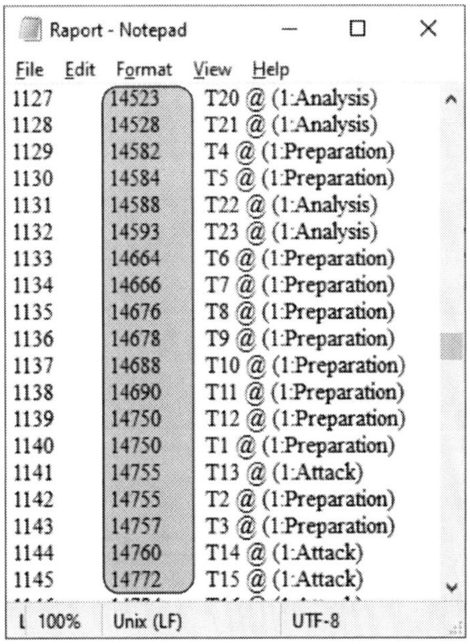

Figure 11. Simulation report. Execution moments of transitions.

IOP Conf. Series: Materials Science and Engineering **898** (2020) 012017 doi:10.1088/1757-899X/898/1/012017

6. Conclusions

Modeling and simulation with colored Petri nets are effective tools for evaluating the performance of automated systems.

The paper shows how colored Petri nets were used for modeling and simulation: modeling an automated system for preparing and analyzing metallographic samples.

The token can be interpreted as a parameter that can take different values. In the analyzed case, two types of samples were considered. The structure of the model is simplified (the number of positions and transitions decreases significantly) and more when using complex colors.

In the case of ordinary timed Petri nets, timings are associated with transitions (T-timed Petri nets) or positions (timed P-Petri nets). In the model described in the paper, the timings were associated with colors and transitions. The number of samples analyzed over an eight-hour time period was highlighted by simulation. The introduction of timings allows the identification of the moments when the analysis of each sample is completed.

By using the model with timed colored Petri nets, the sequences executed by the components of the real system over the considered time interval can be monitored.

The hierarchical organization of the models with colored Petri nets allowed the highlighting of the way in which each subsystem works within the automated system for the preparation and analysis of metallographic samples.

Future research will look at how models with fuzzy colored Petri nets. These will give the models a higher degree of accuracy compared to the real system.

7. References

[1] Madison J, Spowart J E, Rowenhorst D. J., Fiedler J. and Pollock T. M. 2008 Characterization of three- dimensional dendritic structures in nickel-base single crystals for investigation of defect formation *Proceedings of the Superalloys Conference* Roger C. Reed and at.Champion Pennsylvania USA The Minerals, Metals & Materials Society pp 881-888

[2] Vesselényi T, Barabás T and Moga I 2004 *Rob. & Man. Inter. J* **9** 37

[3] Blaga F, Stanasel I, Pop A, Hule V and Buidos T 2014 *A. U. O. Fasc. Man. Tech. En.* **23** 299

[4] Saren S K, Blaga F and Vesselenyi T 2018 IOP *Conference Series: Materials Science and Engineering* Oanta E and at Constanta Romania IOP Publishing 17

[5] Saren S K, Blaga F, Dzitac S and Vesselenyi T 2017 *Procedia Computer Science* Ahuja V Shi Y Khazanchi D Abidi N Tian Y Berg D and Tien M J New Delhi Elsevier B.V. 122 pp 253-260

[6] https://highresbio.com/laboratory-automation/range/ accessed on January 2019

[7] https://www.hseag.com/about-hombrechtikon-systems-engineering-ag/ accessed on January 2020

[8] https://www.picoquant.com/materialsscience accessed on February 2020

[9] http://saxslab.com/materials-science/automation/ accessed on February 2020

[10] Blaga F S, Vesselenyi T, Ursu M P, Hule V and Stanasel I 2019 IOP *Conference Series: Materials Science and Engineering* Raval K R and at. Iasi Romania IOP Publishing 10

IOP Conf. Series: Materials Science and Engineering **898** (2020) 012018 doi:10.1088/1757-899X/898/1/012018

Coordinate Measuring Machine thread position measurement analysis

P E Serban[1], F Peti[2]

[1] Metrology Research, Quality Department, CIE Matricon SA, Gh Doja street, no.155, Targu Mures, Romania

[2]Industrial Engineering and Management Department, Faculty of Engineering and Information Technology, University of Medicine, Pharmacy, Sciences and Technology "George Emil Palade" of Targu Mures, N. Iorga street, no. 1, Targu Mures, Romania

[1] serban.petru@yahoo.com

Abstract. Measuring the true position of a cylinder with a Coordinate Measuring Machine became a simple fact for programmer developer, for this they have to choose between different strategies available on the CMM software to improve measurement accuracy. Everything is changing when we try to measure a true position threaded hole, during serial production this position will pass between some changes that will affect the profile of the thread and the beginning of the threaded hole. Although CMM cylinder strategies methods exist they are not efficiently developed for precise measurement of a thread position. This article will analyse the difference between current methods that are available on the market and will compare the results with an experimental new method, this new method is developed from the method available plus other tools available in the CMM software. The importance of this method is to increase thread position measurement precision during all production cycle time.

1. Introduction

Due to the technological advancement, which accelerates the development of new systems and methods for final product, the design of the finished products has reached a point that increasingly demands for accuracy of the execution and assembly to be better controlled. The impact of these requirements have a direct influence on the development steps of final product that in the end lead to the development of more precise control methods.

Since the final products are made up of assembled parts, by default they will have threaded holes to can facilitate the assembly's processes. The control on the CMM's of these holes is a challenge that increases once with the drawing tolerances. In Automotive Industry we have a variety of parts where we can find more and more threaded holes due to the high machining precision accuracy. The engine blocks can have easily up to 100 threads, or more, that must be controlled during serial production so that the process can be validated and guaranteed.

The current state of the technology offers two different types of approach for this measurement strategy: measuring by using special centering pins and direct measurement on the threaded surface, both types of measurement have pro and cons.

In this article we will present the traditional methods of measurement and a new method developed for the efficiency and accuracy of these types of measurements.

2. Material and methods

For this article we'll use metric threaded holes, this type of thread being the most common thread used in aluminum machined parts that are used in Automotive Industry.

According to the actual level of knowledge, in the measurement process of a thread we always have to take some initial decisions before measuring the threads. Forming threads are cleaner but need to be formed complete and the wear of the tool creates incomplete threads, milling threads have a clear cut true the hole and can present porosity or to have a lot of chips. Threads must be cleaned.

The actual research has a single configuration of the CMM machine so that the results of the measurement to be evaluated only from measurement strategy point of view. The diameter of the probe is very important for thread measurement process. If we use a probe with a small ruby ball we can risk of having measurement errors by not have contact only with the active probe but also with the probe body, if we use a big ruby ball will not be relevant for this research because the focus is to go closest to the pitch diameter.

The tests were made on the same samples threaded holes with the same alignment system and with the same scanning parameters. Our goal is to increase the measurement precision with a combination of parameters that are already available on the software but didn't communicate as a single element.

The tests are divided in four categories, tree of them are traditional measurement methods and the fourth is a combination of all tree initial tests:

2.1. Method 1 (Method 2.1.1 and 2.1.2)

We measure thread True Position (TP) by measuring a cylinder formed by 2 circles.

2.1.1. Normal circles. The 2 circle are scanned at 4 mm between them (figure 1).

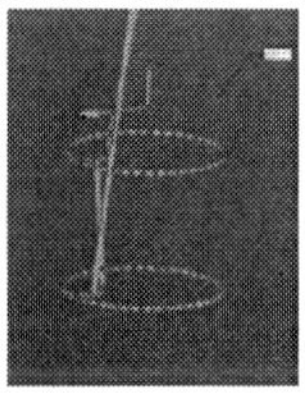

Figure 1. Cylinder created from two circles.

2.1.2. Pitch circles. The 2 circle are scanned at 4mm between them and the rotation will follow the thread rotation direction and pitch (figure 2).

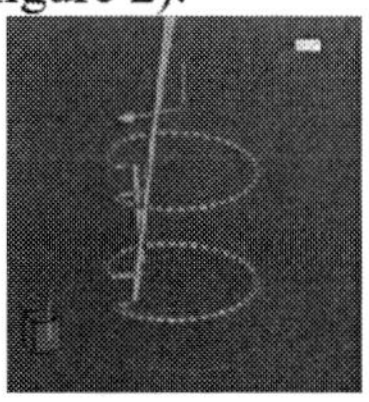

Figure 2. Cylinder created from two circles that are set to have an elliptical move with an increment that follow the pitch of the thread.

2.2. Method 2 (Method 2.2)

We measure thread TP by measuring a cylinder formed by an elliptical revolution. The ellipse will be created from 4 revolutions that will follow the thread rotation direction (figure 3).

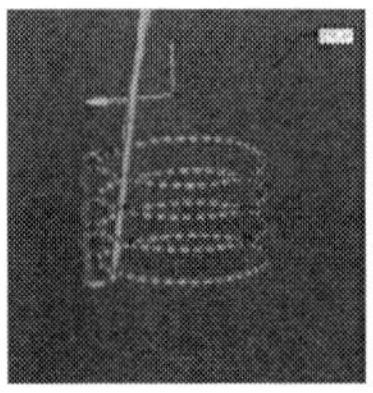

Figure 3. Cylinder created from an elliptical move with an increment that follows the pitch of the thread.

2.3. Method 3 (Method 2.3.1 and 2.3.2)
We measure thread TP by measuring lines.

2.3.1. Cylinder created from 3 lines. The lines will be scanned symmetrical and the length will be the same as in method 2 (figure 4).

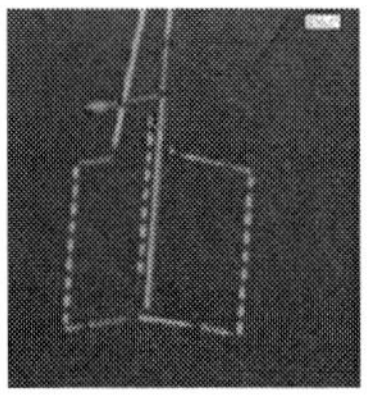

Figure 4. Cylinder created from three symmetric lines.

2.3.2. Cylinder created from 6 lines. The lines will be scanned symmetrical and the length will be the same as in method 2 (figure 5).

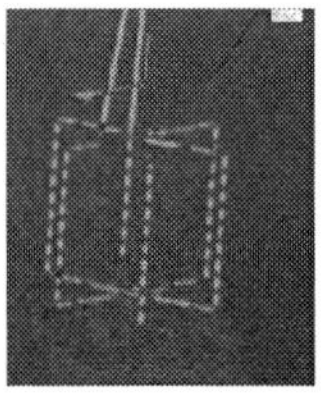

Figure 5. Cylinder created from three symmetric lines.

2.4. Method 4 (Method 2.4)
We measure thread TP by combining the methods described above and by applying a combination of data available in the measuring software. The thinking is to create a cylinder (figure 9) that use pitch elliptical revolution steps as in 2.2, the start point of this cylinder is established in 3 steps: (1) we measure a circle like in Method 1 (figure 6) to determinate the center of the thread, next we will create an evaluation line (figure 7) on the length of the cylinder like in Method 3, the linearity evaluation (figure 8) of the line will indicate the pitch start point (4), the last point is established as the start point for evaluated cylinder (figure 9).

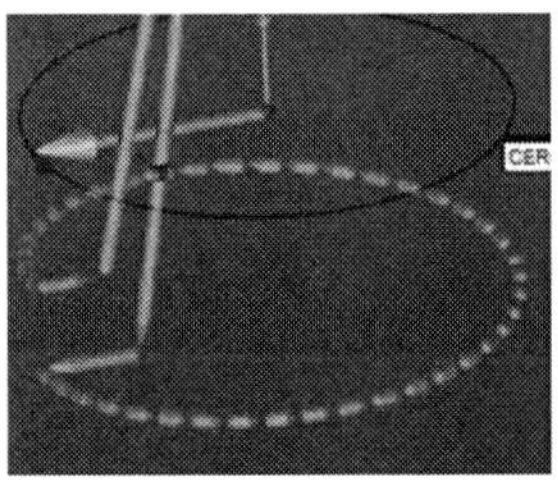

Figure 6. Step 1 represent the initial circle used to centre the measurement.

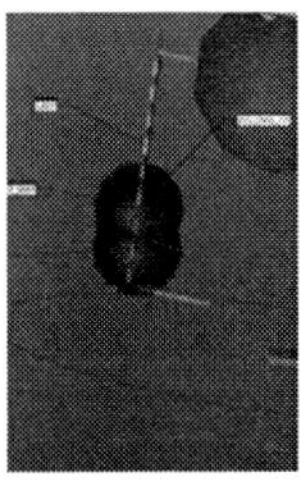

Figure 7. Steps 2 represent a directrix line of the cylinder.

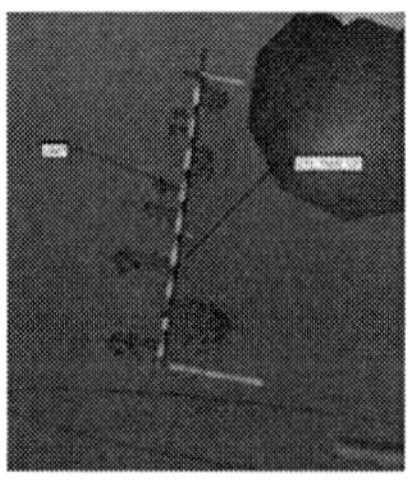

Figure 8. Steps 3 represent the identification of the pitch diameter.

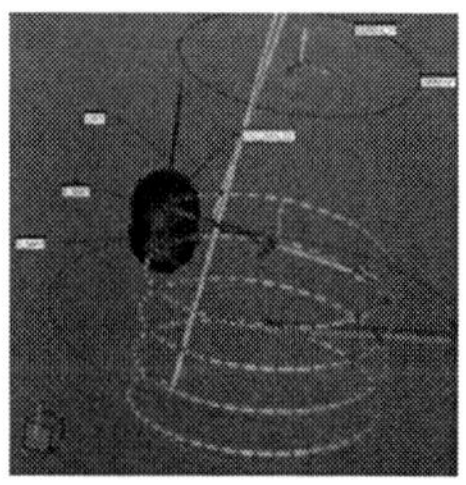

Figure 9. Steps 4 represent the elliptical revolution on the pitch diameter.

3. Results and discussions

The methods described above where evaluated by measuring a device represented in figure 10 that had five M6x1 threads where we measured for tree times same locations.

Figure 10. Device used for evaluate measurement strategy.

3.1. Graphic description for each Zone and each Method

The results are presented by centering all the deviations and by marking each method with a unique color presented in figures: 11, 12 and 13.

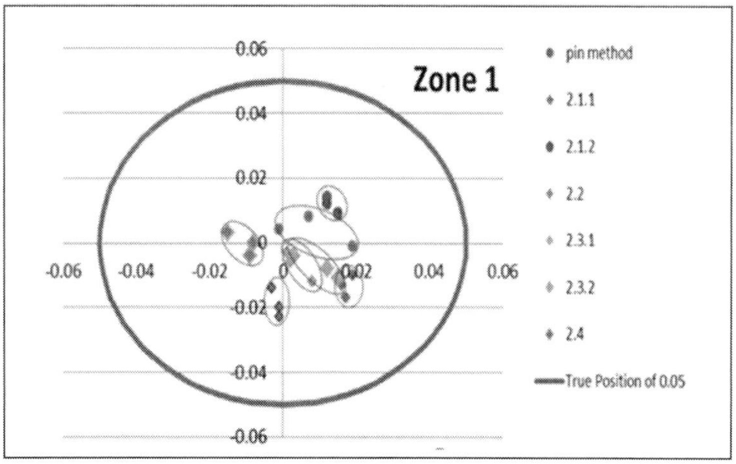

Figure 11. Zone 1 indicate all the measured points deviated from the nominals.

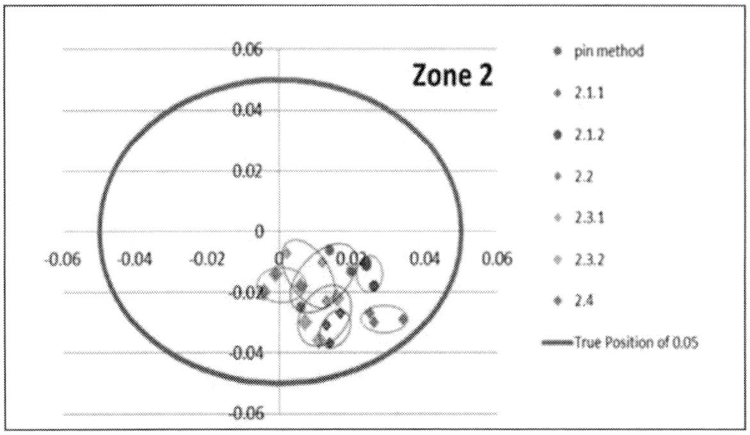

Figure 12. Zone 2 indicate all the measured points deviated from the nominals.

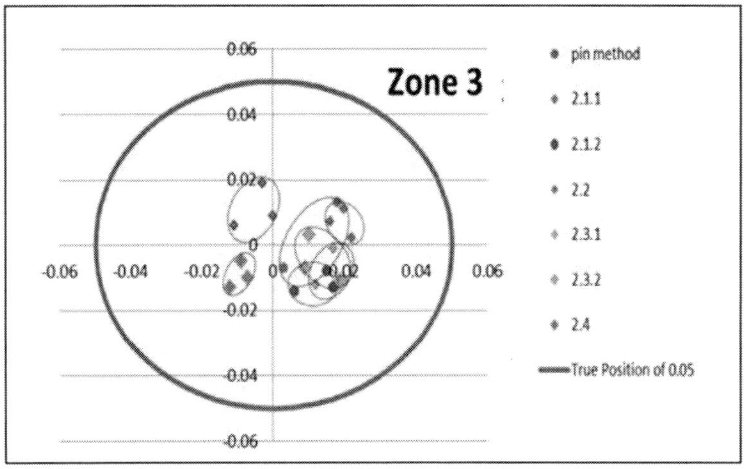

Figure 13. Zone 3 indicate all the measured points deviated from the nominals.

The results are presented by centering all the deviations and by marking each zone with a unique color and are presented in figures: 14 till 19, figure 20 describe each zone marking color. We used 5 zones to can have a better understanding of the distribution of each measurement.

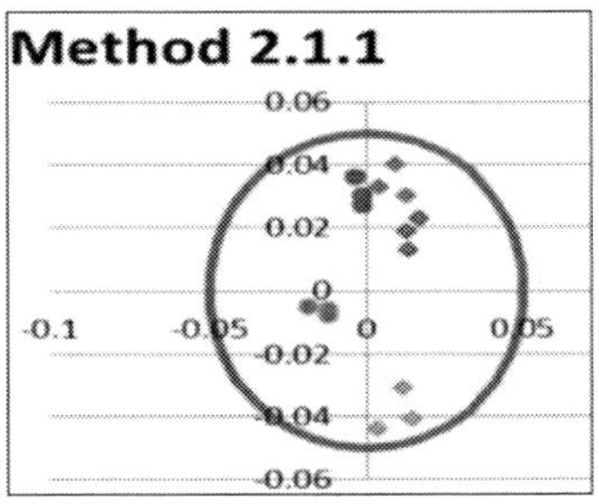

Figure 14. Measurement dispersion for normal circles.

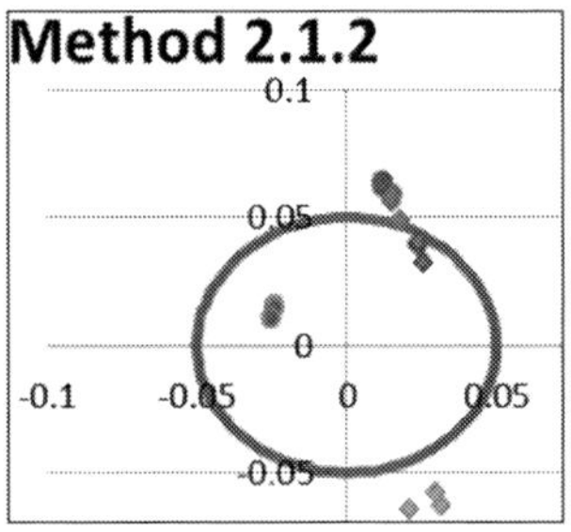

Figure 15. Measurement dispersion for helicoidally circles.

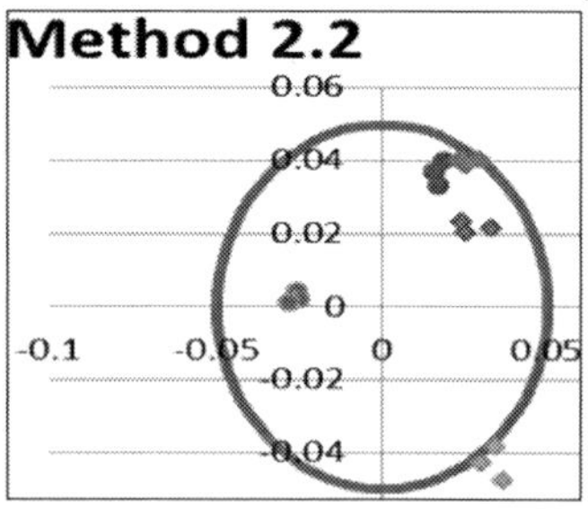

Figure 16. Measurement dispersion for elliptical cylinder.

IOP Conf. Series: Materials Science and Engineering **898** (2020) 012018 doi:10.1088/1757-899X/898/1/012018

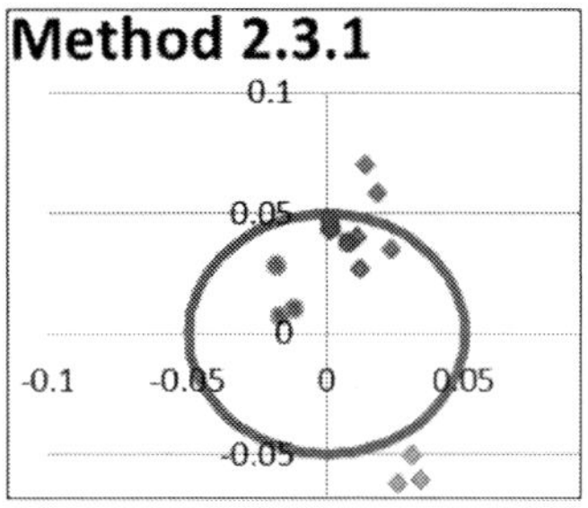

Figure 17. Measurement dispersion for 3 lines cylinder.

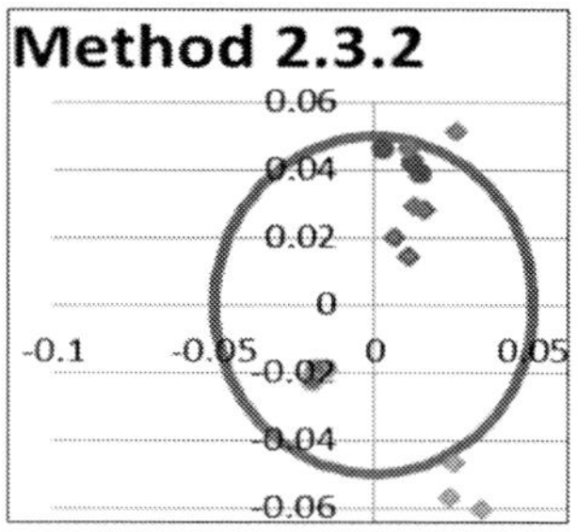

Figure 18. Measurement dispersion for 6 lines cylinder.

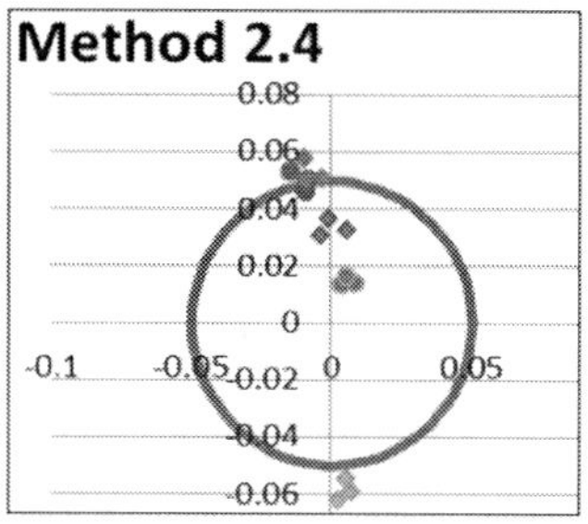

Figure 19. Measurement dispersion for the new cylinder.

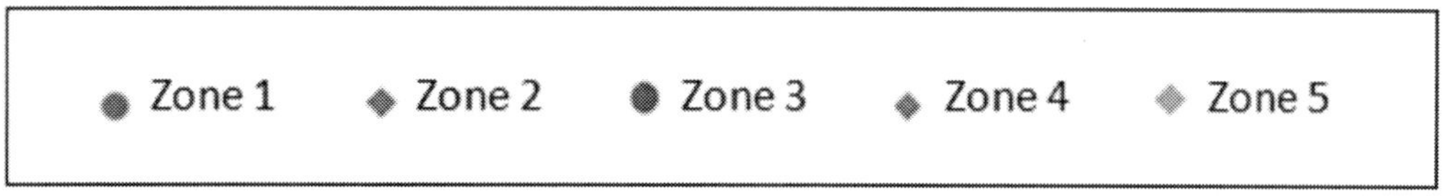

Figure 20. Color legend used to can identify in each Zone de deviations

3.2. Measurement accuracy
The results dispersion from the nominal values shows that each method is located in a different location. This creates a measurement error that influences the results depending on what measurement method was used. The method 2.4 is the most closest to the theoretical values and indicates this for all tree measurement sets.

3.3. Measurement repeatability
The results dispersion graphics made for each method separately identify that even if we use same measurement program the repeatability dispersion exist. The most repeatable methods are 2.1.2, 2.2 and 2.4. We know that all this methods use the spiral measurement strategy with the pitch movement.

4. Conclusions
We start this research with the thinking that the measurements where performed on a part where the machining tool is new, but in serial use, all the tolls will have a cycle time and the measurement must measure in good conditions during all cycle time of the tool.

From the results presented above we have identified that measurements where we use the elliptical revolution movement that follow the pitch step give us the best results. But from our experience we found out that this is not enough, because during the production other factors can influence our surfaces: tool wear, porosity, bad setup between drill ant thread tools, vibrations, etc. This was the start of creating method 2.4 where we are looking to measure always as near as possible the pitch diameter.

Measuring closer to the pitch diameter we'll have, no matter what is the cycle time or the type of the tool, a more accurate measurement then in the other methods. The start of the pitch will constantly be changed during a lifetime of a program, but the measurement program not.

Examples of good and bad surfaces that will have a major impact for each measurement method used are presented in figures 21, 22, 23 and 24.

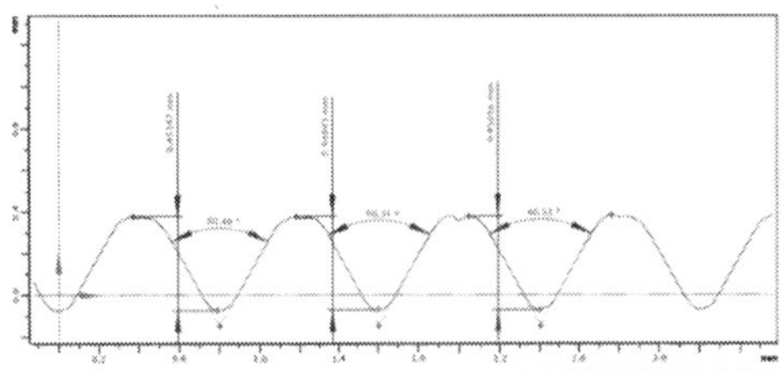

Figure 21. Thread section profile, new tool.

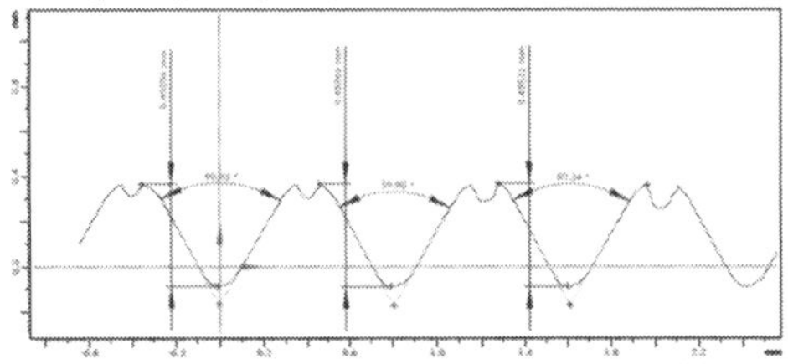

Figure 22. Thread section profile, slightly worn tool.

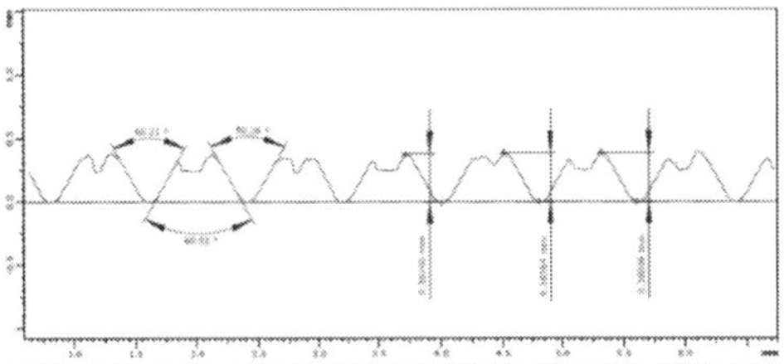

Figure 23. Thread section profile, hard worn tool.

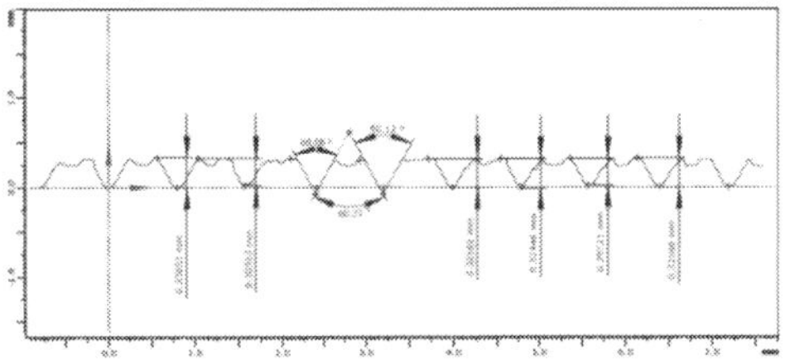

Figure 24. Thread section profile, damaged tool

References

[1] P Martikáň, M Drbúl, J Holubják, J Mrázik, R Joch, *The issue of determining the geometric position deviation of the threaded holes,* Advances in Science and Technology Research Journal, Dec. 2016, pages 47–52

[2] A Yüksel, T O Kılınç, K B Sönmez, and S Ö Aktan, *Comparison of internal and external threads pitch diameter measurement by using conventional methods and CMM's* 19th International Congress of Metrology, 09001 (2019)

[3] www.hexagon.com, data: 03/2020

[4] PC-Dmis 2019R1 Help Manual, data:03/2020

[5] D T Harris, *Screw threads*, Glastonbury Southern Gage, January 2003

IOP Conf. Series: Materials Science and Engineering **898** (2020) 012019 doi:10.1088/1757-899X/898/1/012019

Numerical analysis of the influence of the relative position of the drone's rotor inside the nacelle on the airflow dynamics

A A Sirca[1], F Mariasiu[1], B O Varga[1], S Morariu[2]

[1]Technical University of Cluj-Napoca, Department of Automotive Engineering and Transports, B-dul.Muncii 103-105, Cluj-Napoca, ROMANIA
[2]Inovo Finance SRL, Cluj-Napoca

andreea.sirca@auto.utcluj.ro

Abstract. The use of drones in different economic sectors has become a usual thing at present, being present in a varied and diverse range from a constructive point of view. A special attention is given for agricultural sector where drones can be used for: plant protection and pest control, vegetative crop surveillance, disease and pest estimation, surveillance, etc. However, the big problem that characterizes the performance of a drone remains the flight autonomy, autonomy directly related to the drone's own weight and the configuration of the propulsion system. The paper aims to use computerized numerical analysis methods to study the effect of the relative positioning of a drone's rotor, inside the nacelle, to characterize (and study) the flow characteristics of the airflow. The parameters considered in the study for the comparative evaluation of the research hypothesis are related to the parameters that characterize the dynamic flow of airflow (airspeed and pressure). Based on the results obtained the researches can be further developed by modifying the geometric construction of the rotor nacelle in order to optimize the flow of airflow used for crops' spraying process.

1. Introduction

One of the problems that the agricultural field has faced and continues to face is to limit the losses of agricultural crops due to the effect of diseases and pests. In this sense, chemical solutions are used to reduce and / or eliminate these negative effects. The method of treating agricultural crops is based on the sprinkling process with the necessary liquid of the plants, generally the process being a mechanical one, with agricultural installations and equipment dedicated to this operation.

Unmanned aerial vehicles (UAVs, drones) have been rapidly implemented (and seen in the future) as a new technology in agriculture in the fields of: plant protection and pest control, vegetative crop surveillance, disease and pest estimation, surveillance, etc.

In the field of plant protection and pest control, the most widely used function of UAVs is to automatically perform crop spraying processes. Still these agricultural works are performed on relatively small agricultural surfaces due to the autonomy/total weight of the drone.

However, in estimating the efficiency of the sprinkling process of agricultural crops besides the productivity offered by the use of a drone, it must also be taken into account the important factors that affect the deposition of drops of liquid used on the surface of the leaves: the speed and the height of the drone's displacement, the temperature of the environment. and liquid solution, wind speed, propeller pitch, spray nozzle shape and type, etc. For this purpose, many researches have been made that have studied the influence of these factors on the quality of the primary process of spraying process, both experimental and research carried out by means of numerical analysis [1-6]. Also, lately there are also researches that study the possibility of implementing the Artificial Intelligence techniques in order to increase the efficiency of the use of UAV in the process of sprinkling agricultural crops [7].

The general feature of these researches was that they were made for a classic UAV construction, which consists of a common splash ramp (with a predetermined number of spray nozzles) positioned at the bottom of the UAV. The liquid sprayed through the nozzles is driven to the plants by the random airflow created by the blades of the drone's rotors.

The present paper intends to study the possibility of replacing the spray common ramp with a system embedded in the drone rotor's nacelle. In order to obtain primary data related to the subsequent feasibility of the project, computerized numerical analysis methods were used, simulating the cases in which the drone rotor has 2 and 3 blades in construction. Also, two additional cases of analysis were considered, namely when the rotor is positioned higher respective lower from vertical axe of symmetry in the construction of the nacelle. The quantities considered for the efficiency analysis were the flow velocity and the air pressure to decide further the optimal placement of spray nozzles.

2. Material and Method

For the numerical analysis of the velocity and pressure fields caused by the air flow considering different relative position the rotor of a drone (with 2 and 3 blades in construction), a model was developed using Solid Works software (figure 1). The general geometrical dimensions of the model are: inlet platform diameter 550 mm, output platform diameter 550 mm, platform length 550 mm, blade diameter 450 mm, thickness of the platform walls: upper part 16 mm, lower part 1.58 mm. As main boundary conditions, the density of air (fluid flow) is considered to be 1.205 kg/m^3 (corresponding to ambient temperature of 293.2 K). Drone's rotor speed is set to 2387 rpm for both 2 and 3 blades configuration.

The analysis of the velocity and pressure fields of the air flow was performed for the cases in which the rotor has 2 and 3 blades in design, and the measurements of the obtained values were made both in in the vertical plan of rotor's nacelle.

a) b)

Figure 1. Simulation model for 2-blade case construction of rotor (a-lower positioning; b-upper positioning).

3. Results and Discussions

The computerized numerical analysis (CFD) was applied for following distinctive cases: use of 2 and 3-blade in rotor's construction and upper and lower positioning of rotor related to the symmetry of vertical axe of nacelle. The results obtained from CDF simulation are presented in figures 2-5, analyzed and interpreted from two major parameters point of view, parameters that characterize the air-flow pattern (air speed and relative pressure). Comparative variations on nacelle's vertical axe of air speed and relative pressure for considered cases are presented in figures 6 and 7.

IOP Conf. Series: Materials Science and Engineering **898** (2020) 012019 doi:10.1088/1757-899X/898/1/012019

In Table 1 are presented the results for both configuration of rotor taking into consideration the average and maximum values, and the values registered in horizontal plan at the exit of rotor's nacelle are presented in Table 2.

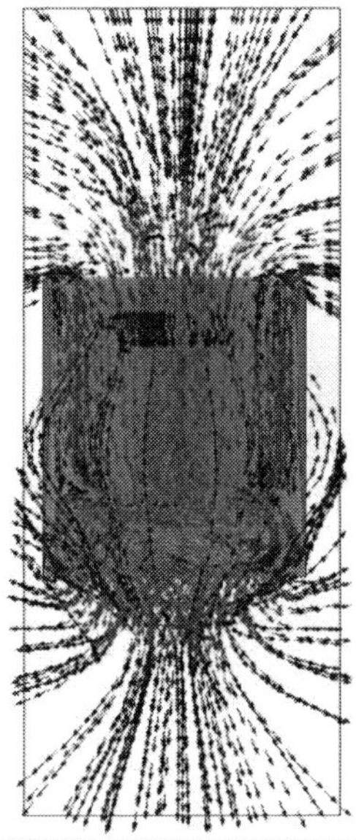

Figure 2. Velocity field in vertical section of nacelle (2-blades rotor case lower positioning).

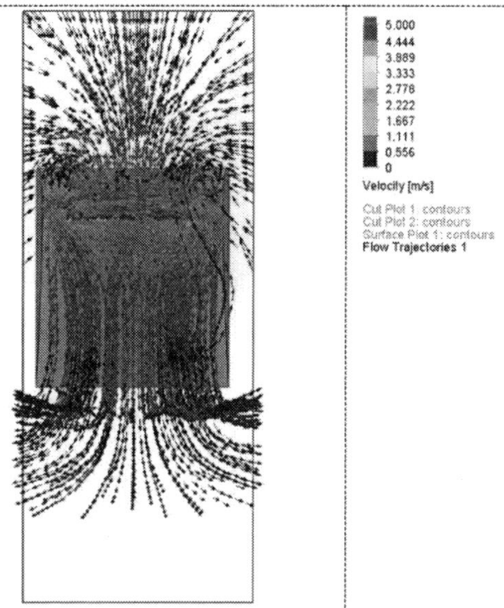

Figure 3. Velocity field in horizontal section of nacelle (2-blades rotor case upper positioning).

Table 1. Air flow characteristics in vertical plan.

	Air speed [m/s]				Air relative pressure [Pa]			
	upper positioning		lower positioning		upper positioning		lower positioning	
Case	2-blade	3-blade	2-blade	3-blade	2-blade	3-blade	2-blade	3-blade
Average	0.78116	1.257491	0.909947	1.819513	101308.2	101312.3	101316.3	101314.2
Maxim	1.895055	2.658613	1.899145	3.548052	101324.6	101334.5	101322.0	101321.4

Table 2. Air flow characteristic at the exit of rotor's nacelle (horizontal plan).

	Air speed [m/s]				Air relative pressure [Pa]			
	upper positioning		lower positioning		upper positioning		lower positioning	
Case	2-blade	3-blade	2-blade	3-blade	2-blade	3-blade	2-blade	3-blade
Value	0.651039	0.76612	0.942191	0.209003	101324.6	101333.7	101322.0	101319.6

Figure 4. Pressure field in vertical section of nacelle (3-blades rotor case lower positioning).

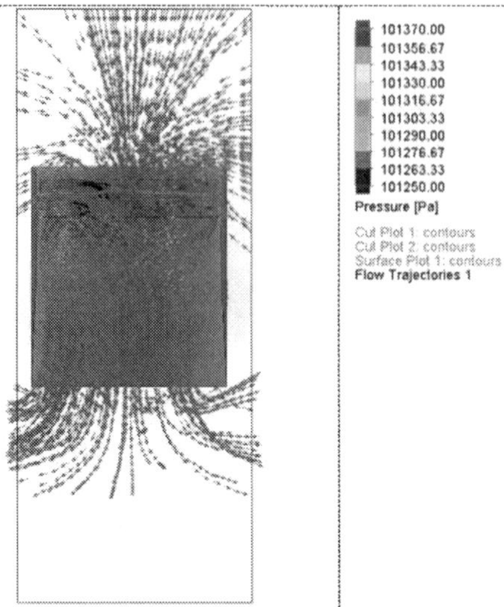

Figure 5. Pressure field in vertical section of nacelle (3-blades rotor case upper positioning).

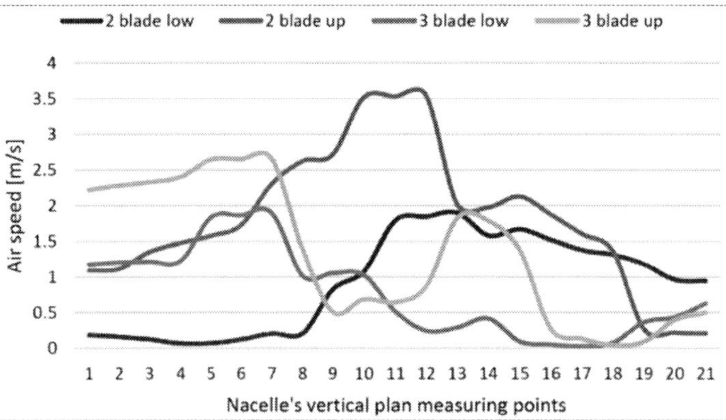

Figure 6. Comparative variation of fluid air speed vs. rotor's positioning and blade configuration

From the CFD analysis, it is observed that there is a characteristic of airflow, directly related to the positioning of the rotor in the vertical plan of the platform. Regardless of the construction version, with 2 or 3 blades, if the rotor is positioned lower, the air flow is concentrated towards the center of the horizontal section of the platform. In exchange for a higher positioning of the rotor, the air flow at the exit of the platform is made towards the circumference (edge) of the horizontal section of the platform.

From the point of view of the measured values (Table 1) it is observed that between the average air speed between the two constructive cases of the rotor considered for the construction with 2 blades the value is with approx. 50% lower than in the case of the construction with 3 blades, for the lower positioning. For the upper position of the rotor the difference is 38% smaller. It should be specified that the maximum speed values are obtained in the blade rotation area.

Similarly, in the case of air pressure (Table 1) the measured values are approximately equal (with relative differences between them less than 1%) both for the constructive variants of the 2 or 3 blade rotor and for the cases related to the relative positioning of the rotor inside. nacelle.

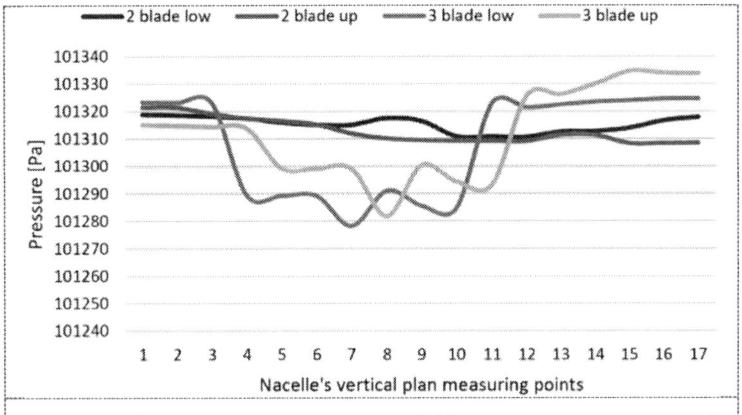

Figure 7. Comparative variation of fluid flow pressure vs. rotor's positioning and blade configuration

When using the model presented for agricultural operations related to crop spraying, it is important to analyse the air flow parameters at exit from the nacelle. Based on the results presented in Table 2 it can be stated that the highest value of air velocity is for the case of 2 lower positioning blades (0.94 m/s) and the smallest value is for the case for the case of 3 blades (0.21 m/s).

The relative difference between these values is 77.6%. In the case of pressure, the measured values are approximately equal (as in the case of speed), the highest value is 101333.7 Pa (lower positioning) and the smallest value of 101319.6 Pa (upper positioning) is recorded for the case of 3 blades.

4. Conclusions

There is a continue development of UAVs applications in agriculture due to the immediate and specific benefits that can be offered. Still, being an emergent technology there are a lot of supplementary researches to be made to optimize their functionality and exploitation.

In the case of using UAVs for plant protection activity (crop's spraying) there are necessary to have a enough autonomy of work (depending of overall UAV's total weight), an uniform fluid flow pattern transmitted to plant surfaces and possibilities to compensate the direct influence of environmental factors (wind, temperature).

The considered hypotheses and specific cases presented in present work shown that the vertical relative positioning of the rotor inside the nacelle have a direct influence on air flow pattern at the exit of nacelle. In both constructions types of the rotor (2 or 3 blades), the spray nozzles will be placed in the center of the horizontal section at the exit of the nacelle when the rotor is located inferiorly, and on the circumference of the horizontal section at the exit of the nacelle when the rotor is located superior, in order to optimize the efficiency of the crop's spraying process.

Acknowledgments

The authors thanks to "POC-A1-A1.2.3-G-2-15 Knowledge transfer partnerships, Project title: Technologies for intelligent urban electric vehicles URBIVEL, ID: P_40_333" research program for help and support.

References

[1] Wang L, Chen D, Zhang M C, Wang Y, Yao Z, Wang S M 2018 *IFAC PapersOnLine* **51-17** 648.
[2] Qin W C, Qiu B J, Xue X Y, Chen C, Xu Z F, Zhou Q Q 2016 *Crop Prot.* **85** 79.
[3] Muraru L, Cardei P, Muraru V, Sfiru R, Codruz P 2019 *Researches regarding the use of drones in agriculture,* Proceedings of 9th International Multidisciplinary Scientific GeoConference SGEM 2019, 30 June - 6 July, 2019, Sofia, Bulgaria.

[4] Yang S, Yang X, Mo J 2018 *Precis. Agric.* **19** 278.
[5] Rose CD, Chilvers J, 2018 *Agriculture 4.0: Broadening Responsible Innovation in an Era of Smart Farming*, Sustainable Intensification and Ecosystem Services, a section of the journal Frontiers in Sustainable Food Systems.
[6] Zhang B, Tang Q, Chen L P, Xu M 2016 *Biosys Eng* **145** 52
[7] Wen S, Zhang Q, Yin X, Lan Y, Zhang J, and Ge Y 2019 *Sensors* **19** 1112

IOP Conf. Series: Materials Science and Engineering **898** (2020) 012020 doi:10.1088/1757-899X/898/1/012020

Research on the behaviour of a $LiFePo_4$ pouch cell under mechanical stress

Szabo I, Kocsis L*, Mariasiu F

Technical University of Cluj-Napoca, Automotive Engineering and Transports Department, Bd. Muncii 103-105, Cluj- Napoca, Romania

levente.kocsis@auto.utcluj.ro

Abstract. In this article, research has been done regarding the behaviour of the pouch cell under the action of mechanical abuse, in order to improve pouch battery cells design. Four types of indentations were performed on pouch cell respectively, three-point bending, lateral bending, indentation with hemispherical head and pinch indentation. Contemporary electric vehicles are a solution in terms of environmental pollution, that is why the large vehicle manufacturers also have different models of electric vehicles. Also, a major problem for electric vehicles customers is the autonomy of a vehicle. That forces major electric vehicle manufacturers to invest in in-depth research into storing electricity in energy cells. Li-ion cells seem to be a viable solution at present, and that is why the paper presented a pouch-type cell subjected to specific mechanical stresses. Also, the internal short circuit can be observed in all four tests performed, and its effects are destructive in the case of lithium-ion pouch cells.

1. Introduction

Pouch cells as an energy source are used in the electric vehicle industry, also in various applications that require portable energy sources. Obviously, there is a global spread of the trend of owning an electric vehicle, being subject to many mechanical stresses that occur during their operation. At the same time, numerous researches perform connection with this important aspect, and some researches will be presented in a brief form in the following chapters.

Extensive research on $LiFePo_4$ pouch type cells were conducted by Wei L. research group and collaborators, in order to establish the behaviour at different loads and mechanical stresses [1]. Experiments were done what included mechanical stresses on the cell with the bending head, the punch was performed with a sharp steel rod head, to determine the short circuit start moment. In parallel, the voltage was monitored with a voltmeter and the cell temperature with a temperature sensor was measures. Also, other researches have also been conducted by Sahraei and collaborators regarding the behaviour of the pouch cell under a certain mechanical stress, among others showing that in some conditions it is enough 1 mm intrusion in the cell and that may introduce soft short circuit, what can lead to serious consequences [4], [5].

In this article, 4 tests were performed with different mechanical stresses on the pouch cell, respectively with the 10 Ah $LiFePo_4$ cathode material, with SOC = 80% (similar cells were used in the manufacture of the Nissan Leaf and Chevrolet Volt [6]).

2. Methods and results

In the realisation of experimental study regarding the structural stresses on the cells that equip the electric vehicles, it was necessary to develop a test stand whose components and devices used were listed in table 1.

The tests were performed with the help of three test heads made especially of steel with S355JR quality. In the figures 2, 3 and 6, 7 the indentation heads are shown. The T bending head has dimension of ϕ 25.10 mm, ϕ 12.5 mm hemispherical head. The pinch head with length of 30 mm of sharp tip. The cylinder speed was set to 4 mm • min^{-1}, and all four experiments were performed under this condition.

Table 1. Devices used in tests.

Experimental stand	Hydraulic press	10 T
	Hydraulic unit	50-600 bar
	Ball valve, hydraulic connections	
	Load cell mounting plate	
	Lighting reflectors	
Measuring instruments	Load cell	0-20 kN
	Multimeter with temperature sensor	0-1300 °C
	AXIOMET Multimeter	0-600 V
	Caliper	0.01 mm
Cell type	4 X Pouch Cell LiFePo4 H - 3210	3.2 V
Used punches	T punch head	25 mm
	Hemispherical head	12 mm
	Pinch indentation	

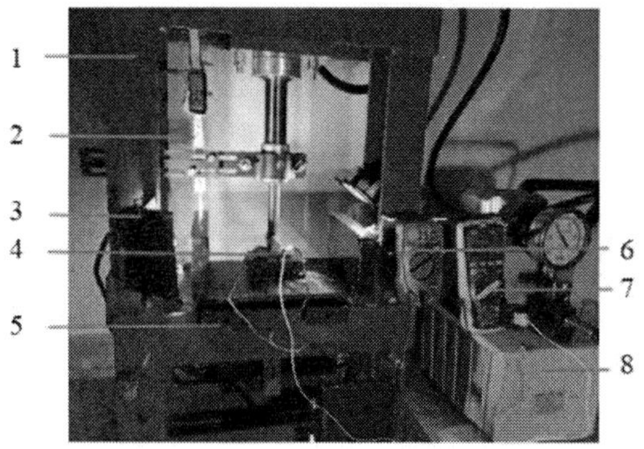

Figure 1. Experimental stand setup. 1 - Hydraulic press; 2 - Calliper; 3- Lighting reflectors; 4 – Punch head; 5 - Load cell mounting plate; 6 - Multitester; 7- Multitester with temperature sensor; 8 – Hydraulic unit.

The auxiliary mechano-welded parts used in the construction of the stand were designed in a CAD design program. Figure 1 includes the configuration of the stand where a 20 kN force load cell was also included, with the help of which the applied force on the cell was monitored. The force was measured and graphically displayed using a dedicated software installed on laptop. The experiments were also filmed using video camera to help in further analyses of obtained results.

Table 2. Summary of the testing parameters and indentation tests

Testing parameters	Intrusion	mm
	Voltage	V
	Temperature	° C
	Load	kN
Indentation tests	Three-point bending	25.1 mm
	Lateral bending	25.1mm
	Hemispherical head indentation	12.5 mm
	Pinch indentation	

Figure 2 shows the configuration of the stand for the three-point bending experiment. On the support with two rollers the pouch cell was placed, with an attached temperature sensor. The three-point bending head with the diameter of Ø 25.10 mm was used as the deformation device. The voltage measurement was performed using multimeter with the terminals attached to the cell terminals.

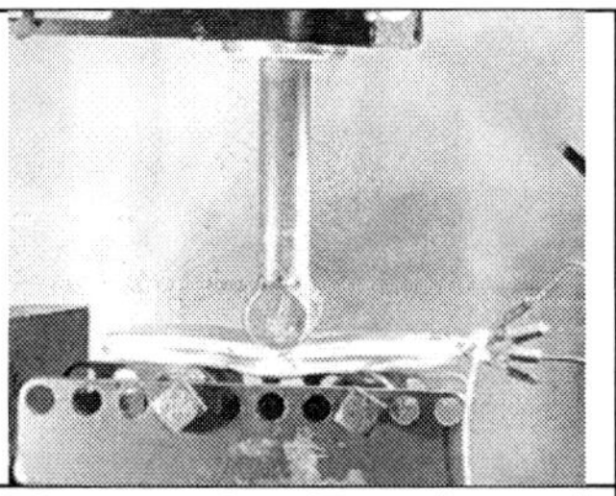	
Figure 2. Experimental stand setup for pouch cell three-point bending.	**Figure 3**. Experimental stand setup for pouch cell lateral bending.

The obtained results were centralized, where the variation of force and deformation appear in figure 4 and the variation of temperature and voltage in figure 4 b.

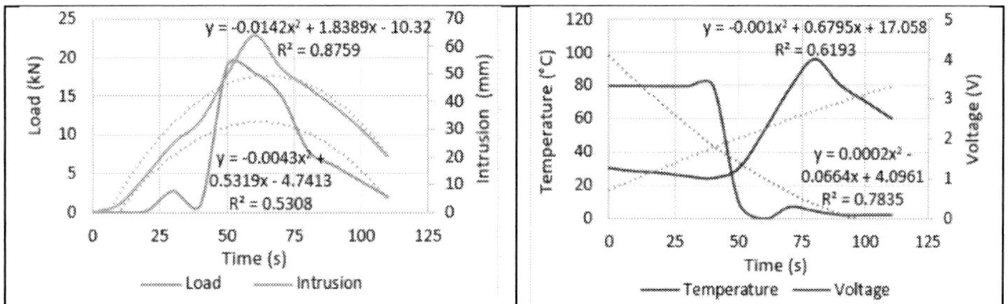

a	b

Figure 4. Pouch cell three-point bending test results: a- variation of force and displacement; b- temperature and voltage variation.

In the case of simple bending, the pouch cell was placed on a flat steel on the stand frame, figure 3 includes the respective configuration. As in the previous experiment, the T head was used in the test. A multimeter was also used to monitor the voltage and a thermometer with a sensor was used to monitor the cell temperature. The obtained results are centralized in figure 5.

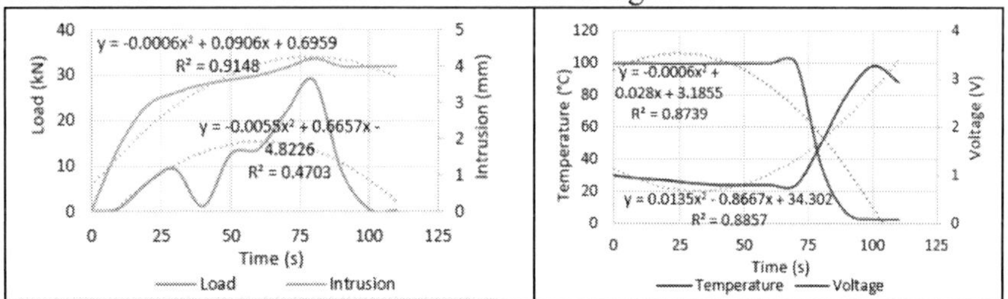

a	b

Figura 5. Pouch cell lateral bending test results: a- variation of force and displacement; b- temperature and voltage variation.

In the tests performed with the hemispherical head, the configuration is found in figure 6, where the cell was placed on the flat surface on the press frame. To measure the voltage variation in Figure 8 b, the terminals on the multimeter were attached to the terminals of the energy cell. The variation of the temperature measured on the cell surface during the experiment is also found in the graph in figure 8 b. The maximum temperature that was measured is approximately 105 ° C.

IOP Conf. Series: Materials Science and Engineering **898** (2020) 012020 doi:10.1088/1757-899X/898/1/012020

Figura 6. Experimental stand setup for pouch cell hemispherical head indentation.

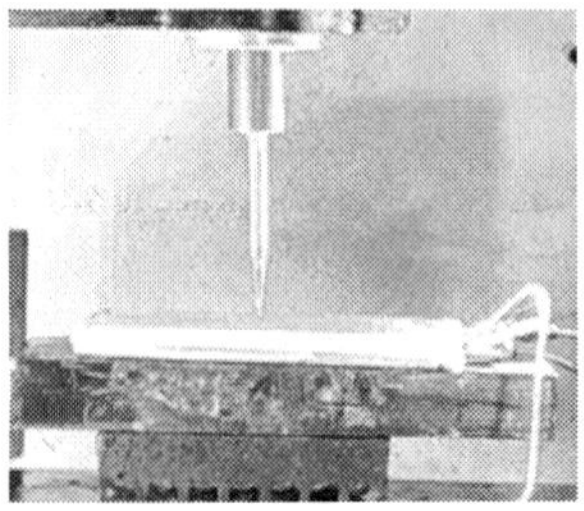

Figura 7. Experimental stand setup for pouch cell pinch head indentation.

The positioning of the pouch cell in the case of the pinch test was done on a flat surface according to figure 7. The penetrating pinch head is mounted on the device support, and the multimeter was connected with electric wires to the pouch cell terminals.

The applied force on the cell when the short circuit is induced by the applied load is 6 kN. The variation of the force appears in the curve represented in figure 8, where the maximum force exerted on the cell was 7 kN.

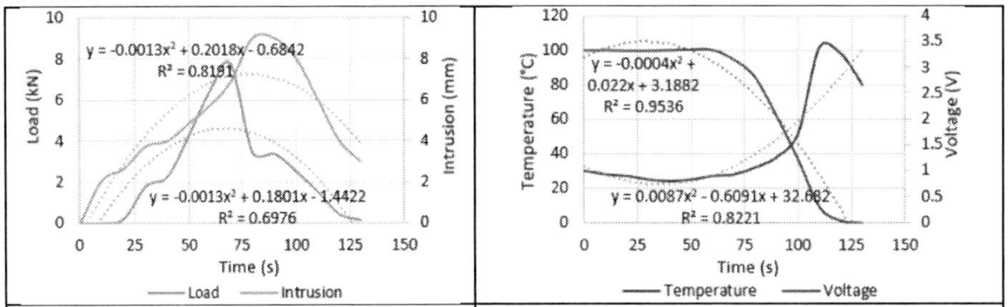

Figura 8. Pouch cell hemispherical indentation test results: a- variation of force and displacement; b- temperature and voltage variation.

The force variation is represented in figure 9, where in the second 20 is the value of the penetration force of 0.2 kN when the short circuit was induced. The variation of voltage is shown in figure 9 a where it is observed that in the second 20 the voltage starts to decrease due to the tripping of the short circuit. After penetration, the temperature rises to 100 ° C in 100 seconds, as shown in Figure 9.

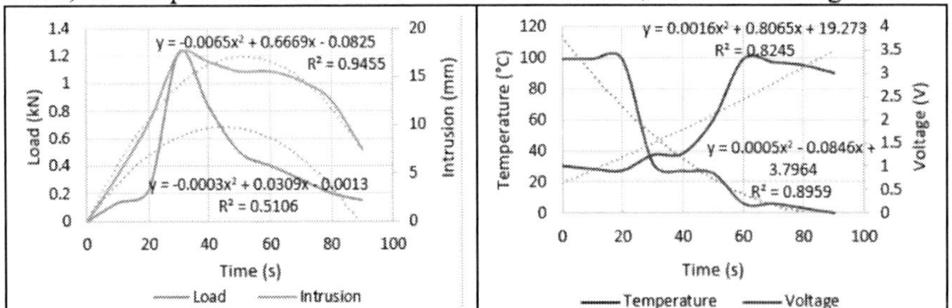

Figura 9. Pouch cell pinch indentation test results: a- variation of force and displacement; b- temperature and voltage variation.

In the case of the three-point bending experiment, the pouch cell case being malleable, the pressing force increases exponentially. The start of the short circuit was at 30 seconds after the start of experiment where the displacement was over 30 mm and the voltage decreases.

The simple bending experiment was also performed for 125 seconds, and the graphs in Figure 5 show that starting with 65 seconds, the short circuit was triggered.

The onset of the short circuit in the case of the hemispherical indentation it was induced from the second 60, and the applied force with 6 kN and with a displacement of 6 mm. From second 75 the temperature rises exponentially according to the graph in figure 8.

3. Discussions

Pouch-type electric cells are by their construction the most fragile type of cells used in batteries (energy storage sources) and therefore special attention must be paid to the possible effects of mechanical tests and stresses that may occur in their operation. Through the tests performed, the primary results obtained are necessary to be applied further in the possible development and construction of a battery housing (consisting of this type of cell and with specificities for this type of cell). Because the determination of the puncture effect of the pouch-type cells is made only on one dimension, it cannot be stated with certainty what is the optimal position of their placement within the construction of a battery. Also, an important role in their positioning has the uniform distribution of the electrolyte in the composition of the bag type cells, which requires additional researches on this issue.

4. Conclusions

The tests performed on the pouch type cell show that the onset of the short circuit does not appear even from the first moments of the application of the global force. By the action of force, the displacement occurs where the deformation is directly proportional to the displacement.

This technical approach can be applied to any battery cell type to study the short circuit starts moment under the action of mechanical stress.

The internal short circuit in the cell is the result of several abuses of a mechanical nature, it is also a topic that must be studied further.

It is also recommended to validate the experiments by the finite element method, and with the help of the results it is possible to obtain and bring cellulite and modules that correspond to the clients' requirements.

References

[1] Li, W., Xia Y., Chen, G. et al. 2018 Comparative study of mechanical-electrical-thermal responses of pouch, cylindrical, and prismatic lithium-ion cells under mechanical abuse. Sci. China Technol. Sci. **61**, 1472–1482.

[2] Hsin W, Edgar L. C, Evan T. R, Clinton S. W, 2017 Mechanical abuse simulation and thermal runaway risks of large-format Li-ion batteries. Journal of Power Sources. **342**, 913-920.

[3] Hailing L, Yong X, Qing Z. 2017 Mechanical damage in a lithium-ion pouch cell under indentation loads, Journal of Power Sources, **357**, 61-70,.

[4] Elham Sahraei, Joseph Meier, Tomasz Wierzbicki,. 2014 Characterizing and modeling mechanical properties and onset of short circuit for three types of lithium-ion pouch cells, Journal of Power Sources, 247, 503-516.

[5] Xia Y, Chen G, Zhou Q, et al. 2017 Failure behaviours of 100% SOC lithium-ion battery modules under different impact loading conditions, Engineering Failure Analysis 82: 149–160

[6] ***, *A Tale of 3 Battery Packs*, https://cleantechnica.com/2016/01/06/a-tale-of-3-battery-packs/ (accessed on 25 April 2020).

Acknowledgments

This research was funded by the Program "POCA1-A1.2.3-G-2-15 Partnerships for Knowledge Transfer, no. financing contract: 6/16.01.2018, project title: Intelligent Urban Intelligent Vehicle Technologies URBIVEL, ID: P_40_333, MySMIS: 105565".

IOP Conf. Series: Materials Science and Engineering **898** (2020) 012021 doi:10.1088/1757-899X/898/1/012021

Research on the behaviour of a $LiFePo_4$ prismatic cell subjected to mechanical stress

Szabo I, Kocsis L *, Mariasiu F,

Technical University of Cluj-Napoca, Automotive Engineering and Transports Department, Bd. Muncii 103-105, Cluj Napoca, Romania

levente.kocsis@auto.utcluj.ro

Abstract. Numerous models of electric vehicles in the automotive market are clear proof of being a solution for the massive reduction of greenhouse gases, caused by the high number of the internal combustion engines in the transport sector. It was observed that by developing and using electric vehicles, the quantities of harmful emissions in large cities can be significantly reduced. The convenience of users of electric vehicles have pushed the researchers to study in more detail and attention the storage of energy in the energy sources used in the construction of electric vehicles. Several researchers have used the resulted data to develop finite element models, therefore it is necessary to study their behaviour. The article presents an experimental research, regarding the analysis of the behaviour and prismatic cell of type $LiFePo_4$ at different mechanical stresses, including three-point bending, lateral bending, indentation with hemispherical punch and pinch indentation. The cell was tested at ambient temperature at 90% SOC.

1. Introduction

Researches regarding the behaviour and stability of electric cells under different and various mechanical stresses is an important issue, given the increase in the number of electric vehicles on the roads. The general aims of researches are to improve energy sources in electric vehicles from the design stage, which correspond to the requirements related to their use and operation in maximum safety and reliability. There are several standard requirements that vehicle manufacturers must take into account regarding the use of energy sources on electric vehicles [1], [2].

The probability that the vehicle's energy source will cause fire risks is high in the event of an accident or damage to the module as a result of improper operation. This phenomenon is directly proportional to the number of electric vehicles in traffic. Therefore, manufacturers can prevent or at least significantly reduce the possibility of damage from the design stage, using CAD design.

Therefore, the article studied the behaviour of a prismatic cell subjected to various mechanical demands. Multiple researches have been conducted by researchers, respectively on different models of lithium-ion cells to study their response to external mechanical abuses [3]. Generally, the results and conclusions shown that always is a correlation between mechanical deformation on cells and short circuit.

In present article, experimental alleged methods included three-point bending on cells, penetration with a hemispherical head, respectively axial and lateral plane compression to determine the moment short circuit starts. Cell voltage and temperature were monitored throughout the experiment.

2. Methods and results

In order to determine the effects of mechanical stresses on the battery cells that equip the electric vehicles, it is necessary to build an experimental stand. The measurements performed also imply the need for measurement and monitoring equipment to immortalize and measure the processes and results of the experimental test. After a detailed analysis of the results obtained, they were passed in a structured form in the paper, respectively in the last part the interpretation of the results. The stand was configured from a 10-tons commercial hydraulic press and a hydraulic unit equipped with an electric motor and the related hydraulic installation. The hydraulic unit is equipped with a simple manual oil flow control valve, to adjust the oil flow to 0.52 - 2.5 L / min. The working pressure of the hydraulic unit varies between 50-600 bar. The entire auxiliary equipment for the test was designed using a CAD software and was welded mechanically. At the same time, this was made according to the execution drawings elaborated according to the necessity and utility within the project.

All the indentation heads were made of steel, where for the bending head was used a ϕ25.10 mm rod and a ϕ12.5 mm for hemispherical head. Also, the pinch head indentation tip has a length of 30 mm. Every loading case was performed with a cylinder speed of 3mm/min.

Table 1. Apparatus and devices used to perform experiments

Experimental stand	Hydraulic press	10 T
	Hydraulic unit	50-600 bar
	Ball valve, hydraulic connections	
	Load cell mounting plate	1 x
	Lighting reflectors	3 x
Measuring instruments	Load cell	0-20 kN
	Multimeter with temperature sensor	0-1300 °C
	AXIOMET Multimeter	0-600 V
	Caliper	0.01 mm
Cell type	4 X Prismatic Battery Cell	3.2 V
Used punches	T punch head	25 mm
	Hemispherical head	12.5 mm
	Pinch indentation	

To reach consistent and relevant conclusion, test parameters and special demands had to be established. The main test parameters are displacement, voltage, temperature and pressing force. The simulated demands are three-point bending, lateral bending, hemispherical sharp pinch head indentation. All the tests were performed on 11 Ah $LiFePo_4$ cathode material pouch cells, with a SOC= 90%.

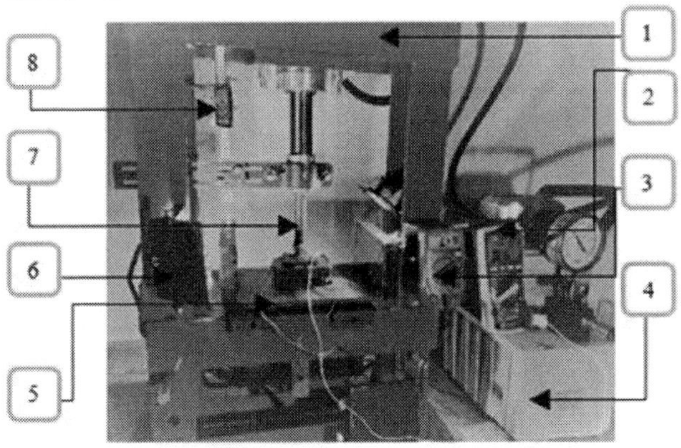

Figure 1. Experimental setup. 1 - Hydraulic press 2 - Thermometer; 3 - Multimeter with temperature sensor; 4 - Hydraulic unit; 5 - Load cell mounting plate; 6 - Lighting reflectors; 7 - Punch head; 8 - Caliper.

Experimental setup for the three-point bending test of the prismatic cell is shown in figure 2. Two support rollers were placed on the stand, over which the cell was placed, and with the bending head attached to the head holder, and the load was applied using this configuration. The cells were larger and relatively more robust compared to cylindrical cells, and the experiment lasted for 200 seconds.

Figure 2. Three point bending experimental setup.

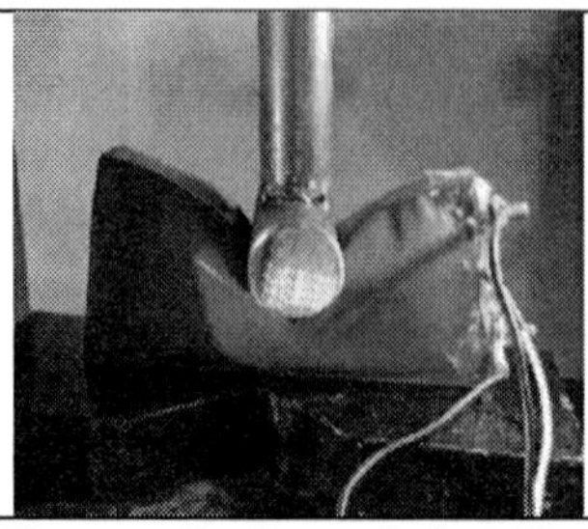

Figure 3. Lateral bending experimental setup.

On the cell terminals, electric wires from the multimeter were attached, to monitor the cell temperature and voltage at a load force of about 15 kN, the cell having a nominal voltage of 3.2 V was short circuited and the voltage dropped to 0 V.

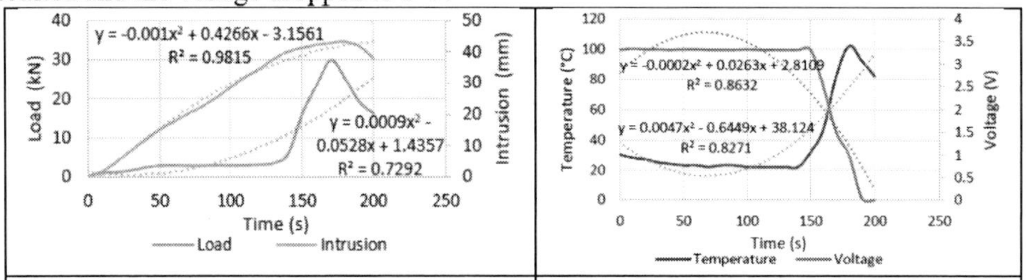

Figure 4. Three-point bending results: a- variation of force and displacement.
b- temperature and voltage variation.

In order to perform the lateral bending experiment on the prismatic cell, it was necessary to mount the bending head in the support, also the prismatic cell was placed on the stand table to perform the experimental test. The connection, to monitor the cell voltage was made using the terminals provided on the cell, and the thermometer sensor was attached to the aluminium casing of the cell. Experimental setup can be seen in figure 3.

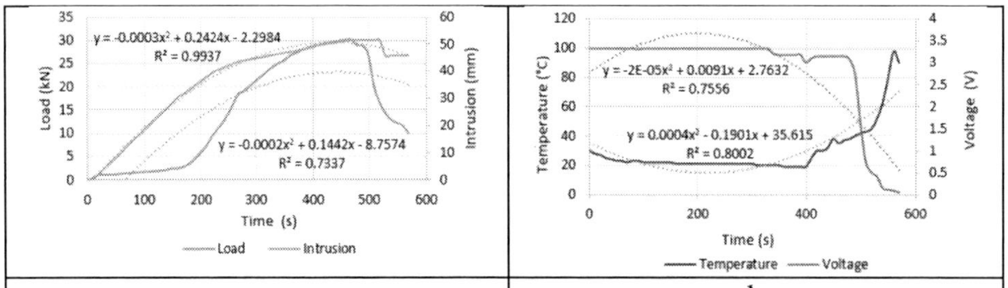

Figure 5. Lateral bending results: a - variation of force and displacement;
b- temperature and voltage variation.

Due to the strength of the cell structure the force exerted by the press was greater, compared to the other cells, according to Figure 5, it reached a maximum of 30 kN. Since this experiment was performed on the side of prismatic cell, given the geometrical width of the cell of 75 mm, the intrusion is more than 66% of the cell (50 mm), as seen in figure 3.

At the same time, the width of the cell also influenced the duration of the experiment, which lasted almost 600 seconds.

The next experiment on the prismatic cell was performed with the hemispherical head with a radius of 6 mm. On the existing terminals the conductors of the multimeter were attached to measure cell voltage and the heat sensor was placed on aluminium casing to monitor the cell temperature as figure 6 shows.

Figure 6. Hemispheric indentation experimental setup

Figure 7. Experimental setup for pinch indentation

As shown in figure 7, by actuating the piston, the pinch head deforms the cell, short circuit being induced. To monitor the voltage, the electric conductors from the multitester were attached to the cell terminals. The results for the hemispherical head indentation experiment can be found in figure 8. The short circuit occurred at a deformation of 6 mm, from which the temperature began to rise as a result of the applied load deformation. In the last experiment on the prismatic cell, we have the same configuration with the flat plate on the frame, respectively the pinch head was mounted in holder. The voltage variation is represented in figure 9. After the cell case protrusion, a short circuit occurred. The voltage variation in figure 9 b shows that from second 40 the voltage drops to close to 0 V and the temperature rises to 120 ° C. The variation of force and deformation can be found in figure 9 a.

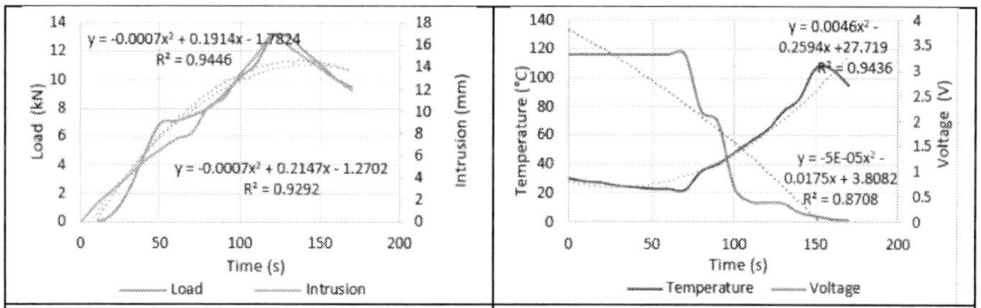

a b

Figure 8. Hemispherical test results: a - variation of force and displacement; b- temperature and voltage variation.

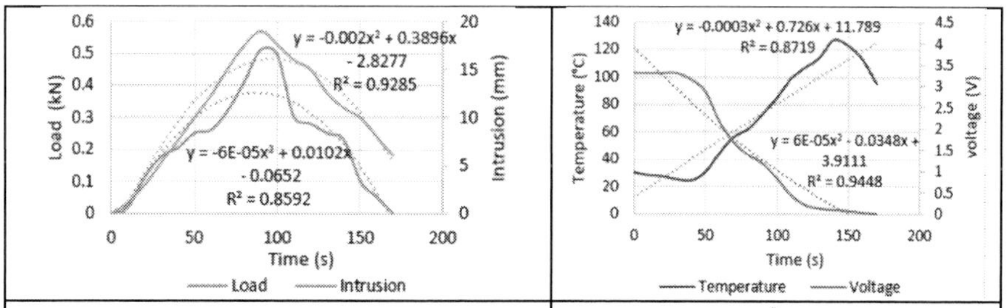

a b

Figure 9. Results of pinch indentation test: a - variation of force and displacement: b- temperature and voltage variation.

3. Discussions

Prismatic cells are widely available in electrical constructions for vehicles and even for testing the effects of different mechanical stresses experimental tests have been performed with different forms of stressing equipment.

From the point of view of the negative effects, it is understood that the most negative effects are obtained for the use of the carcass tightening equipment, for all the mechanical force it is concentrated in a single point and a relatively easy penetration of the carcass is achieved after 38 seconds). However, this test was necessary because there is a high probability that the battery of an electric vehicle will be penetrated by a screw or a thin round profile (as well as debris from the road) during driving / operation.

The other experimental tests performed show the behaviour of prismatic cells in the simulation conditions of a frontal impact accident. By considering the time of the short circuit production as main criteria, it is observed that the best positioning of the prismatic cells inside a battery would be with the smallest side oriented towards the front of the vehicle.

4. Conclusions

Through this configuration of the stand and with the help of testing devices, the test parameters were analysed in terms of the influence of mechanical stresses on the prismatic cell to understand the mechanical behaviour under mechanical loads applied to the cell.

At all four experimental tests at the beginning of the short circuit the voltage decreases, and the temperature increases to values of over 100 ° C.

The results also shows that if the diameter of the indentation head is larger the short circuit it is triggered later, compared with the smaller diameter indentation head tests.

It is recommended that the data obtained in the paper be compared and validated by the finite element method. However, the results can be used as input information for modelling new energy cells used in the construction of electric vehicles.

References

[1] Miller P, Dobedoe T, Duncan G, Pike T, Sharred D and Smout P, Surge Transport and its Role in Technology Transfer of Environmental Awareness in the Transport Sector, IEE Seminar on Automotive Electronic Standards, IET London, Savoy Place, UK, 1999, Ref. No. 1999/206, pp. 4/1–4/8

[2] Miller P, IEEJ Transactions on Industry Applications, 2008, **128**, (7), 880

[3] J. Zhu, T. Wierzbicki, W. Li, A review of safety-focused mechanical modeling of commercial lithium-ion batteries, Journal of Power Sources 378 (2018) 153–168

[4] Sahraei, E.; Campbell, J.; T., Wierzbicki, Deformation and failure mechanisms of 18650 battery cells under axial compression. Journal of Power Sources 336 (2016) 332-340.

[5] Golriz K., E. Sahraei; 2017 Review: Characterization and Modeling of the Mechanical Properties of Lithium-Ion Batteries, Energies 10 (11), 1730

[6] Li, W., Xia Y., Chen, G. et al. 2018 Comparative study of mechanical-electrical-thermal responses of pouch, cylindrical, and prismatic lithium-ion cells under mechanical abuse. Sci. China Technol. Sci. 61, 1472–1482.

[7] Xia Y, Chen G, Zhou Q, et al. 2017 Failure behaviours of 100% SOC lithium-ion battery modules under different impact loading conditions, Engineering Failure Analysis 82: 149–160

Acknowledgments

This research was funded by the Program "POCA1-A1.2.3-G-2-15 Partnerships for Knowledge Transfer, no. financing contract: 6/16.01.2018, project title: Intelligent Urban Intelligent Vehicle Technologies URBIVEL, ID: P_40_333, MySMIS: 105565".

IOP Conf. Series: Materials Science and Engineering **898** (2020) 012022 doi:10.1088/1757-899X/898/1/012022

Aspects of creativity management in the Romanian industry and doctoral research

G M Moraru[1] and L G Popescu[1]

[1] "Lucian Blaga" University of Sibiu, Engineering Faculty, 4, Emil Cioran Street, Sibiu, 550025, Romania

E-mail: gina.moraru@ulbsibiu.ro

Abstract. Creativity management is an extremely vast field, which has preoccupied us for many years. In this paper we present some aspects of the junction between creativity management in academic research and creativity management in industry. Our aim was to compare how creativity management influences the results of the PhD students' researches with the way in which creativity management in industry influences the organizational performances and competitiveness. The research based on questionnaires was carried out in parallel in technical and economic faculties in Romania, as well as in industrial organizations. Even if the variables analysed on the two samples of respondents are different, because they have different aims and interests regarding the use of creativity management, we have made comparisons between the two sets of variables and proved that the impact of creativity management in PhD students' research is high, but the impact of creativity management in Romanian industry is low.

1. Introduction

Creativity management is an extremely vast field. Approaching it even at the level of a single country, such extensive research could be carried out that their simple reading would take years. Therefore, this paper is based only on a sequence of our researches in this field.

Specialists who have analysed workplace creativity highlight two of its essential approaches in the literature: approaching creativity as an individual outcome and as a dyadic or team outcome [1]. Regardless of whether it is about individual, group or organizational creativity, it is capitalized only to a small extent in many fields of activity. For example, Europe is a world leader in science, home of some of the most creative and entrepreneurial minds and some of most innovative ideas anywhere in the world [2]. Even so, Europe is experiencing an innovation deficit in many areas [2].

On other hand, it was proved that many social and cognitive factors can inhibit group creativity. Several social inhibitors of group creativity are social anxiety, social loafing, the tendency to match the ideas of the group members and the downward comparisons inside the group [3]. Among the cognitive interferences related to a low group creativity are ideas production blocking, task-irrelevant behaviours, and cognitive load of each group member [3]. All these are factors that lead to the poor capitalization of creativity in an organization. Therefore, in some cases, the creativity of leaders is more closely linked to individual intra-psychological tasks and abilities than to selling this creativity to the outside world [4]. European Commission highlights the need to invest in skills at all levels and empower universities to become more entrepreneurial and interdisciplinary [2]. So, it is important that academic research be correlated with the skills needs of the industry, and creativity in academia with that in the industrial environment. In this paper we will present some aspects of the junction between creativity management in academic research and creativity management in industry.

2. Creativity management in academia and industry

We have conducted the research in parallel in technical and economic faculties in Romania, and in industrial organizations, too. Starting from the premise that a management model based on strategic planning and product championing can explain the positive effects of leadership on creativity in a highly creative population [5], we must not forget that organizations emphasize the balance between being creative and being productive [6]. Therefore, connections between academic and industrial management of creativity are welcome.

The aim of our paper was to compare how creativity management influences the results of the PhD students' researches with the way in which creativity management in industry influences the organizational performances and competitiveness. Regarding our aim, we must first clarify that "creativity is related to, yet different from, discovery and invention" [7]. This study continues our research work in the field of creativity management started years ago [8]. A premise of the study is our proven belief that group and organizational creativity have enormous potential for development if creativity management uses benchmarking processes [8].

2.1. Methodology of the research

A series of questionnaires were completed by 50 PhD students in management or industrial engineering and another series of questionnaires by 82 managers. The two series of questionnaires were different, because PhD students have certain interests related to the use of creativity management in doctoral training, and managers have other interests, related to creativity in their organizations.

The PhD students were 56.0% female and 44.0% male. 40.0% of them graduated a technical faculty, 28.0% an economic one, 20.0% another kind of faculty, and 12.0% graduated two faculties. The managers were 30.5% female and 69.5% male. 11.0% of them graduated a high school, 64.6% an engineering faculty (technical, food, chemistry, textile, etc.), 17.1% an economic one, 2.4% a low faculty, and 4.9% graduated two faculties. Other characteristics of respondents are shown in figure 1.

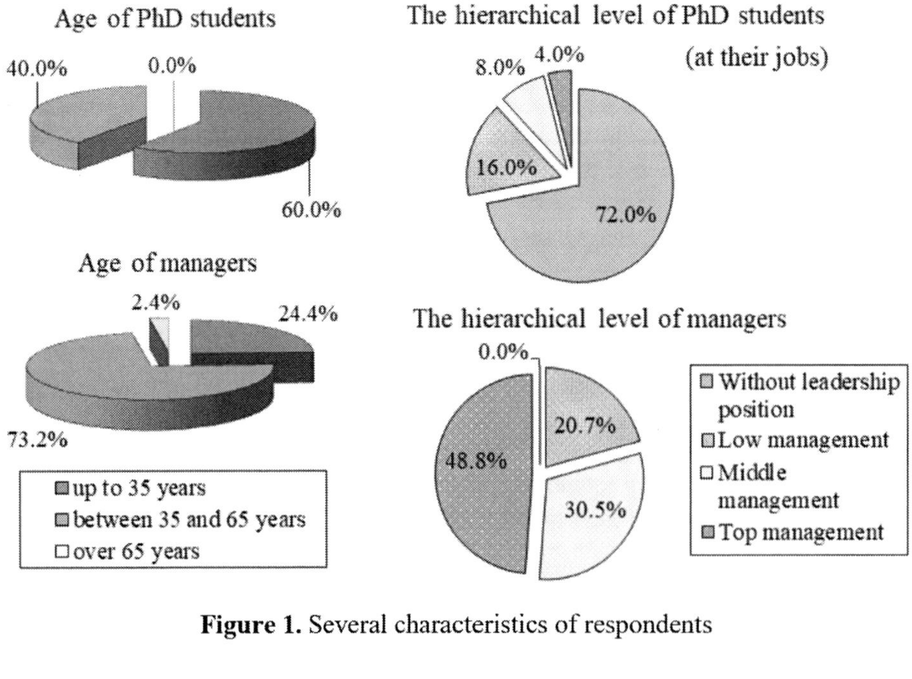

Figure 1. Several characteristics of respondents

2.2. *Main results*

The doctoral students were asked a question that tested the degree to which their doctoral training was oriented towards the management of creativity. They were asked to tick the agreement with a series of statements, using the following evaluation scale: 0 points – "I do not know", 1 point – "total disagreement", 2 points – "partial disagreement", 3 points – "neutral", 4 points – "partial agreement", and 5 points – "total agreement". The statements were therefore assimilated to the following variables:

- Variable a: "In my doctoral training I used notions of creativity management, especially when writing reports";
- Variable b: "The topic of my doctoral thesis is directly related to the management of creativity";
- Variable c: "Managers will find in the thesis useful guidelines for the use/development of creativity in companies";
- Variable d: "The PhD research will also be based on the results of a benchmarking process";
- Variable e: "The PhD thesis will make a junction between modern management, the intangible values of a company and performance".

Figure 2 shows the percentages of respondents who expressed their agreement/disagreement with these statements.

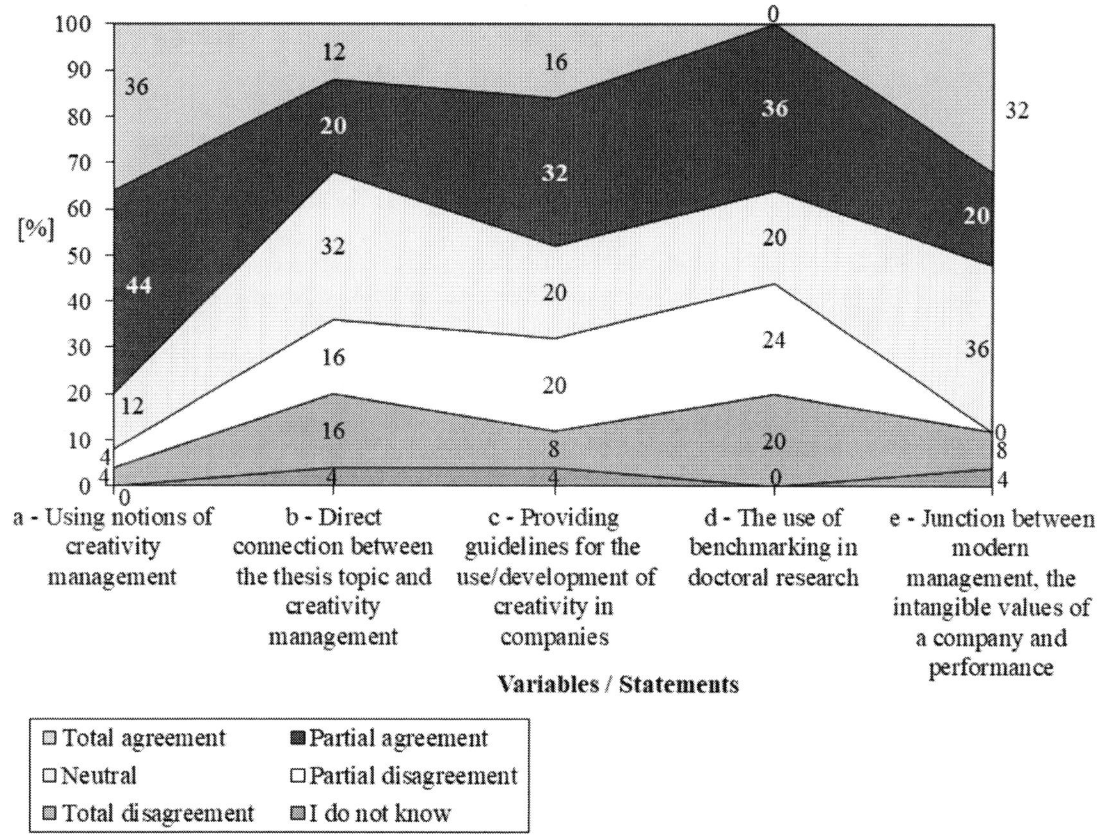

Figure 2. PhD students' agreement with the sentences regarding their doctoral training and thesis

Because we suspected that an agreement with the first two variables (a, and b) increases the chances of an agreement with variable c, we developed a three-dimensional graph, using the STATISTICA software, to highlight the connection between the three variables (figure 3). For the sample of PhD students analysed, the use of notions of creativity management in the doctoral training (variable a) seems to have little influence on the issuance of creativity management guidelines (variable c). On the one hand, it is normal for PhD students to use methods and tools of creativity management in their doctoral training, being focused on solving the main problem of the thesis, not on the development of creativity management guidelines. On the other hand, the connection of thesis subject with creativity management leads to recommendations for the use/development of creativity in companies. So the variable b greatly influences the variable c (figure 3).

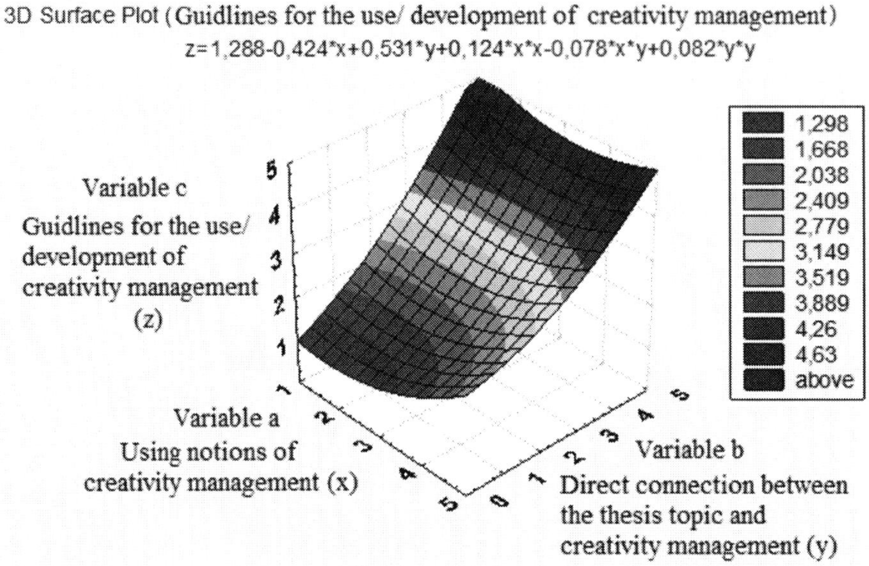

Figure 3. The possibility of elaborating some creativity management guidelines in the PhD thesis (the correlation between variables a, b and c)

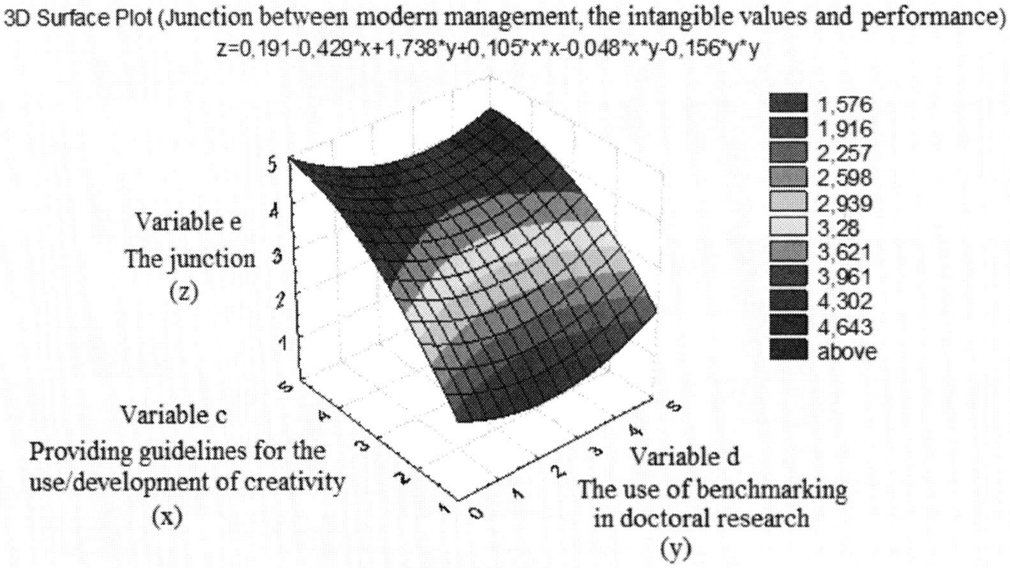

Figure 4. The junction between modern management, the intangible values of the company and performance favoured by two other factors (the correlation between variables c, d and e)

We have also demonstrated that the junction between modern management, the intangible values of the company and performance (variable e) can be favoured, in a doctoral thesis, by issuing guidelines for the use/development of creativity in companies (variable c) and the fact that the PhD research is also based on the results of a benchmarking process (variable d). Figure 4 shows this correlation.

In order to compare the variables analysed in the PhD students' questionnaires (a, b, c, d, and e) with the managers' answers, we have calculated the averages of the variables and translated them on the scale 0 ... 100 (table 1).

Table 1. Averages of the variables analysed in the PhD students' questionnaires

Variable	a	b	c	d	e
Average on the scale 0 … 5	4.04	2.84	3.16	2.72	3.56
Average translated on the scale 0 … 100	80.80	56.80	63.20	54.40	71.20

Unfortunately, asked if they hope for a future implementation in the Romanian industry or economy of the discoveries made in the doctoral thesis, only 24.0% of the PhD students gave an affirmative answer.

As we mentioned before, the results of creativity and the use of creativity management in industry are part of a different category from the results pursued in doctoral research. However, both areas are favoured by the tandem use of creativity management with modern benchmarking, which stimulates creativity at all levels: individual, group and organizational [8]. One of the most important questions addressed to Romanian managers was the request they quantify the impact that creativity management and benchmarking have on several factors that define the prosperity of an organization: organizational performance, competitiveness and success of the company on the market, increasing turnover, company value and employee job satisfaction. We mention that the managers' opinions are found only in a proportion of 69.5% in figure 5, because to a previous question only they knew how to give a definition, complete or not, of benchmarking.

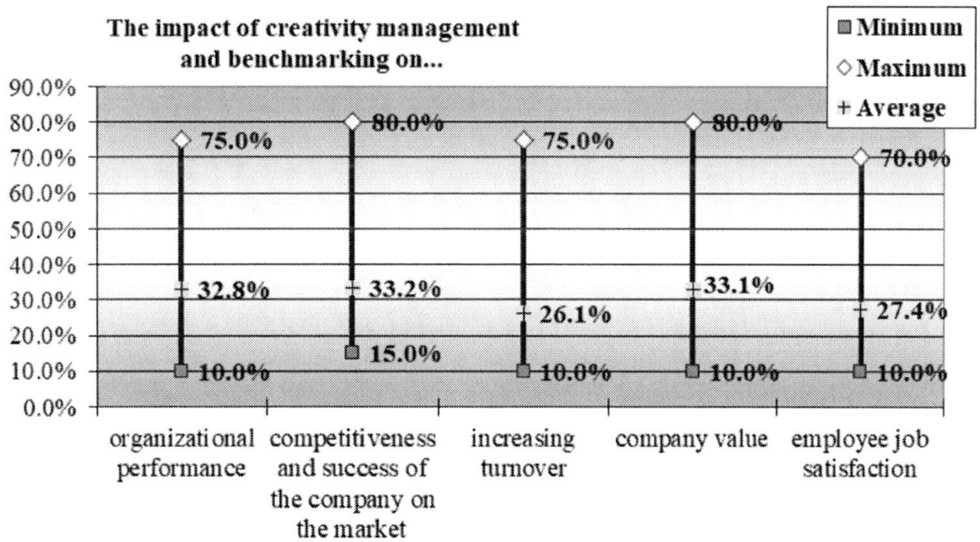

Figure 5. The impact of creativity management and benchmarking on five factors that define the prosperity of an organization

Comparing averages in figure 5 with those of the variables analysed on the doctoral students' sample, we observe that, on the whole, the effects of creativity management in industry are much weaker than in doctoral research. By example, we can consider that providing guidelines for the use/development of creativity in companies in a doctoral thesis (variable c) is a performance of the academic research which is quoted at 63.20 on the scale 0 … 100 (table 1). But the organizational performance is influenced by creativity management in Romanian industry only at an average weight of 32.8%, which is the largest of the averages shown in figure 5. Moreover, let's not forget that only 24.0% of doctoral respondents hope that the results of their own theses will be implemented in practice.

3. Conclusions

Coordinating creativity at the societal level requires first awareness of its importance as a resource, and then the application of effective creativity management in all sectors of activity. The efficiency of using creativity is also given by a correlation of its effects in academic research with its effects in industry. We have conducted in this paper only an exploratory research on the junction of these two fields, distributing different questionnaires to a sample of PhD students in engineering or management and, respectively, to a sample of industry managers.

Even if the sets of variables analysed on the two samples of respondents are different, because their aims and interests regarding the use of creativity management are different, we can say that the use and the impact of creativity management in PhD students' research is high (table 1), but the impact of creativity management in Romanian industry is low (figure 5).

In the future, descriptive, explanatory and predictive research is needed, to follow the way in which the results of doctoral theses can be implemented in the Romanian industry, in order to efficiently correlate the creativity in academic research with that in industry. These in-depth studies will need to be conducted on each industry and even on different categories of topics/issues addressed in PhD students' research.

References

[1] Zhou J Hoever I J 2014 Research on Workplace Creativity: A Review and Redirection *Annu. Rev. Organ. Psychol. Organ. Behav.* **1** 333-359

[2] European Commission 2018 *A renewed European Agenda for Research and Innovation - Europe's chance to shape its future* COM(2018) 306 final (Brussels)

[3] Paulus P B 2000 *Applied Psychology: an Int. Review* **49**(2) 237-262

[4] Mayer C H Oosthuizen R M 2020 *Creativity Studies* **13**(1) 21-40

[5] Vessey W B Barrett J D Mumford M D Johnson G Litwiller B 2014 *The Leadership Quarterly* **25**(4) 672-691

[6] Vele C L 2017 Employee creativity in Romanian construction companies. An empirical research *Review of Management and Economic Engineering* **16** No 2(64) 267-278

[7] Gabora L 2020 Creativity *Oxford Research Encyclopedia of Psychology* forthcoming Jan-Mar 2020 Subject History and Systems of Psychology (Oxford: UK/OUP) *Preprint* https://arxiv.org/ftp/arxiv/papers/1912/1912.00091.pdf

[8] Moraru G M 2011 *Creativity management and benchmarking promoting performance and competitiveness in modern organizations* PhD Thesis in Management ("Lucian Blaga" University of Sibiu: Romania/unpublished by a publishing house)

IOP Conf. Series: Materials Science and Engineering **898** (2020) 012023 doi:10.1088/1757-899X/898/1/012023

Study on the improvement of the vehicle deceleration signalling system

G M Moraru[1] **and L G Popescu**[1]

[1] "Lucian Blaga" University of Sibiu, Engineering Faculty, 4, Emil Cioran Street, Sibiu, 550025, Romania

E-mail: gina.moraru@ulbsibiu.ro

Abstract. Our paper presents a study regarding the improvement of the vehicle deceleration light signalling system. Based on global statistics on the number of road accidents caused by non-compliance with the safety distance and a market research on the availability of purchasing an extra option to allow drivers to be informed about the speed reduction of the vehicle in front, we have applied the Product Life Cycle Management (PLM). For this purpose, we made two market researches. Based on the results of the primary market research and the technical solutions currently in place, a new market research was carried out through an opinion poll. All the results of the opinion polls were discussed in a brainstorming session, which was attended by representatives of automotive companies, representatives of academia and professional drivers. At the end, we have demonstrated the possibility of developing a new product that aims to reduce road accidents caused by rear-end collisions. The working method, proposed has led to the need to find new constructive solutions and ways to apply them.

1. Introduction

The driver of one vehicle driving behind another is obliged to keep a sufficient distance from it to avoid a collision, in the event that the one in front slows down or brakes suddenly. It goes without saying that drivers must always adapt their speed to road conditions so that they can perform any maneuver safely. Drivers' attention must be constantly focused on the road [1]. In most European countries, the general rule is that each driver should keep a sufficient distance between his vehicle and the front vehicle to avoid an accident in the event that the front vehicle suddenly stops or slows down. This rule is not always imposed by specific distances to be observed, such as the minimum distance or time required between vehicles. In cases where minimum distances or times are required, the figures vary, from country to country. The second rule or the half-distance rule can be applied. Different specific rules can be applied to heavy goods vehicles. The second rule is often used as a rule of thumb and taught in driving schools. Some devices, such as road markings, are used to help drivers follow this rule.

Acording to Asociation for Safe International Road Travel approximately 1.35 million people die in road crashes each year, on average 3,700 people lose their lives every day on the roads and an additional 20-50 million suffer non-fatal injuries, often resulting in long-term disabilities [2]. Unless action is taken, road traffic injuries are expected to become the fifth leading cause of death by 2030. One of the most common causes of road accidents is non-compliance with the safety distance.

2. Methodology and problem description

The proposed methodology is that of the product life cycle management (PLM). Thus, when analyzing the customer's need, a stage prior to conception, two phases of market research can appear: a first overall research, aiming at the opportunity to design a new product or improve an existing one, and another more detailed research, mostly qualitative, to determine in detail the needs and demands of consumers towards the product [3].

If a driver does not respect the safety distance, he cannot accurately estimate how aggressively he brakes the vehicle traveling in the same direction and thus road accidents occur.

The brake lamp lights come on several times in traffic. The cases in which they can ignite are the following:

- There is a need for braking to stop the vehicle or avoid obstacles;
- Press the brake pedal lightly (preventively);
- Voluntarily or involuntarily supports the foot on the brake pedal;
- There are technical problems with the drive system - the lights are on permanently or not at all.

It is often very difficult to estimate how aggressively the vehicle in front brakes. The question arises: How much will the vehicle in front do? Will it stop? Slow down? Or is the driver just resting his/her foot? Or maybe something else? In such circumstances, fuel consumption is also increased because the driving style is not very fluid.

2.1. Market research - the primary source

In order to discover the need for a product to help reduce road accidents caused by rear-end collisions, a market research was conducted using a questionnaire. The survey was conducted on a sample of 100 people from urban areas, 80 men and 20 women with a driver's license, aged between 21 and 60 years.

According to the answers in the questionnaire, 17% of respondents stated that they were involved in a traffic accident while driving, and 23% that they were involved in a traffic accident but not in the driver's position. Of those involved in traffic accidents, 20% said the accidents were caused by non-compliance with the safety distance.

Regarding the questions related to the automatic safety systems and the degree of confidence in the autonomous braking system, the majority of respondents (65%) answered that they trust the modern automatic safety systems and 45% that the degree of confidence in the autonomous braking system, is medium. Only 30% said there was a high degree of trust in these systems. According to 20% of respondents, the human role in traffic can be replaced by automated systems, with a majority of 70% stating that the human role in traffic can only be partially replaced by automated systems.

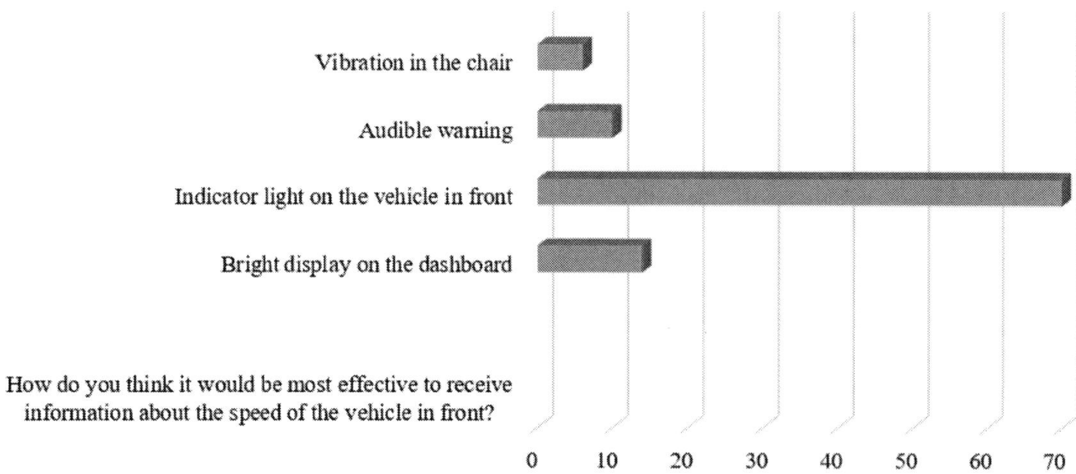

Figure 1. Results of the exploratory research

Given that 50% of respondents stated that in certain situations they would find useful in traffic information about the degree of speed reduction of the vehicle in front, and 55% would be available to purchase an extra option to inform drivers of rear vehicles about the degree of braking (regardless of price), there was the issue of how to receive information and the degree of reduction of the speed of the vehicle in front. The answers to this question are shown in figure 1.

Analyzing the results of the first questionnaire, it was concluded that it is appropriate to implement a light signaling system to warn the driver of the rear vehicle about the degree of deceleration, when the need for braking occurs. It will be referred to as the "Flash Deceleration System".

2.2. *Market research - current market solutions*

The regulations of the Convention on Road Traffic from Vienna suggested the activation of red lighting lamps with a higher light intensity than the rear position lamps, when the driver applies the brakes of the vehicle [4]. These are officially called stop lamps and are mounted in multiples of two, symmetrically at the left-right edge of the rear of each vehicle.

As regards the Center High Mount Stop Lamp (CHMSL), in Europe since 1998 United Nations Regulation No. 48 has been applied: a mounted central stop lamp (brake) is higher than the left-right lamps. CHMSL is sometimes informally called "central brake lamp", "third brake light", "eye level brake lamp" [5]. CHMSL can use one or more filament or led bulbs or a neon tube strip as a light source.

The CHMSL module is intended to provide a warning, but due to certain overlaps, left or right brake stops cannot always be observed. Also, in the event of a stop lamp failure, they provide a redundant stop light signal.

2.2.1. Autonomous Emergency Braking - AEB

From 2014, all car manufacturers are required to offer on newly launched cars an autonomous emergency braking system entitled AEB - Autonomous Emergency Braking [6]. All vehicles sold in the European Union must be equipped with autonomous automatic braking systems. These systems generally provide a warning for the driver before acting on the brakes.

As of mid 2022, all new cars put on the EU market will have to be equipped with advanced safety systems. Following an agreement with the European Parliament on lst of March 2019, the Council has adopted at the end of 2019 a regulation on the general safety of motor vehicles and the protection of vehicle occupants and vulnerable road users in a bid to significantly reduce the number of road casualties [7].

The European New Car Assessment Programme – Euro NCAP provides the consumer with information on the safety of new cars. The Safety Assist score is determined from tests to the most important driver assist technologies that support safe driving to avoid and mitigate accidents. In these tests, Euro NCAP tests system functionality and/or performance during normal driving and in typical accident scenarios [8].

The autonomous emergency braking system uses a radar, lasers or a video camera to detect an imminent danger, namely a collision with the car in front. The software alerts the driver and prepares the braking system. If the driver does not react, the technology intervenes on the brakes and actuates them. Twenty carmakers have agreed that their cars will have a very important option as standard: automatic braking if the danger of collision is imminent. The transition will be gradual, starting in 2022, depending on the class they belong to, but it is expected that in 2025 absolutely all cars to be built will benefit from this equipment [9].

2.2.2. Emergency stop signal (ESS)

Regulation No. 48 of the Economic Commission for Europe of the United Nations (UNECE) - uniform provisions concerning the approval of vehicles with regard to the installation of lighting and light-signalling devices requires that the lamps supplying the ESS flash at 4 Hz when a passenger car decelerates by more than 6 m/s^2 or a truck or bus decelerates by more than 4 m/s^2 [5].

Emergency Stop Signal, or ESS, was taken over by companies such as Toyota, Mercedes-Benz, Volvo, BMW that launched equipped vehicles that transmit a special light signal when the vehicle is braked quickly and severely. This is officially called the "emergency stop signal". Mercedes vehicles flash the stop lamps for the ESS, while vehicles in the Volkswagen manufacturer's group (VW, Audi, SEAT, Skoda) light up the flashes. Other methods of indicating severe braking have also been implemented; some Volvo models make the brake lights brighter, and some BMW cars have an "adaptive brake lamp" that effectively increases the size of the brake lights in high braking conditions by illuminating the brake lights at higher intensity than normal [10].

There is currently considerable debate about the measurable increase in safety performance of the system, as studies on in-service vehicles have not shown significant improvements. The idea behind these emergency braking indicator systems is to draw attention to the fact that it is a special emergency.

2.2.3. Emergency Brake Blinker - EBB

Another solution that can be used as an alternative to ESS systems pre-installed on some newer models from the factory, is the electronic module "Emergency Brake Blinker" - intermittent signal to brake. The EBB electronic module is an electronic module based on an electronic accelerometer assisted by a microcontroller that detects sudden braking and quickly flashes the 3rd stop on the brake and in case of very violent braking also activates the breakdowns, making the car much easier. to be noticed by the one behind.

There are currently several safety systems on the car in terms of braking systems, but even the most advanced automated safety systems cannot yet replace the human input from traffic. Most have limitations on speed or visibility. That is why it is necessary for the driver of the vehicle to have as much traffic information as possible. The information must be available in real time, and be as easy to observe and understand as possible.

2.3. Identification of product requirements and proposed solutions

Based on the results of the primary market research and the technical solutions currently in place, the question arises of identifying a solution that would create a brake light of variable and/or gradual intensity that would respond to the intensity of the deceleration. This will increase traffic safety by preventing rear-end collisions and reducing fuel consumption.

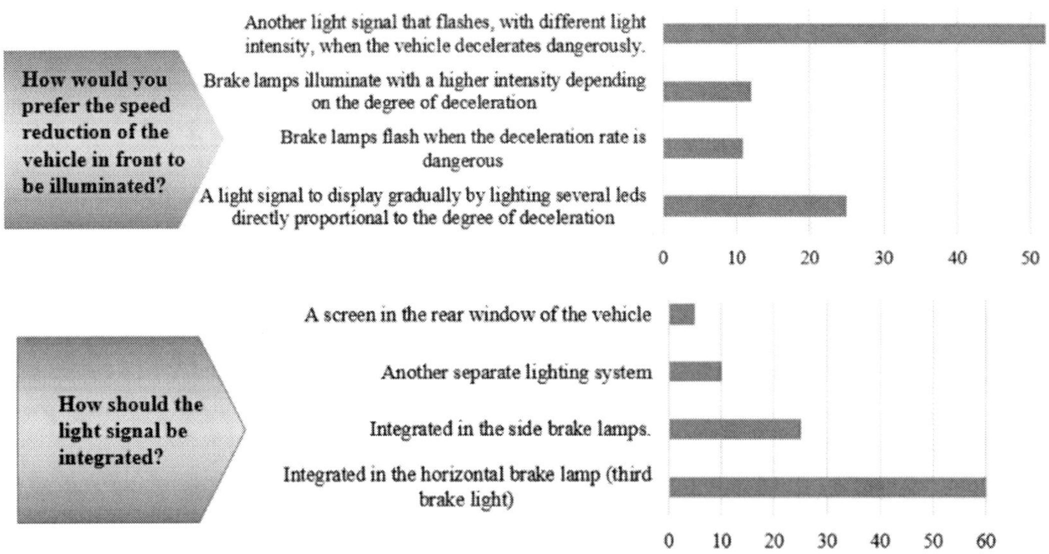

Figure 2. The answers to the first two questions in the questionnaire

In this sense, a new market research was carried out through an opinion poll. The questionnaire distributed to a number of 100 drivers had as its theme the characteristics related to the new product (light system for decelerating the vehicle). Figure 2 shows the first two questions in the questionnaire and the corresponding answers.

In order to determine customer preferences for product features, the following questions were asked to choose the preferred feature depending on how useful/appropriate they consider them. The questions and answers are shown in figure 3.

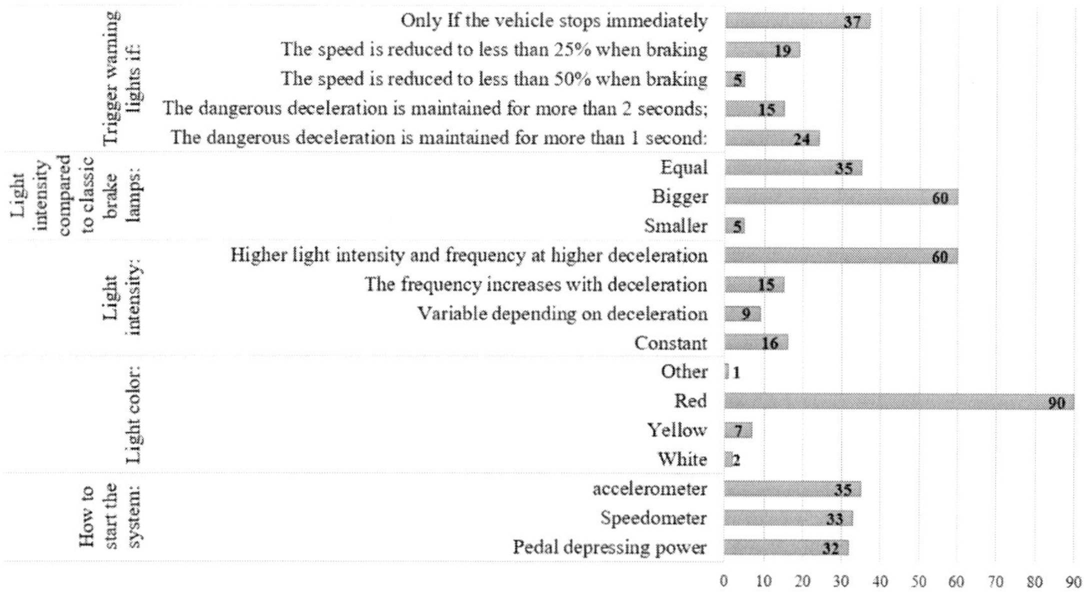

Figure 3. Customer preferences for product features

After analyzing the results of the second questionnaire, the following requirements for the proposed product were reached:

- Another light signal that flashes, with different light intensity, when the vehicle decelerates dangerously;
- The light signal should be integrated in the horizontal brake lamp (third brake light);
- The triggering mode should take into account the information received from both the accelerometer and the speedometer;
- The light should be red;
- The lights should be of higher intensity and frequency at higher deceleration, as at low intensities they may not be observed, and in certain situations, especially in daylight, the light intensity may not be properly assessed;
- The system should automatically trigger the fault signal (flashing) if deceleration is dangerous.

All the results of the opinion polls were discussed in a brainstorming session, which was attended by representatives of automotive companies, representatives of academia and professional drivers. Implementation, at the end of which a new list of requirements is drawn up. The requirements have been ranked according to cost, utility and field of use. The requirements and the way of their implementation are presented below:

- The product must provide information on the braking degree of the vehicle in front. This information is provided by displaying the degree of deceleration by means of the central (horizontal) brake lamp. At a deceleration of more than 6 m/s^2 in motor vehicles and 4 m/s^2 in trucks and coaches respectively, the leds in the middle lamp will flash, with a frequency similar to the hazard warning lights and will gradually increase with increasing deceleration.

- The product must have low manufacturing costs, so that it can also be mounted on low-cost vehicles. In addition to the ideas resulting from the opinion poll, the development of a communication system (radar, satellite, etc.) with the car on the same direction of travel was also discussed. It is hoped that this "Flash Deceleration System" will have a low cost to create the preconditions for a large-scale implementation in all cars regardless of the range they belong to.
- The product must comply with European and global automotive regulations.The provisions of the Romanian road legislation are aligned with the European standards, so the product will comply with the legal norms only if it will be integrated in the factory car, and will not be available as an attachable extra option. Due to the legislation, the product is not intended for physical customers, but is addressed to automotive manufacturers.
- Information to be available in real time. Due to the fact that the product will be electronic, the information will be available instantly.
- The information must be easily intelligible, visible from a sufficient distance, in adverse weather conditions, or in various light conditions. To meet these requirements, the leds in the lamp will be red and will have at least as much intensity as the classic brake lamps.
- The information should be unmistakable. The location of the central brake lamp, the horizontal orientation, as well as the ignition mode will provide unmistakable information.
- The product can be easily integrated functionally, regardless of the make or model of the vehicle. The product will be easy to integrate, functional because it does not require bulky components.
- The design of the product must be easily integrated into the design of any vehicle. Due to the fact that no new components are added to the exterior of the vehicle it will not affect the design.
- The identity element (main feature) of the product should derive from the way it solves the problem.The main feature of the product is the unique way it provides information in traffic, so the identity element is strong.
- The product has the potential to become a standard in the automotive industry. It is recommended to use a standard number of leds to avoid confusion in interpreting the data.
- The product to comply with the automation trend of vehicles. The product is also integrated with modern automatic braking systems, due to the fact that it does not receive data from the brake pedal, but from the speedometer and accelerometer. So, whether the car brakes "alone" or the driver brakes, the data will be the same.
- The product should be eco-friendly - environmentally friendly. Due to the fact that it will help to use a more fluid finish style, with less braking, the product will also contribute to fuel economy. Also the components from which it will be built will be 95% recyclable (plastic and metal).
- Abbreviated name - the brand should be easy to remember and inspire. The name must be easy to pronounce in any language. The English abbreviation for "Flash Deceleration System" is the most appropriate (FDS).

2.4. Product concept

In order to attract the attention of drivers in traffic and even more on the very strong braking, the proposed system should fave the light intensity of the leds can also increase by flashing 2 rows of leds intermittently. The first row will illuminate normally when braking with deceleration below 6 m/s^2, respectively 4 m/s^2, and above this value the second row will also light up, both rows flashing. Thus, the middle lamp, located in the rear window area, is also visible by the drivers in the column. The system does not disturb traffic because it only works in extreme situations, thus avoiding the risk of chain jamming. In this case of high deceleration, when the vehicle is stopped this system automatically controls and the emergency system starts.

This system also operates in the event of an accident (when the driver of the vehicle may have problems and cannot operate the emergency system), thus informing other road users, of the existence of a special situation forcing them to stop to avoid an accident. This system called Flash Deceleration System works independently of the braking system, based on a deceleration sensor that controls the leds and the car's emergency system via an electronic circuit.

The product itself will consist of 2 components: one hardware and one software. Regarding hardware components the product will be composed of the following physical components:

- Speedometer interface - a module that must take data from the speedometer or on-board computer about the vehicle's speed per unit of time and turn it into percentage deceleration data.
- Sensor accelerometer that measures acceleration based on vehicle inertia.
- Microcontroller the data processing center coming from the speedometer interface and the accelerometer using special software. It will calculate how many leds will need to receive current depending on the degree of deceleration.
- Led display - a component that must be easily configurable depending on the design of the car model on which the system will be mounted.
- Electrical cables - connect the physical components of the product.

The software component will have the role of calculating if one or both rows of leds light up, of course with the help of software. The software must take into account the degree of deceleration (based on the information received from the speedometer) from the moment the brake pedal was depressed, but provided that the accelerometer records a certain value for a certain period of time. The values will be defined after performing several tests. Figure 4 shows schematically the logical scheme of the operation of the proposed system.

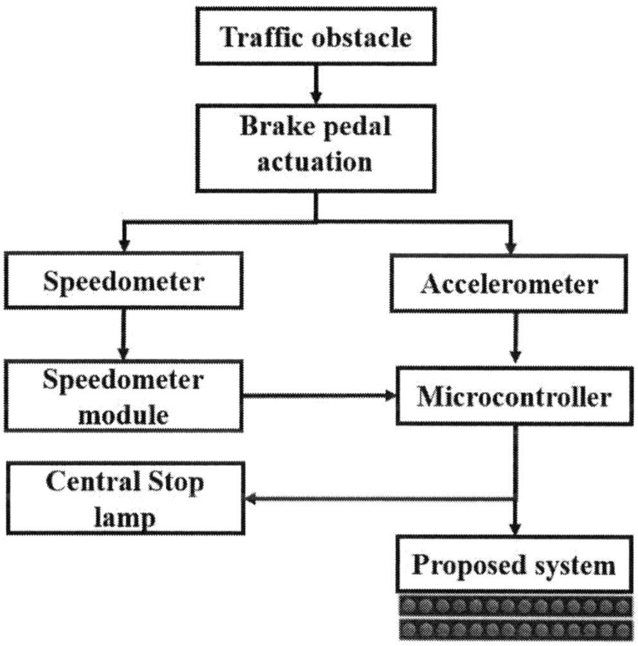

Figure 4. Logical scheme of the operation of the proposed system

- When the need arises to brake, either modern automatic braking systems come into operation or the driver actuates the brake pedal;
- The speedometer records the reduction in travel speed, and sends the data to a data processing module;
- The accelerometer records the deceleration of the vehicle;
- Data related to deceleration, speed and time are processed by a microcontroller;
- The microcontroller calculates with the help of a software what is the degree of deceleration and controls how many leds will receive electricity.

- When the deceleration percentage reaches 100%, or a critical level established after the tests, the second lamp will flash intermittently to draw even more attention to the danger. The emergency lights shall be operated under the same conditions.

3. Conclusions

The approach through the PLM prism of the light signaling system of the vehicle deceleration facilitated the detection of the shortcomings, and the studies clarified the needs and highlighted the ways to improve it. This working method has led to the need to find new constructive solutions and ways to apply them.

References

[1] Trufasu V 2018 The distance to be kept between the machines while driving (Distanta care trebuie pastrata intre masini in timpul deplasarii - Romanian) *Adevărul* 25 October https://adevarul.ro/locale/galati/distanta-caretrebuie-pastrata-masini-timpul-deplasarii-zeci-metriopreste-autovehicul-circula-60-kmh-1_5bcf8f37df52022f75cc0f16/index.html

[2] Association for Safe International Road Travel 2020 *ASIRT website* https://www.asirt.org/safe-travel/road-safety-facts/

[3] Popescu L G Brindasu P D 2015 *Aspects regarding automotive products (Aspecte privind produsele din domeniul auto - Romanian)* (Sibiu: Romania/Lucian Blaga University Press) pp 58-73

[4] United Nation 1968 Convention on Road Traffic *Treaty Collection* (Vienna)

[5] Economic Commission for Europe of the United Nations 2016 Regulation No 48 of the Economic Commission for Europe of the United Nations (UNECE) — Uniform provisions concerning the approval of vehicles with regard to the installation of lighting and light-signalling devices [2016/1723] *Official Journal of the European Union* L265/126

[6] European Commission 2014 *Commission Staff working document on the implementation of objectives 4 and 5 of the European Commission's policy orientations on road safety 2011-2020 – deployment of vehicle technologies to improve road safety* SWD (2014) 297 final (Brussels)

[7] Council of the EU 2019 Safer cars in the EU *Press release* 8 November

[8] Euro NCAP 2020 *Safety Assist* https://www.euroncap.com/en/vehicle-safety/the-ratings-explained/safety-assist/

[9] Mearian L 2016 All cars will have automatic braking in six years *Computerworld* (IDG Communications Inc) https://www.computerworld.com/article/3045553/all-cars-will-have-automatic-braking-in-six-years.html

[10] Linhong Led Auto Lights 2020 (Dongguan: China/Lin Hong Electronic Technology Co LTD) http://ro.linhongautoled.com/info/emergency-stop-signal-29250076.html

IOP Conf. Series: Materials Science and Engineering **898** (2020) 012024 doi:10.1088/1757-899X/898/1/012024

The influence of ethanol and biodiesel blends on diesel engine performance and emissions

B C Benea[1]

[1]Department of Automotive and Transportation, Mechanical Engineering Faculty, Transilvania University of Brasov, Brasov, Romania

b.benea@unitbv.ro

Abstract. This paper presents the variation of energy parameters and pollutant emissions of the diesel engine for different types of diesel-biodiesel-ethanol blends. The biofuels added in fossil fuels influence the mentioned characteristics. Ethanol is a renewable fuel, which is obtained from plant biomass, sugar and starch. The tests were performed on a 4-cylinder engine, water cooled with pure diesel (D), 10% ethanol (D90E10), 10% ethanol 10% biodiesel (D80B10E10), and 15% ethanol 10% biodiesel (D75B10E15). The comparative results of the engine operation with these fuels for the evaluation of performance and pollutant emissions were presented. Nitrogen oxides (NOx) decreased moderately for fuel blends compared to pure diesel. A slight increase in hydrocarbon (HC) and carbon monoxide (CO) emissions was observed with fuel blends due to low combustion temperature inside the engine cylinder. It can be concluded that the engine power has a small variation compared to diesel operation, and the pollutant emissions are lower when the engine runs on fuel mixtures.

1. Introduction

Compression ignition engines are widely used in industry, agriculture, transportation and as stationary engines. They have high reliability and low fuel consumption. Among the disadvantages presented by CI engines are: high emissions of NOx, particulate matter and other polycyclic aromatic hydrocarbons. Due to the increasingly demanding pollution norms have been developed: high-performance injection systems, gas recirculation systems (EGR), systems to reduce NOx emissions (AdBlue), the use of biofuels.

Biofuels have the advantage of being renewable, reducing dependence on oil resources [1]. At the same time, being oxygenated compounds, mixed with diesel can reduce pollutant emissions. Alternative fuels that mix with diesel must be soluble in diesel and have high cetane numbers, and the raw material for obtaining them must be cheap, easy to produce and the production process must be clean [2]. Currently, the most common biofuel used for diesel engines is biodiesel (alkyl ester of fatty acids from plants and animal fats) [3]. Biodiesel is a renewable fuel that can be produced from vegetable oils, waste oils, and animal fats. Biodiesel has about 10% oxygen in its structure, contributing to a more complete combustion [4].

Ethanol is less toxic, has a higher volatility and a higher oxygen content, which can offer a high potential to reduce pollutant emissions from the engine. Ethanol is a renewable fuel and can be obtained at a fairly low price on the market, which makes it an attractive alternative fuel. It is widely used in ethanol / gasoline blended fuel mode for gasoline engines. There are some problems with applying ethanol-diesel mixtures to the diesel engine. The solubility of ethanol in diesel is poor [5], which makes it difficult to mix ethanol in diesel with a high ethanol ratio. Ethanol can be used in diesel engines in several ways such as mixing ethanol and diesel before injecting it into the combustion chamber. Another method is fumigation [6]; in this case, the ethanol will be injected into the intake manifold and mixed with the intake air. A third technique is to inject fuel (diesel and ethanol) through a separate injection system into a combustion chamber. In the case of this work, the first method was used, blending ethanol with diesel because it does not require any modification of the engine equipment.

2. Method

The tests were made on a Renault K9K engine. A schematic diagram of the engine test bed is presented in figure 1. The engine specifications are presented in table 1.

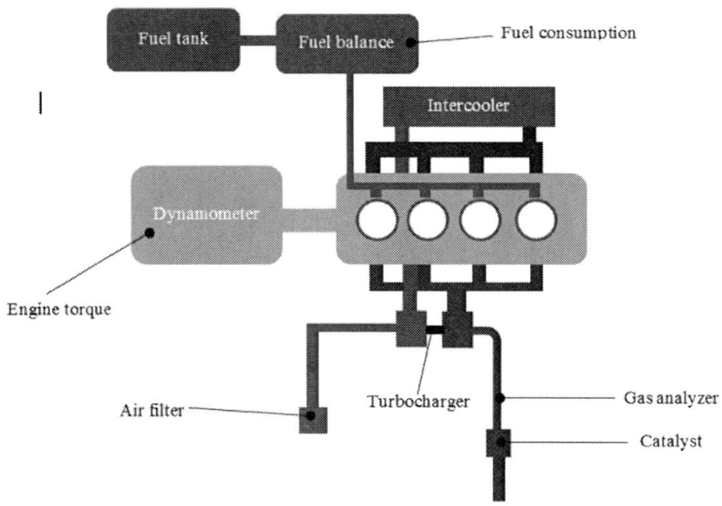

Figure 1. Schematic diagram of the engine test bed.

Table 1. Engine properties.

Engine type	Renault K9K four stroke
Number of cylinders	4
Bore (mm)	76
Stroke (mm)	80.5
Total displacement (cm^3)	1451
Compression ratio	15.3
Maximum power	72 kW at 3700 rpm
Maximum torque	200 Nm at 2700 rpm
Fueling	Common-rail direct injection

The engine was mounted on a Horiba Titan 250 test bench. The engine test bed is equipped with an electric Dynas3 LI250 dynamometer, which is designed for operated within a range of 0-8000 rotations per minute. It can measure engine power up to 250 kW with an accuracy of ±2%.

The engine was fuelled with a blend from mineral diesel and 10%/15% of ethanol and 10% of biodiesel obtained from waste oil. The characteristics of fuel are presented in table 2.

The Pierburg Hermann HGA 400 analyser was used to measure pollutant emissions. The analyser specifications are presented in table 3.

Table 2. Fuels properties.

Properties	Diesel	Biodiesel	Ethanol
Density (at 20°C), kg/m^3	840.2	862.3	796.3
Viscosity (at 20°C), mm^2/s	5.34	6.49	3.53
Cetane number	51.1	65.2	8.4
Flash point, °C	67	171	17.5
Net calorific value, MJ/kg	43.16	38.56	24.89

Table 3. Fuels properties.

Components	Measuring range	Accuracy
Carbon monoxide (CO)	0 – 10% vol	<1,2%vol±0,06% vol
Hydrocarbon (HC)	0 – 20000 ppm	< 220 ppm ±11 ppm
NO_x	0 – 5000 ppm	< 450 ppm ±4 ppm

The tests were done at an ambient temperature of 23°C and a relative humidity of 62%. The tests were performed at an engine speed of 3900 rpm (maximum engine power speed) for 25%, 50%, 75% and 100% engine load.

3. Results

3.1 Brake Power

Figure 4 shows the increase in brake power when the engine is running with a mixture of ethanol and biodiesel, at higher engine loads, while at low loads the power differences developed by the engine are insignificant

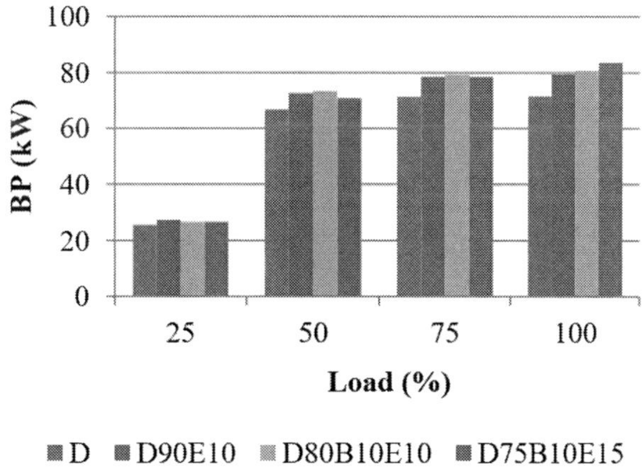

Figure 2. Brake power at different loads.

3.2 Carbon monoxide emissions (CO)

As shown in figure 5, CO emissions with combined fuels are higher than those of pure diesel. Increased emissions are caused by inefficient combustion, which is also observed by the variation of HC emissions (figure 6). Usually, any engine is most prone to CO emissions when the engine is running on a rich mixture or when the proper air-fuel mixture is not being made.

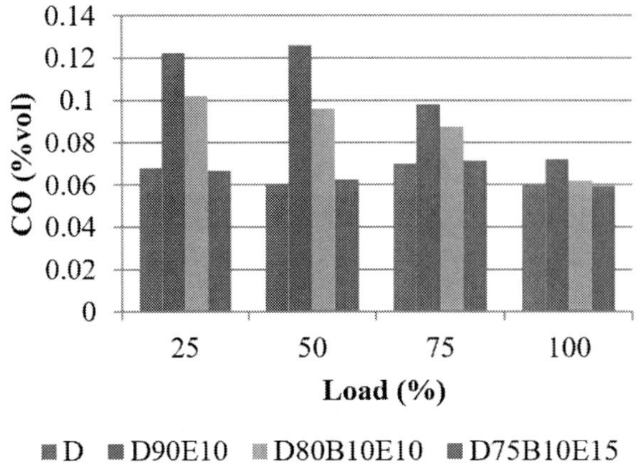

Figure 3. CO emissions.

3.3 Hydrocarbons emissions (HC)
Hydrocarbon (HC) emissions are considerably higher at all operating modes, as shown in figure 6. Unburned hydrocarbon emissions are due to improper injection process. Due to the low cetane number of ethanol, the ignition delay increases, producing an accumulation of unburned fuel in the cylinder. Due to this, areas with a rich mixture appear which burn incompletely, leading to an increase in hydrocarbon emissions.

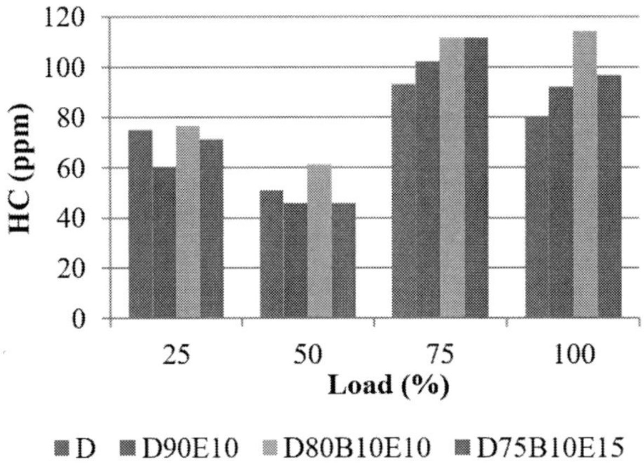

Figure 4. HC emissions.

3.4. NO_x emissions
The NOx formation rate decreased with the operation of the engine with ethanol-based mixtures compared to the operation with simple diesel. This can be attributed to the addition of ethanol, which possesses high latent heat of vaporization. During vaporization ethanol absorbs amount of heat from cylinder and drops the in-cylinder temperature. The formation of NOx emissions is favoured by the temperature and the presence of nitrogen in the cylinder. By decreasing the temperature in the cylinder, the rate of NOx formation also decreased.

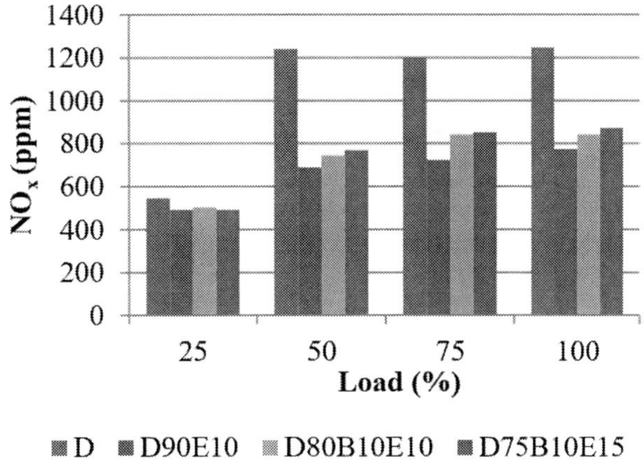

Figure 5. NO_x emissions.

4. Conclusions

In this study, the power generated, CO, HC and NOx emissions resulting from the addition of ethanol and bioethanol in a concentration of 10% and 15% to diesel fuel were investigated. For this purpose, a 4-cylinder engine in line was used at loads of 25%, 50%, 75% and 100%, at maximum power speed.

At higher engine loads, the engine's brake power (BP) increased with the ethanol-diesel-biodiesel mixture. A decrease in the power at low loads of the engine powered by an ethanol mixture was observed.

Hydrocarbon emissions (HC) are considerably higher at all engine tests regimes. Carbon monoxide (CO) emissions are higher than diesel fuel due to incomplete combustion, which certainly reflects an inhomogeneous air-fuel mixture.

The rate of formation of nitrogen oxides (NOx) emissions decreased when the engine was supplied with ethanol mixtures compared to the operation with mineral diesel. This is mainly due to the addition of ethanol which possesses high heat of vaporization.

5. References

[1] Shahir S A, Masjuki H H, Kalam M A, Imran A and Ashraful A M 2015 Performance and emission assessment of diesel – biodiesel – ethanol/bioethanol blend as a fuel in diesel engines: a review. *Renew. Sustain. Energy Rev.* **48**, 62–78

[2] Zoeldy M 2011 Ethanol-biodiesel-diesel blends as a diesel extender option on compression ignition engines. *Transport* **26**, 303-309

[3] Park S H, Cha J and Lee C S 2012 Impact of biodesel in bioethanol blended diesel on the engine performance and emissions characteristics in compression ignition engine. *Appl. Energy* **99**, 334-343

[4] Ileri S E 2013 Experimental investigation of the effect of diesel-cotton oil-n-butanol ternary blends on phase stability, engine performance and exhaust emission parameters in a diesel engine. *Fuel* **109**, 503-511

[5] Hansen A C, Zhang Q and Lyne P W 2005 Ethanol-diesel fuels blends – a review. *Bioresource technology* **96**, 277-285

[6] Chauhan B S et al 2011 Experimental studies on fumigation of ethanol in small Diesel engine. Energy **36,** 1030-38

IOP Conf. Series: Materials Science and Engineering **898** (2020) 012025 doi:10.1088/1757-899X/898/1/012025

Wear of three-treads uncoated and coated integral hob milling tools

S Sovilj-Nikić[1], B Sovilj[2], G Varga[3] and N Ungureanu[4]

[1] Iritel a.d. Beograd, 11 000 Belgrade, Serbia
[2] University of Novi Sad, Faculty of Technical Sciences, 21 000 Novi Sad, Serbia
[3] University of Miskolc, Faculty of Mechanical Engineering and Informatics, 3515Miskolc, Hungary
[4] North University of Baia Mare, Engineering Faculty, 430311 Baia Mare, Romania

E-mail: sandrasn@eunet.rs

Abstract. Tribological Processes in the gear cutting of cylindrical gears on cutting elements of hob milling tools develop under specific conditions. The particular requirements under which the contacts between the hob milling tools and the workpiece material are made, make it difficult to become familiar with the tribological processes on them so that today there is more knowledge about their consequences on the machined surface of the teeth and the cutting elements of the hob milling tool, and less about their nature. In this paper, based on the experimental investigations, the wear analysis of three-treads uncoated and coated integral hob milling tools in the gear cutting of cylindrical gears is given.

1. Introduction

Tribomechanical systems for transmitting the energy of mechanical movement at a distance, as well as for modifying its parameters, also carry rotational movements. The elementary functions of the gearbox are energy distribution, reduction or increase of speed, change of type of action, start, stop, change of direction and more.

The functionality of toothed gears is based on a specific form of tribological- elements and appropriate conjugation laws of the contact surface.

When gear cutting as a significant element of tribomechanical systems, it is very important to consider all tribological aspects, since errors made during machining cannot repair even perfectly organized maintenance.

2. Tribological processes, deterioration and wear

Tribological processes take place with the continuous destruction of parts of the surface of elements of the tribomechanical system and the emergence of new physically or topographically similar or completely different from the previous ones [1,2]. The destruction and formation of parts of the contact surface is temporally uneven and spatially discrete. Contact surfaces machined by hob milling always have irregularities. The connection between the contact surface topography and the development of tribological processes is extremely complex [3,4,5,6]. The change of topography in the development of tribological processes can be represented by the model as in figure 1 [2,3].

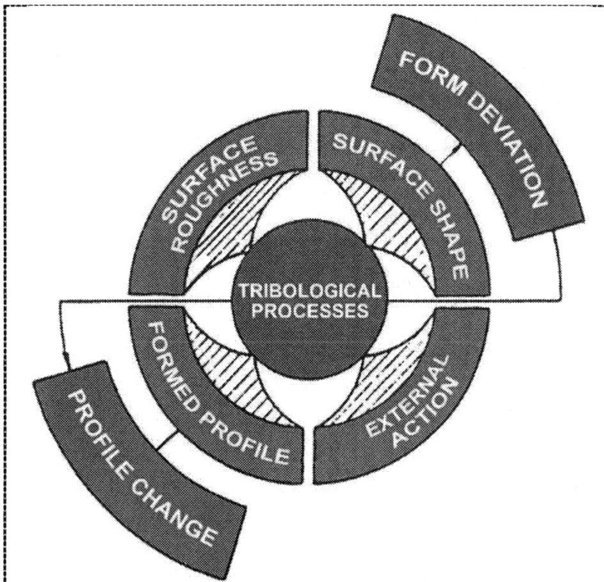

Figure 1. Change of topography in the development of tribological processes

The deterioration of elements of tribomechanical systems in the process of gear cutting is one of the negative phenomena. The basic deterioration patterns of the hob milling tool are given in figure 2.

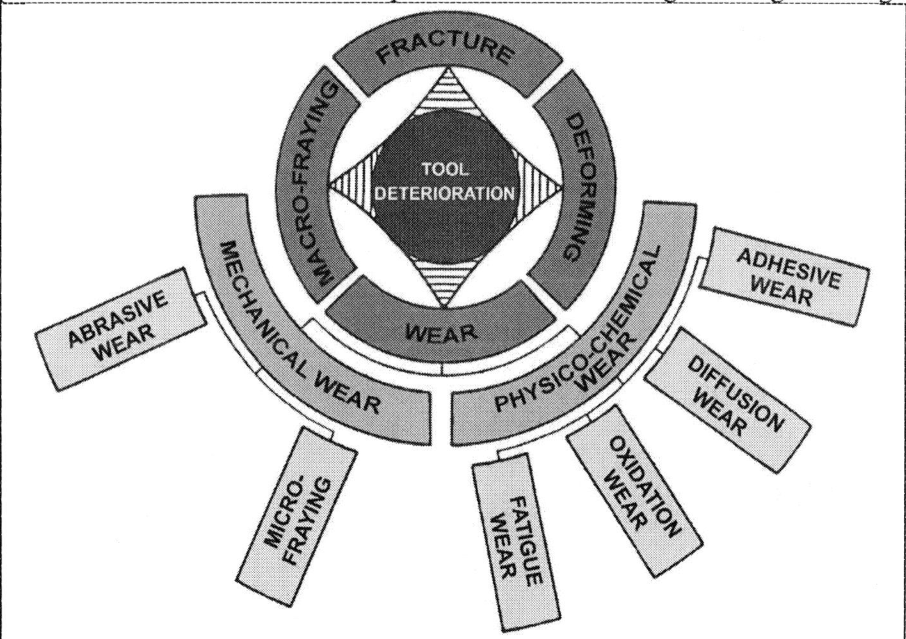

Figure 2. Basic deterioration forms of hob milling tools

As can be seen in figure 2, wear is one form of deterioration of the hob milling tools.

In hob milling, all cutting elements do not play the same role in the process of gear cutting, that is, they do not remove the same amount of material from the workpiece [7]. For this reason, the wear of the cutting elements or the teeth of the hob milling tool is different. The teeth that take away most of the material when gear cutting wear the most.

3. Results of experimental research

This paper presents a part of the results of several years of research on the process of hob milling of cylindrical gears serration and the possibility of introducing into the production process of coated hob milling tools, which is related to the wear of three-treads uncoated and coated integral tools. Based on the experiment plan, tests were carried out, using hob milling tools from the same batch of HS-6-5-2-5 material, uncoated and coated with TiN.

During the test, the gear cutting of cylindrical gear was performed on the workpiece of the same material and characteristics with three-treads uncoated and coated integral hob milling tools. For these studies, an original form was developed to monitor the development of tribological processes in the gear cutting by hob milling. On this form, it is possible to enter information about: workpiece, hob milling tool, hob milling machine, technological process, sketch of workpiece in gear cutting, tooth wear, development of the wear process, wear distribution and more.

The paper provides data for the experiment with uncoated three-treads hob milling toll 1U (U-uncoated), and for all other experiments changes only the data for the tool mark 6U (U-uncoated), IVC (C-coated), VC (C-coated).

Workpiece data: outer diameter d_k = 305 mm, number of teeth z_2 = 100, module m, standard module 3 mm, angle of contact line $\alpha = 20°$, angle of inclination of teeth $\beta = 0°$, tooth width b = 90 mm, material HS-6-5-2-5, hardness HB = 180.

Hob milling data: integral 1U, diameter of hob milling tool d_g = 120 mm, length of hob milling tool l_g = 115 mm, module m = 3 mm, basic profile II, normal pitch t_n = 9,42 mm, coil angle ω = 10,4°, coil pitch $T = 91{\cdot}2$ mm, direction of coil tread L, number of treads z_1 = 3, number of grooves i = 10, material of hob milling tool HS-6-5-2-5, hardness HRC = 65, accuracy class A, hole diameter d = 22 mm.

Machine data: hob milling machine MODUL-ZFWZ-250x5A, Cooloing agent REZANOL EP40
Process data: number of revolutions n = 200 rev/min, axial feed S_a = 6 mm/rev.

The primary processes of wear are manifested in two basic ways as a flank wear and as a crater wear. Numerous technological parameters, i.e. conditions, influence which type of wear will be primarily dominant when gear cutting. During the experiment, only a part of which is presented in this paper, flanh wear was the primary wear zone. In figure 3 wear process on inlet lateral flank, top flank and outlet lateral flank is shown. The wear of oulet lateral flank for pretreatment VB_0 = 0,5 mm is adopted as a wear criterion.

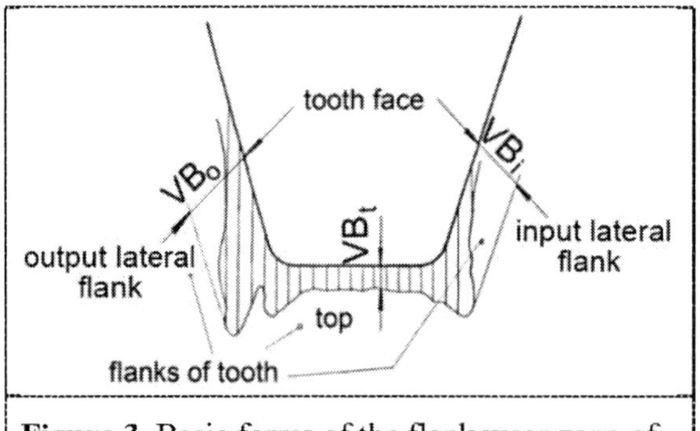

Figure 3. Basic forms of the flank wear zone of hob milling tool

In figures 4, 5, 7 and 8 characteristic examples of wear forms for two uncoated and two coated three-treads integral hob milling tools are given The tooth wear forms of the uncoated three-treads integral hob milling tool 1U of HS-6-5-2-5 material on three combs (1,5,9) are given in figure 4. The wear distribution for this tool on the input lateral flank and output lateral flank is shown in figure 6. The tooth wear forms of the uncoated three-treads integral hob milling tool 6U of HS-6-5-2-5 material on three combs (1,5,9) are given in figure 5. The tooth wear forms of the three-treads integral hob milling tool

IVC of material HS-6-5-2-5 coated with TiN on three combs (1,5,9) are shown in figure 7. Figure 8 shows the tooth wear forms of the three-treads integral hob milling tool VC of material HS- 6-5-2-5 coated with TiN on three combs (1,5,9).

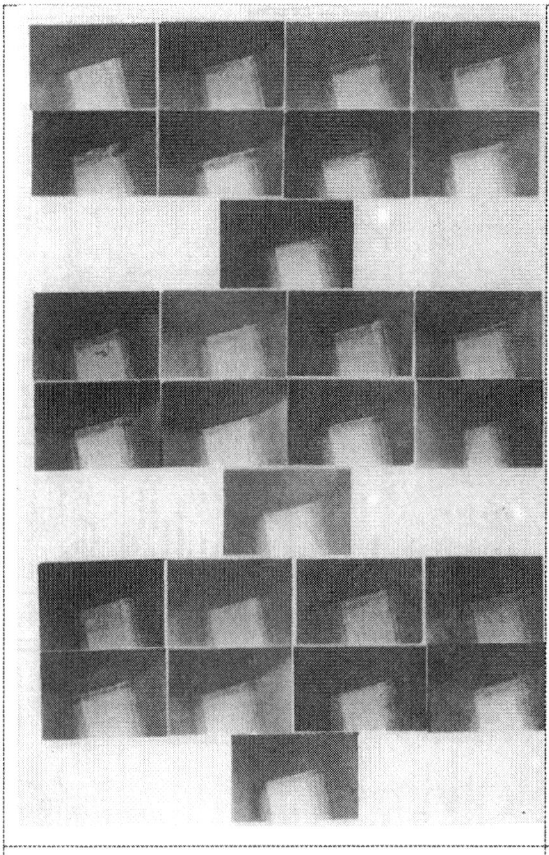

Figure 4. Records of the tooth wear form of the uncoated three-treads integral hob milling tool 1U

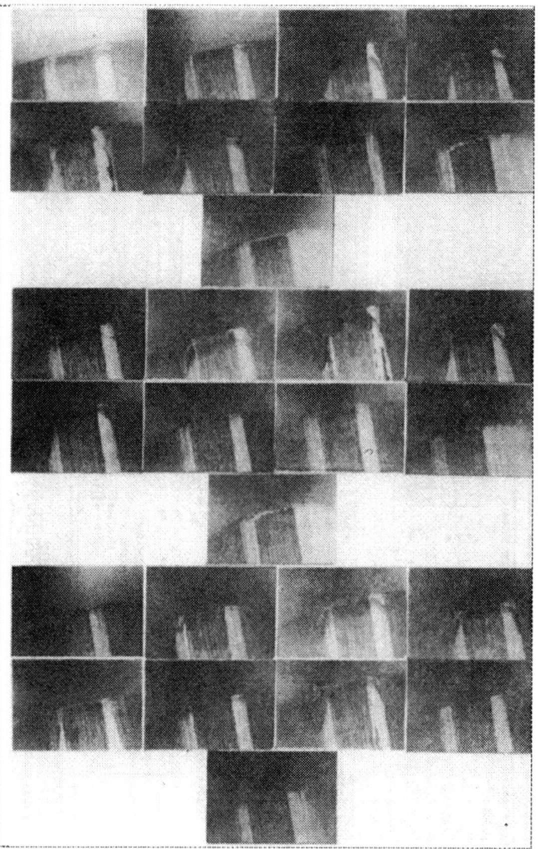

Figure 5. Records of the tooth wear form of the uncoated three-treads integral hob milling tool 6U

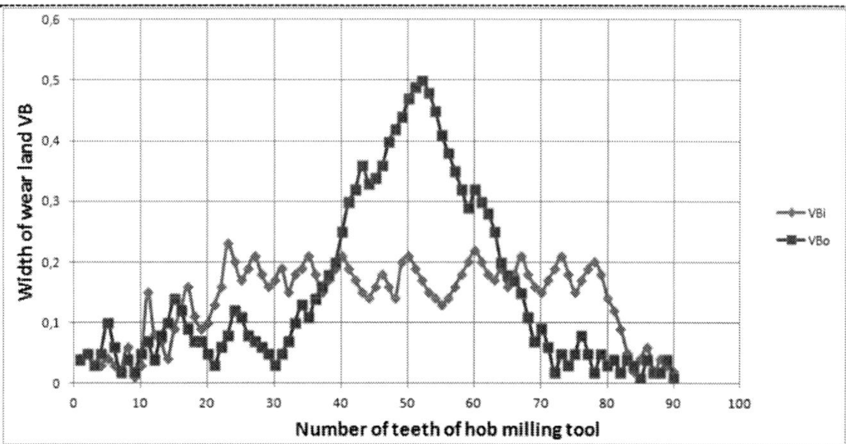

Figure 6. Wear distribution at the input and output lateral flank for uncoated three-treads integral hob milling tool 1U

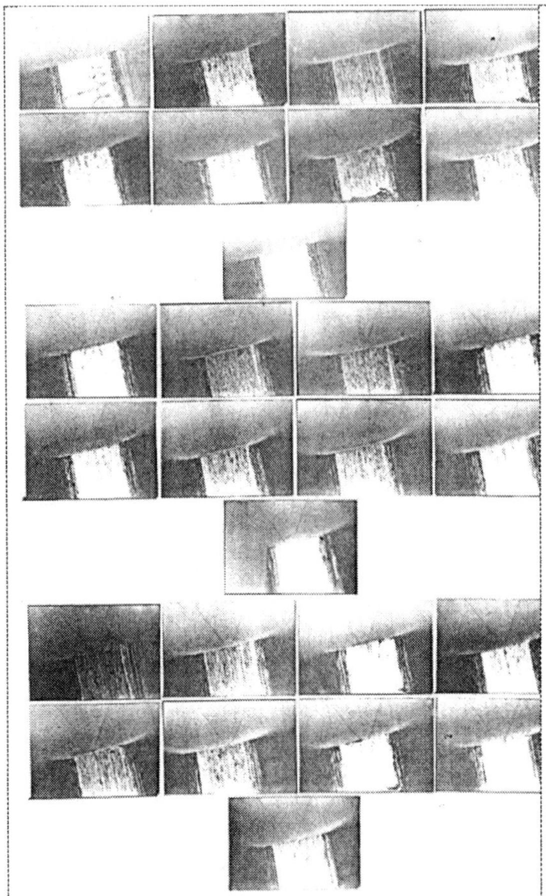

Figure 7. Records of the tooth wear form of coated three-treads integral hob milling tool IVC

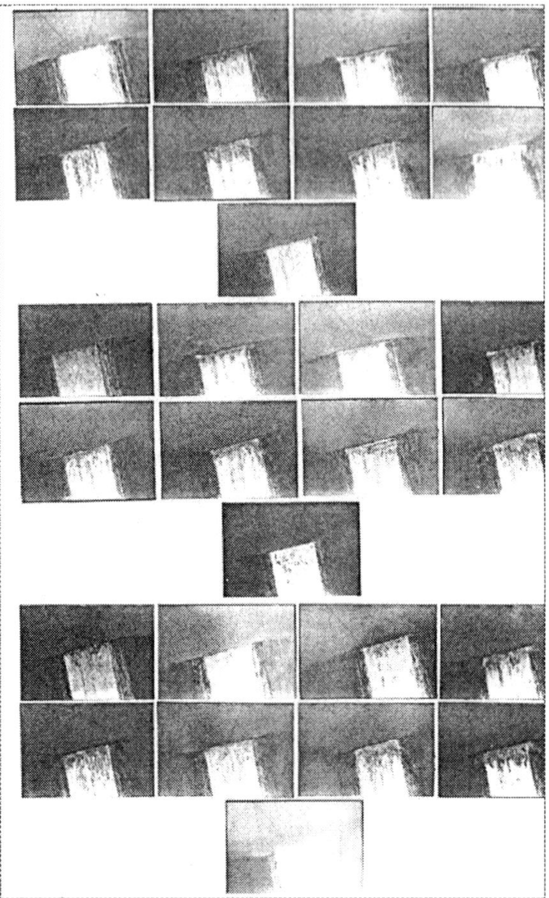

Figure 8. Records of the tooth wear form of coated three-treads integral hob milling tool VC

Comparative tests of uncoated three-treads integral hob milling tools and those coated with TiN were performed under real production conditions. Experimental studies have confirmed that there is concentrated wear of the cutting teeth of the hob milling tools. The morphology of the flank wear is characterized by the appearance of most often one groove on the input lateral flank and two grooves on the output lateral flank. The grooves are located in the zones of the transition of the top flank into the input and output lateral flank of the cutting elements of the hob milling tool.

Experimental studies have shown that three-treads integral hob milling tools process 2.46 more cylindrical gear teeth than three-treads uncoated integral hob milling tools.

4. Conclusion

Based on the presented in this paper, it can be concluded that the introduction of three-treads coated integral hob milling tool into the production process of gear cutting of cylindrical gears is justified, because significantly higher productivity is achieved and also better quality of machined surface is obtained.

5. References

[1] Bouzakis K-D, Skoridatis G and Michalidis N 2008 Innovative methods for characterizing coatings mechanical properties, *Proceedings of International Conference 'The Coatings 2008'*, October 2008, Kallithea, Greece

[2] Sovilj-Nikić I *Modeling and optimization of hob milling process (Modelovanje i optimizacija procesa odvalnog glodanja)*, Fakultet tehničkih nauka, Novi Sad, Serbia, Doctoral Thesis (unpublished)

[3] Sovilj B 1988 *Identification of tribological processes during hob milling (Identifikacija triboloških procesa pri odvalnom glodanju)*, Fakultet tehničkih nauka, Novi Sad, Serbia

[4] Sovilj-Nikić I, Sovilj-Nikić S, Sovilj B and Đokić R 2017 Analysis of roughness parameters of the tooth-face of model hob milling tool, *Machine design*, **9** (3), 99-106

[5] Sovilj-Nikić I, Sovilj-Nikić S, Sovilj B and Blanuša V 2017 Topography of characteristics surfaces of model hob milling tools for machining of gear serration, *Proceedings of International Scientific Conference IRMES 2017*, 7-9 September 2017, Trebinje, Bosnia and Herzegovina, 241-246

[6] Sovilj B, Sovilj-Nikić S and Juliana Javorova 2019 Tribological researches of triboelements topography of hob milling process of cylindrical gear serration, *Power Transmissions, MATEC Web of Conferences* ***287***, 05003 (2019), p.13

[7] Sovilj B and Sovilj-Nikić S 2018 Tribological researches of gear cutting processes of cylindrical gears, invited paper in *Proc. of 13th International Scientific Conference MMA 2018*, Novi Sad, Serbia, 107-116

Acknoledgements

The research was funded by the Ministry of Education, Science and Technological Development of the Republic of Serbia within the project III 43008, and it is also the result of the cooperation within CEEPUS projects CIII- RO-0058-07-1920 and CIII-RS-0304-10-1920.

IOP Conf. Series: Materials Science and Engineering **898** (2020) 012026 doi:10.1088/1757-899X/898/1/012026

Overcoming the barriers of consignment stock policy implementation in a manufacturing company

M Faur[1], A Sipos[2], C Bungau[1] and C I Gherghea[1]

[1]Faculty of Management and Technological Engineering, University of Oradea, 1 Universitatii street, Oradea, Romania

[2]Faculty of Agricultural Sciences, Food Industry and Environmental Potection, Lucian Blaga University of Sibiu, 7-9 Dr. Ion Ratiu street, Sibiu, Romania

monica.faur@csud.uoradea.ro

Abstract. The present study represents a part of a wider research, centered on consignment stock program adoption and implementation in a supply chain governed by lean and agile strategies. The paper attempts to explore the challenges and the barriers which may occur prior or during the implementation stages of the project, to identify and mitigate the risks and also to establish safe approaches to overcome project failure. The work includes a brief overview on consignment stock model and its benefits, along with the implementation risks perceived by the internal stakeholders, mostly viewed as a human factor reaction to change. The methodology consists of a survey addressed to the business functions directly affected by the project, in order to identify the main barriers and risks, and of an assessment of the risk severity and its distribution within and among different groups of stakeholders, using One-way ANOVA statistical method. A solutions framework to the identified risks is proposed, in order to overcome implementation barriers.

1. Introduction

In manufacturing companies, the levels of inventories at all stages, as raw material, work-in-process and finished goods inventories, may indicate the company's competitive positioning. Lowering the inventory in terms of level and value, especially for the purchased items as raw materials and packaging, is an essential objective in any business environment [1]. In this view, 'consignment stock' (CS) represents a particular case of the vendor managed inventory approach, where the vendor maintains a stock of materials at the buyer's plant, which in the present case is the manufacturing company. The idea is based on a close collaboration between the vendor and the buyer, aiming to create a win-win situation through revenue sharing [2-3]. In accordance with a general CS policy, the vendor guarantees the buyer to continuously keep a stock between a minimum and a maximum level, limits that have been agreed between the two parties [4].

Under a CS approach, the inventory location is with the buyer, along with the holding costs for stocking items and ordering responsibility, while the inventory ownership is with the vendor, until the stock is consumed [5]. At that moment, the inventory which corresponds to the consumed amount is transferred to the buyer, via invoicing, usually once or twice a month, depending on the consignment agreement concluded between the parties. As a consequence, there are no financial implications for the buyer concerning the consigned stock, meaning that the overall raw materials or packaging inventory is diminished with the value of the consigned materials. There are several benefits associated with CS implementation, as 'zero cost' for the consigned inventory, 'zero' procurement lead-times, no out of stock situations, improved cash-flow, no more disruptions and forecast accuracy problems, just to name a few of them. All of these contribute to enhanced agility, while shortening the chain downstream towards the final customer. From the benefits point of view, CS model seems to be a very attractive proposal, however every significant change is linked to challenges and brings elements that can't be neglected and need to be identified and deeply assessed prior to the decision making process, in order to find out at what extent is a supply chain or an organization ready to adopt it, and how to proceed in order to avoid an implementation failure. There have been discussions and studies that revealed the fact that supply chains under certain management strategies, as lean and agile practices for example, are more suitable to support innovative initiatives and transformation processes, improving overall performance [5-8].

Lean and Agile strategies seem to be a very common approach that companies choose to adopt, in order to get enhanced efficiency and effectiveness in their supply chains [9]. Lean technics are mostly used to improve effectiveness [10], while agile tools enable the chain to become more flexible, increasing its adaptability, towards greater responsiveness and customer satisfaction [5].

2. Problem formulation

The present research is based on exploratory case studies, represented by real attempts of CS implementation in two different subsidiaries of a multinational company. The two facilities, A and B, producing FMCG goods, are based in different countries, with certain differences in respect of the economic and legal environments. The medium size plant A was chosen as a pilot of CS adoption project, resulting in a successful implementation, discussed by Faur & Bungau (2019) [1]. Based on the positive and valuable outcome of plant A, the group management recommendation was to implement the same project in a larger size plant, B. Despite the fact that the two plants are part of the same organization, governed by the same strategies and rules, and having almost the same organizational culture, the idea of adopting CS in plant B generated several challenges to the project team, raised by the stakeholders. At a first sight it looked like a usual resistance to change, which is somehow in human's nature, but the things escalated and generated real barriers to the implementation process. In this respect, the present research aims to investigate and assess the risks associated with the challenges, and propose a solution framework, in order to overcome the barriers towards achieving the benefits of a CS program.

3. Research methodology

The data collection has been conducted by one of the authors, being part of the buying organization project team. A questionnaire survey has been addressed to the internal and external stakeholders, in order to find out their concerns and their risk perception regarding the CS system adoption. Once data consolidated, the main risks have been identified and all stakeholder categories have been asked through another questionnaire to assign scores between 1 and 5 to each identified risk, 1 being the score for the lowest risk and 5 for the highest risk. The data has been further analysed through one-way ANOVA statistical method, run in Matlab program. One-way (or single factor) analysis of variance, abbreviated ANOVA, is a statistical model that can be used to compare means of two or more samples, using the F distribution [11]. Developed by statistician and evolutionary biologist, Ronald Fisher, the method allows the estimation of the relative significance of different parameters within and among the groups [12].

4. Barriers and perceived risks identification and assessment

When it comes to acceptance of a new technology, a new business model adoption, or a new policy implementation, perceptions of risk and benefit are generally important, if not crucial [13-15].

Decision making needs to also consider risk perceptions and these perceptions are based on a frame of reference and sometimes on incomplete knowledge on the subject [16]. Provided information related to a new project can be insufficient or ambiguous, generating discomfort and sometimes a fake perception of risk. Perceived risks and benefits of an innovative approach are constantly subject to change, depending on the information shared and on the problem understanding degree. A new technology, a new policy may be perceived as risky, but, when it is embedded in routine behaviour the perception of it can change to not risky [17]. Furthermore, perceived risks and benefits also influence each other; the level of perceived benefits can have an impact on the acceptability levels of the perceived risks [18]. Slovic (1999) notes that risk is subjective, "from the initial structuring of a risk problem, to deciding which end-points or consequences to include in its analysis, identification and estimation of exposure, and so on"[19].

The perceived risk may consist of psychological, physical, financial, social, and performance factors [20] and can affect a decision implementation by bringing negative consequences [21]. Some authors also mention functional risk, and time risk [22]. The idea is to identify all type of perceived risks prior to a project implementation in order to prevent a negative outcome. Therefore, further research into this area is required. However, the perceived risk can have a negative relationship with the variable perceived benefit, generating barriers to a project implementation. Perceived risks might also have substantial impacts on the decisional processes for innovations [23-24]. From the above categories, the following risk types are most related to the present research: financial, functional, psychological, performance and time-related. Financial risk relates to the potential negative financial outcomes which are associated with new system adoption [25]. Psychological risk can be defined as anxiety and/or uncomfortable feelings arising from anticipated post-behavioural emotions such as worry and tension [26-27]. Functional and performance risk origin from the insufficient information shared, misaligned procedures, lack of knowledge, lack of training, or lack of understanding the functionality of the new proposed approach (how it works). Time risk relates to the perception that the adoption and use of an innovative idea will take too long [28], or will be a perceived as waste of time [29-30].

The carried-out surveys by the project team on B plant' stakeholders revealed the main barrier categories, challenges and risks associated with the CS adoption and implementation process, presented in figure 1.

Barrier Category	Challenges	Perceived Risks
Internal processes	Extra resources not included in business plan (BP), additional workload and missing competences	- Inaccurate reports due to increased complexity - Extra financial resources for training requirements - Increased time per operation (more attention required) - Additional human resources vs BP - special aprovals needed, along with financial resources - Extra working time - overtime constraints
Organizational capabilities	Storage space limitation	- Reshape warehouse material flow due to extra storage place required by CS policy (material delimitation per each vendor)
Regulation	Legal constraints	- Lack of local legislation regarding CS concept
Financial	Additional reports and workload	- Weekly reconciliations with vendors - time consuming activities
Technical	Integrated information systems	- Lack of coordination and correlation in case the vendor uses a different system than SAP
Internal procedures	Clear procedures	- Misaligned procedures across functions
Vendor selection	Vendors evaluation	- Questioning suppliers' reliability and capabilities

Figure 1. Barrier categories, challenges and related risks that limit the implementation of CS program.

The collected data is influenced by people's perceptions or interpretations. The consolidated data from the second questionnaire addressed to the stakeholders is reflected by table 2.

Risk no.	Perceived Risks	Allocated scores by the internal stakeholders					
		Planning	Finance	Operation	Procurement	RM Wh.	Legal
R1	Inaccurate reports due to increased complexity	3	4			5	
R2	Extra financial resources for training requirements	4	3	5	2	4	
R3	Increased time per operation (more attention required)	3	4	4		4	
R4	Additional human resources vs BP - special aprovals needed, along with financial resources		4			5	
R5	Extra working time - overtime constraints	3				4	
R6	Reshape warehouse material flow due to extra storage place required by CS policy (material delimitation per each vendor)			3	2	4	
R7	Lack of local legislation regarding CS concept		3	2	2		4
R8	Weekly reconciliations with vendors - time consuming activities	4	4	1		4	
R9	Lack of coordination and correlation in case the vendor uses a different system than SAP	4	4	1	3		
R10	Misaligned procedures across functions	2	3	3	2	4	3
R11	Questioning suppliers' reliability and capabilities		4		3		2
		Scores between 1-5 (1 lowest risk - 5 highest risk)					

Figure 2. Score allocation for the perceived risks.

Data analysis is further required in order to evaluate the risk dispersion on each steakholder category and the risk relationship among different categories.

5. Risk assessment using one-way ANOVA

ANOVA determines whether the groups created by the levels of the independent variable (the perceived risks, in our case) are statistically different by calculating whether the means of the treatment levels are different from the overall mean of the dependent variable (represented by the stakeholder groups). The test statistic, F, assumes independence of observations, homogeneous variances, and population normality. The one-way ANOVA tests the null hypothesis: $H_0 : \mu_1 = \mu_2 = ... = \mu_s$, where μ = group mean and s = number of groups, which states that samples in all groups are drawn from populations with the same mean values. If, however, the one-way ANOVA returns a statistically significant result, we accept the alternative hypothesis ($H_\alpha : \exists \mu_i \neq \mu_k, i \neq k$), which is that there are at least two group means that are statistically significantly different from each other.

Following the data processing in Matlab, ANOVA model provided the information highlighted in figure 3.

ANOVA Table

Source	SS	df	MS	F	Prob>F
Groups	16.7095	5	3.3419	4.8	0.002
Error	23.6905	34	0.69678		
Total	40.4	39			

Figure 3. Processed data through One-way ANOVA, in Matlab.

where 'Groups' represents SSA (Sum of Squares Among groups), defined by the following relation:

$$SSA = \sum_{k=1}^{s} n_k (\overline{x}_k - \overline{\overline{x}})^2 \tag{1}$$

$\overline{x}_k$ being the mean of group k; n_k is the number of values corresponding to group k and s is total number of groups;

'Error' (Matlab notation) represents SSW (Sum of Squares Within groups), calculated by the below relation:

$$SSW = \sum_{k=1}^{s} \sum_{i=1}^{n_k} (x_{ik} - \overline{x}_k)^2 \tag{2}$$

where x_{ik} is value i from group k and n is total number of values $n = n_1 + n_2 + \ldots + n_s$

'Total' means SST (Total Mean Square), representing the variance between groups;

'SS' = the numerical value of SSA and SSW;

'df' = the number of freedom degrees

'MS' reflects MSA (Mean Square among Groups) and MSW (Mean Square within Groups);

'F' = Fisher test - representing the ration of the mean squared errors (*MSA*/*MSW*);

'*p*' is the probability that the test statistic can take a value greater than or equal to the value of the test statistic, i.e., $P\ (F > 4.8)$. The small *p*-value of 0.002 indicates that differences between group means are significant (figure 2).

The synthesis of ANOVA application is presented in table 1.

Table 1. Synthesis of One-way ANOVA application

Variation source	Degrees of freedom	Sum of Squares	Mean Square (the variance)	*F* test
Among Groups	s-1	SSA	$MSA = \frac{SSA}{s-1}$	
Within Groups	n-s	SSW	$MSW = \frac{SSW}{n-s}$	$F_{calculat} = \frac{MSA}{MSW}$
Total	n-1	SST	$MST = \frac{SST}{n-1}$	

6. Results and interpretations

The performed statistical analysis revealed that signifficant differences exist in at least two groups from the compared ones. The differences refer to the severity degree of the risks.There are also differences between pairs of groups acting as a majority, and a third group: e.g. Finance and Raw Material Warehouse (RM Wh.) groups compared to Procurement – figure 4.

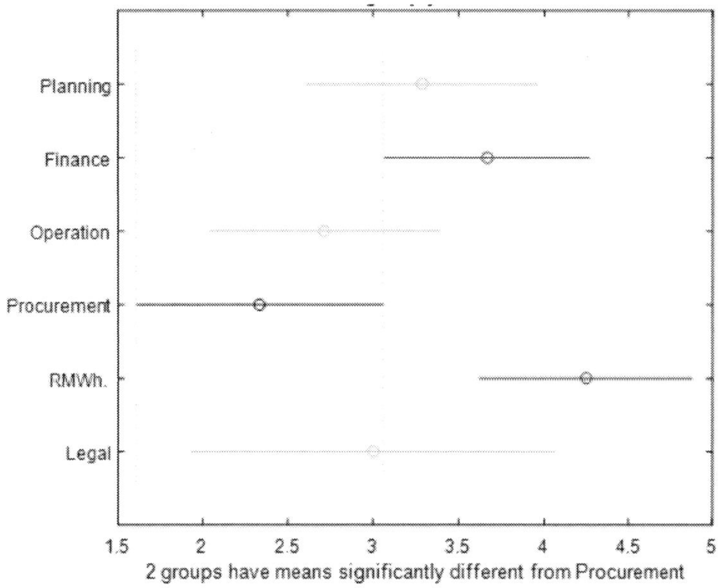

Figure 4. Risk severity degree between groups (Matlab generated).

ANOVA method also provided the output shown in figure 5, concerning risk dispersion within stakeholders.

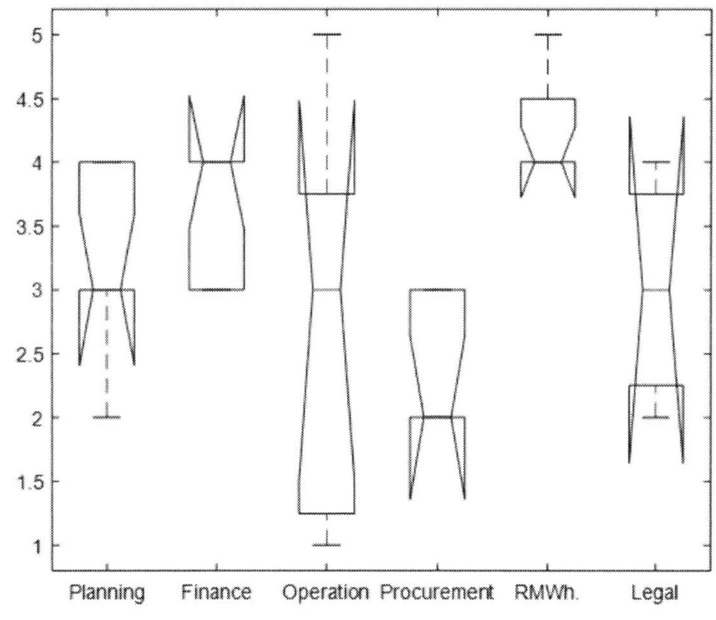

Figure 5. Perceived risk dispersion within groups (Matlab generated).

The output shows that there are significant differences among Procurement function and RM Wh. group in terms of risk severity. It can be seen that RM Wh. category is the most affected by the CS policy adoption, which is true, because the most activities connected with the new project are executed by this department. The scores assigned to the risks that are relevant for this compartment are concentrated at relative high level, compared to Procurement function, which has a risk level below average. Operation (Production) has a wide risk dispersion, which is somehow understandable, as almost all the other departments operate as support functions for the Production group and at the end all the perceived risks have an impact on the manufacturing process. Finance and RM Wh. have quite similar average risk levels and the same situation can be observed when comparing Planning, Operation and Legal functions. Procurement is the less impacted function and this is correct, as once the consignment agreements are concluded, Procurement responsibility is only to monitor the suppliers and the other groups to stick to the agreements. Legal compartment identified a reduced number of risks, but they cover a large range in terms of severity, almost like the Operation function. From the risk type perspective it has been found that finance and functional risks have the highest severity level (e.g. Finance group and RM Wh.).

The analysis provided a general view of the risk perceptions among and within the stakeholders. This approach helped to further more deeply assess the causes that generated the risks and the people perceptions. As a next step, a solution framework has been proposed, having the purpose to mitigate risks and bring them to an "acceptable level", make the stakeholders understand the outcome of the CS program and its great benefits for the business, in order to give the project an implementation chance. The solution framework is presented in figure 6.

Perceived Risks	Proposed solutions for "acceptable level" of risk
R1 - Inaccurate reports due to increased complexity	Resources allocation for new competences develpoment
R2 - Extra financial resources for training requirements	Organizing workshops with SAP experts on CS module
R3 - Increased time per operation (more attention required)	Skills development, motivating incentives
R4 - Additional human resources vs BP - special aprovals needed, along with financial resources	FTE (full time equivalent) re-calculation, additional resources allocation
R5 - Extra working time - overtime constraints	
R6 - Reshape warehouse material flow due to extra storage place required by CS policy (material delimitation per each vendor)	Re-thing stock dimensions for all materials in place; implementation plan / time-frame to be extended in order to properly prepare the storage space
R7 - Lack of local legislation regarding CS concept	Legal and Financial functions research in the field; finding a compromise solution; carrefully drafting vendor agreements
R8 - Weekly reconciliations with vendors - time consuming activities	System integration with vendors, for data visibility and transparency; ease-up reconciliation reports
R9 - Lack of coordination and correlation in case the vendor uses a different system than SAP	Technical know-how and expertize from consultants
R10 - Misaligned procedures across functions	Service level agreement on procedures across all implied functions
R11 - Questioning suppliers' reliability and capabilities	Accurate and transparent evaluation of vendors

Figure 6. Risk solution framework.

7. Conclusions

The case study reveals that even though a project implementation has a successful outcome in one facility of an organization, the same positive result is not warranted for another subsidiary of the same company, acting in a different economical and legal environment and having a larger size.

In case the challenges are addressed and the identified risks are assessed and brought to a level of acceptance, the chances of a positive implementation will significantly increase.

Many risk analysts assume that people opposition to innovative projects is mostly due to unfounded fears of their perceived risks. The one way ANOVA provided a useful output to analyse the risk dispersion within stakeholder groups and risk impact among different groups, which further permitted to investigate whether the perceived risks are real and have a significant degree of severity.

The assessment of the perceived risks demonstrates that people perception can be changed when the issues are properly addressed, problems are clarified, pertinent information is provided and targeted solutions are proposed.

The barriers arisen in a project implementation can be overcome by bringing the risks to an acceptable level, which could allow the acceptability of the project that has been judged or declined.

Further research objectives on the perceived risks might be in the range of grouping the risks by different criteria and exploring which specific risks are most predictive of acceptance.

References

[1] Faur M and Bungau C 2018 Supply chain 'leagility' through adopting consignment stock strategy in manufacturing companies *Proceedings of the 6th RMEE Int. Mng. Conf.* (Cluj-Napoca) 623-63

[2] Battini D, Gunasekaran A, Faccio M, Persona A and Sgarbossa F 2010 Consignment stock inventory model in an integrated supply chain, *Int. J. Prod. Research* **48** 477-500

[3] Chen J M, Lin I C and Cheng H L 2010, Channel coordination under consignment and vendor-managed inventory in a distribution system, *Transp. Res. Part E: Log. & Transp. Rev.* **46** 831-843

[4] Valentini G and Zavanella L 2003 The consignment stock of inventories: industrial case in performance analysis *Int. J. Prod. Ec.* **81–82** 215–224

[5] Faur M and Bungau C 2019 Exploring the insights of a consignment stock program implementation in a leagile supply chain *Annual Sess. Scientific Papers- IMT Oradea* 291-295

[6] Draghici A 2007 Adaptive technologies and business integration: social, managerial and organizational dimensions *IGI Global* 211-243

[7] Avasilcai S 2007 Performance evaluation in Romanian industrial organisations *Proc. Conf.: Intelligent Manuf. & Autom., Annals of Daaam for 2007 The 18th Int. Daaam Symp* 27-28

[8] Gherghea IC and Bungau C 2018 Poka yoke application synthesis in manufacturing engineering *Proc. 6th Rev. Manag. Econ. Eng. Int. Manag.* (Cluj Napoca) 564–571

[9] Faur M and Bungau C 2019 Outsourcing towards greater agility by investigating decoupling points in leagile supply chains *Int. Manufact. Science and Ed. Conf.* MATEC Web of Conferences (Sibiu) 290, 07006

[10] Gherghea I C, Bungau C and Negrau D C 2019 Lead time reduction and increasing productivity by implementing lean manufacturing methods in cnc processing center *IOP Conf. Ser. Mater. Sci. Eng.* **568** 1

[11] Howell D 2002 *Statistical Methods for Psychology*, Duxbury ISBN 0-534-37770-X p. 324–325

[12] Moore D S, McCabe G P 2003 *Introduction to the Practice of Statistics* (4th ed. W H Freeman and Co.) p 764 ISBN 0716796570

[13] Hurlimann I C 2007 Is recycled water use risky? An urban Australian community's perspective *Environmentalist*, **27** 83-94

[14] Otway H J and Von Winterfeldt D 1982 Beyond acceptable risk: on the social acceptability of technologies *Policy Sci.* 14 247-256

[15] Van Dijk H, Fischer A R H, Marvin H J P and Van Trijp H C M 2017 Determinants of stakeholders' attitudes towards a new technology: nanotechnology applications for food, water, energy and medicine *J. Risk Res.* 20 277-298

[16] Weisenfeld U and Ott I 2011 Academic discipline and risk perception of technologies: an empirical study *Res. Pol.* **40** 487-499

[17] Flynn R, Bellaby P and Ricci M 2006 Risk perception of an emergent technology: the case of hydrogen energy *Forum Qual. Soc. Res.*, **7** 1 https://doi.org/10.17169/fqs-7.1.58

[18] Fishoff B, Slovic P, Lichtenstein S, Read S and Combs B 2000 How safe is safe enough? A psychometric study of attitudes towards technological risks and benefits *The Perception of Risk, Earthscan*, London 80-103

[19] Slovic P 1999 Trust, emotion, sex, politics, and science: Surveying the risk-assessment battlefield *Risk Anal.* **19** 689–701

[20] Jacoby J, Kaplan I B 1972 The components of perceived risk *SV Proc 3rd Annual Conf. of the Association of Consumer Research*

[21] Solomon M R 2016 Buying, owning and being *Consumer behaviour 11th Ed – Bookman Ed*

[22] Hsu T H and Lin I Z 2006 Using fuzzy set theoretictechniques to analyse travel risk: an empirical study *Tour Manag.* **27** 5 968-981

[23] Janssen J, Marell A and Nordlund A 2011 Exploring consumer adoption of a high involvement eco-inovation using value-belief-norm theory *J. Cons. Behaviour* **10** 1 51-60

[24] Pletchnig M, Heidenreich S and Spieth P 2014 Innovative alternatives take action–Investigating determinants of alternative fuel vehicle adoption *Transp. Re. Part A: Policy & Practice* **61** 68-83

[25] Stone R N and Gronhaug K 1993 Perceived Risk: Further Considerations for the Marketing Discipline *Eu. J. of Mkt.* **27** 3 39-50

[26] Dholakia U M 2001 A motivational process model of product involvement and consumer risk perception *Eu. J. of Mkt* **35** 11/12 1340-1362

[27] Hirunyawipada T and Paswan A K 2006 Consumer innovativeness and perceived risk: implications for high technology product adoption *Eu. J. of Cons. Mkt* **23** 4 182-198

[28] Forsythe S, Liu C, Shannon D and Gardner L 2006 Development of a scale to measure the perceived benefits and risks of online shopping *J.of Interactive Mkt.* **20** 2 55- 75

[29] McGuire K A, Kimes S E, Lynn M, Pullman M E and Lloyd R C 2010 A framework for evaluating the customer wait experience *J. of Service Mngm.* **21** 3 269–290

[30] Roselius T 1971 Consumer Rankings of Risk Reduction Methods *J.of Mkt.* **35** 1 56- 61.

IOP Conf. Series: Materials Science and Engineering **898** (2020) 012027 doi:10.1088/1757-899X/898/1/012027

Data acquisition for medical devices

R Veres[1] B C Feier[2] S Ilea[1] G Bohm-Revesz[1]

[1]Faculty of Managerial and Technological Engineering, Doctoral School of Industrial Engineering, University of Oradea, Romania
[2]Faculty of Medicine and Pharmacy, University of Oradea, Romania

ralph.veres@catedral.ro

Abstract. Following the path of the powerful evolution in many fields, engineering and medicine are converging to a healthy collaboration. Engineering, as a profession, is a practising one, and it is based on modifying and harnessing the three fundamental resources of the known world: information, energy and materials. Delivering solutions using the latest technologies and also trying to innovate for a better world, the continuous advancement of methods and many other improvements are focusing on helping and developing new ways to have better results. Medicine, with its many fields, is going hand in hand with this exponential growth. One method used in engineering to study and characterize different processes or phenomena is the acquisition and digital processing of data. With the development of computational techniques, it became possible to capture biological data, radiography and tomography being two examples that allow the reconstruction of images that are useful for diagnosis and establishment of treatment. In the following paper, methods known in the industry are presented and related to different procedures used in this collaboration between engineering and medicine.

1. Introduction
The first element used to work with is information. The central and key element for research is data and the need for acquisition it. New methods are coming to the market with improvements, considering the predecessors, being also more efficient, cost-effective and precise, this component is mandatory for present applications.

As presented in [1] and [2], there are many amounts of information in the surrounding world, the intent being to acquire and process this kind of data for useful applications. Acquisition data is done by processing the sampling signal that shows us surrounding conditions and then turning the resulted data into digital or numeric values that can be manipulated by computer-aided programs. In the evolutive field of medicine, the basis for ethical practices is represented by information samples that can lead to precise and useful analysis.

The need for capturing three-dimensional models has become a necessity in the last years, mainly due to rapid evolution on 3D technologies and also for the need of evolution the imagistically zone of medicine.

2. Methods
Very common and a beneficial method in our days in represented by Magnetic Resonance Imaging, also known as MRI, and uses strong magnetic fields and radio waves to generate images of organs in the human body.

Modern magnetic resonance imaging techniques can provide high-quality anatomic details, even too small organs and extremities such as feet and hands. The benefits of MRI are represented by real-time visualization of anatomic structures, like significant blood vessels and also for the display of the soft tissues, sub-structures and surrounding organs and components, as the authors suggest in [3] and [4].

MRI can provide a big amount of data and information resulting from that the analysis of these becomes a complex task for clinicians, being time-consuming and can inquire errors. Nowadays, computerized methods can assist clinicians for this data diagnostic—segmentation method most used in the field of MR data processing.

Segmentation comes hand in hand with 3D imaging and design programs used to reproduce full-scale models of the organs that are scanned (figure 1). Programs like 3D Slicer and Mimics that can transform DICOM files resulted from an MRI scan to many other formats, like STL are stepping forward, furtherly being capable to modelling and optimize different organs or structures in the computer-aided programs.

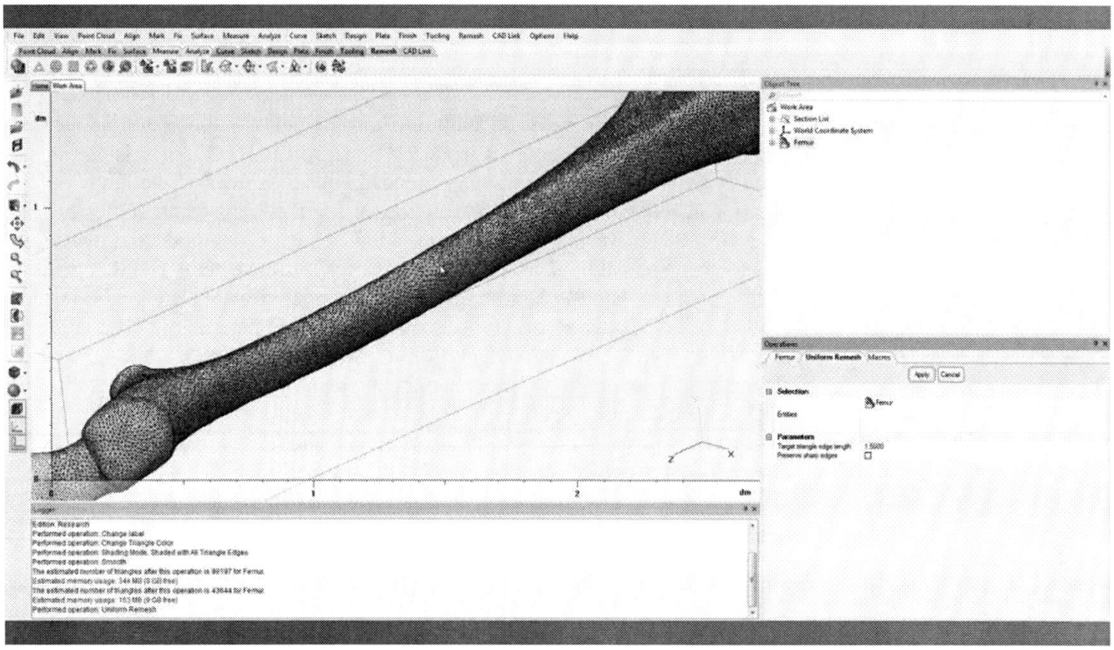

Figure 1. A femur reconstruction in MIMICS, after an MRI scan segmentation.

There are also a few possibilities of capturing 3D models, as presented in [5],[6],[7], the most known being 3D scanning. This technique is spreading in new domains and makes a powerful appearance in the medical sphere. 3D scanning brings advantages, the most common one being the volumetry. To be able to observe parameters of the human body is one of the most important factors in diagnosis and also for evaluations that can result after it.

3D scanning has an advantage compared to other methods: the spatial resolution of models that can also be enlarged by one dimension and can be registered in relation to other parameters.

On the present market, we can find different 3D scanner (figure 2) with different principles, but the main feature of them is they can have modules implemented. The first module is the raw data capturing, where most of the scanners measure the distance of the surface from the sensor. Another module is represented by the sensor movement, that allows building an elaborate 3D model by positioning the scanner's sensor in several viewpoints from which all the details of the object and surface are visible. Another module is represented by computation of the 3D point position. Mainly, 3D scanners use geometric transformations to build a 3D model, and can be used to model and optimize furtherly the objects.

Figure 2. 3D scanner.

3D scanning has an advantage compared to other methods: the spatial resolution of models that can also be enlarged by one dimension and can be registered in relation to other parameters.

On the present market, we can find different 3D scanner with different principles but the main feature of them is they can have modules implemented. The first module is the raw data capturing, where most of the scanners measure the distance of the surface from the sensor. Another module is represented by the sensor movement, that allows building a complex 3D model by positioning the scanner's sensor in several viewpoints from which all the details of the object and surface are visible. Another module is represented by computation of the 3D point position. Mainly, 3D scanners use geometric transformations to build a 3D model and can be used to model and optimize furtherly the objects.

Advantages of 3D scanners are undeniable are consistent in comparison with other methods. Here are some of them, as the authors help us to consider them in [1]:

- Low operational cost – the ease of use and operation is not as difficult as MRI/CT equipment
- Simple manipulation – is a simple system, mainly with a very accessible user interface, and also with very simple handling, it can be facile to use by the personnel
- High speed – related to time efficiency, 3D scanning is fast and time-saving
- Accuracy – Modern 3D scanners are very accurate, this being the essential parameter for good results and the main parameter to distinguish changes of human body
- Harmless operation – there is no harm for both the personnel or the patient, as it is in other methods, like CT scans, Radiography, etc.
- No limitations – it can be used in many cases, being no limitations regarding different parts in the human body (stents, pace-makers, etc.)

3. Conclusions

In terms of advancement, medical imaging is aided with technological improvements that bring us new methods to develop new products, new techniques and brings us viable solutions in the sphere of research. For us, in the zone of PhD research, this kind of methods can help us raise the quality of our research and also defines us the primary three needs of engineering: cost optimization, high quality and reasonable time.

Compared to other methods, 3D scanning brought us advantages, mainly in cost optimization zone, but also in quality and time essential parts, compared to more known CT/MRI scans and also helps us improve basic knowledge of the nonconventional technologies applications in the field of medicine.

References

[1] Chromy, Adam. (2016). Application of High-Resolution 3D Scanning in Medical Volumetry. International Journal of Electronics and Telecommunications. 62. 10.1515/eletel-2016-0003.

[2] S. H. Ridner, L. D. Montgomery, J. T. Hepworth, B. R. Stewart, and J. M. Armer, "Comparison of upper limb volume measurement techniques and arm symptoms between healthy volunteers and individuals with known lymphedema," Lymphology, vol. 40, no. 1, pp. 35–46, Mar. 2007, PMID: 17539463.

[3] Rachida, ZEGOUR & Belaid, Ahror & Ben Salem, Douraied. (2018). A Segmentation Method of Skin MRI 3D High Resolution in vivo. Medical Technologies Journal. 2. 255-261. 10.26415/2572-004X-vol2iss3p255-261.

[4] Tomei, Ernesto & Marini, Mario & Stagnitti, Andrea & Sartori, Alessandro & Bertana, Luca & Ansari, N & Passariello, Roberto. (2011). Determination of bone age using MRI of hand/wrist: a pilot study. 10.1594/ecr2011/C-0963.

[5] A. L. Reyes, J. M. Cervantes, and N. C. Gutirrez, "Low-cost 3D scanner by means of a 1D optical distance sensor," Procedia Technology, vol. 7, pp. 223–230, 2013.

[6] K. Liu, Y. Wang, D. L. Lau, Q. Hao, and L. G. Hassebrook, "Dual-frequency pattern scheme for high-speed 3-D shape measurement," Optics Express, vol. 18, no. 5, p. 5229, Mar. 2010.

[7] Veres, R. (2018). RAPID PROTOTYPING AS AN AID IN MEDICINE. Nonconventional Technologies Review, 22(3). Retrieved from http://revtn.ro/index.php/revtn/article/view/38

IOP Conf. Series: Materials Science and Engineering **898** (2020) 012028 doi:10.1088/1757-899X/898/1/012028

Validation of an algorithm for predicting the remaining useful life, for a model with linear degradation

G Grebenişan[1,3], N Salem[2], S Bogdan[1], D C Negrău[1]

[1]University of Oradea, Romania
[2]Zarqa University, Jordan

[3]grebe@uoradea.ro

Abstract. This paper aimed to validate a working tool, component of the Predictive Maintenance Toolbox ™, produced by Matlab (MathWorks), in the case of a procedure for monitoring the operation of mechanical systems, in order to diagnose a failure of the process and to estimate the remaining useful life (RUL). This toolbox provides toolsets, materialized in function files, for labeling data, designing condition indicators, and estimating a parameter named the remaining useful life of a machine. You can analyze and label machine data imported from local files, cloud storage, and distributed file systems. The algorithm suggested by Matlab (software owned by MathWorks) was used in detail to process part of the data set provided freely by NASA through The Prognostics Data Repository, The Prognostics Center of Excellence (PCoE) at Ames Research Center. Of the 4 data sets, only one was used for this paper. Each data set is composed of 3 working files, in text format, for training, test and algorithm validation, and solution statement, respectively. The results obtained confirm the validity of the computer-assisted training system, diagnostics, prognosis, and validation tools, on a statistical basis, in the case of consistent databases.

1. Introduction

Monitoring the operation status of industrial equipment provides data on their condition, [1], [2]. Any failure, or deterioration of the state of the installations, can be detected, and preventive measures can be taken within an appropriate cycle to avoid catastrophic failures. This is done by monitoring parameters such as vibration, solid wear in the oil, noise emission, etc. Changes to these parameters help detect the spread of defects, diagnose the causes of the problem, and anticipate failure. Maintenance can be supported, so corrective actions can be planned accordingly. Applying condition monitoring in machines and installations leads to savings in maintenance costs and improved availability and safety, [3], [4].

The main function of monitoring the operating condition of a mechanical system, machine, or installation is to provide an almost accurate diagnosis of the condition of the machine and its rate of change, so that preventive measures can be taken at a given time. Knowledge of the condition of the machines can be obtained by selecting a parameter that indicates the deterioration of the condition of the machine. The value of this parameter can be measured periodically or continuously. In some cars, the deterioration of the conditions develops so fast that it is possible that there are only a few seconds between the detection of faults and the total failure. In such cases, continuous or online monitoring with an automatic shutdown of the system or machine is recommended.

2. Fault detection and diagnosis

When a process error occurs, it must be detected as soon as possible. The fault detection system must indicate that something is wrong in the process. After detection, the fault is diagnosed, then is isolated, and an attempt is made to detect the cause of the fault. Typically, the techniques used to detect and diagnose defects are divided into two general categories: estimation methods and pattern recognition methods, [1], [3], [5], [6].

Fault detection and diagnosis is currently a significant issue in process automation, [5, 6]. Screening and diagnostic methods based on pattern recognition and expert systems, respectively, have been suggested to solve the problem of singularity (anomaly, [6]), as defined by a process failure. Various methods and techniques have been suggested in the literature, such as model-based techniques such as pattern recognition, neural networks, with a lot of possible architectures for diagnosing errors. The neural network was the first automatic technique for learning the various situations that define a system failure.

Usually, an anomaly is considered that non-compliant state that does not respect the development of expected behavior, therefore, the problem of detecting an anomaly is reduced to find, in the available data set, those data models that do not follow the rules, which are out of general description process. As a general approach, these anomalies, or non-compliant models, are often referred to as exceptional values, discordant observations, exceptions, aberrations, surprises, peculiarities, or contaminants in different fields of application, [7]. Of these, anomalies and outliers are two terms most commonly used in the context of anomaly detection, sometimes interchangeably. Examples of anomaly detection are found in a wide range of applications, such as credit card fraud, insurance or health care detection, cybersecurity intrusion detection, critical security breach detection, and military surveillance of enemy activities. The importance of anomaly detection is since data anomalies translate into relevant, often critical, information that can be used in a wide variety of fields of application.

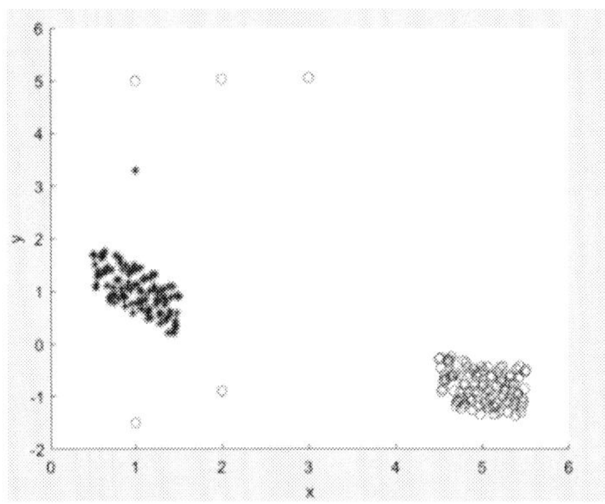

Figure 1.- Normal regions and anomalies data, [3]

Figure 1 illustrates the scattering of anomalies in a simple set of data represented in a two-dimensional graph. There are two well-defined regions, in which the data are grouped, each representing the normal behavior of a system, or systems, while the scattered, discordant data, sufficiently distant from the two typical regions are singularities, or anomalies of this data set, [8]. It is essential that these anomalies are not treated as "noisy". Noises are generally defined as obstacles in the work of analysis of phenomena and processes and must be treated as component parts of the mathematical model of the process, [1, 2, 3, 4, 9, 10, 11]. Considering a linear model, which expresses the relation between the time series of the inputs, X_t, and the time series of the outputs, Y_t, in the form:

$$Y_t = \upsilon(B)X_t + N_t \tag{1}$$

where, $\upsilon(B) = \upsilon_o + \upsilon_1 B + \upsilon_2 B^2 + \dots$, is the transfer function, expressed polynomially, $\upsilon(B) = \dfrac{\Omega(B)}{\delta(B)} = \delta^{-1}(B)\Omega(B)$ respectively, represented as a ratio of mathematical operators, and which denotes the dynamic relationship between outputs and inputs, in which **B** is a ***backward shift operator*** (given an observable time series z_t, with the components: : $z_1, z_2, \cdots, z_N$, then $Bz_t = z_{t-1}, B^m z_t = z_{t-m}$, while N_t, represents the filtered component of the noise superimposed over the input signal, a_t, and which influences (transforms) the z_t component as follows:

$$z_t = \mu + a_t + \psi_1 a_{t-1} + \psi_2 a_{t-2} + \dots = \mu + \psi(B)a_t \tag{2}$$

where μ is a parameter that generally expresses the "level" of z_t, and a_t is a sequence of random components, weighted, after the weighted operator $\psi(B) = 1 + \psi_1 B + \psi_2 B^2 + \dots$, called the filter transfer function (this is different in definition and mathematical model, of anomalies, singularities, etc .: $N_t = \psi(B)a_t$, so that the analysis of the behavior of a process, or system, imposes the determination, both of the transfer function, $\upsilon(B)$, and of the transfer function of the filter, $\psi(B)$, [1], [2], [9], [10], An iterative numerical procedure is presented in [12], starting from the analysis of two classes of outliers, generated by dynamic models of exceptional interventions, at unknown moments of time: innovative outlier (IO) and an additive outlier (AO). Starting from a stochastic process model, x_t, following an autoregressive-integrated-moving average (ARIMA) model (possibly with the characteristics **p**- a positive integer indicating the degree of the nonseasonal autoregressive polynomial, **d**- a non-negative integer indicating the degree of nonseasonal integration in the linear time series, **q**-a positive integer indicating the degree of the nonseasonal moving average polynomial known), [1], [13], [14]:

$$\phi(B)\alpha(B)x_t = \theta(B)a_i \tag{3}$$

where **B** is the ***backward shift operator***, defined above; $\phi(B) = (1 - \phi_1 B - \phi_2 B^2 - \dots - \phi_p B^p)$; and $\theta(B) = (1 - \theta_1 B - \theta_2 B^2 - \dots - \theta_q B^q)$; there are two polynomials whose roots lie outside the circle with unit radius $\alpha(B) = (1-B)^{d_1}(1-B^s)^{d_2}; d = d_1 + sd_2$; and $a_t\{\}$ has been defined above, has the definition as a function of $N_t = \psi(B)a_t$. The model of an exceptional external intervention, [10], [12], is mathematically represented by a dynamic model:

$$z_t = \frac{\omega(B)}{\beta(B)}\zeta_t^{(T)} + x_t$$
where
$$\zeta_t^{(T)} = \begin{cases} 1 & \text{for} \quad t = T \\ 0 & \text{otherwise} \end{cases} \tag{4}$$

The significance of the parameter $\zeta_t^{(T)}$ is: the moment when the intervention takes place on the process x_t, and the respective operators $\omega(B) = (\omega_0 - \omega_1 B - \omega_2 B^2 - \ldots - \omega_s B^s)$; are two polynomials depending on **B**, whose ratio $\omega(B)/\beta(B)$ characterizes the dynamic behavior of the intervention. Proving that these interventions can cause pronounced bias in the procedure for calculating and estimating correlations, partial autocorrelations, and autoregressive moving average (ARMA) parameters, it is obvious the need to find a method to identify and determine these interventions, respectively eliminate their effects. Assuming that the two classes of interventions in the time series model, innovational outlier (IO) and an additive outlier (AO), are characterized, each by the following models (the operators used here, $\theta(B), \phi(B), \alpha(B)$ have been previously defined), [10], [12], [15]:

-a dynamic model for the innovative outlier (IO)

$$z_t = \frac{\theta(B)}{\phi(B)\alpha(B)}\omega\zeta_t^{(T)} + x_t \tag{5}$$

- a dynamic model for the additive outlier (AO)

$$z_t = \omega\zeta_t^{(T)} + x_t \tag{6}$$

If the last relations are rewritten, in terms of a random sequence, a_t, we obtain:

- for innovational outlier (IO)

$$z_t = \frac{\theta(B)}{\phi(B)\alpha(B)}\left\{a_t + \omega\zeta_t^{(T)}\right\} \tag{7}$$

- for additive outlier (AO)

$$z_t = \frac{\theta(B)}{\phi(B)\alpha(B)}a_t + \omega\zeta_t^{(T)} \tag{8}$$

Thus, analyzing the last two models, it is found that additive outlier (AO), affects the time series model only at the level of observation **T**, while innovational outlier (IO), is an exceptional intervention, that means an unusual event, at time **T**, but which affects all observations, z_T, z_{T+1}, z_{T+2} ..., which follows, from that moment, through the $\frac{\theta(B)}{\phi(B)\alpha(B)}$ term.

3. Experimental setup

Datasets are delivered, [11], [16], [17], [18], in directories and files consisting of several time series. Each set of data is further divided into training and testing subsets. Each time series comes from a different engine - that is, the data can be considered to be from a group of engines of the same type. The experiment begins with each engine having a certain degree of initial wear and different levels of manufacturing accuracy classes that are not known to the user. This wear and the level of manufacture are considered normal; i.e., it is not considered a state of error. There are three condition settings that have an important effect on engine performance and functional characteristics. These operational settings are included in the data (operational setting 1; operational setting 2; operational setting 3. The data is contaminated with sensor noise.

Each engine runs typically at the beginning of each time series and, according to the experimental scenario, develops a failure at some point during the series. In the training set, the defect increases in amplitude, eventually generating system failure. In the test set, the time series ends shortly before the system fails. The aim of the research is to predict the number of operating cycles remaining before the failure of the test set, i.e., the number of operating cycles after the last cycle in which the engine will continue to run. A vector of true remaining useful life (RUL) values, which was also provided for the test data. The data is provided as a text file structured as tables with 26 columns of numbers, separated by spaces, cell arrays, and column vectors, in txt files. Each row is a snapshot of the data taken during a single operational cycle, and each column is a different variable. The columns correspond to the variables, to the table header of the data structure:

1) unit number
2) time, in cycles
3) operational setting 1
4) operational setting 2
5) operational setting 3
6) sensor measurement 1
7) sensor measurement 2
...
26) sensor measurement 26

The experimental scheme and the appropriate scenario belong to NASA, The Prognostics Center of Excellence (PCoE) at Ames Research Center, Turbofan Engine Degradation Simulation Data Set, [11], [16].

4. **Results and Discussions**

Data used here were received from NASA's database, The Prognostics Center of Excellence (PCoE) at Ames Research Center, Turbofan Engine Degradation Simulation Data Set, [16], were preprocessed in Matlab(all the file functions designed for this algorithm was used in original form, same as definitions and statements were used in original form, [17], [18]), using the specific *loadData* helper function, which converts the text file, data in tables, in cell array files as well as in vector files. The basic file, "*degradationData*", is a cellular structure, with 100 cells, arranged vectorially in a single column. Each cell represents a table of numerical data, arranged in different numbers of rows (samples) and 26 columns. Rows represent a time sequence, i.e., a set of 26 values collected at a time, value according to the second column, "time". The columns represent the variables defined in Section "3. Experimental setup ": 1) unit number; 2) time, in cycles; 3) operational setting 1; 4) operational setting 2; 5) operational setting 3; 6) sensor measurement 1; 7) sensor measurement 2; ...; 26) sensor measurement 26. Preprocessing in Matlab is a continuous process, so it is inappropriate to call it "pre", as definitions of variables, their processing, are performed throughout the workflow.

The converted data, from the table to the cell structure, "*degradationData*" with dimensions: 249x1 cells, is partitioned (using the *cvpartition* function) into data needed for training (200 of the 249 cells of the *degradationData* time series). A partition of this primary data (49 cells from *degradationData*) will be used for the validation process, to evaluate the performance of the procedure/algorithm.

In figure 2 and figure 3, one can analyze the plots of data evolution for a group of 7 sensors/graph, on 2 samples, and 5 samples, respectively. This analysis is not very performant or efficient because there

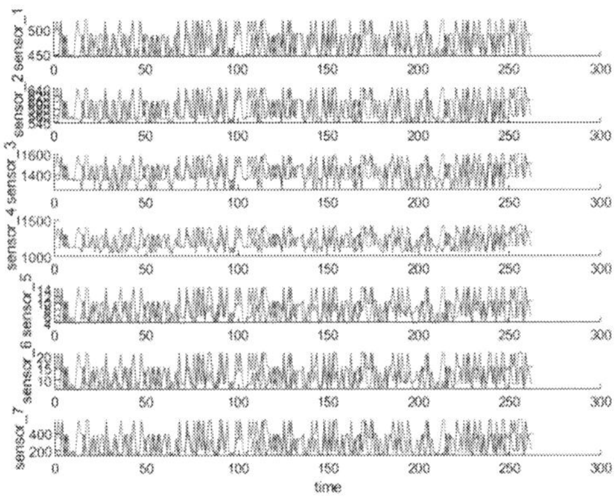

a)-sensors #1 to #7

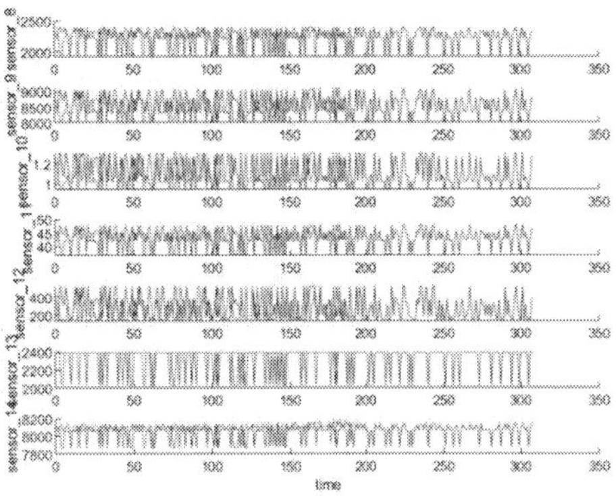

b)-sensors #8 to #14

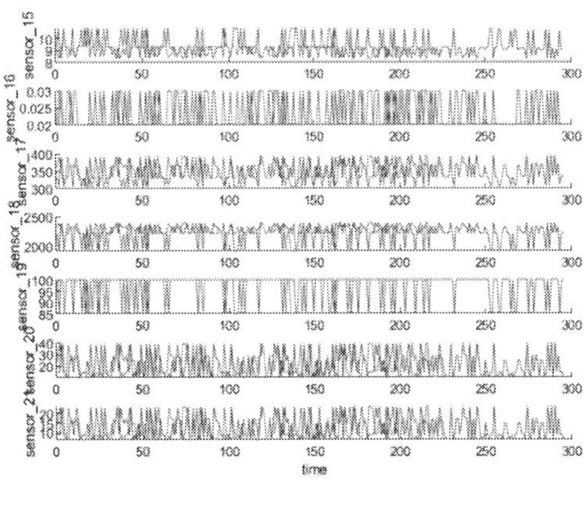

c)- sensors #15 to #21

Figure 2. Graphical representation of data evolution for a group of sensors, on 2 samples

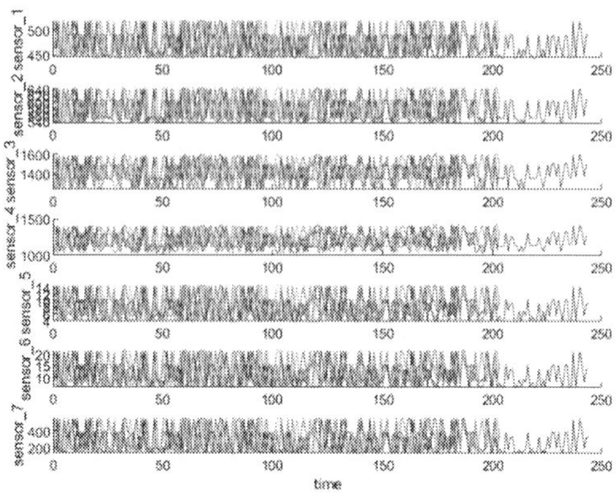

a)-sensors #1 to #7

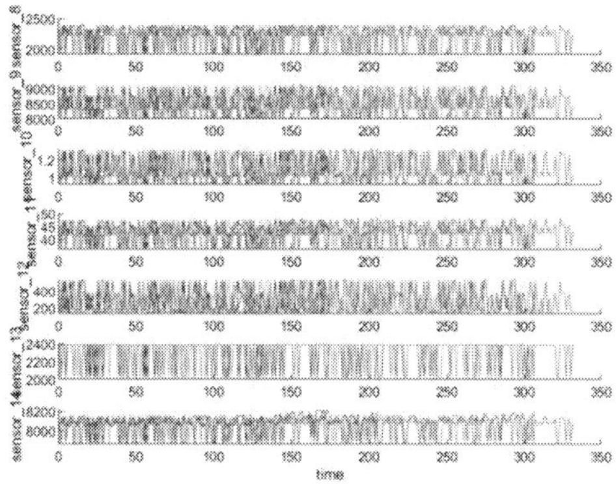

b)-sensors #8 to #14

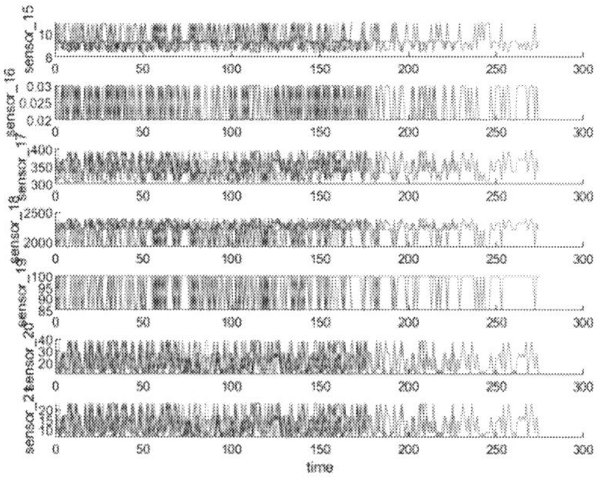

c)- sensors #15 to #21

Figure 3. Graphical representation of data evolution for a group of sensors, on 5 samples

are no obvious trends in data measurements. To have a clearer image, in mind, of the evolution/trend of the chart, to be able to highlight and extract clear trends of degradation, respectively marking the occurrence of any trend of failure, the three settings of the operations will be used (different, for each sensor). It should be noted that each member of the ensemble contains 3 operating conditions:

"op_setting_1", "op_setting_2" and "op_setting_3". First, the data will be extracted from each cell of the *degradationData* structure (249 cells, each cell is a table with 49473 rows and 26 columns), as column vectors, then these vectors will be concatenated into an extended table with 43352 rows and 26 columns. Then, from this table, with 49473 rows (samples) the data corresponding to the 3 columns of the operating conditions: "op_setting_1", "op_setting_2" and "op_setting_3" are extracted, and next will be grouped in an array with 49473 rows and 3 columns, by vertical concatenation.

Considering two working regimes: "clustering" and "normalization", there are three graphical representations captured in figures 4, 5, and 6, respectively, which can simplify the analysis of the evolution trend of the data collected from each sensor. Thus, figures 4 and 5 are valid for data processing in clustering mode, and the last of these two shows that the small distances between the different operating points, the centroids, coincide with the operating points (6 regimes, respectively, distinct operating points). The K-means algorithm is used to automatically locate the 6 clusters. Repeating the algorithm 5 times, and the results are identical: 0.377212.

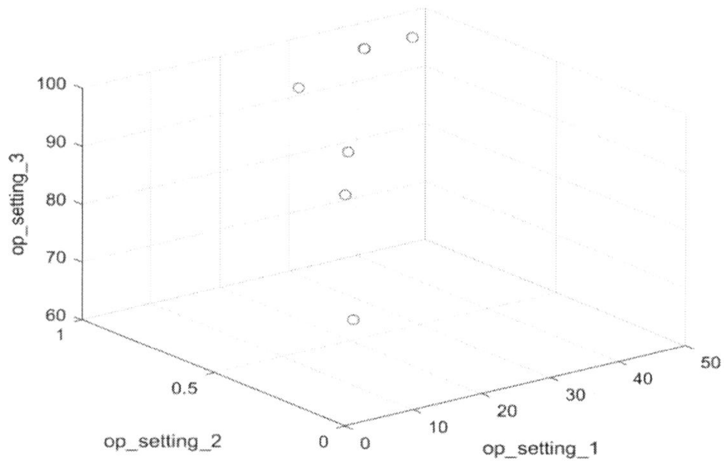

Figure 4. The 6 operating points in clustering mode

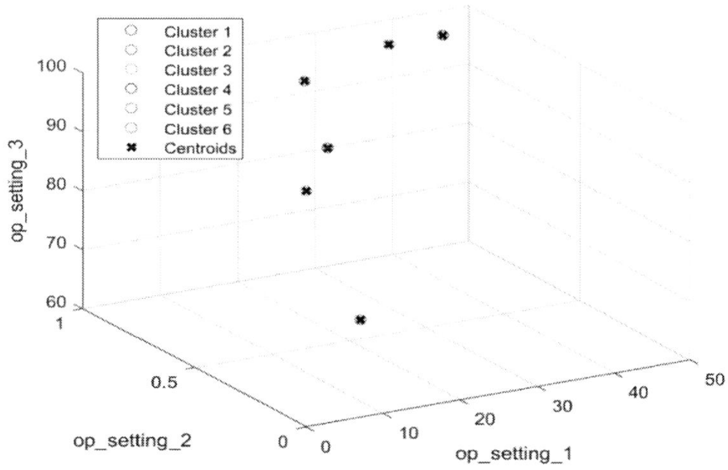

Figure 5. The simultaneous plot of the clustering results and the identified cluster centroids (with "x" marker)

Figure 6 shows the trend of evolutions, mostly positive (increasing graph), of data collected from sensors # 2, # 3, # 4, #8, #9, #11, #13 and # 17. Graphical representations are now, after data normalization, more expressive and easy to be analyzed. This is the most important gain of the normalization operation.

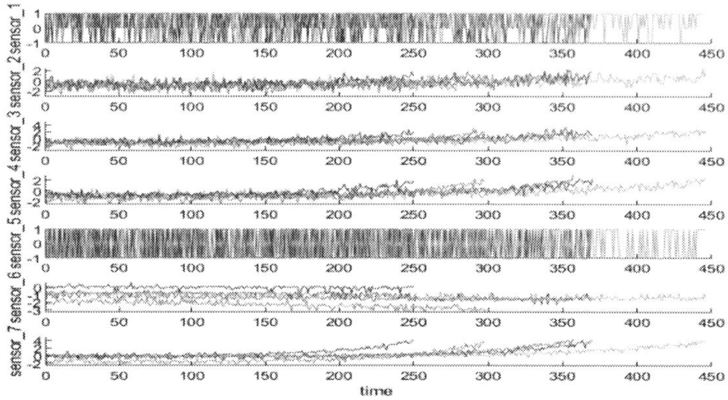

a)-sensors #1 to #7

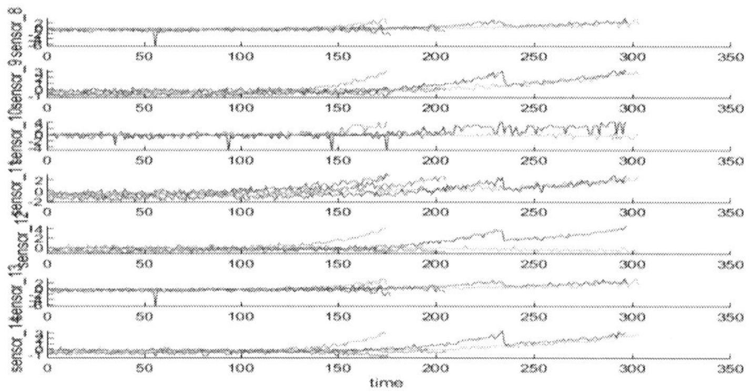

b)- sensors #8 to #14

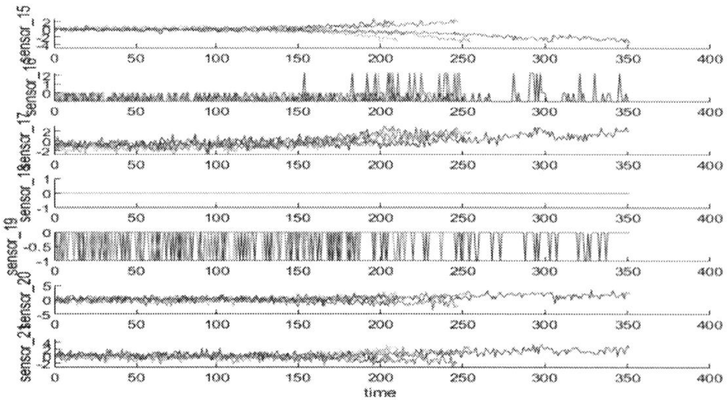

c)- sensors #15 to #21

Figure 6. Using the data normalized by the working regime, the degradation trends for some sensor measurement s

IOP Conf. Series: Materials Science and Engineering **898** (2020) 012028 doi:10.1088/1757-899X/898/1/012028

Figure 7 concentrates on two subfigures, the most trendable measurements, which in fact show a strongly increasing trend, in the evolution graph. Once the data has been normalized, the graphical representation is more obviously expressive.

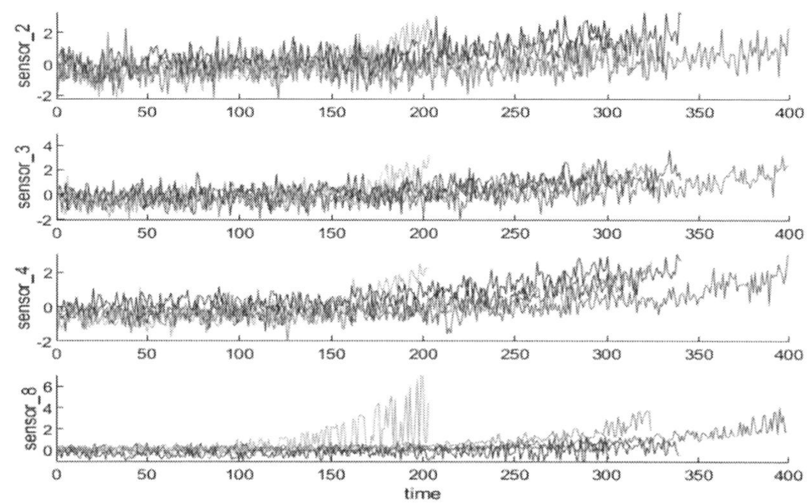

a)-sensorsTrended # 2, # 3, # 4, #8

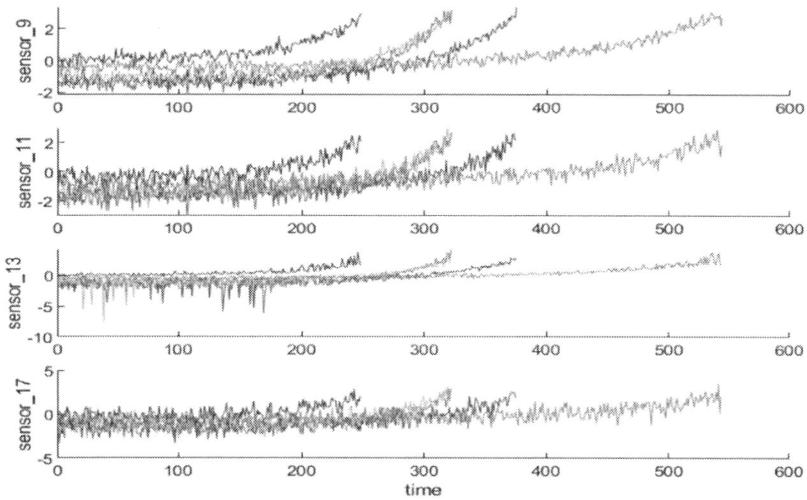

b)- sensorsTrended #9, #11, #13 and # 17

Figure 7. The selected most trendable sensor measurements

Next, we'll go to the creation of a health indicator, from fusing the measured data. A system is supposed to start behaving poorly, starting from a state of operation in accordance with the rules within which it was designed, called a state of "health". If one considers the state of health a function, the value 1 is assigned to the state of health at the beginning, and to the failure, the value zero is assigned to it. It can be considered that the health condition has a linear evolution, the degradation being linearly between 1 and 0. The measured and collected data will be fused with the 21 sensors in a so-called health indicator, based on similarity. Several models and fusing techniques can be used in this case, [11], [13], [17], [18]. Figure 8 shows the linear representation of evolutions in three cases: 5 samples, 25 samples, and 100 samples.

IOP Conf. Series: Materials Science and Engineering **898** (2020) 012028 doi:10.1088/1757-899X/898/1/012028

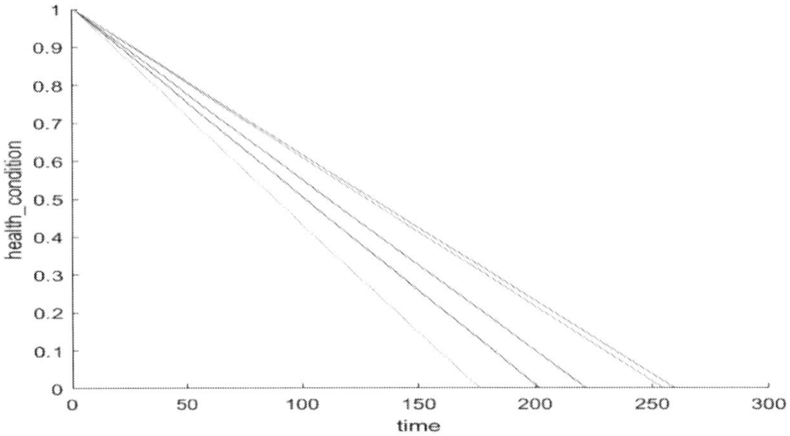

a)-5 samples

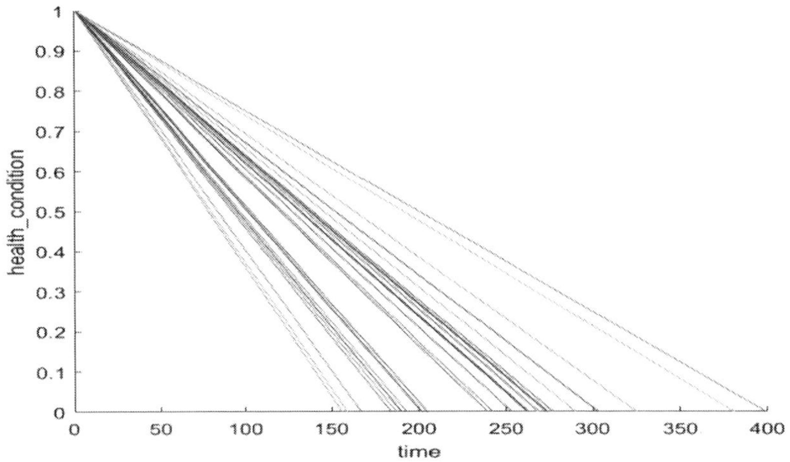

b)-25 samples

Figure 8. Three cases of the health condition evolution trend

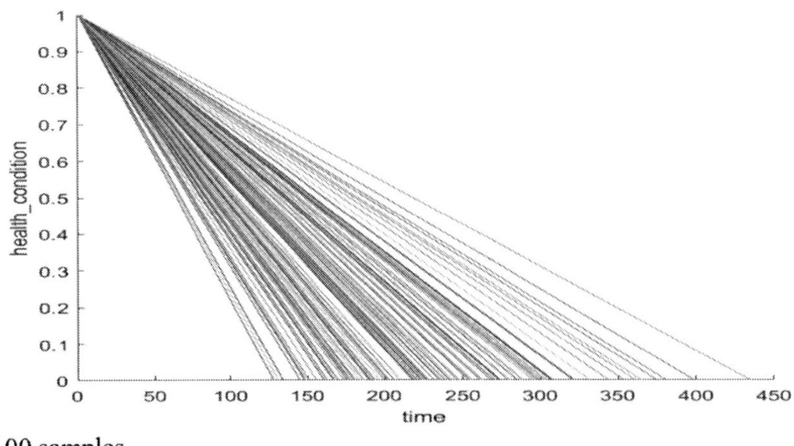

c)-100 samples

The slope of the lines in Figure 8 is consistent with the rate of degradation of the systems represented by the data collected from the sensors. Linear regression, therefore, can be chosen as a model for Health

Condition Indicator, in which the data for characterizing the regressors are those corresponding to the sensors with the highest trendability: # 2, # 3, # 4, #8, #9, #11, #13 and # 17, plots in Figure 9.

Figure 9. Plots of the fused Health Condition Indicator for training data in three cases: 5 samples; 25 samples and 50 samples

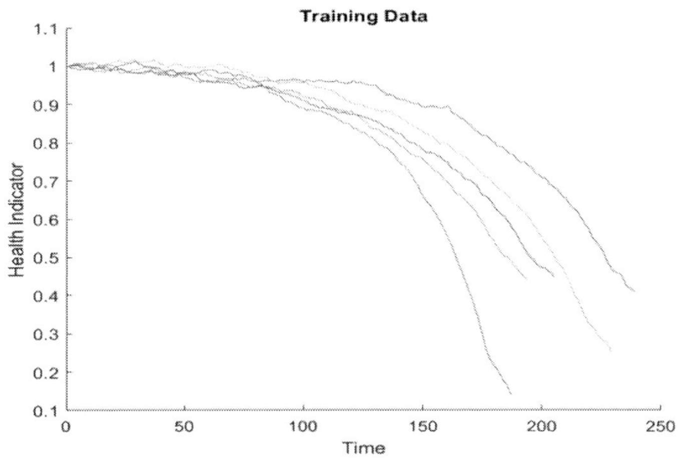

a)-5 samples

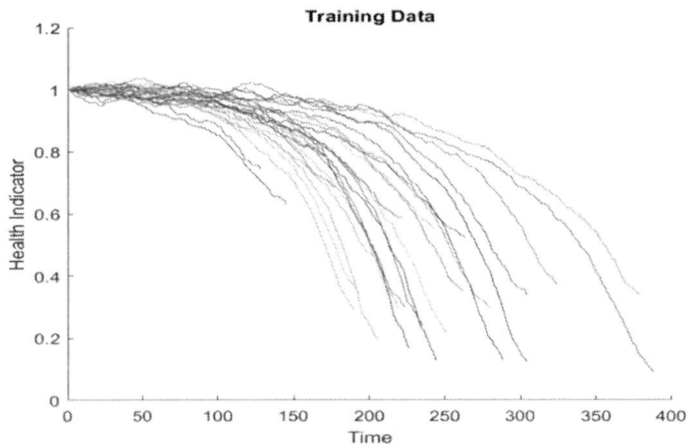

b)-10 samples

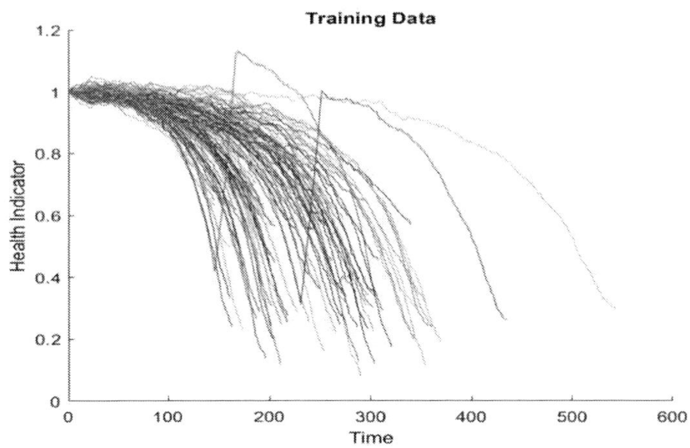

c)-50 samples

One may repeat the data normalization process and the sensor fusion process for the validation data set, figure 9.

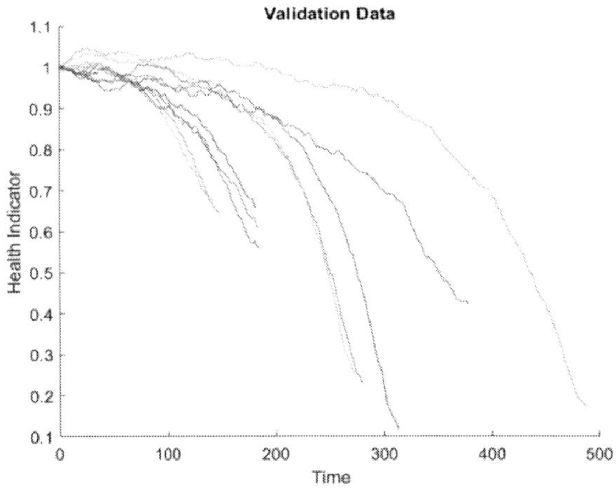

a)-10 samples

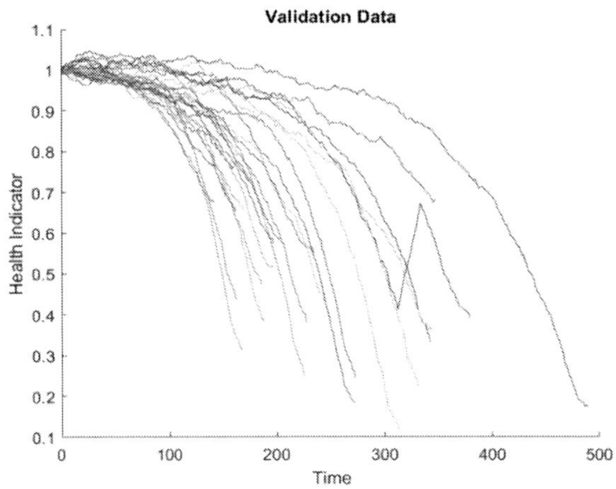

b)-25 samples

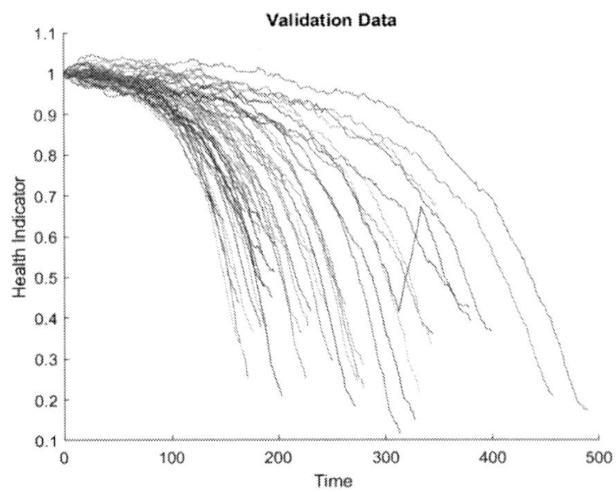

c)-49 samples

Figure 10. Plots of the fused Health Condition Indicator for validation data in three cases: 10 samples; 25 samples and 49 samples

IOP Conf. Series: Materials Science and Engineering **898** (2020) 012028 doi:10.1088/1757-899X/898/1/012028

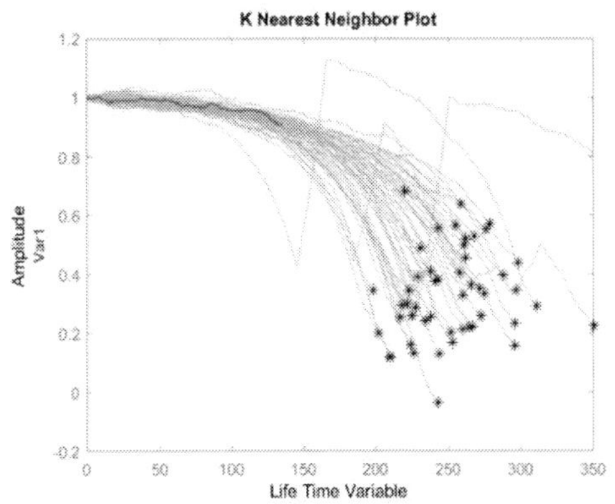

Figure 11. The RUL validation data truncated at 50% (red curve)

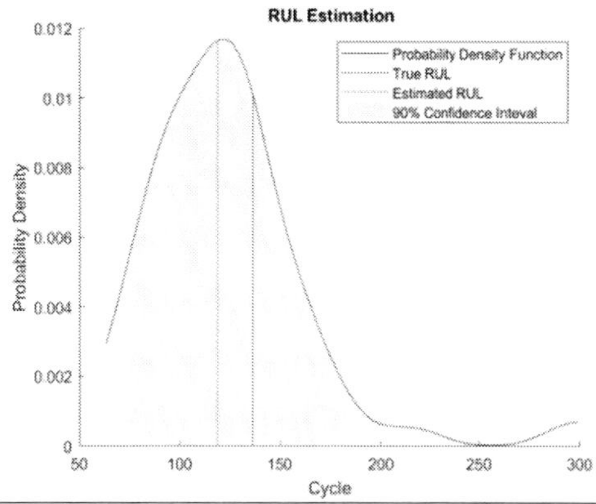

Figure 12. The predicted RUL(green vertical) compared to the true RUL(red vertical) and the probability density function (black curve) of the estimated RUL_50%

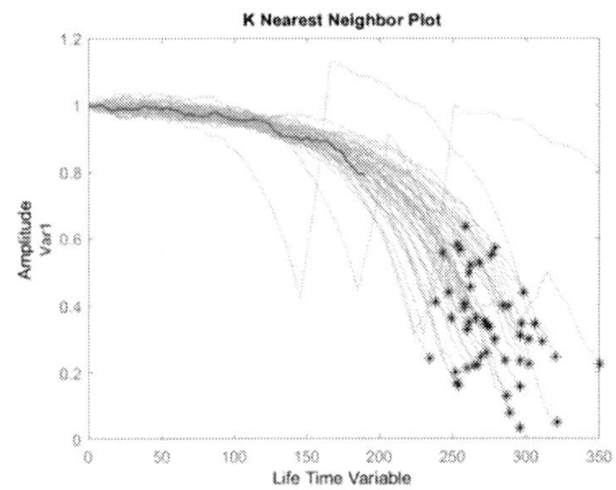

Figure 13. The RUL validation data truncated at 70% (red curve)

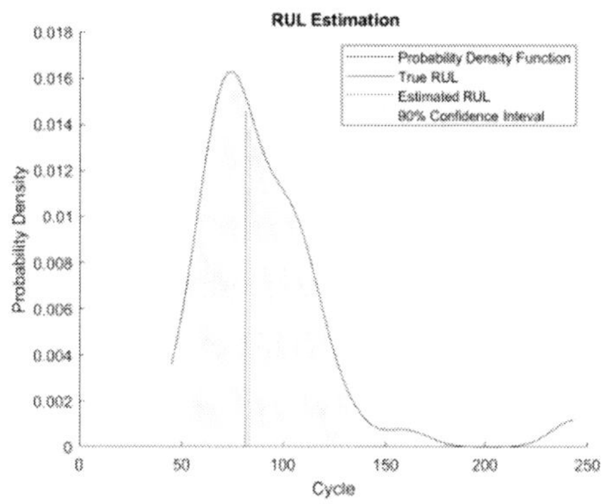

Figure 14. The predicted RUL(green vertical) compared to the true RUL(red vertical) and the probability density function (black curve) of the estimated RUL_70%

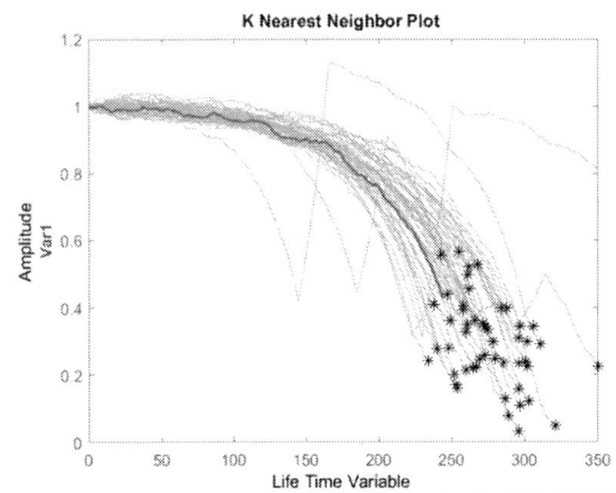

Figure 15. The RUL validation data truncated at 90% (red curve)

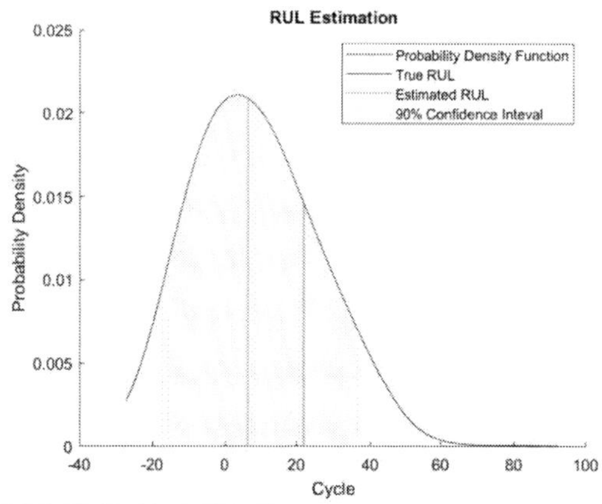

Figure 16. The predicted RUL(green vertical) compared to the true RUL(red vertical) and the probability density function (black curve) of the estimated RUL_90%

Using the training data, one can now build a model of Remaining Useful Life with residuals-based that is compared for fit compared to fused data, mathematically using a second-order polynomial curve. The fit error is calculated as the difference between the training data in the structure of the health indicator of the machine and the estimation data of the same health indicator of the machine. The *residualSimilarityModel* Matlab function is used to estimate the remaining RUL of a component with a similarity model based on the residual comparison. This method can be applied in this case because it has data sets that characterize degradation profiles for a set of similar components, with the same specifications, and the dynamics of the degradation process are known. This stage of the algorithm consists of two distinct steps: configuring the model (residualSimilarityModel) and comparing the data obtained in this model with the fused trained data. To similarity RUL model may evaluate in three partial data sets will be used, i.e., samples of 50%, 70%, and 90% of the previously determined validation set, to predict its RUL, figures 11 to16.

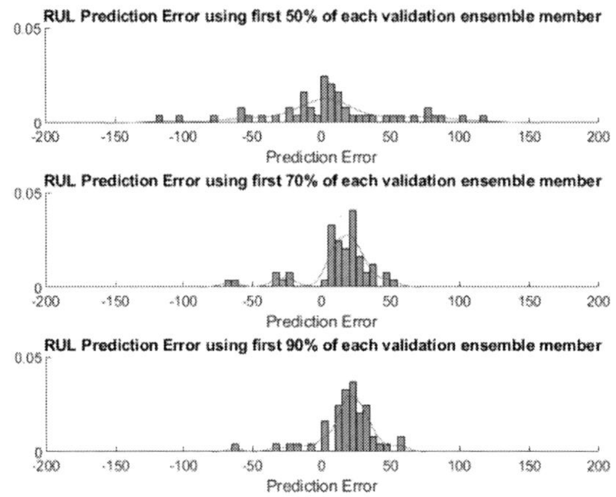

Figure 17. The histogram of the error versus probability distribution.

In figure 17 is represented the histogram of the error between estimated RUL and true RUL for each breakpoint with a probability distribution when the evaluation was made for all validation data.

5. Conclusion

In practice, a lot of applications and calculation systems, algorithms, and elaborate schemes for predicting the remaining useful life are used, [3], [4], [11], [16], [17], [18]. By definition, widely accepted, degradation models extrapolate past behavior to predict a future state. This type of RUL calculation fits a linear model, most often, or exponentially adapted to the degradation profile of a condition indicator (also called health indicator), taking into account the degradation profiles of the user as a whole. Then, for the training validation, the degradation profile of the test component is used to statistically calculate the time remaining until the indicator reaches a prescribed threshold. These models are most useful when there is a known value of the status indicator indicating a fault. Of the two available types of degradation models: the linear degradation model (*linearDegradationModel*) and the exponential degradation model (*exponentialDegradationModel*), the first model was used in the paper, as the collected data have a linear degradation profile (using statistical procedures) and the degradation type is not cumulative. The computational algorithm used here briefly states that after the degradation model is created, or the data are preprocessed so that it meets the degradation profile, the model is initialized using historical health data of an assembly with similar components, such as many machines, or equipment, manufactured to the same specifications, to do this, a specific fit function is used, after which appropriate calculations can determine it, predicted RUL of similar components using *predictRUL*.

References

[1] Isermann R, Fault-Diagnosis Systems. An Introduction from Fault Detection to Fault Tolerance, ISBN-10 3-540-24112-4, Springer, Berlin, 2008.

[2] Wang L and Robert X. Gao (Eds.) Condition Monitoring and Control for Intelligent Manufacturing, ISBN 978-1-84628-269-0, 2006, Springer-Verlag.

[3] Grebenişan G, Bogdan S, Salem N and Negrău D C, A brief assessment of outliers and malfunctions detecting techniques with an application on lubricant condition monitoring, Mat. Sc. and Eng. 568 (2019) 012039 IOP Publishing doi:10.1088/1757-899X/568/1/012039.

[4] Grebenişan G, Bogdan S, Salem N, and Negrău D C -A neural networks approach of process fault diagnosis using time series collected data through oil condition monitoring, Mat. Sc. and Eng. 568 (2019) 012079 IOP Publishing doi:10.1088/1757-899X/568/1/012079.

[5] Ortiz D, Byington C, Patrick R, Ture C, Farnach J, Moffatt J, Combined Lubrication Monitor for On-Line Gearbox Health Assessment, doi: 10.1109/AERO.2011.5747563, 2011, Aerospace Conference, IEEE Xplore.

[6] Teng H S, Chen K and Lu S C-Y, Adaptive Real-time Anomaly Detection Using Inductively Generated Sequential Patterns, Conference Proceedings on Research in Security and Privacy, 1990.., 1990, IEEE Xplore DOI: 10.1109/RISP.1990.63857

[7] Chandola, V., Banerjee, A., and Kumar, V. 2009. Anomaly detection: A survey. ACM Comput. Surv., 2009, 58 pages. http://doi.acm.org/10.1145/1541880.1541882

[8] Hodge V J. and Austin J, A Survey of Outlier Detection Methodologies, Artificial Intelligence Review 22: 85–126, 2004, Kluwer Academic Publishers, doi: 10.1023/B:AIRE.0000045502.10941.a9

[9] Box G E. P., Jenkins G M., Reinsel G C., Ljung G M., Time Series Analysis: Forecasting and Control, Fifth Edition, 2015, John Wiley & Sons.

[10] Box G E. P. & Tiao G. C. (1975) Intervention Analysis with Applications to Economic and Environmental Problems, Journal of the American Statistical Association, 70:349, 70-79, DOI: 10.1080/01621459.1975.10480264.

[11] Saxena A., Goebel K., Simon D. and Eklund N., "Damage propagation modeling for aircraft engine run-to-failure simulation," 2008 International Conference on Prognostics and Health Management, Denver, CO, 2008, pp. 1-9, doi: 10.1109/PHM.2008.4711414.

[12] Chang, Ih, George C. Tiao, and Chung Chen. "Estimation of Time Series Parameters in the Presence of Outliers." *Technometrics*30, no. 2 (1988): 193-204. doi:10.2307/1270165.

[13] Statistics and Machine Learning Toolbox™ User's Guide, last accessed on February 2020, https://www.mathworks.com/help/releases/R2018a/pdf_doc/stats/stats.pdf,.

[14] Abraham B, Box G. E. P., "Bayesian Analysis of Some Outlier Problems in Time Series", 1979, Biometrika 66(2):229-236, DOI: 10.1093/biomet/66.2.229

[15] Rousseeuw P J., Leroy A M., "Robust Regression and Outlier Detection", John Wiley & Sons.

[16] The Prognostics Center of Excellence (PCoE) at Ames Research Center, Turbofan Engine Degradation Simulation Data Set, https://ti.arc.nasa.gov/tech/dash/groups/pcoe/prognostic-data-repository/#turbofan, May 2020 last accessed.

[17] Predictive Maintenance Toolbox™, May 2020 last accessed, https://www.mathworks.com/products/predictive-maintenance.html.

[18] Similarity-Based Remaining Useful Life Estimation, May 2020 last accessed, https://www.mathworks.com/help/predmaint/ug/similarity-based-remaining-useful-life-estimation.html?s_tid=srchtitle,.

IOP Conf. Series: Materials Science and Engineering **898** (2020) 012029 doi:10.1088/1757-899X/898/1/012029

A brief overview of Additive Manufacturing

D C Negrau[1], G Grebenişan[1], C Gherghea[1]

[1] University of Oradea, Faculty of Managerial and Technological Engineering, Universităţii Street No. 1, Oradea, Romania
E-mail: dan.negrau@yahoo.com

Abstract. In this paper, a brief overview is presented, resulting from a recent literature review of some representative books or papers regarding various research on additive manufacturing. Some basic terms are presented, in the context of founding the existence of several terminologies for some specific expressions of this subject, as well as different definitions for them materials used at that process, areas of using, problems appeared during process of manufacturing, heat treatments used, advantages and disadvantages. This research is carried out to identify a solution for controlling the parameters during the additive processing (AM) process, parameters with which to improve the quality of the parts obtained by AM

1. Introduction

A new industrial revolution is the implementation of additive manufacturing (AM), designed to replace totally or partially technologies for manufacturing processing and subassembly in all industrial fields. Industrial analysts are considering PAs new rights or technologies mentioned to currently influence and change industrial technologies. PA is often referred to as printing (3D printing), which may be similar to laser printing (figure 1). PA builds solid objects by superimposing material layers; each layer is drawn by the computer [1,2].

For manufacturing of parts with AM process is necessary to use a powder as:

- Carbon steel
- Aluminum
- Stainless steel
- Titanium
- Another material

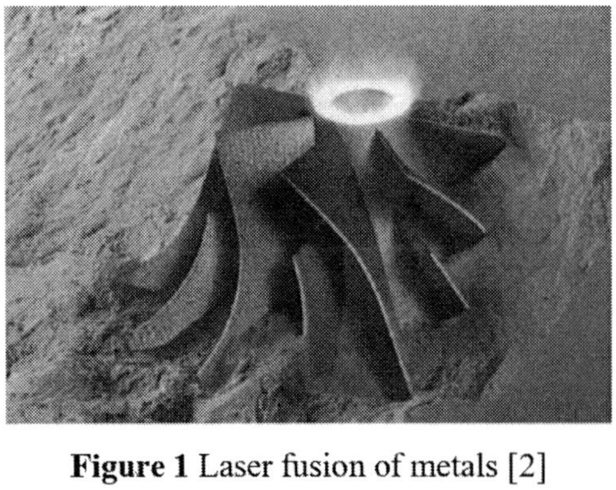

Figure 1 Laser fusion of metals [2]

At manufacture of parts from metalic materials, the parts is built by metalic powder and laser light (figure 2), and the direction of deposition is diferent with de milling operation (figure 3). The parts are built layer by layer using a laser, according to a 3D model. Thus, additive processes have a different approach than milling, turning and deformation processes, used predominantly in industrial production.

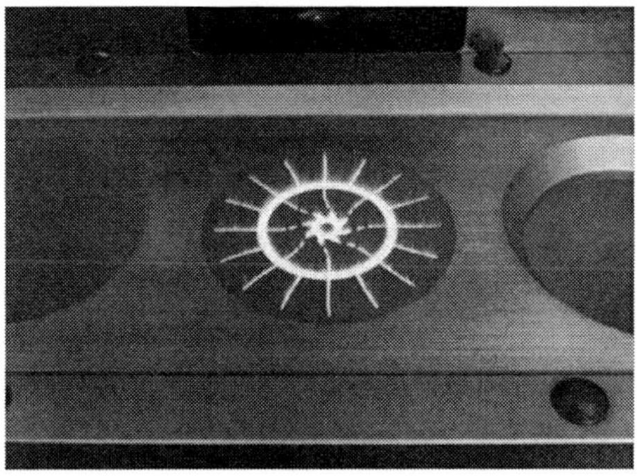

Figure 2 Laser metal fusion (LMF)[3]

Figure 3 Laser metal deposition (LMD)[4]

Research has shown that hybrid design systems have potential and efficient use for high-functional molds in production, almost smooth nest shape manufacturing, repairs and coating (LMD). There are a number of companies that have built hybrid machines that incorporate selective laser melting (SLM) technology and milling function as a solution for finishing the part obtained through SLM [5].

Laser additive processing equipment makes it possible to manufacture parts of very high shape and complexity, only in terms of roughness and dimensional accuracy; the parts manufacturing by laser fusion or loaded by laser welding do not correspond, thus being necessary subsequent finishing operations (figure 4).

a) LMD deposition b) Milling part after LMD deposition

Figure 4 LMD Proces and milling (on -LASERTEC 65 3D hybrid) [5]

For a very good accuracy and dimension, it is necessary that the part is milling on the same machine (mixed machines) as shown in figure 4 a), where the part was obtained by laser melting deposition (LMD) or by LMF or on another machine. The tools which are used at this manufacturing process is, a laser melting head, the head through which the metal / non-metal powder is also evacuated, and for the finishing of the resulting part, we use cylinder-frontal mills, spherical, profiled cutters, drills (figure 4b).

2. Materials used at Additive Manufacturing AM and areas of use

Additive processing (PA) builds solid objects by superimposing layers of materials, and as the raw material used by PA is the powder that can be made of thermopolymer, or metal, which can be aluminum, stainless steel, titanium or other materials. The layers can contain even more types of materials [7].

With additive manufacturing systems could be produced in the future, better aircraft engines with lower fuel consumption. To do this, engineers need to improve today's industrial 3D printers so that these machines can process extremely heat-resistant and obviously high-strength alloys [8].

In the medical field for to find an alternative solution to edentulism problems, a procedure for the development and manufacture of a personalized dental appliance, less expensive and adequate are the implants, avoiding an initial surgery and allowing immediate intervention [9]. This procedure, in addition to direct 3D digital modeling of the custom implant [9], uses a combination of the two advanced additive manufacturing technologies, SLM and SL [10,11].

This combination allows the photopolymerization of a polymer in the gaps of a 3D housing of the metal core previously produced by SLM. These biocompatible metal housings, which have the necessary mechanical strength, have been optimized based on the models, favoring the regeneration and growth of rigid and soft tissues. The basic structure is then placed on the SL platform and the additive process begins with the first polymerization of the resin layer until the complete manufacture of the tooth crown, the crown that will be built based on the 3D CAD model (figure 5)

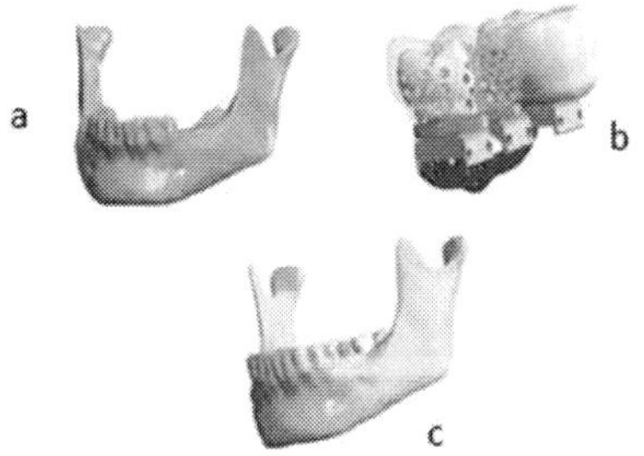

Figure. 5 3D CAD Model: a) the jaw of the patient with edentulism problems; b) generation according to a Voronoi model and a topological and geometric optimization plan for establishing the dental points; c) resulting from the assembly between the jaw and the personalized bio-inspired dental implant. [10]

After the 3D modeling, the additive deposition will be performed and a dental crown will be presented, shown in figure 6, resulting in the dental bridge (figure 7).

Figure 6 Production by SLM of dental cores and dental bridge [11]

Figure 7 Multimaterial and personalized physical model of the dental bridge [11]

There are also other areas such as the automotive industry where it is suitable for manufacturing prototype parts (figure 8) for new cars, parts that after optimization will be mass-produced or if it is a very small or unique series cars, they can be obtained entirely by 3D printing, parts made of different types of metals or alloys [12,13,14].

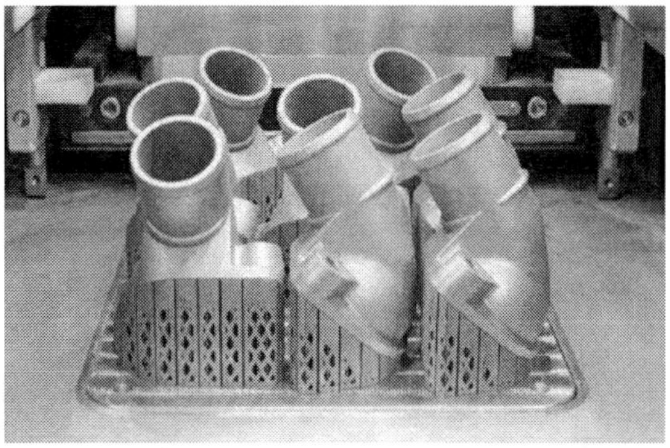

Figure 8 Mercedes Benz thermostat cover with technological addition [13]

3. Problems encountered at laser fusion of metals

Among the various materials processed by selective laser melting, steel is one of the most used materials for manufacturers due to its wide range of properties and applications [15].

Working parameters during deposition have an influence on the fatigue resistance of steel parts obtained by SLM [16,17], and from here results in the apparition of cracks. Current studies show the importance and the effects of deposition, directions of deposition, heat treatment, surface quality, and the energy density of the beam during deposition by SLM (figure 9) [18].

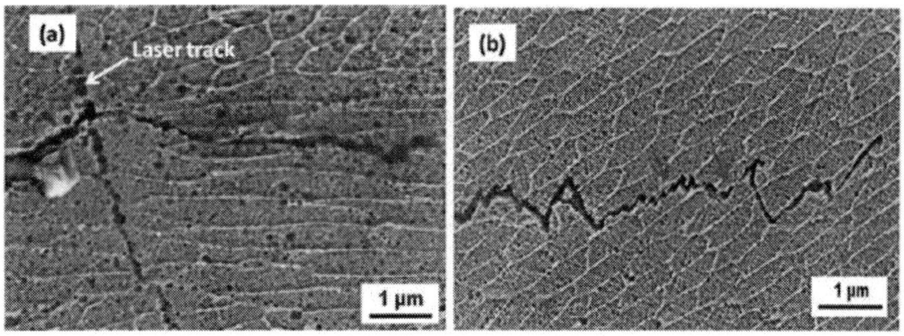

Figure 9 Crack propagation after mechanical test part (a) horizontal (b) vertical steel 316L [18].

Therefore, the appearance of cracks is due to the fact that the working parameters of the machine are not adjusted correctly (melting temperature, deposition speed, the distance between layers), parameters that are adjusted according to the type of powder material, to avoid overheating (figure 10) [19].

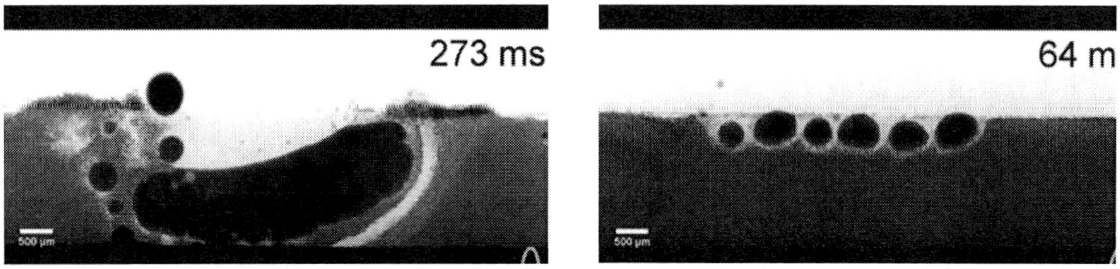

Figure 10 Appear of the inclusions at 3D printing [19]

Have also been made to improve the printing process, so that no further operations are required as specific heat treatments (isostatic pressure) [20, 21].

4. The influence of heat treatment applied to different types of metals

In general, there are four types of post-treatments recommended for LPBF from IN625: annealing of normalization of internal stresses, normalization of recrystallization, treatment with a solution, and hot isostatic pressing [22,23].

Carrying out the heat treatment on samples of nickel superalloys (IN625), heating them to 700 C, 1000 C, and 1150 C and keeping them one hour before air cooling, it was found that the microstructure of the IN625 parts at the annealing state at 700 ° C was the same as that of a state of material obtained by casting.

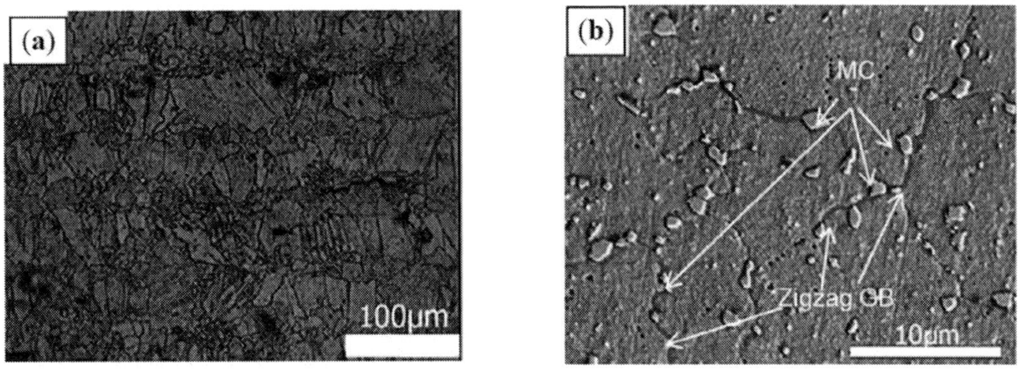

Figure 11 LPBF plane microstructure of LPBF of IN625 samples after annealing at 1150 C. (a) Metallographic optical image, (b) Boundary structure and carbide distribution [24]

Comparison of different types of recycling temperatures can be concluded that recovering the high temperature allows recrystallization (figure 11a) of a sample built IN625, High Z contrast deviations are dissolved and MC carbides precipitate and granule delimitation becomes stable due to carbide fixation MC (figure 11 b) [24].

5. Conclusion

The purpose of this article is to present a brief overview, resulting from a recent literature review on the additive manufacturing process, and in future identify a solution for controlling the parameters during the additive processing (AM) process, parameters with to improve the quality of the parts

References

[1] Ebert-Uphoff, Imme & Gosselin, Clément & Rosen, David & Laliberte, Thierry "Rapid Prototyping for Robotics." Rapid Prototyping for Robotics. 10.5772/4639 (2005).

[2] Linkan Bian, Nima Shamsaei, and John M. "Laser-Based Additive Manufacturing of Metal Parts" Usher-by Taylor & Francis Group, book title Modeling, Optimization, and Control of Mechanical Properties 2018

[3] W Meiners,K Wissenbach &R Poprawe "Laser additive manufacturing of metallic components: materials, processes and mechanisms" Institute of Materials, Minerals and Mining and ASM International Published by Maney for the Institute and ASM International DOI 10.1179/1743280411Y.0000000014 2012

[4] Yamazaki Mazak Corporation, Oguchi, Aichi. "Development of A Hybrid Multi-tasking Machine Tool: Integration of Additive Manufacturing Technology with CNC Machining" 480-0197, Conference on Electro Physical and Chemical Machining (ISEM XVIII) Japan Procedia CIRP 42 (2016) 81 – 86 18th CIRP

[5] Ian Gibson , David Rosen , Brent Stucker" Additive Manufacturing Technologies 3D Printing, Rapid prototyping,and Direct Digital Manufacturing",book title Additive Manufacturing TechnologiesVerlag New York, Springer-2015
[6] ***https://en.dmgmori.com/products/machines/additive-manufacturing/powder-nozzle/lasertec-65-3d-hybrid*** accessed at 2020.03.04)
[7] ***https://www.ttonline.ro/revista/fabricatie-aditiva/prelucrarea-aditiva-pa-ii***Accessed at 2020.04.04
[8] ***https://www.ttonline.ro/revista/tehnologii/sistemele-de-fabricatie-aditiva-descopera-superaliaje*** Accessed at 2020.04.04
[9] O. Ciobanu and G. Ciobanu *"The use of 3D scanning and rapid prototyping in medical engineering*". Academica Brancusi (2013) 241-247.
[10] W. Gao, Y. Zhang, D. Ramanujan *"The status, challenges, and future of additive manufacturing in engineering ",* , Computer-Aided Design. 69 (2015) 65–89.
[11] P. Li, Z. Wang, N. Petrinic, C.R. Siviour, Materials Science & Engineering A 614*" Deformation behaviour of stainless steel micro lattice structures by selective laser melting"* (2014) 116–121.
[12] ***https://robbreport.com/motors/cars/father-and-son-3d-printing-lamborghini-2857428/*** accessed at 2020.04.06
[13] ***https://formlabs.com/blog/3d-printed-car-how-3d-printing-is-changing-the-automotive-industry/*** accessed at 2020.04.06
[14] ***https://www.sme.org/technologies/articles/2019/august/automotive-industry-warms-to-3d-printing/*** accessed at 2020.04.08
[15] Rui Li, Yong Seok Kim , Hoang Van Tho , Young Jin Yum , Won Jun Kim and Soon Yong Yang, *"Additive manufacturing (AM) of piercing punches by the PBF method of metal 3D printing using mold steel powder materials"* * Manuscript Received April 24, 2018; Revised September 14, 2018; Accepted October 20, 2018
[16] Shahriar Afkhami, Mohammad Dabiri, S. Habib Alav, Timo Björk,*" Fatigue characteristics of steels manufactured by selective laser melting"* Antti Salminen International Journal of Fatigue 122 (2019) 72–83
[17] Carelyn E. Campbell Sudha CheruvathurEric A. Lass" *Additive Manufacturing of 17-4 PH Stainless Steel: Post-processing Heat Treatment to Achieve Uniform Reproducible Microstructure JOM",* The Minerals, Metals & Materials Society (outside the U.S.) SUDHA CHERUVATHUR ERIC A. LASS and CARELYN E. CAMPBELL Vol. 68, No. 3, 2016 DOI: 10.1007/s11837-015-1754-4 _ 2015
[18] Suryawanshi J, Prashanth KG, Ramamurty U. Mater *". Mechanical behavior of selective laser melted 316L stainless steel"* Sci Eng, A 2017;696:113–21.
[19] **** https://www.oxmet-technologies.com/abd-xam*** accessed at 2020.04.08
[20] **** https://www.oxmet-technologies.com/abd-xam***
[21] Mater Kreitcberg, A.; Brailovski, V.; Turenne, *"S. E_ect of heat treatment and hot isostatic pressing on the microstructure and mechanical properties of Inconel 625 alloy processed by laser powder bed fusion"*. Sci. Eng. A **2017**, 689, 1–10.
[22] Fang, X.Y.; Li, H.Q.;Wang, M.; Li, C.; Guo, Y.B. "Characterization of texture and grain boundary character distributions of selective laser melted Inconel 625 alloy" Mater. Charact. 2018, 143, 182–190.
[23] Mater Kreitcberg, A.; Brailovski, V.; Turenne, *"S. E_ect of heat treatment and hot isostatic pressing on the microstructure and mechanical properties of Inconel 625 alloy processed by laser powder bed fusion"*. Sci. Eng. A **2017**, 689, 1–10.
[24] Zhihua Tian, Chaoqun Zhang , Dayong Wang, Wen Liu, Xiaoying Fang, DanielWellmann, Yongtao Zhao 2 and Yingtao Tian "*A Review on Laser Powder Bed Fusion of Inconel 625 Nickel-Based Alloy"* 17 November 2019; Accepted: 18 December 2019; Published: 20 December 2019

IOP Conf. Series: Materials Science and Engineering **898** (2020) 012030 doi:10.1088/1757-899X/898/1/012030

Theoretical and experimental research on the development of an analysis system of the hand tremor movement for patients with Parkinson's disease

St Borta and M Baritz

Design Product and Environment Faculty, University Transilvania Brasov

email: stefania.borta@yahoo.com

Abstract. Theoretical studies indicate the existence of the tremor in the human body, normally physiological, barely visible, and which occurs when the upper limbs are stretch to the front or when very high precision is required in carrying out an activity. Different experimental researches seek the theoretical fundament of the tremor movements or, develop evaluating systems for this type of movement, and last but not least are creating systems for behavioural help for people who develop forms of tremor with effects on comfort. Thus, in the first part of the paper, the specific clinical tremor forms are analyzed, as well as the manifestation forms and the parameters quantifying methods of this movement form developed in the human body. In the second part of the work is reviewed the substantiation of the analysis models of tremor type oscillations. In the third part of this paper, the experimental setup developed is presented, focusing on finding the best behavioural help solution for patients with pathological tremors and presents some considerations related to the device constructive aspects. In the final part, the approach conclusions of developing an assistive system for patients with pathological tremors are presented, on the highlight of the parkinsonian tremor patients.

1. Introduction

The manifestations of tremor can be classified into several groups according to specific neurological characteristics. Thus, specialized studies [1] classify tremor status into: rest tremor, postural tremor, action tremor, kinetic tremor, intention tremor, and a type of tremor which involves customizing the actions, named tasks specific tremor.

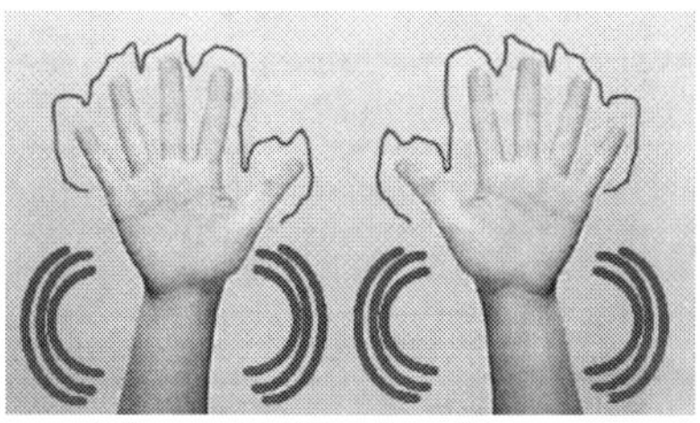

Figure 1. The hand tremor

The first two types of tremor occur when the affected body part is not active and is supported to overcome gravitation, for example when the affected upper limbs are voluntarily held in a different position (stretched to the front) than the normal one (orthostatic position). The variants of action and kinetic tremor are manifested in the case of voluntary or directed movements performed to evaluate various neurological examination tests (approaching and moving away the hand from the nose, etc.).

The other variants (intentional tremor and personalized task tremor, respectively) are characterized by an increase in tremor amplitude as the target is approached or are manifested during particular actions, such as writing. Besides, the same variants of tremor movement can be classified according to other criteria such as: state (normal or pathological), activation conditions (rest, postural, kinetic/intentional) and tremor frequency (setting limits of variation that divides the oscillation motion into low, medium, or high-frequency tremor) [2].

Among these forms, Parkinsonian tremor stands out due to the diversity of forms in which it can manifest (rest, postural or kinetic) or the low and medium frequency levels (3-7 Hz) of the oscillations during the action. Therefore, in order to perform a clinical evaluation of the different forms of tremor, it is necessary to identify in the first place the location of the tremor, its activation state and its frequency, and then to add the amplitude of movement.

Thus, this important feature of the oscillation-amplitude motion is taken into account for the evaluation of clinical forms and the creation of evaluation scales concerning this parameter. Clinical tremor assessment scales include the Fahn-Toulouse-Marin (FTM) scale, which assigns 0 to 4 points for tremor amplitude levels in different conditions and respectively 0 to 4 points for degrees of severity in daily activities. Unified Parkinson's Disease Rating Scale (UPDRS) [4] assigns 0 - 4 points for the pair of parameters amplitude and severity of rest and forms of postural or kinetic tremor. The ratings of the evaluation scale are on average proportional to the logarithm of the amplitude of the displacement as shown in the research [3].

The physiological variants of the tremor state were differentiated by identifying the basic and/or pathological forms and were indicated several different mechanisms to establish the origin of the tremor. In most research on this issue [5] several types of tremor-generating mechanisms are identified, such as:

- mechanical mechanism (each segment of the limbs involved or the entire upper or lower limb has a certain resonance frequency, which depends on the task to be performed);
- sensory reflex mechanisms or central oscillating mechanisms, as a group of oscillating neurons located in a specific structure of the brain or which are manifesting as a network or loop of several different structures.

As it shown in various studies, the tremor" associated with Parkinson disease (PD) is one of the most widely studied and the second most common pathological tremor, with prevalence of 102-190 cases per 100,000 population in Western countries. Age at disease onset is usually after 60 and incidence increases with advancing age. [6]

Resting tremor is present in 80% of patients with autopsy-proven PD. Asymmetrical onset of tremor is commonly observed, and tremor onset may be coincident with other parkinsonian symptoms of rigidity and slowness of movement (bradykinesia). As PD progresses the severity of tremor may diminish, but the tremor is accentuated by performing mental tasks or contralateral voluntary movements and during ambulation. In a sample of PD patients, resting tremor may be inhibited by voluntary movement and up to 20% of PD patients also exhibit postural or kinetic tremor". [7]

2. Theoretical aspects of tremor mathematical model and software applications

For fundamental research studies, especially in the field of tremor level assessment, a series of mathematical models have been developed based on the study of oscillations. A first differential nonlinear model is constructed as a Van der Pol oscillator [8] defined by equation (1) in which p represents the position coordinate (as a function of time), and $\mu \geq 0$ is a scalar parameter indicating the nonlinearity and the degree depreciation.

$$p'' - \mu(1 - p^2)p' + p = 0 \tag{1}$$

The parameter μ is calculated according to the equation (2) where F = the rate of emotional arousal, and I = inhibition rate. The parameter μ can take zero values ($\mu = 0$), when the excitation rate is equal to the inhibition rate (simple harmonic oscillator) or non-zero values ($\mu \neq 0$), where equation (1) is transformed into equation (3) by entering the parameter b.

$$\mu = \frac{F}{I} - 1 \tag{2}$$

$$p'' + bp' + p = 0 \tag{3}$$

As is mentioned in [8] "the coefficient b is interpreted as damping (with $b < 0$ corresponding to anti-damping behaviour where solutions gain energy over time). In the case of the Van der Pol equation, b is replaced by a nonlinear term which is negative when $|p| < 1$ and positive when $|p| > 1$."

Based on this tremor motion oscillation model and along with other approaches that take into account the Deep Brain Stimulation (DBS) process, open-source platforms have been developed with applications for calculating the parameters needed to be evaluated in different forms of tremors.

These applications include: the online platform for tremor analysis that uses the "R" language-called "TREMOROTON". To validate this application, the researchers compared using an intraclass correlation coefficient, the tremor frequency estimate obtained with Tremoroton, with the results obtained from commercially available software, obtained from a sample of 20 patients (10 with essential tremor and 10 with Parkinson's diagnosis). The experimental system (figure 2) of this research was based on the analysis of the activity recorded with an accelerometer and electromyographic sensors (EMG) positioned on the arm muscles. [9]

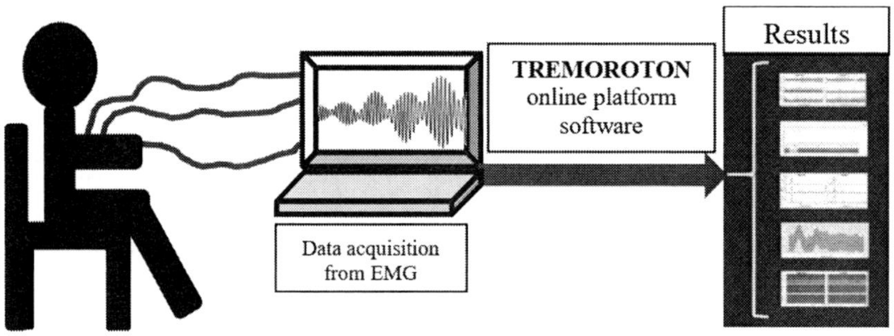

Figure 2. The TREMOROTON software platform

Another experimental structure developed with a software application that supports the analysis of tremor movements is TREMBAL, a system "that quantifies tremor using electromagnetic motion tracking sensors with four sensors (trakSTAR), a remote control and a standard notebook computer with custom software" (figure 3). [10]

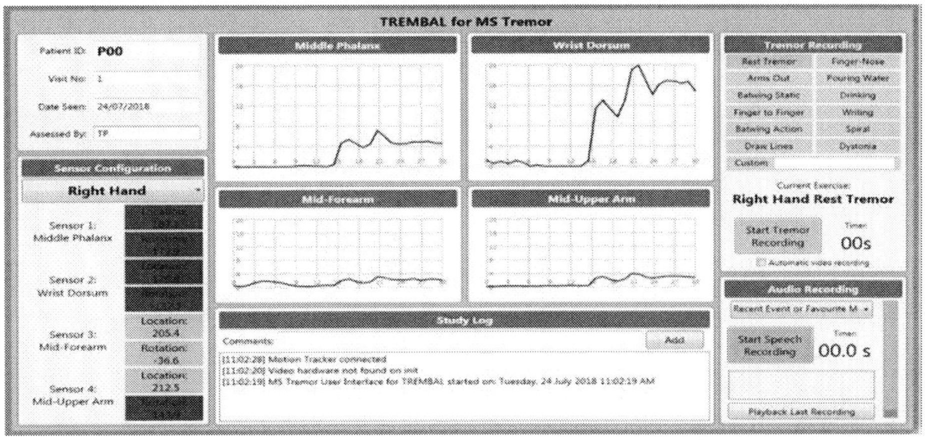

Figure 3. The graphical user interface of the TREMBAL software application

In order to make these systems for assessing and measuring the level of tremor more practical in terms of use in experimental research, mobile software applications have been developed for tablet or smartphone devices [11]. Thus, the researchers developed "a novel open-source mobile app for tremor quantification TREMOR12", offering "a low-cost tremor quantification method only for research purposes and algorithm development, and may help to improve treatment evaluation". [12] The TREMOR12 application was developed in *Xcode 7* using the Swift version programming language. 2.0 and through it, 4 parameters can be extracted, with 3D representation.

These parameters are: acceleration, rotation angle, rotation speed and gravitational acceleration, the first three parameters providing values of quantification of tremor, and the fourth can be used to calculate a standardized 3D space to make comparisons between measurements. The TREMOR12 application together with the TREMOR 12P version (data processing mode) runs on iPhone and iPod Touch and requires iOS8 or a newer version. Although not yet a certified and widely used application, it is a useful tool for further research that can be further developed.

3. Experimental setup

There is also an important aspect of research on tremor-type movements, in addition to evaluations and measurements of kinematic and dynamic parameters. It is about the identification of solutions to help patients with this pathological form, to be able to carry out daily personal or professional maintenance activities, providing them solutions to increase and improve postural comfort.

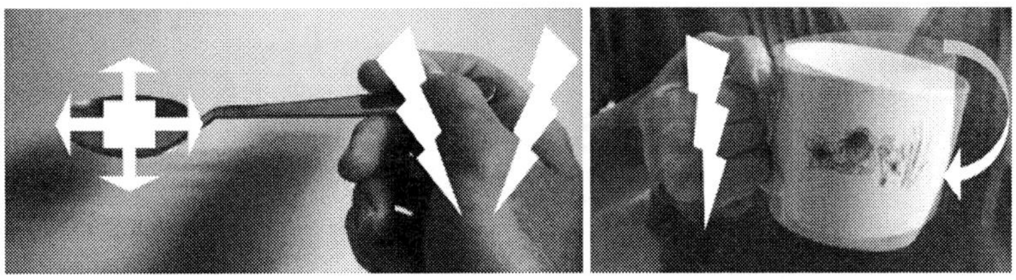

Figure 4. The handling of objects in case of tremor

These devices are addressed primarily to patients with a tremor in the upper limbs and who have to perform actions to handle various objects in the ambiental environment (figure 4).

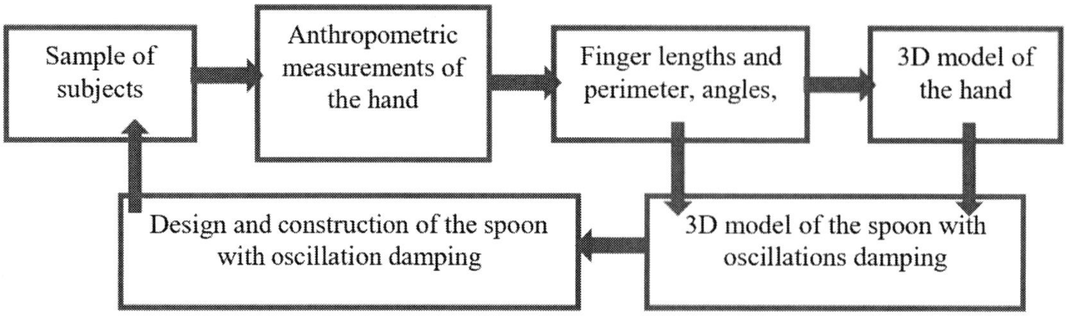

Figure 5. Block diagram to develop an attenuation system (spoon) of tremor movement in the hand

The movements due to the tremor can take place in vertical or horizontal direction regardless of how the object is caught, but the effects are the same because the object is becoming an oscillating exciter that amplifies the movement both in frequency and amplitude.

Therefore, instruments containing damping systems for these tremor oscillations are useful for patients with such manifestations. As such, the design and construction of such systems require in the first stage, an assessment of the anthropometric dimensions of the sample of subjects for which a tremor attenuation system is developed.

These measurements are part of the procedure for developing an attenuation system (spoon), for these movements (figure 5) and respectively for the overall analysis of the tremor-type movements at the level of the upper limbs.

The tremor-type motion analysis system developed in this research is based on image acquisition from a system (Contemplas), consisting of 3 high-speed video cameras (250 frames/sec), which synchronously captures images in three different directions for as then they are processed with the help of the Templo software application (figure 6).

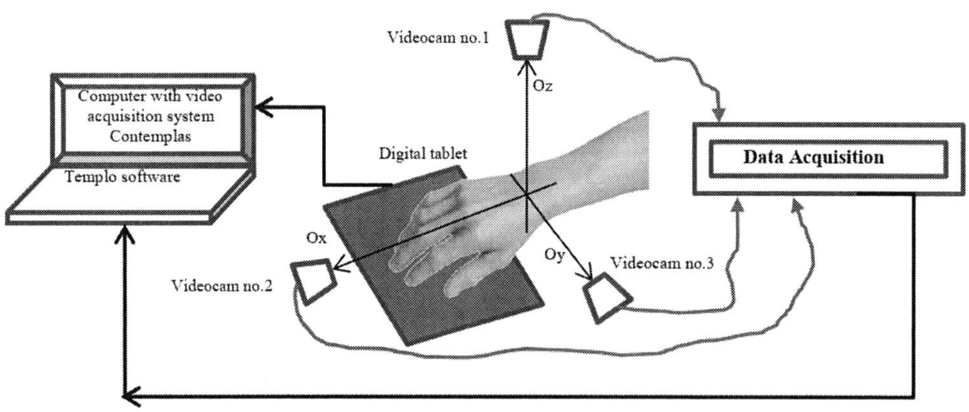

Figure 6. Block diagram to record the hand movements in tremor

The obtained images are synchronized and processed to identify the parameters of the tremor movement (frequency, amplitude) and also to determine the direction and the way of the movements in parallel with the use of the online application [13] for the 2D recording of tremor movement on a digital tablet with a touchscreen, as a path displayed on the touch screen.

4. Results and conclusions

The analysis procedure is designed to contain several work steps to be able to verify the initial hypotheses. Thus, the recordings on the video system were made in three stages of work, namely:

- recording the tremor movements without any attenuation device and comparing with the routes recorded on the digital tablet, then,
- in the second stage the subjects are recorded holding a regular spoon in their hand (video image acquisition) and respectively
- the third stage in which the subjects are recorded with the video cameras but also on the tablet, holding in their hand the tremor attenuation system, spoon with damping (figure 7a).

Following the construction of this experimental setup, the development of the image acquisition activity and the correlation with the routes recorded on the touchscreen tablet, a series of aspects were found that will determine changes and optimizations of the activity modules (positioning of video cameras, constructive modification of the system damping, modification of the spoon-grip system with damping, etc.).

The need for a calibration stage of the video cameras about the activity space and a resizing of the actuation systems of the cushioning structure in the spoon in order to obtain an ergonomic and accessible variant for patients was also identified.

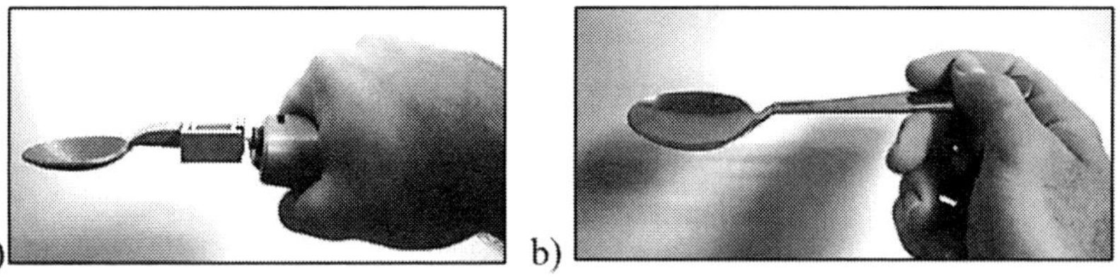

a) b)

Figure 7. Grip modification of the damping spoon system (a), in relation to the usual catching (b)

The configuration thus made requires a large volume of image acquisition and processing, for which the experimental system is provided with a memory hard-drive dedicated to these operations, to avoid loading the computer and reduce the working speed. Preliminary results of the use of this analysis system lead us to the conclusion that through a development and diversification of records, the information that can be obtained leads to the improvement of manoeuvring aid systems (figure 7) for patients with Parkinson's tremor.

5. References

[1] http://www.scholarpedia.org/article/Tremor; [Accessed April 2020];
[2] R. Elble, Essential Tremor (ET), *International Essential Tremor Foundation (IETF),* 2012;
[3] R. Elble, The Essential Tremor Rating Assessment Scale, *J Neurol Neuromedicine*, 1(4); 34-38; 2016;
[4] Ch.G. Goetz, The Unified Parkinson's Disease Rating Scale (UPDRS) Status and Recommendations, State of the Art Review, *Movement Disorders* Vol. 18, No. 7, 2003, pp. 738–750
[5] G. Deuschl, J. Raethjen, M. Lindemann, P. Krack, The pathophysiology of tremor. *Muscle Nerve* 24: 716-735; 2001;
[6] SK. Van Den Eden, CM. Tanner, AL. Bernstein, RD. Fross, A. Leimpeter, DA. Bloch, LM. Nelson, Incidence of Parkinson 's disease: Variation by Age, Gender and Race/Ethnicity. *Am J Epidemiol* 157: 1015-1022; 2003;
[7] AJ. Hughes, SE. Daniel, L. Kilford, AJ. Lees, Accuracy of clinical diagnosis of Parkinson's disease: a clinico-pathological study of 100 cases. *J Neurol Neurosurg Psychiatry* 55: 181-184; 1992;
[8] A. Harikrishna, D. Effah Osei, M.W. Kaminska, S. Majumdar, On numerical study of Parkinson tremor, (preprint and has not been certified by peer review), 2016;
[9] F. Vial, P. McGurrin, T. Osterholt, D. Ehrlich, D. Haubenberger, M. Hallett, Tremoroton, a new free online platform for tremor analysis, *Clinical Neurophysiology Practice* Volume 5, , Pages 30-34, 2020;

[10] T. Pereraa et al., Validation of a precision tremor measurement system for multiple sclerosis, *Journal of Neuroscience Methods* 311, pp.377–384; 2019;
[11] J-F Daneault, B Carignan, CÉ Codère, AF Sadikot, C Duval, Using a smart phone as a standalone platform for detection and monitoring of pathological tremors. *Front. Hum. Neurosci.* 6:357.2013;
[12] P.L. Kubben, M.L. Kuijf, L.P.C.M. Ackermans, A.F.G. Leentjes, Y. Temel, TREMOR12: An Open-Source Mobile App for Tremor Quantification, *Stereotact Funct Neurosurg*. Aug; 94(3): 182–186; 2016;
[13] https://www.parkinsonsmeasurement.org/toolBox/tremor.html [Accessed April 2020];

6. Acknowledgments

In these experiments we've developed the investigations with equipment from Applied Optometry and Medical Engineering Laboratories and the research is part of Ştefania-Gabriela Bortă's student diploma project works that was funded by the Transilvania University of Brasov.

IOP Conf. Series: Materials Science and Engineering **898** (2020) 012031 doi:10.1088/1757-899X/898/1/012031

Theoretical and experimental research on the use of the blinking reflex for command and control of human body movement

D Drăguşin and M Baritz

Design Product and Environment Faculty, University Transilvania Brasov

email: deliadragusin@yahoo.fr

Abstract. The blink reflex is a behavioural motor response that is normally found in the process of seeing at newborns and human adults. As with the corneal reflex, the supraorbital branch electrical stimulation of the trigeminal nerve causes a facial nerve bilateral response, namely eyes blinking. Therefore, this sensitive process can also be a neurological control element for performing movements, related to the natural blinking process amplitude, speed and frequency. In the first part of the paper, a series of blinking process physiological aspects are analyzed, in order to identify the movements command and control components. In the second part of the paper, it is analyzed the devices constructive variants for evaluating the blinking process and the connecting possibility with separate sensory elements that can collect information from the visual system neuro-motor level. In the third part of the paper, there are mentioned some aspects related to the experimental installation design and the command and control mode for setting in motion a system using the coordinated ocular and neurosensory blinking mechanism. The final part presents the conclusions of this approach to the development of a system with the use of oculomotor and neurosensory stimuli.

1. Introduction

The blinking mechanism of the eye in response to environmental stimuli is a typical defensive response to potential threats to the body [2] and is usually manifested as a variety of somatosensory and non-somatosensory stimuli. When brought into the state of stimulation by a series of intense somatosensory stimuli, the blink reflex is analyzed as a manual and non-automatic reflex. A series of studies [3] have shown that the blinking reflex has two components (figure 1), a component that determines an anterior-initial response R1 and a component with a posterior-late response R2.

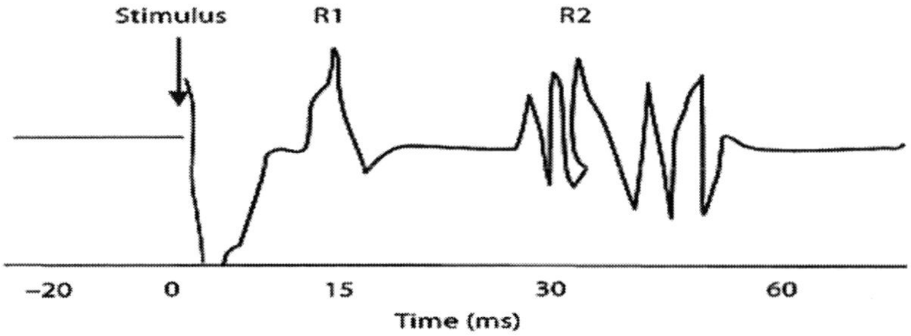

Figure 1. Diagram of the blinking reflex with the two components: R1 - anterior-initial component and R2 - late component [4]

The R1 response is usually present on the stimulated side, while the R2 response is usually present bilaterally. The R1 response is defined as representing the reflex pathway between the main sensory nucleus and the facial nucleus on the stimulated side. Instead, R2-type responses are mediated by a multi-synaptic pathway between the spinal cord nucleus and the inter-neurons that form connections with the facial nuclei on the laterally and counter-laterally stimulated side.

According to the results obtained from recent studies [3], the previous R1 response is usually stable and reproducible, with a biphasic or three-phase morphology, noting that in a small percentage of normal subjects, it cannot be reliably generated by both parties. Posterior R2 responses are polyphasic and variable from one stimulation to stimulation, so that through repeated stimulation, R2 responses tend to stabilize and become normal. A series of neurological-ocular pathologies can present the lack of the blink reflex as a form of manifestation, which leads to obtaining a permanently open eye situation, exposed to the risk of infection, irritation or even pain. At the same time, the loss of the blinking may lead to permanent damage to the cornea due to ulcers or infections caused by the drying process of the corneal surface or the penetration of particles from the environment. As shown in experimental research on the analysis of phenomena in cases of facial paralysis [5] in addition to the functional deficiency of the blinking process and other facial movements, the various forms of facial paralysis are also a major psychological barrier to a healthy social life. A number of researchers have shown in the laboratory that a possible way to restore the blinking process in the case of an eye neurologically affected by paralysis is to place a closed-loop neural prosthesis (facial stimulation device) that can detect normal blinking on one side of the normal face. Depending on this, it can simultaneously generate the stimulation of blinking on the paralyzed side. This is possible because the blinking process (as an automatic reflex), being symmetrical (synergic) process and using the normal state of the healthy eye, can trigger prosthetic assisted blinking on the facial area affected by unilateral paralysis. As shown, [5] "one of the challenges of using healthy blinking as a trigger for induced blinking in facial paralysis is the rapid and accurate detection of healthy blinking in a non-invasive and non-disruptive manner."

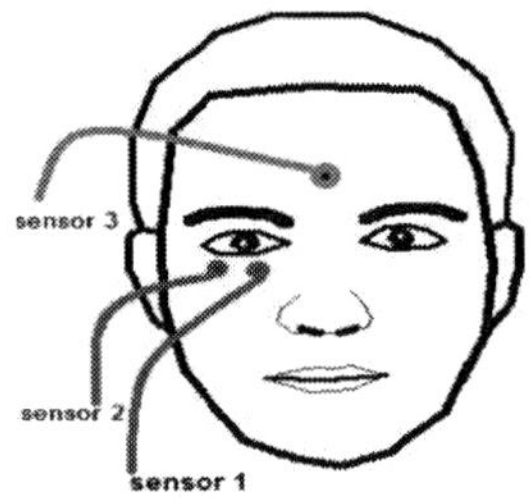

Figure 2. The placement of the sensors in electromyography recordings of the blinking process

Thus, the electromyography (EMG) method of the facial muscle structure in the ocular area can be considered the most frequently proposed and used method for noninvasive detection of blinking on the intact side, followed by the detection of tissue movements in the area of the ocular orbital periphery, by using accelerometers. But even this method, even if it has a high level of accuracy, cannot be used constantly because it requires placing electrodes on the facial surface, in the eye area and this can influence the dynamics of eyelid movements in the process of normal blinking (figure 2). Therefore, in order to increase the comfort and efficiency of the determinations, the detection of the blinking process can be performed without the interaction of the sensors with the skin surfaces in the eye area, and this can be done with the help of video acquisition systems consisting of infrared radiation (IR) lighting sources. This type of illumination of the facial-orbital area makes it possible to record eye movements without taking into account the photo-motor reflex at the pupil level due to light radiation with wavelengths in the visible range (400-700 nm). The video methods for detecting eye movements (the blinking process) are based on the analysis of images purchased with a high-resolution video camera and acquisition frequency, being much more convenient to use and with multiple possibilities to determine the position and shape of the eyelids and the blinking mode (continuous, sequential or complete/incomplete). This camcorder can be fixed on a chin rest for laboratory testing or on a special pair of glasses for dynamic research in different environments. A very important application of this method is the protection system for drivers against the phenomenon of falling asleep while driving, made with a mobile IR video camera system, mounted on the arm of the glasses with the system with *Raspberry Pi Zero*. Despite the fact that the IR video camera has solved the low performance at night of the cameras with visible radiation, this system is still considered less robust during the day due to the compound white light radiation that interferes with the reflections of the IR radiation. [6,7]

2.Theoretical aspects of blinking process calculus

For fundamental research studies, especially in the field of performance, or the prevention and detection of neurological pathologies, a series of devices and systems for recording and processing images that can indicate the behavior of subjects in different situations related to the orientation of the direction of gaze (convergence, fixation and focus) have been developed in international research centers.

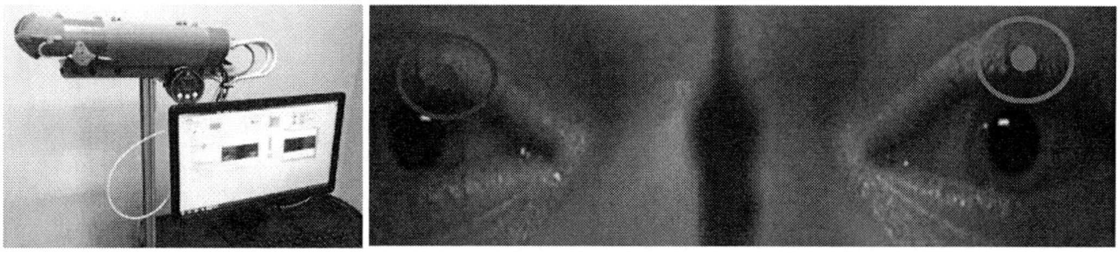

a) b)

Figure 3. a) Blink reflexometer housing unit and software interface, b) Location of the upper right (red dot) and left (blue dot) eyelids are located in each image acquired during a blink. [8]

These systems although high-performance and high-precision, with real possibilities to obtain quantifiable results, have not yet entered the clinical sphere and are still in various stages of development (figure 3). [8] The theoretical determinations that were the basis for the development of these devices, take into account, in the process of the blinking cycle, the opening of the eyelid slit in the form of the area or the distance between the upper and lower eyelid. Thus, a contour of the orbital area between the upper and lower eyelid is defined and analyzed in the three extreme positions of the blinking process (maximum open, half open and minimum open), according to figure 4.

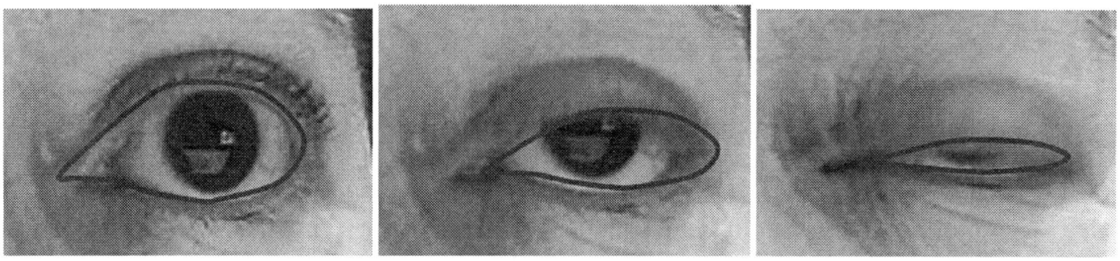

a) b) c)

Figure 4. a)Blinking process- maximum opening, b) medium opening, c)minimum opening

Also, in order to be able to calculate the area and the height of the eyelid slit, respectively, 4 main points (upper/lower 1-4) and 4 other cardinal points (up/down, nasal/temporal 5-8) are chosen on the contour of the opening profile, according to figure 5. The calculation of the opening area of the eyelid slit is determined, in ideal format, using the perimeter between points 1-4 together with points 5 and 7 (figure 5), and the distance between the upper and lower eyelid, using points 6 and 8 (figure 5)

The size of the EAR (eye aspect ratio), "defined as the ratio between the height and the width of the eye contour" [9], can be calculated using multiple video recordings of eyeball images and the determination of mediated values of distances A, B and C (figure 6), as Euclidean distances, as presented in equations (1) and (2).

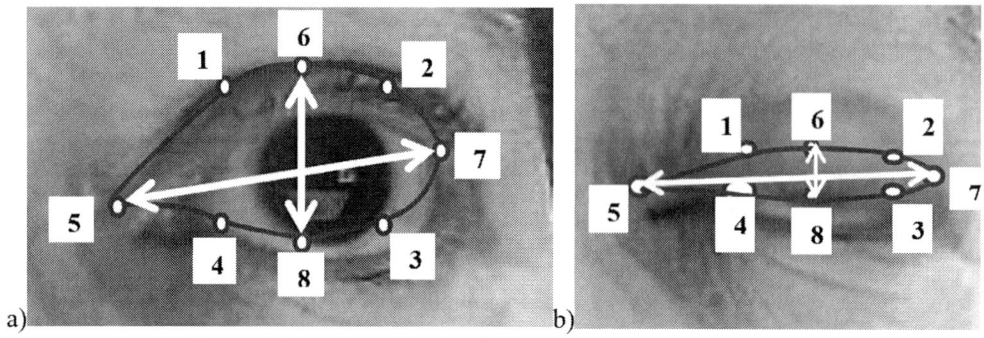

Figure 5. a) Eye open maximum, b) eye open minimum

$$d(p,q) = d(q,p) = \sqrt{(q_1 - p_1)^2 + (q_2 - p_2)^2 + \cdots \dots (q_n - p_n)^2} = \sqrt{\sum_{i=1}^{n}(q_i - p_i)^2} \qquad (1)$$

$$EAR = \frac{A+B}{2*C} \qquad (2)$$

From equation (2) it can be seen that the values of the EAR coefficient decrease as the sizes of distances A and B decrease (the eyelid slit decreases) and the upper eyelid approaches the lower one in the blinking process, size C having a small variation.

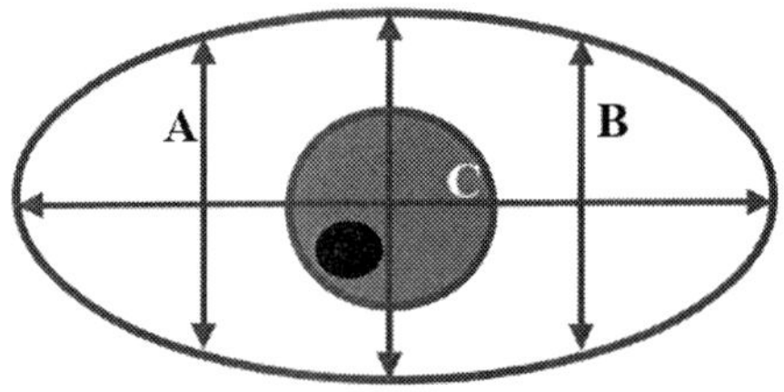

Figure 6. Ideal format to calculate *Eye Aspect Ratio (EAR),* defined as the ratio between the height and the width of the eye contour [9]

In terms of the area of the eyelid opening, there will always be a T_{eye} threshold (which can be determined experimentally) of this value below which the eyeball is considered closed (the eyelid slit no longer allows light radiation to enter the eyeball to form a whole image on the retina) and therefore, the half-cycle of the blinking process can be considered completed. This opening/closing threshold is determined experimentally and therefore requires a careful analysis of the type of visual system of the subjects participating in the recordings, of the mode of reaction at the ocular level and last but not least, of the refractive state of the visual system (emmetropic or ammetropic). In this way, it will be possible to determine the state of the eyeball, from the point of view of the blinking reflex, comparing the EAR value with the opening/closing threshold established experimentally on a group of subjects as different as possible in relation to the visual system

$$Eye\ state = \begin{cases} open, & EAR \geq T_{eye} \\ closed, & EAR < T_{eye} \end{cases} \qquad (3)$$

The closed/open state of the eyeball is thus analyzed in different situations of activity (driving, tracking eye movement for various activities, control of mobile systems with the help of eye movements, medical tests related to neuromotor health, positioning of the head and eyes, ocular ergonomics, etc.)

The optimal value of this threshold of 0.25 has been identified from the literature [9-11], which means that if the eyelid slit is only 25% open, then the eye can be considered closed and the blinking half-cycle achieved, even if the eye has not closed completely.

3. Experimental setup

In order to capture and process the information regarding the control of a mini-car (information recorded in parallel with a neurosensory headset and a video camera integrated in special glasses), an experimental system was designed and developed, using the blinking process (physiological and controlled) to track the movement of the eyeballs, in order to convert it into command and control signals. The experimental system developed in this research consists of a pair of glasses (realized by 3D printing) integrated with a video camera, a NeuroSky type neurosensory headset, both devices being dedicated to capturing blinking eye movements, then an electronic acquisition and compatibility system in real time of the signals coming from the two sensory devices and respectively a computer with dedicated software applications (fig.7). The element that takes over the computer commands according to the information entered and coming from the sensory devices is represented by a mini-automobile with all-wheel drive systems

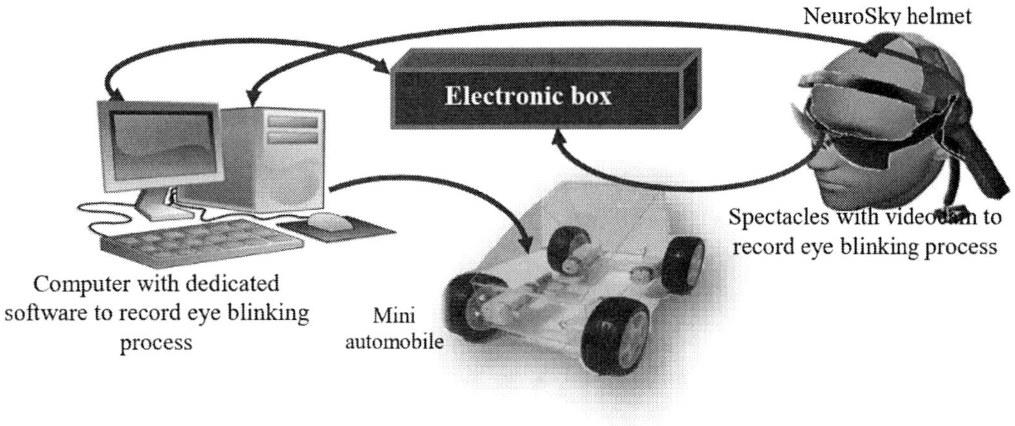

Figure 7. Experimental setup

The experimentation procedures contain a series of modules through which the acquisition of signals from the NeuroSky headset are analyzed in real time and then compared with the commands of the software application developed on the Arduino board for the acquisition of the sequences of the complete blinking cycles.

According to research in the field of visual function, the blinking process takes place, on average, with a frequency of 17 blinks / min, which indicates that the average duration of a full cycle (open-eye-closed-open) of physiological blinking may take between 300-550 ms, with an average interval between two blinking cycles of 2.8-3.1 sec.

Therefore, a complete and controlled blinking cycle, consciously performed to trigger a motion control of the mini-car must last on average at least 5 times longer than the average physiological blinking cycle.

From the value of the duration obtained of this complete blinking cycle, the open-closed and closed-open transitions of the target eyeball must take place within a range of approximately 40% of this estimated duration and the closed eyeball period must be about 60%. This accentuated, controlled and conscious form of the blinking process is necessary to enable the optoelectronic system to capture the image, to transmit it to the computer, to be analyzed and transformed, through the software application on the Arduino board, into a control signal for mini-car's motors.

The control of the mobile must also take into account the travel requirement, in relation to the space and trajectory required. As such, blinking commands will be quantified, based on a programming code, so that they can be recognized, analyzed and transformed into appropriate commands (forward, left, right, etc.) for the mini-car used. If it is necessary to travel with a larger and heavier vehicle, the design difference will only be related to the power of the engines and gear transmissions

4. Results and conclusions

Following the construction of this experimental setup and the development of the command and control activity of the mini-car movement through the blinking process, a series of aspects were found that led to the modification and optimization of the action modules. Thus, it was identified a need for a step of calibrating the video camera in relation to the activity space and a reprogramming of the maneuvers to start, go and stop from the same commands of the blinking process. It was also found during the experiments that there is a need to use a second computer with Bluetooth to communicate with the NeuroSky headset in order not to create a conflict of information transmission in parallel with the acquisition of images from the video camera. This can be done quickly and for the next stage, it is planned to create an external module for the compatibility of signals from both sensory systems.

The design and development of this experimental system proves to be useful also in making records for the analysis of different behavioral states of subjects or for the control of interactive activities with the computer. [12-15]

5. References

[1] J. M. S. Pearce, Observations on the Blink Reflex. *Eur Neurol* 2008;59:221-223.;

[2] R.J. Bufacchi, S. Ponticelli, G. Novembre, M. Kilintari, Y. Guo, G.D. Iannetti, Muscular effort increases hand-blink reflex magnitude, *Neuroscience Letters,* 2019, May 29; 702: 11–14;

[3] D. Preston, B. Shapiro, *Electromyography and Neuromuscular Disorders, Clinical-Electrophysiologic Correlations, 3rd Edition,* ISBN: 9781455726721, November 2012, pp.664,

[4] https://www.karger.com/Article/Fulltext/114053 [Accessed 25 April 2020].

[5] A. Frigerio, T. A Hadlock, E. H Murray, J. T Heaton, Infrared- Based Blink Detecting Glasses For Facial Pacing: Towards A Bionic Blink, *JAMA Facial Plast Surg.* 2014 May-Jun; 16(3): 211–218.

[6] Medium. 2017.*Wearable Vision Assistance Device With The Raspberry Pi Zero.* [online] Available at: <https://medium.com/hacksters-blog/wearable-vision-assistance-device-with-the-raspberry-pi-zero-68eb7a67c58e> [Accessed 25 April 2020].

[7] Instructables. *Poor Man's Google Glass/Aid For Those With Tunnel Vision.* [online] Available at: <https://www.instructables.com/id/Poor-Mans-Google-GlassAid-for-Those-With-Tunnel-Vi/> [Accessed 1 May 2020].

[8] N. T. Tsai, J. S. Goodwin, M. E. Semler, R. T. Kothera, M. Van Horn, B. J. Wolf, D. P. Garner Development of a Non-Invasive Blink Reflexometer, *IEEE J Transl Eng Health Med 2017; 5: 3800204*

[9] A Z Mohammed, E A Mohammed, A M Aaref, Real-Time Driver Awareness Detection System, *IOP Conf. Series: Materials Science and Engineering*, 745 (2020) 012053;

[10] https://www.hindawi.com/journals/cmmm/2020/1038906/[Accessed 1 May 2020].

[11] D. Dou, Z. Zhang, Blink Detection Based on Pixel Fluctuation Ratio of Eye Image, 2020 *J. Phys.: Conf. Ser.* 1453 012073;

[12] W. Liu, J. Qian, Z. Yao, X. Jiao, J. Pan, Convolutional Two-Stream Network Using Multi-Facial Feature Fusion for Driver Fatigue Detection, *Future Internet* 2019, 11, 115; doi:10.3390/fi11050115,

[13] N. Irtija, M, Sami, M. A. R. Ahad, Fatigue Detection Using Facial Landmarks, *ISASE-MAICS* 2018;

[14] V. Karthikeyan; B.P. Kumar; S.S. Babu; R.Purusothaman; S. Thomas, A Narrative Vehicle Protection Representation for Vehicle Speed Regulator Under Driver Exhaustion-A Study, 2014;

[15] A Z Mohammed et al, Real-Time Driver Awareness Detection System, 2020 *IOP Conf. Ser.Mater. Sci. Eng.* 745 012053;

6. Acknowledgments

In these experiments, we have developed the investigations with equipment from Applied Optometry Laboratory and the research is part of Delia Dragusin's student diploma project work that was funded by the Transilvania University from Brasov

IOP Conf. Series: Materials Science and Engineering **898** (2020) 012032 doi:10.1088/1757-899X/898/1/012032

New research direction on performance and co-creation: a literature review

M B Tudose[1], G Agafitei[1] and S Avasilcai[2]

[1] PhD candidate, Department of Engineering and Management, FDIMA, Gheorghe Asachi Technical University, 29 Dimitrie Mangeron Bd., Iaşi, 700050, Romania
[2] PhD Professor, Department of Engineering and Management, FDIMA, Gheorghe Asachi Technical University, 29 Dimitrie Mangeron Bd., Iaşi, 700050, Romania

brindusa.tudose@gmail.com

Abstract. The purpose of this study is to review literature on topics of performance and co-creation. The study has two premises: performance includes economic and behavioural aspects; co-creation produces direct and indirect effects on performance. In order to validate these two, we took into consideration all studies available on WoS platform. Applying different selection criteria, we obtained a sample of 14 studies published between 2016 and 2020. Through observation, comparison and logic deductions we were able to appreciate the authors` interest in identifying new research directions. Thus, we show that: a) co-creation has a direct impact on performance, competitiveness and value for stakeholders; in this context, performance is regarded as a multidimensional concept with wide implications in various fields such as management, financial, marketing, innovation etc.; b) co-creation has an indirect influence on performance through elements like: brand, product-service systems and cross-sector partnerships. These aspects are empirically validated in different context. The results of our research prove to be useful from two points of view: theoretical (because the study presents in an authentic manner the state of the knowledge on performance and co-creation) and methodological (because it offers support to all those interested in finding new research directions).

1. Introduction

Performance topic is subject to great debate and it includes all social sciences and engineering fields of activity. Performance is a particular result in management, economy and marketing, results that allow assessing competitiveness, organisational efficacy and efficiency and its structural and procedural components [1]. For a clear understanding, performance needs to be approached from the point of view of those interested in attaining and measuring performance but also in the context in which performance needs to be time related in order to be measured [2].

Performance includes all economic and behavioural aspects of individuals and companies. Economic aspects are translated in terms of results (financial and non-financial), most of them measurable. Behavioural aspects may be approached in terms of results but they imply complex measurement and assessment procedures. This is the context in which interests regarding performance assurance includes aspects that depend on stakeholders' behaviour, where consumers/clients have a dedicated role. Some authors [3, 4] have concluded that communication process and high level of interactions with consumers/clients are the most important success determinants of a product and consequently of a business.

Communication is a key element in the co-creation process and it facilitates an active involvement of the client. Co-creation activities determine consumers to contribute in a significant way in designing, developing, producing, marketing and distribution of an existing product or a new one, which in turn reduces the cost of innovation and initiates the success path for a business [5]. Not least, co-creation reduces lost time on irrelevant routines, allowing companies to focus on solving their emerging difficulties [6].

The main objective of this paper has been to evaluate the advance in terms of research from the perspective of a vision that approached co-creation as a determinant factor of performance. This objective is assumed in the context in which modern performance management systems have extended the area of performance assessment towards non-financial aspects, by forcing companies to improve product quality, develop their flexibility and reliability and direct their strategy towards innovation. From this point of view, co-creation has the aim of increasing clients' involvement in product development processes and growing company and shareholders' value [7].

For the present research objective, the paper has been structured as it follows: first section represents current state of the research on performance and co-creation; second section details on the research methodology; next section represents the outline on the most recent research on performance and co-creation; last section is a synthesis of the conclusions and possible future research.

2. Review of literature

2.1. State of the art regarding performance

Conceptualisation efforts on performance topics have marked the advance in terms of research; these efforts have not been downsized by negative heuristic (based on denial, rejection, critique), they were crowned by positive heuristic that allowed for theoretical foundation improving.

During first debates, performance shows up as a conditioning element, context in which it is stated that continuous performance improvement is the vital objective of a company [8]. These first attempts have been criticised for neglecting two important aspects (the structure and availability of resources within a company). Subsequently, new elements have been included in order to define the concept of performance [9]: efficient organising, higher productivity rate (associated with high levels of satisfaction and motivation for organisation members), high growth rate for turnover (associated with reduced costs) and absence of work conflicts.

About ten years distance, performance has been defined as the extent to which an organisation, as a social system, takes in consideration its means and goals [10]. This way, performance it's not just an efficiency related concept of an organisation but also an indication for the manner in which an organisation achieves its objectives [11]. Although there have been multiple initiatives to define performance, some authors invoke incomplete conceptualisation [12] or lack of consistency in defining the concept [13]. These signals have determined a new construct; performance is associated with capacity of an organisation to efficiently use its available resources to obtain results according to the objectives that were set, taking into consideration their relevance for stakeholders [14]. Also, the performance topic coverage extends, including quality management assessment and determining value for clients and other interested parties [15].

At the corporate level, the essence of performance translates in growth and survival on the long term [16]. More recent research mention performance as a barometer that reflects the actual business and development trend [17], respectively as an artefact depending on which success of an organisation is measured in the context of a free, competitive and global market [2]. In this new context, corporate performance focuses on measuring what has been realised by a company which represents good conditions for a certain period of time [18].

2.2. State of the art regarding co-creation

Co-creation term has been presented for the first time in the beginning of 2000 to explain the relation between companies and consumers [19]. In return getting users (beneficiaries) involved as a source of innovation has been mentioned since the 18th century. Over time there has been interest in explaining this concept. For example, some authors [19] made a comparative analysis to better explain what is and what is not co-creation; others [20] performed a comparative analysis between classic strategy and strategy based on co-creation, in which they detail the stakeholders for whom value is created, the goals pursued and benefits of co-creation. Modern management dictionaries (such as Business Dictionary) place co-creation as a business strategy that substantiates on clients experience and dynamic relations.

Co-creation represents a way of getting clients directly involved in the process of developing a product. Today, companies try more and more to integrate this feedback in designing new products and services. Lusch et al. [21] support the idea that co-creation with clients has the purpose to innovate, being a fundamental part of modern marketing. In a digital era, co-creation implies social learning processes through which people learn from each other, they create and collaborate by means of social networks [22]. Co-creation has facilitated the interaction between brands, employees and clients and using online platforms allowed implementing standardised working processes [23].

Co-creation implies sharing activities with clients in order to make innovations closer to clients' preferences, with a higher probability of commercial success [5] [24]. Although co-creation may be considered important, many companies have difficulties in using and transforming these knowledge in significant results [25].

3. Methodology of research

In order to evaluate the extent to which researchers show interest for analysing correlations between performance and co-creation we revised research in the past five years. By using the search term 'performance management' we sum up 277110 studies (published in the last 45 years), among which 15% are business related and 8% management related. When using keyword 'co-creation' we get 5959 studies, out of which 1847 business related and 1564 management related (out of which 70% are published between 2016 and 2020). On the other hand, search using both 'performance' and 'co-creation' revealed 873 studies published between 2016-2020 out of which 354 business related and 342 management related. Among these 873 sources selecting the *open access* option has generated 198 results, out of which 68 business related and 59 management related. At the first evaluation we concluded that some studies are made on public or financial entities so we made a new search by integrating the criteria 'performance company'. The final results of these searches were 14 articles (published between 2016 and 2020).

4. Landmarks of recent research focused on performance and co-creation

According to the research methodology we have identified 14 studies that approach the concepts of co-creation and performance in an integrated manner. In this section we will go over the main research ideas they convey.

Although not discussing the topic of co-creation in an explicit manner, Nisar et al. [26] base their study on relevant research in this field. Their research emphasises the following essential aspects: a) information diversity and meaning have a positive impact on income (sales generated), considered a relevant indicator for firm performance; b) reputation is positively influenced by users' involvement; c) both 'firm-generated content' and 'user-generated content' are important for involving users and firm performance; d) cause effect relation between social media and firm performance are evident (confirmed by previous research).

Omar et al. [27] showed that, in order to offer value added services, firms focus on consumers. This approach (based on dialogue, participation and involvement) has a strong impact on firm performance (through brand). In their research authors integrate co-creation in the innovation field of expertise, considering it an extension of marketing. DART model (dialogue, access, transparency and referred risk) is considered a part of the co-creation model based on innovation. According to the authors, customer value co-creation behaviour represents a voluntary behaviour that offers value to the company and consumers.

Dahlin et al. [25] have made a research studying the relation between the absorption, co-creation and innovation performance and showed that involving clients (and importance of this involvement for innovation processes) vary not only among companies but also among countries and context. Using a questionnaire on a sample of 1102 companies (Swedish and Norwegian multi-group SEM), results showed that: a) there are differences between countries regarding the role of co-creation, because the business logic are different from country to country; b) co-creation (appreciated through the utility given by the clients knowledge) may play a vital role in the relation between the capacity of absorption and innovation performance; c) independent of the national context, companies with a weak capacity of absorption cannot rely on co-creation as much as other large companies; d) national context may affect clients involvement in the innovation processes.

Mulyana et al. [28] showed that value co-creation (understood as a commitment of clients in developing new products) places clients' behaviour and attitude in the centre of marketing strategies. Analysing data collected by applying a questionnaire (using structural equation modelling), authors have shown the following: a) the higher the brand commitment the more willing the consumers will be to 'sacrifice' for companies in creating value co-creation; b) in turn, value co-creation influences perceived quality (which - in turn - affects the brand advantage and contributes to creating competitive advantages); c) brand advantage may be influenced only by perceived value and cannot be influenced straight by co-creating value; d) high advantage for the brand represents a warranty for long term sustainability. Based on these results it has been confirmed that a strong consumer-brand commitment will be perceived as pride when consumers receive acknowledgement for being partners in creating value (context in which they will be voluntary doing sacrifices for the company if the value they create represents fulfilling their needs and wishes).

Based on a conceptual research (with focus on clients' involvement behaviour, respectively, on particular types of behaviours of the customers engagement), Beckers et al. [7] investigated, in premier, the financial consequences of firm-initiated customer engagement. The results of this research reveal the following: a) initiatives showing involvement of companies' clients may have a negative impact on market value; b) there is a significant variation in shareholders' answers to different involvement initiatives; c) initiatives focusing on word-of-mouth are considered to be less negative than initiatives expressed through social media; d) customer engagement initiatives are more efficient on competitive markets that require more activities focused on clients and a strong differentiation; e) in certain conditions initiatives coming on behalf of the firm inviting customers to contribute can increase the efficiency and effectiveness of value creation f) initiatives launched by firms with the aim to engage clients are less beneficial for firms with a strong reputation.

By performing a chronological analysis of the literature regarding client participation, Boaventura and Brito [29] propose a model that takes into consideration organisations strategy to invite customers to control and act in order to contribute. They show that in highly relational services, clients' participation is intrinsic and mandatory but this can change and participation may become voluntary when companies promote these opportunities.

Reim et al. [30] made a multiple exploratory study on two Swedish manufacturing companies that offer product-service systems, showing that offering these services certainly results in economic, environmental and social benefits but they also involve significant challenges regarding the relational dynamic between supplier and client. This is due to the fact that offering product-service systems on global markets with different types of clients raises the risk of uncertain behaviour (and also potentially adverse) of clients. Consequently, understanding the way clients may react in the context of product-service systems represents a major challenge. To face this challenge, authors propose mechanisms that diminish the negative client behaviour (monitoring, sharing and trust mechanisms). Although (at global level) these mechanisms present advantages and disadvantages, individually approach, each one proves to be useful for solving certain agency problems (e.g. sharing mechanism is recommended when clients focus on costs, while trust mechanism is usually necessary when customers are opponents of the contract).

In the attempt to offer new perspectives on using social media as a means of building relations with consumers and energise the collective synergy for co-creation, Choi and Burnes (2017) [31] recommend to the producers two different manners of involvement: bonding (meaning that they preserve the cultivating emotional connection with clients) and spreading (which involves clients' creativity). By making an extensive research on music industry, authors provide proof that confirm the following: a) the increase of social media has permitted to make connection whit clients that have their own social network; b) prosocial attitude of consumers towards a certain type of music is based on trusting the labels; c) the two characteristics (bonding and spreading) have an impact on each other. They recommend firms to build strong and distinct identities around brands and their products in order to communicate with their consumers and offer them something in which they can actively get involved.

Ercsey [32] has analysed the way in which the level of involvement influences two types of behaviour: the consumers' participation behaviour and the consumers' citizenship behaviour. Based on a quantitative research they showed that: the level of involvement influences both mandatory and voluntary behaviour of clients (in the context of value co-creation); b) clients' feeling for co-creation may be investigated as a multidimensional construct; c) value co-creation doesn't only refer to co-production through firm-client interaction but also to value co-creation through client-client interaction; d) social involvement related to cultural services has led to greater client engagement in co-creation.

To give meaning to the impact of cross-sector partnerships on co-creating dynamic capabilities for stakeholder orientation, Dentoni et al. [33] have taken a theoretical approach (based on an inductive research) and showed that cross-sector partnerships were created as new forms of organisation that allow members to co-create resources and capabilities to move towards their set sustainability objectives. Due to the fact that this result is valid only for companies, authors suggest that there are no data to evaluate the way in which cross sector partnerships have an impact on dynamic capabilities of non-business partners.

The other studies in the sample exceed the objectives of this study since they refer to: non market strategy of large companies and its correlation with corporate social responsibility [34], developing business model frameworks in order to better show service logic principles [35]; predetermined structures for differentiating services [36]; activities that managers may perform in order to improve the competitiveness of a company and that of a service network [37].

5. Conclusions

The main goal of the paper has been to carry out a review of research on performance and co-creation. The results show that we are being witness to scientific development (materialised in new and original research ideas). The paper brings twofold contributions: a) that of conceptualisation and determining cause-effect relation between co-creation and performance; b) that of research by identifying new research directions.

The advance of theoretical research (although not to the same extent shared in practice) allowed for a better understanding of co-creation role in meeting corporate performance. Although there is no consensus in defining concepts, researchers' opinions converge towards a dominant direction: a) performance reflects the capacity and capabilities of an organisation in trying to efficiently use its available resources according to the objectives and taking into consideration the relevance of these for the stakeholders; b) co-creation is the process through which consumers contribute significantly to designing, developing, producing and distributing a product. First, the direct effect of co-creation results in reducing innovation costs. Second, decreasing cost of innovation has a positive impact on performance.

Summarising the results of the most recent studies that approach the interrelation between performance and co-creation we showed that:

a) co-creation has a direct impact on financial performance measured through sales volume [26], innovation performance [25], service performance [29] marketing performance [28], shareholders value [7] and competitiveness [32].

b) co-creation has an indirect impact on performance through variables such as: brand [27] [31], product-service systems [30] and cross-sector partnership [33].

These results are empirically validated in several contexts: that of companies producing goods [26] [28] [31] [33]; that of companies offering services [27] [29] [32]; that of companies offering service systems for a product [30]; that of SMEs [25], respectively of listed companies [7].

The results confirm that in the field of scientific research new research directions are shaped and they validate two premises on which research has been built (performance includes economic and behaviour aspects; co-creation produces direct and indirect effects on performance). The reasons that led to the development of research were diverse and included: hypotheses that were the basis of research; validity of previously formulated ideas; transformations at global society level.

Limitations and future research directions. Because research focused on performance and co-creation have taken amplitude along the years we have to mention that for the elaboration of the research we considered a selection of the most recent studies. For this reason we admit that the study is not intended to be exhaustive. We are considering extending the research with the aim to identify the extent to which different modern methods of measuring performance include co-creation as a determinant factor of performance.

References

[1] Verboncu I and Zalman M 2005 *Management and performances* (Management şi performanţe-Romanian) (Bucharest: Editura Universitara) p 106
[2] Tudose B M, Avasilcăi S and Golban R 2019 Assessing financial performance of companies manufacturing industrial goods. Evidence on performance dynamics in the period before and after the crisis (*Proc. Int. Conf. European Union's Structural Challenges: The Way Forward* vol. 6) ed. Ţigănaşu R, Simionov L, Alupului C, Cărbune A (Iasi: Editura Universităţii „Alexandru Ioan Cuza") pp 335-55
[3] Gruner K and Homburg C 2000 Does Customer Interaction Enhance New Product Success? *Journal of Business Research*, **49**(1) pp 1-14
[4] Lundkvist A and Yakhlef A 2004 Customer Involvement in New Service Development: A Conversational Approach *Journal of Service Theory and Practice* **14**(2/3) pp 249-57
[5] Li P Y and Huang, K. F. 2019 The antecedents of innovation performance: the moderating role of top management team diversity, *Baltic Journal of Management* **14**(2) pp 291-311
[6] Zhang M, Zhao X, Voss C and Zhu G. 2016 Innovating through services, co-creation and supplier integration: Cases from China, *International Journal of Production Economics* **171** pp 289-300.
[7] Beckers S, van Doorn J and Verhoef P 2018 Good, better, engaged? The effect of company-initiated customer engagement behavior on shareholder value, J. *of the Acad. Mark. Sci.* **46** pp 366-83
[8] Chandler A D 1962 *Strategy and structure* (Cambridge: MIT Press)
[9] Lupton T 1977 *Organisational Behavior and Performance* (London: The Macmillan Press)
[10] Robbins S P 1987 *Organisational Theory: Structure, Design, and Application* (San Diego: Prentice-Hall)
[11] Cherrington D J 1989 *Organisational behavior: The management of individual and organisational performance* (Boston: Allyn&Bacon)
[12] Cameron K S 1986 Effectiveness as paradox: Consensus and conflict in conceptions of organisational effectiveness *Management science* **32**(5) pp 539-53
[13] Kirby J 2005 Toward a theory of high performance *Harvard Business Review* **83** pp 30-39
[14] Peterson W, Gijsbers G and Wilks M 2003 An organisational performance assessment system for agricultural research organisations: concepts, methods, and procedures *ISNAR Research Management Guidelines* **7** pp 1-81

[15] Moullin M 2007 Performance measurement definitions: Linking performance measurement and organisational excellence *International Journal of Health Care Quality Assurance* **20**(3) pp181-83
[16] Bercovitz J and Mitchell W 2007 When is more better? The impact of business scale and scope on long-term business survival, while controlling for profitability. *Strategic Management Journal* **28**(1) pp 61-79
[17] Peleckis K, Krutinis M and Slavinskaitė N 2013 Multi-criteria evaluation of the efficiency of the main activities of alcohol industry companies (Daugiakriterinis alkoholio pramonės įmonių pagrindinės veiklos efektyvumo vertinimas-latvian). *Verslo ir teisės aktualijos* **l**(8) pp 1-16
[18] Batchimeg B 2017 Financial Performance Determinants of Organisations: The Case of Mongolian Companies *Journal of Competitiveness* **9**(3) pp 22-33
[19] Prahalad C K and Ramaswamy V 2004 Co-creation experiences: The next practice in value creation *Journal of Interactive Marketing* **18**(3) pp 5-14
[20] Ramaswamy V and Gouillart F 2010 The Power of Co-Creation. Build It with Them to Boost Growth, Productivity, and Profits, New York: Free Pres
[21] Lusch R F, Vargo S L and O'Brien M 2007 Competing through service: Insights from service-dominant logic *Journal of Retailing*, **83**(1) pp 5-18.
[22] Hartley J, Sorensen E and Torfing J 2013 Collaborative innovation. A viable alternative to market competition and organisational entrepreneurship *Public Administration Review* **73** pp. 821-30
[23] McIntyre D P and Srinivasan A 2017 Networks, platforms, and strategy: Emerging views and next steps *Strategic Management Journal* **38**(1) pp 141-60
[24] Kazadi K, Lievens A and Mahr D 2016 Stakeholder co-creation during the innovation process: Identifying capabilities for knowledge creation among multiple stakeholders, *Journal of Business Research* **69**(2) pp 525-40.
[25] Dahlin P, Moilanen M, Ostbye S E and Pesämaa O 2019 Absorptive capacity, co-creation, and innovation performance. A cross-country analysis of gazelle and nongazelle companies, *Baltic Journal of Management* **15**(1) pp 81-98
[26] Nisar TM, Prabhakar G, Ilavarasan P V, and Baabdullah A M 2020 Up the ante: Electronic word of mouth and its effects on firm reputation and performance J*ournal of Retailing and Consumer Services* **53** pp 1-30.
[27] Omar N A, Kassim A S, Shah N U, Alam S S, and Wel C A C 2020 The Influence of Customer Value Co-Creation Behavior on SME Brand Equity: An Empirical Analysis *Iranian Journal of Management Studies (IJMS)* **13**(2) pp 165-96
[28] Mulyana D, Rudiana D and Taufiq A R 2019 The role of value co-creation based on engagement to develop brand advantage *Polish Journal of Management Studies* **20**(1) pp 305-17
[29] Boaventura P S M and Brito E P Z 2018 Leveraging Highly Relational Service Performance Through The Participation of Empowered Customer *Brazilian Journal of Marketing BJM* **17**(3) pp 314-28
[30] Reim W, Sjödin D, and Parida V 2018 Mitigating adverse customer behaviour for product-service system provision: An agency theory perspective, *Industrial Marketing Management* **74** pp 50-61
[31] Choi H and Burnes B 2017 Bonding and Spreading: Co-Creative Relationships and Interaction with Consumers in South Korea's Indie Music Industry, *Management Decision* **55**(9) pp 1905-23
[32] Ercsey, I 2017 The Role of Customers' Involvement in Value Co-creation Behaviour is Value Co-creation the Source of Competitive Advantage? *Journal of Competitiveness* **9**(3) DOI: 10.7441/joc.2017.03.04
[33] Dentoni D, Bitzer V and Pascucci S 2016 Cross-Sector Partnerships and the Co-creation of Dynamic Capabilities for Stakeholder Orientation *J Bus Ethics* **135** pp 35-53

[34] Nana, R K H M and Beddewela E 2019 The politics of corporate social responsibility in the mining industry in Burkina Faso, *Africa Journal of Management* **5**(4) pp 358-81

[35] Ojasalo J and Ojasalo K 2018 Service Logic Business Model Canvas *Journal of Research in Marketing and Entrepreneurship* **20**(1) pp 70-98

[35] Roaj De Francisco L, Botero L M B and Valencia C F I 2016 Analysis of service management structures in companies from the service sector *AD-MINISTER* **29** pp 121-46

[37] Alexander A, Teller C and Roggeveen A 2016 The boundary spanning of managers within service networks *Journal of Business Research* **69** pp 6031-39

IOP Conf. Series: Materials Science and Engineering **898** (2020) 012033 doi:10.1088/1757-899X/898/1/012033

Analysis and improvement of freight transport using software products

M Inţă[1], M Bădescu[1] and C Purcar[1]

[1] "Lucian Blaga" University of Sibiu, Engineering Faculty, 4, Emil Cioran Street, Sibiu, 550025, Romania

E-mail: marinela.inta@ulbsibiu.ro

Abstract: The purpose of the paper is to provide certain solutions to improve freight operations within a company. Nowadays, information technology occupies an important place and computer programs facilitate work in all areas of activity.
With the help of these tools that record important data related to the company's activity and operations management clearly and concisely, the company's management easily obtains answers to essential questions. After a critical analysis of the main problems faced by the transport company, the four most important problems to be solved are selected. Based on these, the most suitable software is selected for the company after customized consultation for the best solutions of these problems. There are presented the structure of this software and also the advantages brought after its implementation in the company. Among the most important are improve communication within the company, cost reduction through efficient management of truck supplies and rigorous organization of space in trucks. The software can be tested and supplemented at any time with other suitable sections.

1.Introduction

According to Philip Kotler, logistics involves the planning, implementation and control of the physical flow of materials, finished products, information between the point of origin and the point of consumption, in order to meet customer needs and obtain an appropriate profit. In a modern conception, logistics considers not only the flow of goods from the producer to the customer, but also the flow of products from suppliers to producers [1].

Logistics involves, in fact, the management of the entire supply chain, built from the flows that contribute to the formation and addition of value by the participants in this chain. Under these conditions, the logistics activity involves the administration of the entire physical distribution system of the marketing channel, respectively the activity of suppliers, supply agents, market operators, channel members and customers. These activities include forecasting, procurement, production plans, order recording, inventory management, warehousing and transport organization, all of which are in fact components of logistics [2-3].

The efficiency of logistics implies the correlation of the three categories of activities carried out (supply, production support and transport), activities carried out both within the company and at the interface with the sequences downstream and upstream, within the distribution and purchassing channels. The integration of the three categories of logistics activities is possible through information flows over the supply chain.

At present, the importance and role of logistics, considered as a supply chain, is growing, due to at least four considerations [2-3]:

1. First of all, distribution is a key element in serving customers, in meeting their increasingly demanding requirements. Thus, efficient logistics focuses on attracting and retaining customers.
2. Secondly, logistics is a cost element in the activity of any company. Efficient logistics requires improving the efficiency of the distribution system, so as to visibly reduce costs, with positive implications for both the company and customers.
3. Thirdly, developments in the field of information technology have a decisive influence on logistics, helping to improve its processes.
4. Fourthly, the increase of the variety of products, of their assortment range imposed the need to improve the logistics activity.

Some of the main objectives of logistics are: to ensure the planned level of services at the lowest cost, to reduce the time and space distances between the supply and demand of goods and to reduce the costs of storage and distribution [4].

Since the 1960s, a paradigm shift has appeared in managerial thinking, the promoters of which have been the engineers from Toyota plants, respectively from producing as much as possible to produce only as much as needed [4-5]. This new organizational perspective concerns the quality in terms of operational activity and processes.

2.The importance of transport in the SCM

It can be said that transport is the "blood" of an economy through which the "oxygen" it needs to exist is transported. Transport is one of the major components of the logistics system, due to its contribution to the fulfilment of the logistics mission. The main decision-making issues addressed relate to the evaluation and selection of carriers, transport scheduling and route setting.

Thus, it can be said that transport is defined as the physical movement of persons and goods between two points [6]. Logistics services include a range of various activities. Of these, transport has a significant share both in terms of presence and especially in terms of costs. The company's options for transport are numerous, but different in implications: to have their own means of transport, to rent, to use specialized companies, to choose a certain mode of transport. All this must be analyzed in a specific context and no template solutions can be given in advance [7].

Transport can be defined as the activity by which goods are moved over different distances between different points. The main modes of transport are: road, sea, rail, air and pipeline. Of these, road transport is a flexible mode of transport in terms of route and period of operation. The goods can be delivered directly to the customers' premises or to a place designated by them. Its disadvantages of road transport include the fact that restrictions on customs controls (for international transport) can be time consuming. Also, long distances and the need to make water crossings reduce the attractiveness for road transport. In addition, in some parts of the world, especially in underdeveloped countries, road infrastructure is poor.

The limited perspective on space travel will lead to the choice of the carrier company that applies the lowest transport tariff. However, the user is interested in getting the desired services, in addition to the spatial movement at the best price. The user is interested in a transport company that will bring the goods safely and in terms of the contract. Each transport service user will select the service or combination of services that provides the most convenient quality-to-cost ratio. The selection of transport services is based on the cost and performance characteristics of the transport modes.

The most important features considered by specialists are the following [7-8, 11-13]:

- *Costs* - depending on the mode of transport there are differences in costs. According to studies conducted on the US market, road transport is on average 7 times more expensive than rail, and rail is about 4 times more expensive than water or pipeline, while air transport costs twice as much as the road.

- *Transit time / Speed* - is one of the most important features for users of transport services. Transit time is the average time required to reach the goods from origin to destination. This aspect is all the more relevant as the movement of goods involves the use of several modes of transport. A long transit time has as a direct effect a high level of costs with basic and safety stocks, to ensure business continuity.
- *Consistency* - refers to the ability of a mode of transport to maintain the duration of transit over time. Transit time variation is a measure of uncertainty about the performance of transport modes. A high degree of consistency means a potentially low variation in transit time.
- *Availability* - this characteristic refers to the ability of the mode of transport to move goods between any pair of points, consisting of a point of origin and a point of destination. Of all the means of transport, the road is characterized by the highest availability. The means of transport can move the goods directly from the source to the destination.
- *Flexibility* - reflects the ability of a mode of transport to meet the special requirements of users of transport services
- *Frequency* - indicates the number of scheduled deliveries in a certain time interval
- *Safety* - is one of the most important features of modes of transport. It refers to the ability to maintain the quality of products during transit and to avoid the loss and deterioration of products. A low degree of safety leads to increased costs. The main categories of costs they generate are the following: cost of lost goods, cost of replacement or repair of damaged products, cost of stopping the production process, due to lack of raw materials, materials, necessary components, cost of sales lost due to unavailability of products, administrative costs regulating the situation between the transport service user and the transport company, the cost of the necessary safety stocks, the cost of insurance.

3.Transport in global and local vision

Global business changes have forced many organizations to use strategic management, and one of the processes is, of course, transportation management. In order to survive in business and do business successfully, an organization must find a way to provide value-added products or services that differentiate it from the competition [6, 11].

One of the solutions that can meet the transport companies is the rapid development of information technology that allows the realization of the business strategy by improving the processes within it. Transport problems are part of every economy, especially if we followed the importance of the market in the modern economy [12]. Currently, there are numerous software packages that solve minor theoretical transport problems and highlight the advantages of computer-processed data in transport management.

Because transport connects production and consumption, transport taxes are a significant part of the price. The transport load is a special case of general load related to linear programming. Currently, this IT field belongs to operational research and has developed rapidly recently.

4.Analysis of freight transport within the company Transauto S.R.L.

4.1. Company presentation

Transauto SRL is a medium-sized company, with 72 employees, of which 26 drivers, 10 TESA employees, 1 manager, 1 unskilled worker and owns 30 cars for transporting goods and people. The company has the main activity of freight transport but also provides passenger transport services.

Within the company, the freight forwarding activity is performed according to the following flow chart (figure 1):

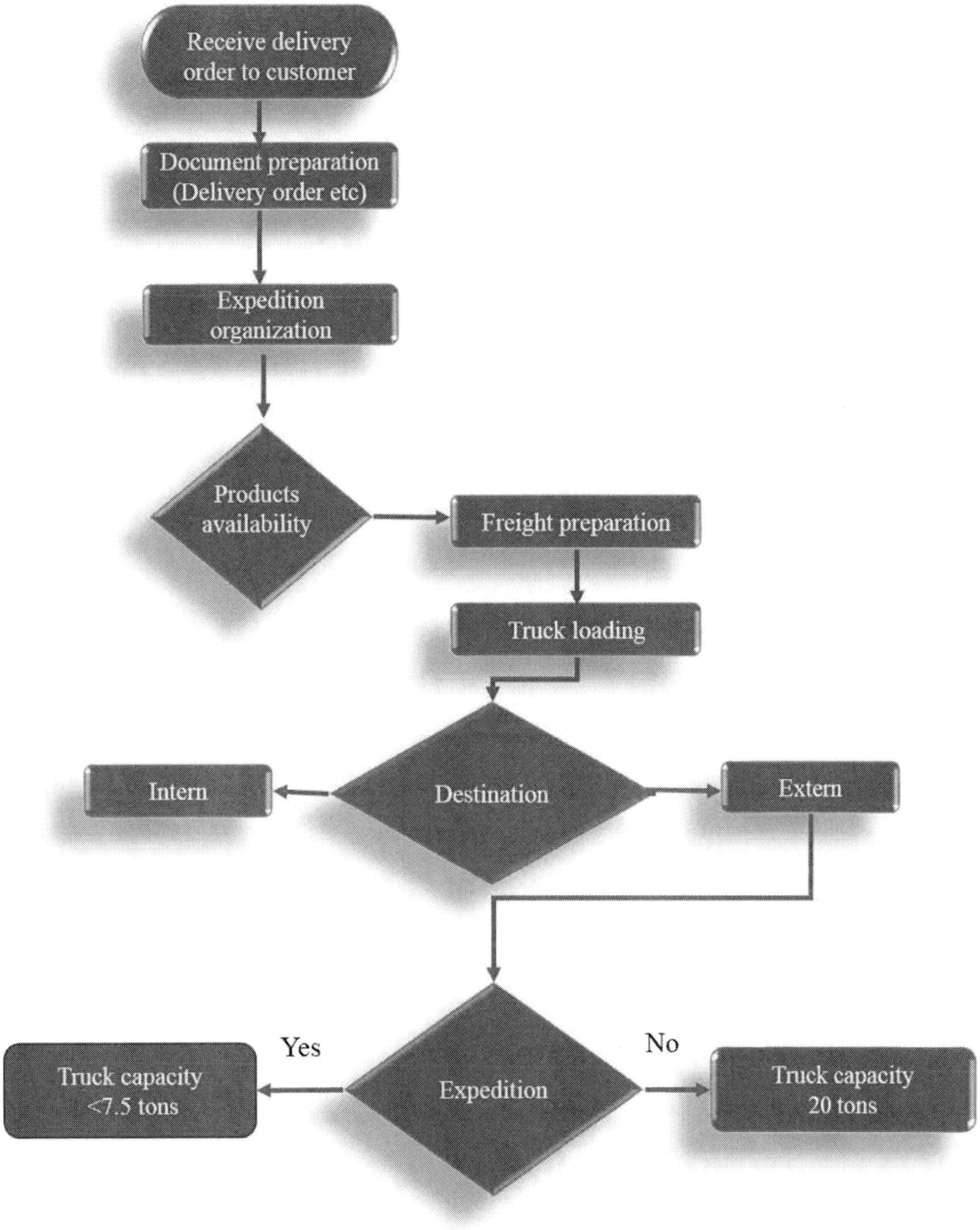

Figure 1. Flow chart of Transauto shipping business

Thus, the delivery order is received by Transauto staff 2-3 days before the shipment of the goods, either by mail or fax and includes information on the type of goods, length, width, weight and height of the goods and the place, the date and time from which the goods must be picked up. An important aspect is the exact date of delivery of the products to the final beneficiary, the delays meaning penalties for

Transauto in terms of existing contracts. After receiving the order, the optimal route in terms of costs and distance is established so that the goods reach the consignee in a timely manner. The main countries of destination of the products are: Czech Republic, Germany, Belgium and France; Transauto sends about 5 cars a week to these destinations (figure 2).

Figure 2. Map of the main destinations Transauto

4.2. Problems in freight transport

Following the analysis, several problems were identified in the transport of goods. A problem identified in the transport of goods is the delay in the delivery of products to the customer.

This problem has its cause in the deficient supply of raw materials but also in various unforeseen events such as: technical problems, calamities, long waiting in customs queues, etc.; thus, the supply is not made on time and the production department cannot finish the products on time. However, over 75% of delivery orders are honored on time.

Another problem identified is related to the impossibility of loading the parts in the truck due to some defective lifting installations.

In order to find a solution for solving the problems, a problem evaluation matrix was created to focus on the most important ones and on solving them.

The following criteria have been identified for the elaboration of the problem evaluation matrix:

- chronicity - the project must correct a problem that occurs frequently, not a recent one;
- duration - projects must have a duration of less than one year;
- urgency - the project is urgent if it addresses issues that make the organization vulnerable in relation to the external environment;
- possible resistance to change - we choose the project that will probably meet the lowest resistance;
- the problem must be measurable - the project does not start if the necessary data is not.

For the evaluation of the problems, scores will be given on a scale from 1 to 5 (table no. 2) where:

- 1 - unimportant problem;
- 5 - very important problem

The selection criteria are weighted according to importance available [14-15].

Table 1. Problem evaluation matrix

Identifyed Problems	Cronicity 30 %	Duration 20 %	Urgency 30 %	Possible resistance to change 10 %	The problem must be measurable 10 %	Total
Failure to meet the deadlines for delivery of products to customers	2	2	5	1	3	13
Transport costs too high	5	4	5	5	5	24
Truck load issues	2	2	3	3	3	13
The supplied products do not						
Lack of communication between departments	3	2	1	1	1	8

Following the completion of the evaluation matrix, it was concluded that the most serious problem is related to transport costs.

4.3. Implementation of Soft Transport software

In order to solve the problems identified after consulting with the company's staff, the aim was to find an optimal solution to improve the current situation. It was concluded that the problems that can be solved can be solved by implementing an innovative software product. After conducting a market study on the software products used by the transport companies, we chose to test the Soft-Transport software, specially designed for freight companies and which includes many elements to help management make the best decisions [16]. The program greatly facilitates the activity of dispatching. This software has been tested for 1 month within the company, observing over time the advantages of use. Soft-transport software (figure 3) includes several sections that allow the development of a database with valuable information for management.

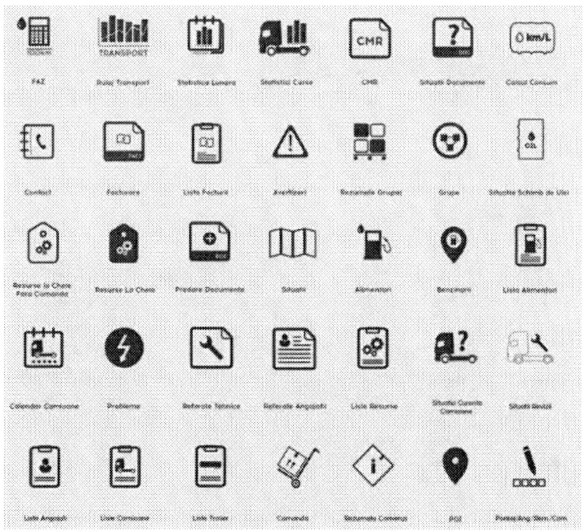

Figure 3. Sections of Soft Transport (Transport roll; Monthly statistics; Race statistic; CMR; Document situation; Consumtion calcul; Contact; Billing; Invoice list; Warn; Group summaries; Groups; Oil change; Key resources without order; Key resources; Handling documents; Situation; Fuelling; Gas stations; Fuelling list; Truck calendar; Problems; Technical references; Referred employees; Resource list; Current track situation; Situation revisions; Employees list; Truck list; Trailer list; Orders; Orders summary; PO; Clocking/ Employees/Commands

The main advantages of testing are:

- The **Current Trucks Situation Section** found the answers to the following questions:
  - What is the number of trucks working for a particular customer?
  - What is the situation of the truck with the number ...?
  - What is the city / country where the goods are to be unloaded?
- The **Unlocked Turnkey Resources** section was able to find out as soon as possible how many trucks are out of order.

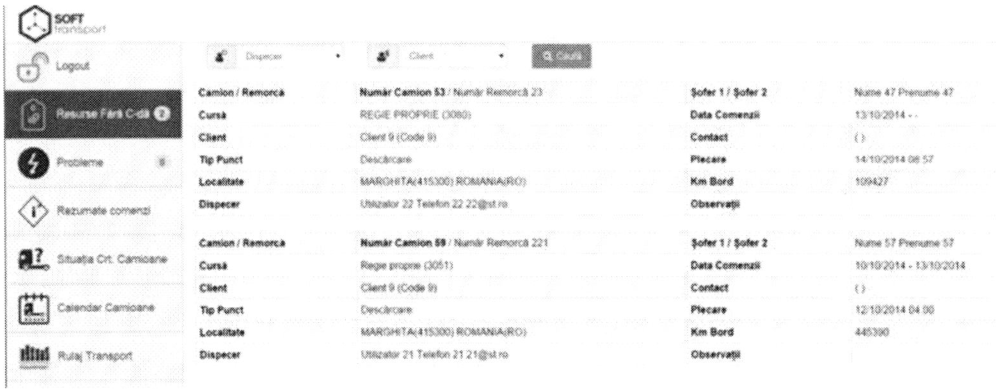

Figure 4. Soft Transport Sequence - Resources without command (Key resources without order; Problems; Order summaries; Current track situation; Truck calendar; Turnover transport)

- The exact record of the available trucks was kept (figure 5) and it was possible to find out in the shortest time if other orders can be honored, besides the already planned ones (figure 6). This tool has played an important role in meeting delivery deadlines to customers.

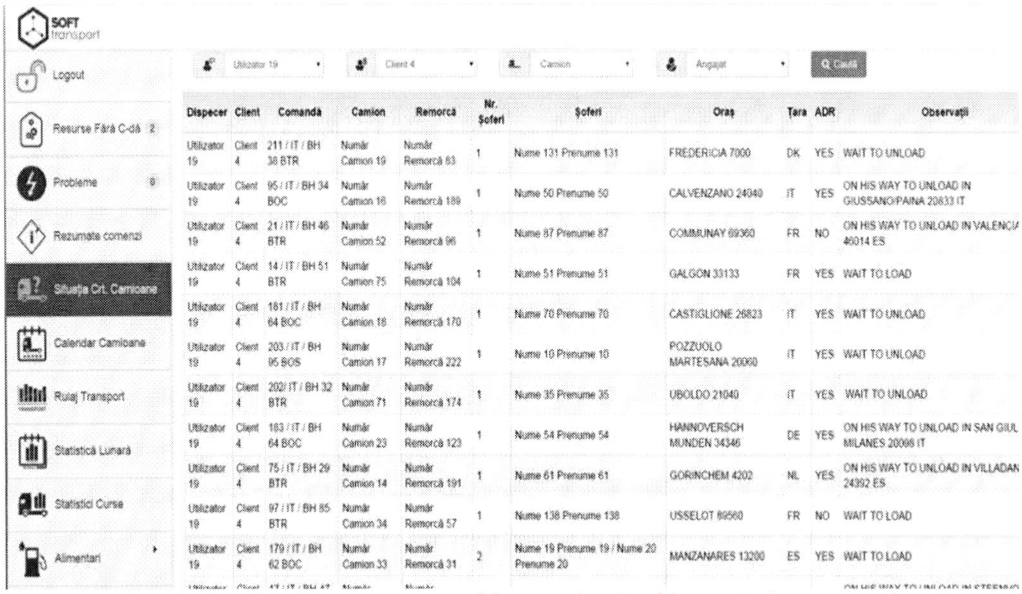

Figure 5. Truck situation (Key resources without order; Problems; Order summaries; Current track situation; Truck calendar; Turnover transport; Monthly statistics; Race statistic; Fuelling)

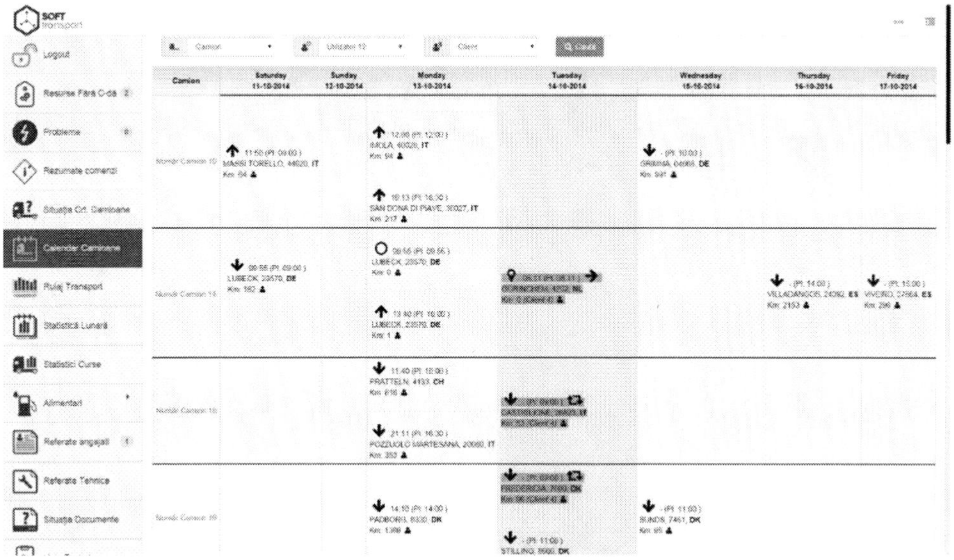

Figure 6. Truck Calendar (Key resources without order; Problems; Referred employees; Technical references; Document situation)

- Problems encountered during transport are also monitored in a special Problems section (figure 7). For example, the nature of the technical problem can be recorded - the tires, kilometers, etc., the observations related to the problem, but also if an order has been placed for a certain truck part.

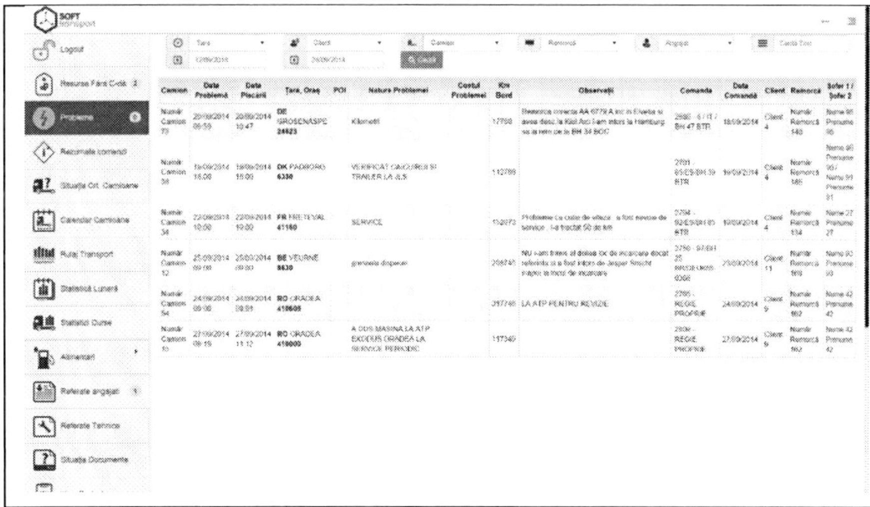

Figure 7. Soft Transport Sequence – Problems (Key resources without order; Problems; Referred employees; Technical references; Document situation)

- Or,the situation of the documents (figure 8) handed over can be followed in the section with the same name. Documents can also be uploaded here. This section also answers the questions: when and to whom were the transport documents handed over to the customer and what non-delivered documents are on board a truck.

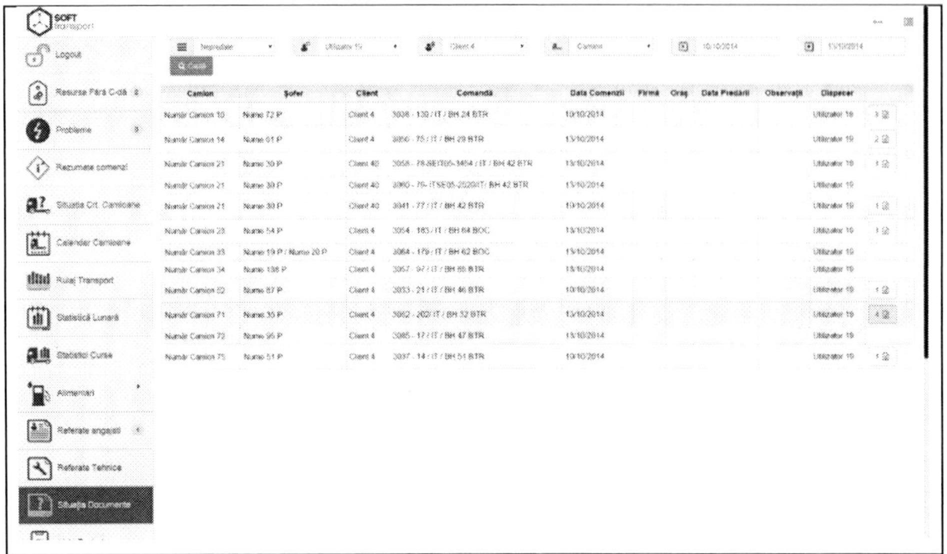

Figure 8. Soft Transport Sequence - Document situation (Key resources without order; Problems; Referred employees; Technical references; Document situation)

So, testing the software application brought many advantages to Transauto, namely (table no. 3:

- Strengthened control within the company;
- The problems in carrying out the transport activity were reduced by 20%;
- An 8% reduction in operating costs was observed;
- Documents and human resources have been managed more efficiently;
- Rigorous advance planning is also one of the control elements that helps to meet delivery deadlines to the customer.

After the acquisition and implementation of the software product within the Transauto company, an improvement of the entire transport activity is observed. (figure 9).

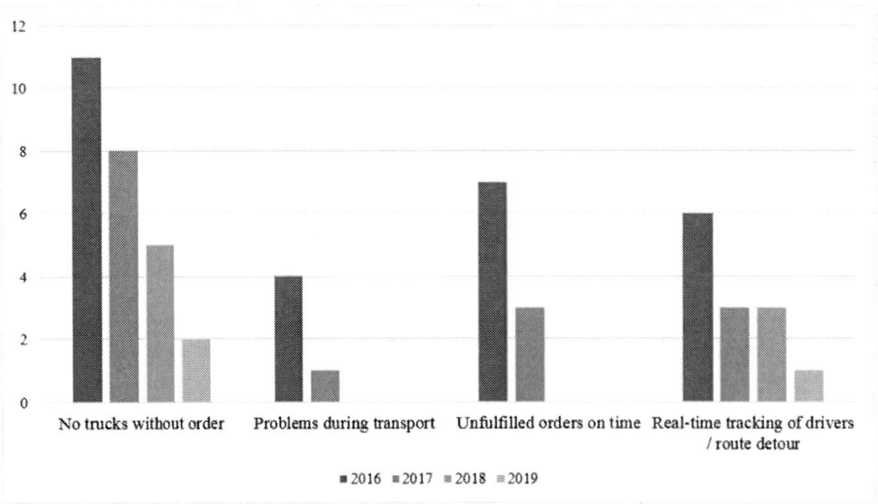

Figure 9. Comparative analysis in testing and after implementation

Even if testing Soft-Transport software has proven to be very useful, it could be improved or supplemented with other sections that bring even better control within the company. The directions that management should focus on after purchasing the Soft-Transport package are the following:

- Defining optimal routes for racing;
- Applications to help optimally arrange products in the truck.

5. Conclusions

In conclusion, logistics services comprise a range of diverse activities. Of these, transport has a significant share both in terms of presence and especially in terms of costs. Global business changes have forced many organizations to use strategic management, and one of the processes is, of course, transportation management. In order to survive in business and do business successfully, an organization must find a way to provide value-added products or services that differentiate it from the competition.

One of the solutions that can meet the transport companies is the rapid development of information technology that allows the realization of the business strategy by improving the processes within it.

Within Transauto SRL, the implementation of IT applications, namely Soft Transport, solved the problems encountered: non-compliance with product delivery deadlines, problems related to truck loads, poor communication within the company and cost reduction through efficient management of truck supplies and rigorous organization of space in trucks.

References

[1] Kotler P 2000 *Marketing Management* (Bucuresti Ed. Teora)
[2] Whiteing T 2001 *Manufacturing Logistics* (Pergamon: Brewer AM Button KJ Hensher DA in the book Handbook of logistics and supply-chain management)
[3] Christopher M 1998 *Logistics and Supply Chain Management – Strategies for reducing cost and improving service* (London)
[4] Meyr H 2004 *Supply Chain Planning in the German automotive industry OR Spectrum* Vol 26 pp 447-470
[5] https://www.investopedia.com/terms/s/scm.asp
[6] Wood D Johson J 2000 *Contemporary Transportation* (Prentice Hall, Upper Saddle River)
[7] Ballou R 2004 *Business Logistics Management* (Pearson Prentice Hall)
[8] D'este G 2001 *Freight and Logistics Modeling*, in the book Handbook of logistics and supplychain management, edited by Brewer, A.M., Button, K.J. & Hensher, D.A., Pergamon, Amsterdam. pp. 521-534.
[9] Davis M 2003 *Operational benefits of the long creek-weight-in-motion system* (Departament of civil Enginering, University of New Brunswick)
[10] Saifizul AA Karim MR Yamanaka H 2010 *Prospect of using weight in motion based system for enhancing vehicle weight enforcement- a case study of Malaysian roads*. (Busan, Korea: Word Congress, Intelligent Transport System)
[11] Beaumelle B Jacob VF 2010 Improving truck safety: potential of weight technology. *IATSSS Research* pp 9-15
[12] *The competitive route to sustainability and safety* http://ec.europa.eu/research/transport/road/index_en.htm..
[13] Shapou Y Shaohua L, Yongjie L 2010 Investigation on dynamical intersection between a heavy vehicle and road pavement. *Vehicle System Dynamic* Vol 48 pp 923-944
[14] Kamran M Sajid A 2010 *Critical analysis of Six Sigma implementation. Total Quality Management*
[15] ISO 13053-1:2011 *Quantitative methods in process improvement Şix Sigma - Part 1: DMAIC methodology* https://www.iso.org/standard/52901.html
[16] https://www.soft-transport.com/about.php

IOP Conf. Series: Materials Science and Engineering **898** (2020) 012034 doi:10.1088/1757-899X/898/1/012034

Material handling based on object recognition with the help of FANUC delta-parallel industrial robot

Gy Korsoveczki[1], G Husi[1]

[1]Department of Mechatronics Engineering, University of Debrecen, Faculty of Engineering, Debrecen, Hungary

korsoveczki.gyula@gmail.com

Abstract. The task was made that recognizes objects with the help of delta-parallel constructed pick & place FANUC robot, that can be found at the Mechatronics Department, Engineering Faculty, University of Debrecen. The detection of the different objects was realised by a SONY XC-56 typed industrial monochromatic camera and using the own image processing software component of the robot. To accomplish the material handling a linear drive, an electropneumatic gripper and the digital output system of the robot were used. During the project, the implementation of a material handling operation was the aim that is based on the machine vision of the robot. As a result, the robot used its vision and object recognizing ability based on iRVision environment, completed object recognition and material handling procedure with the help of a linear drive controlled by robot program code.

1. Introduction

At the Cyber-Physical and Intelligent Robot Systems Laboratory is a special laboratory was created, where new industrial solutions are showcased. The laboratory can be found at University of Debrecen's building mechatronics research center [1]. Nowadays we are at the beginning of the new industrial revolution, which is the Fourth Industrial Revolution, or briefly Industry 4.0. One of the results of the new technologies is that they increase the effectiveness of the industrial production processes. The Machine Vision is very important due to the automatization processes. In the industry three essential adaptability goals have been determined, such as object recognition, determination of the size and the position [2], [3]. Related to the importance of the technology and the fact that at the Mechatronics Department it was possible to deal with a robot that is able to use machine vision I studied the material handling based on object recognition with the help of the mentioned industrial robot.

2. Technical overview of the robot

2.1. FANUC M-1iA 0.5A robot and its appliances

The robot that was used, a M-1iA serial 0.5A typed delta-parallel structured robot produced by FANUC Corporation. It is ideal for precise assembly procedures. The number of moveable axes is 6, its maximal reach is 280 mm, the maximal speed is 4000 mm/s, maximal repeatability is +/- 0,02 mm, mechanical weight is 23 kg, and the load capacity of the 6. axis is 0,5 kg [4]. The robot has a R-30iB Mate typed Open Air control unit that can be found on the base construction of the robot. The operating system that the robot uses is the FANUC's own developed operating system, without frame system. It has IP67

protection against the damage caused by the industrial environment [4]. The primary control possibility of the robot is the Teach Pendant that name is iPendant Touch that provides command structured online-indirect programming [4].
The machine vision of the robot the XC-56 type monochromatic camera produced by SONY is responsible for. The resolution is VGA, the maximum scanning speed is 120 pixel/sec, the imaging mode is CCD, maximum illumination is 0,5 lx, and it has IP66 protection [5]. The camera has also a 12VC1040ASIR objective produced by Tamron, that can produce a suitable image quality in the field that is near the infrared (under 850nm). The imaging size is 1,2", it has variable lens type, manual iris type, the focal length range is 10-40 mm and the operating temperature is (-20) – (+60) °C [6]. The machine vision of the robot is existed by the iRVision image analyser developing environment. It has a direct connection with the imaging device, that is the SONY XC-56 camera. The image analyser system and the developing environment can be reached from the Teach Pendant of the robot, or from the Microsoft Internet Explorer browser, the 192.168.0.1. address.

Figure 1. The robot, the control unit, the Teach Pendant, the camera and the objective [4], [5], [6]

3. Optimization of the workspace

3.1. The digital output system of the robot

As a result of my earlier project, the digital output system of the robot was developed. Due to it, controlling different devices by digital signal through the output system of the robot became possible. A switching circuit was made to the robot and its task is to receive the 24 V reference signal voltage that comes from the output ports of the robot. It contains photo-Darlingtons to realise the galvanic isolation and to activate the associated Finder 55.34.9.024.0040 typed miniature industrial relays. For this reason, with the help of the relay connection, it is possible to operate devices from program code. It contains an electropneumatic vacuum gripper that are controlled by the RO1 relay (Robotic Output 1) [7].

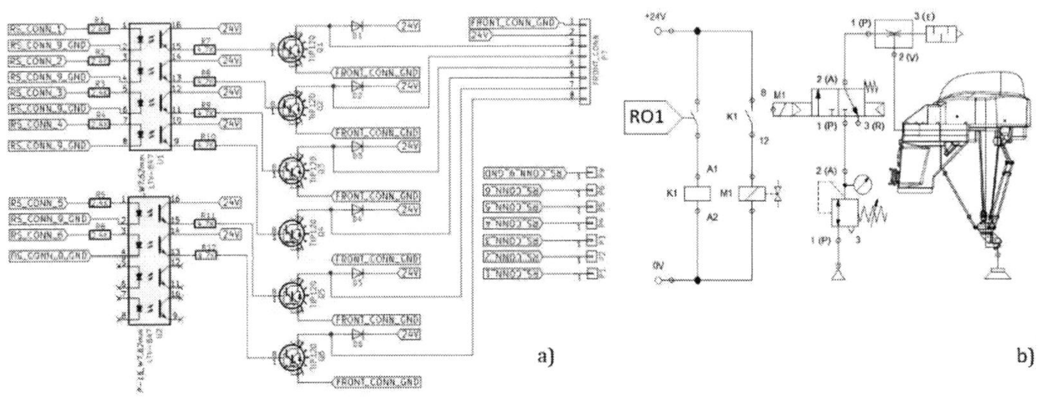

Figure 2. The schematic of the switching circuit (a) and the electropneumatic gripper (b) [7]

3.2. Providing of the optimal lightning

In the interest of the lightning, EGLO typed lamps was equipped on the frame construction of the robot to give enough and well-positioned light against the light pollution. The lamps operate RGB LEDS. They have GU10 typed connection, their performance is 4W one by one. Their minimum lightning intensity is 220 lm, they can be operated via 220-240V AC supply voltage and their current use is 40 mA. Their operation is possible with the help of 6. relay via program code. The lamps were equipped with 2 beholders that were designed printed in 3D.

3.3. Connecting of the linear drive to the robot

During the project, an ISEL LEZ1 type linear belt drive was connected to the robot in its workspace. Its task was the carrying the objects to help the material handling process. The belt spacing is 3 mm, the belt width is 9 mm, the repeatability is +/- 0,2 mm, in case of 1:1 transfer the belt feed is 60 mm/revolute, maximum speed is 1,5 m/sec and the step angle if the stepper motor is 1,8° [8]. To control of the belt drive moving stepper motor a Pololu A4988 type motor controller device was chosen, besides Arduino Nano is the applied microcontroller. The operation of the linear drive is in connection with the digital output system of the robot. Due to it, 3 fixed positions were defined for the stepper motor. These 3 positions connect of the 3 output ports of the robot (RO2, RO3, RO4). An RO5 output relay and a potentiometer were defined as well to position the linear drive proportional according to the potentiometer.

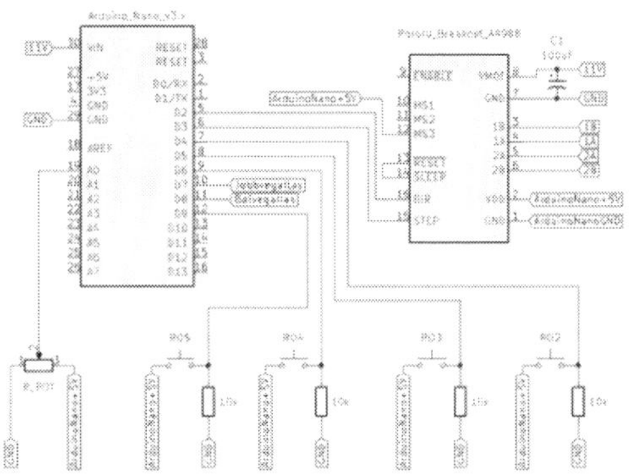

Figure 3. The schematic circuit for controlling the linear drive

Figure 4. The modified workspace

3.4. Creating new worktable for the object recognition

The stability of the worktable is a basic demand as the procedure works with Z offset = 0 value. In this case the X, Y values and the rotation value to the reference position are detected. Otherwise the displacement along the z line are not detected, because the process claims flat. That is why the vertical displacement of the work table can result the failure of the procedure. For the same reason it is necessary to provide the parallel position with the x and y lines of the worktable. One more important thing that the camera view needs to be perpendicular to the flat of the worktable. In the lack of it, the robot program that recognizes the objects can be inaccurate. It is appropriate, if the background is different from the objects that are recognized. As the mentioned objects are white, the surface of the worktable has been covered with black pasteboard that provides the contrast.

4. The performance of the necessary calibrations

4.1. The tool frame calibration

Determining of the Tool Center Point (TCP) is important when the robot calculates its motion between 2 points itself. In this case the unit controller has to take into consideration the geometrical parameters of the tool avoiding the collision. To calibrate the tool, the Three Point Method was chosen. The calibrated tool was an electropneumatic operated vacuum gripper. However, due to the geometry of the gripper it was not possible to rotate it accurately around the chosen reference point. To solve this problem a calibration tool was designed in form of 3D model and 3D printing. The geometrical parameters are equals with the vacuum gripper, but it has spike forming which is suitable for the calibration and the calibrated parameters are accurate in case of the vacuum gripper.

4.2. The calibration of the user frame and the camera

During the calibration of the user frame, that user frame is taught for the robot where the target operation will be done. To accomplish that the more accurate Four Point method was used. In the course of this method, four points was thought for the robot. These points are a point along the X line, a point along the Y line, the Orient Origin point, and the System Origin point. As the tool frame calibration was associated with object recognition procedure as well, so it was necessary to accomplish it together with the camera calibration. During the camera calibration, the real physical coordinates and the direction of the user frame is associated with the pixels that represent the image of the camera. Implementation of it happens with calibration grid plate. The calibration grid plate is the multitude of the points that are situated in regular distances. These points are reference and orientation points. The orientation points sign the certain directions. These directions have to be the same with directions and points thought during the calibration of the user frame. The software part of the camera calibration was accomplished in the iRVision environment.

5. The object recognizing procedure

During the material handling the task of the robot is to detect the taught objects that can be found in is workspace using object recognition and then to detect the shaped holes of the object panel that belong the objects. The robot has to determine offset values during the detection, which contain the X and Y coordinates of the found objects and the R rotation around the Z line referred to the reference position. According to the determined offset values, the robot has to fit the objects into the suitable hole that was determined in the robot program. As there are 3 types of objects and 3 types of shaped hole belongs to them, it was necessary to implement 6 different object recognizing procedure that detects the suitable object and shaped hole in the interest of success. The process provides 2-dimensional vision with 1 camera.

Figure 5. The recognition of the pentagon object

6. The material handling process

It became possible to create the robot programs after the implementation and testing the object recognizing procedures. These programs carry out the desired process of the material handling based on object recognizing. With the help of the subprograms, detection of the objects and fitting them into the right hole happens. The task of the main program is to control the lightning, the linear drive and the subprograms.

IOP Conf. Series: Materials Science and Engineering **898** (2020) 012034 doi:10.1088/1757-899X/898/1/012034

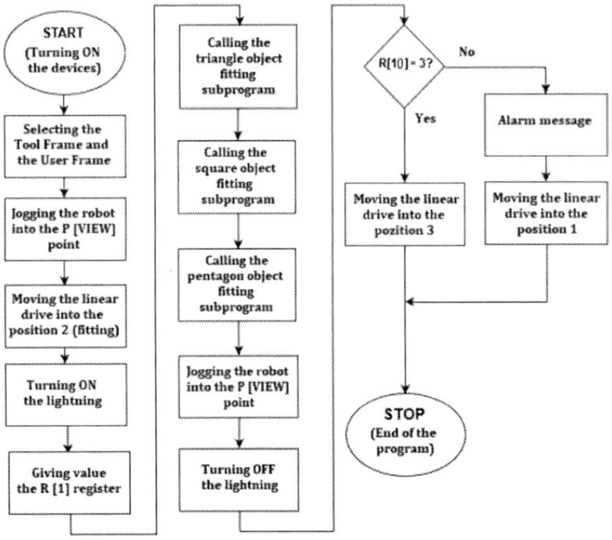

Figure 6. The flowchart of the main program

Conclusion

During the project, the implementation of a material handling operation was the aim that is based on the machine vision of the robot. As a result, the robot used its vision and object recognizing ability based on iRVision environment, completed object recognition and material handling procedure with the help of a linear drive controlled by robot program code. Other developing possibilities are enlarging the robot with other equipment, realising image recognition procedures form more camera standing and developing the digital input and analogue output system of the robot.

References

[1] G Husi, Zs Molnár, N C Obinna, and T I Erdei 2017 A novel Design an Augmented Reality Based Navigation System & it's Industrial Applications 15th IMEKO TC10 – Technical Diagnostics in Cyber-Physical Era Budapest, Hungary

[2] "Robot controlling method for tracking moving object using a visual sensor", T. Arimatsu; T. Jyumonji; K. Otsuka; H. Kubota, 2018.

[3] "Kinetic sensor implementation in Fanuc robot manipulation", M. Pajor; K. Miadlicki; M. Saków, 2018.

[4] FANUC Hungary Ltd., „M-1iA/0.5A," [Online]. Available: http://www.fanuc.eu/hu/hu/robotok/robotsz%c5%b1r%c5%91-lap/m1-sorozat/m-1ia-05a. [Accessed: 29.11.2019.].

[5] SONY, „SONY XC-56 camera," [Online]. Available: https://pro.sony.com/bbsc/ssr/product-XC56/. [Accessed: 29.11.2019.].

[6] "Tamron., „Tamron 12VC1040ASIR," [Online]. Available: http://www.hoodi.net.tw/tamron/down/TAMRON_ALL.pdf. [Accessed: 29.11.2019.].

[7] Gy Korsoveczki, G Husi, T I Erdei, „Development and testing of the pneumatic and output system of the FANUC Spider robot", Papers on Technical Science vol. 11. (2019) 121-124. 2019.

[8] "ISEL Hungary Ltd., „LEZ1," [Online]. Available: http://www.isel.hu/index.php/klub-7/piac/1886/mechanic/belt-linear-unit. [Accessed: 29.11.2019.].

Acknowledgement

The publication was supported by the project EFOP-3.6.1-16-2016-00022. This project was co-funded by the European Union, from the European Social Fund.

IOP Conf. Series: Materials Science and Engineering **898** (2020) 012035 doi:10.1088/1757-899X/898/1/012035

Artificial Intelligence Applications in Autonomous Vehicles: Training Algorithm for Traffic Signs Recognition

C Dursun[1], T I Erdei[1], G Husi[1]

[1] Mechatronics Department, University of Debrecen, Faculty of Engineering Debrecen, Hungary

candrsun@gmail.com

Abstract. As the last point of autonomous vehicle development, these vehicles are being evolved by most of the car manufacturing companies, which increases the need for novel image recognition and automation solutions. This paper presents autonomous driving for vehicles at the stage of level 2 (partial Automation; the vehicle can perform steering and acceleration, though the human driver still monitors all tasks and can take control at any time). YOLO (You Only Look Once), is a Python-based image processing algorithm that was used to achieve the goals. This algorithm is extremely useful due to its real-time capabilities. A new data set was gathered using miniature signs, and a Python script was developed to label these newly trained objects. The algorithm draws bounding boxes around the object with object name and the rates of accuracy. The created system performs well, with further plans to improve it, and apply it in a simulated/modelled environment.

1. Introduction

The world automotive engineers evaluate autonomous driving at six levels, "0" is completely under the control of driver without any assist, "1" the vehicle features a single automated system, for instance, controlling the speed by cruise control, "2", the vehicle can control the steering wheel and the acceleration, but the human can monitor all the tasks and takes the control if needed. Levels "3-6" are the automated systems which monitor the driving environment [1].

The algorithm You Only Look Once (YOLO) is an image processing algorithm which is very popular lately. Although other algorithms are very accurate, none of them is as fast and precise as YOLOv3 as it can process images in real-time [2]. In this article, I am going to present how to collect custom data to create a dataset and implement it for YOLOv3 algorithm.

2. Background

2.1. Understanding the YOLO algorithm

The COCO (Common Objects in Context) is large-scale object detection, segmentation, and captioning dataset [3]. YOLO is much faster compared to other algorithms [4]. We can make changes in accuracy or speed, but if we increase the speed-accuracy will decrease, and vice-versa. R-CNN (Convolutional Neural Network) based object detection algorithm primarily determines the areas where objects are likely to be found then in these areas, processed for classification [5]. Although this method gives good results, one image is processed under two different operations which increases the count of the process and as a result gives a low FPS (Frames per second).

IOP Conf. Series: Materials Science and Engineering **898** (2020) 012035 doi:10.1088/1757-899X/898/1/012035

On figure 1, the structure of R-CNN can be seen. R-CNN and Faster R-CNN algorithms can process the image in one operation, but when it comes to real-time image processing, the algorithm cannot perform more than 7 FPS. YOLO can estimate the classifications and the coordinates by crossing images to deep convolution neural network to detect the object in real-time which can reach more than 40 FPS with an Nvidia CUDA supported GPU (Graphical Processing Unit) [7].

Namely principal of the estimate is recognition an image in a single process. To do this process firstly, it separates the image into an SxS (S number by S number) grid. But this also can be 3x3, 5x5, 19x19 etc. [8]. According to this process output of the image can be seen in the following image.

Each grid inspects if there is an object inside them, if yes then check if the object midpoint of the grid, if yes again then check its length, height and find which class it belongs to.
To be more specific, for example, in the presented image, 7th grid is responsible for recognition because the midpoint of the car is on the 7th grid.

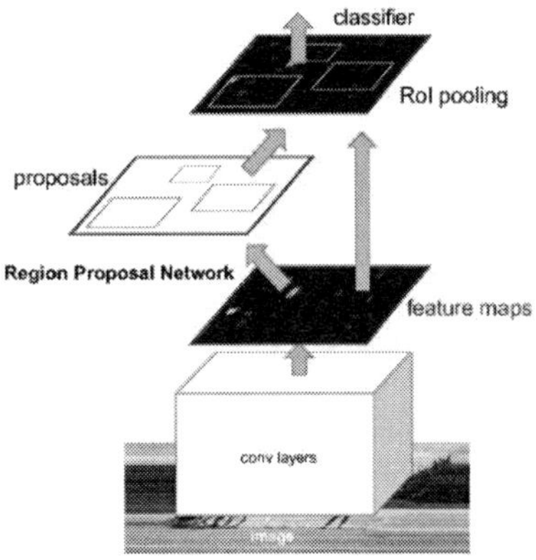

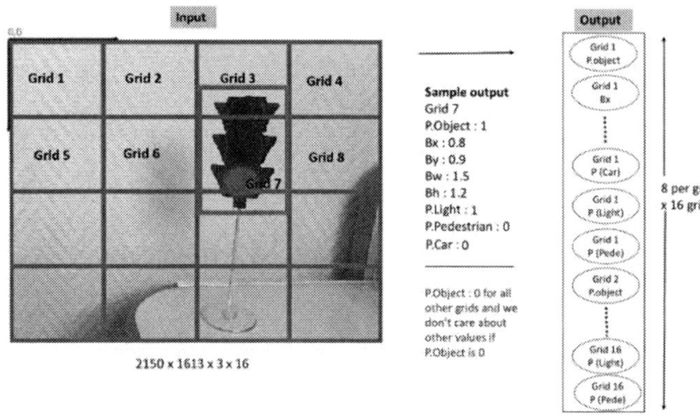

Figure 1. R-CNN Architecture [6] **Figure 2.** YOLO algorithm input-output

According to this YOLO creates prediction vector for each grid and inside of them: Confidence score: This score presents when the object inside the grid and its accuracy, at "0" it's absolutely outside of the grid, at "1", absolutely inside. The operating parameters are: Bx: X coordinate of the object's midpoint; By: X coordinate of the object's midpoint; Bw: Width of the object; Bh: Height of the object.

Probability of classes: Algorithm estimate of all classes. For instance, in the presented image when we look to grid 7, if there is a car, car would be "1", light would be "0" and pedestrian also "0". Calculation for confidence score (IOU = intersection over union, the value that is estimated between boxes): confidence score = P(Object) x IOU. P(Object). This tells us if "P" is within the grid. [9].

According to output vector each grid can recognize only one object. For instance, if we use 3x3 grid YOLO could recognize 9 objects. Well, if there are more than one object in a grid what would happen? Or even midpoint of two object is inside of one grid. This was the one of the problems that could not be answered with the first version of YOLO. With the second version of YOLO algorithm was extended with Anchor Boxes which solves this issue [10]. The Anchor Boxes method was firstly used on R-CNN algorithm and it recognizes the boxes around the object. With two anchor boxes, each object in the training image is assigned to a grid cell that contains object's midpoint and anchor box for the grid cell with highest IOU [11].

After adding the anchor boxes, the output vectors can be mathematically expressed. In figure 3, two anchor boxes can be seen. In this image midpoint of the Traffic light and the firefighting truck are exactly inside the same grid. Our goal is to draw a bounding box around the object and at the same time know the height and the width of the boxes.

With the second version of the YOLO we have the anchor boxes we can directly find the anchor box that is most similar to the object box and we can estimate the width and height difference of that box relative to the anchor box, so we can handle the problem that we have faced with the first version of

YOLO. Now we can predict Tx, Ty, Tw, and Th (where Tx, Ty, Tw, and Th are the distance from the given grid border on each axes) for every box by neural network. Thus, we know which grid cell we are working on which is expressed by Cx and Cy (where Cx and Cy are the distance from the top left point, expressed in X, Y coordinates.) Finally, the last items are Pw and Ph which represent the height and width of anchor boxes. On figure 4, we can observe as the dotted rectangle represents the anchor while, and the blue box the predicted boundary box.

Figure 3. Anchor boxes

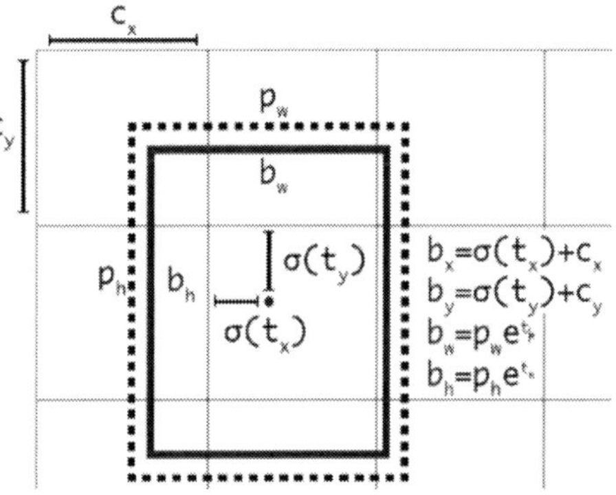

Figure 4. Anchor boxes prediction [12]

In this presented method we are normalizing the parameters between "0" and "1" as a result our neural network is being more stable which makes the training process easier and precise. In the following section data collection for YOLO v3 will be introduced.

3. Dataset training and testing

3.1. Collecting data for training

The purpose of the project is training the algorithm for autonomous driving therefore we are going to collect traffic sign (speed limit and stop sign) and traffic light images for identification. One of the feature plans of project is implementing YOLO v3 real-time image processing algorithm into a robot for a real-world demonstration. A track line will be drawn on the floor, then the miniature traffic signs, and traffic lights will be placed around the track. The robot will follow the line, and YOLO will check the signs and make decisions (for example, control speed based on speed limit signs). The traffic signs must be captured for a custom dataset as the miniatures slightly differ from full-sized ones.

Traffic sign figures were found online and printed out. The traffic sign and traffic light figures are presented on the following image. We need to provide hundreds of images that will be used to train new detection classifier, YOLO v3. Many pictures were taken in different conditions, to teach the algorithm different disturbances it may encounter on an image. These are as follows:

- All photos must be in 1:1 scale
- Different angles (changing the position of the figure after each capture)
- Random backgrounds (putting the figure to different background after each capture.)
- Different light sources (capturing photos under sunlight or in a low light source.)
- Different visibility (for example half of the figure is visible while half of it not
- Capture the signs together in one photo and separately

As seen on figure 6, one sign was captured front of completely different backgrounds. I took pictures of them in one area then moved them to a different area and repeated the process in harder and harder visual environments with more random objects and busier backgrounds. I tried to include similar objects in some pictures to help the detector avoid falsely identifying things that look like traffic signs. In total, 700 images were taken (100 per sign).

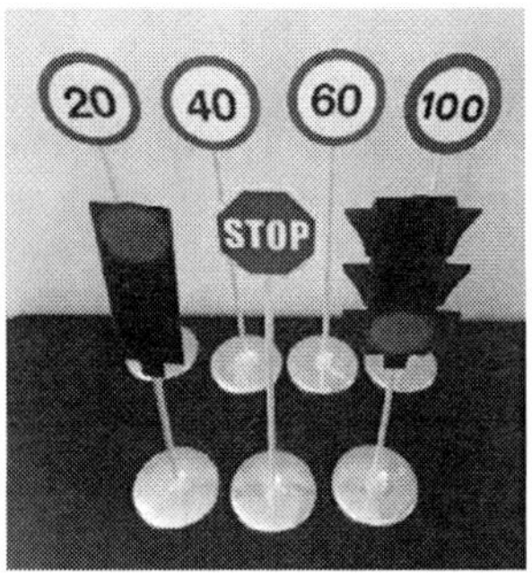

Figure 5. Traffic signs and traffic lights for training

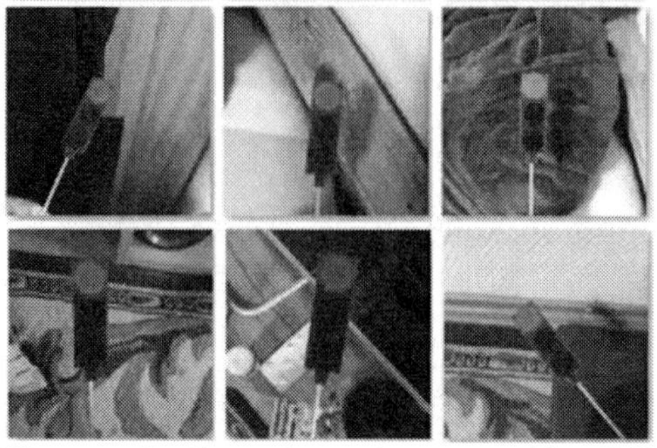

Figure 6. Capturing pictures with different backgrounds

3.2. Image labelling (Annotation), model training and testing

After collecting the images next step is labelling every single image one by one to train, what the objects are and where they are located. The "Labelimg" free labelling tool was used, which is a graphical image annotation tool. With the tool all that was needed is opening the training directory click the "create Rectbox" button and draw a rectangle around the objects in the image, then type in a label for each. The saved XML file contains the size of the image and where the labelled object is using an X-Y coordinate system.

Moreover, we need to create CLASSES.txt and CLASSES_annotation.txt files before starting to train the model. Classes exact meaning is separating object from each other for instance, "traffic light green" or "speed limit 100" As CLASSES.txt presented on the following image, it contains name of the classes while CLASSES_annotation.txt contains location of the image and coordinates of the bounding boxes for every single pictures that we collected. As for content of CLASES_annotation.txt, we are inserting each XML files content into a single TXT file. This concludes the model training.

The YOLO v3 algorithm is a one-shot learner which means that to predict the object, it sends each image to the network only once. To train the model we will use Nvidia GTX 1070 GPU with 8gb of ram, because YOLO can use the full potential of CUDA cores. An important point before start training is to be sure there is no space between any folder name for instance and of there are, they must be substituted with underscores. It took 10 hours to complete the training process. At this time final trained weight is also created, which we are going to use to test the model in the next step.

To test the model, we used a webcam which is connected to a PC. As a result, model is working successfully, and can predict the traffic signs and lights in real time, performing at about 13-14 FPS, which is not too fast, but satisfactory for our model. Detecting object at a long distance is at times unreliable, but the system works well otherwise, even in low-light conditions as seen on figures 7-8.

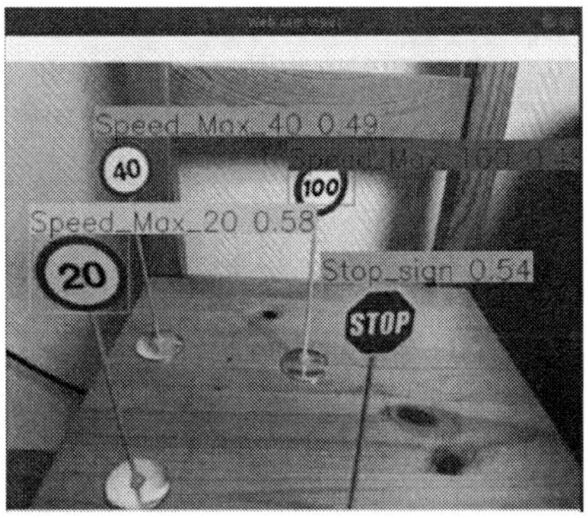

Figure 7. Testing the model under well-lit conditions

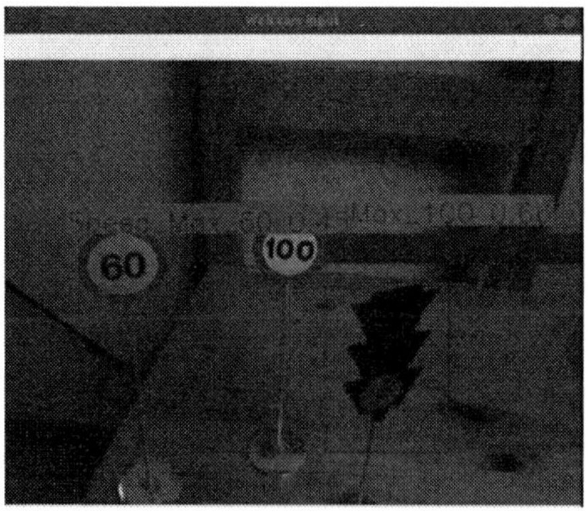

Figure 8. Testing the model in low-light conditions

4. Conclusion

YOLO v3 has demonstrated remarkable performance gains while running at real-time detection. Results can be used many applications if it can be trained with a proper dataset. I classified 7 classes for an existing image recognition algorithm called YOLO to use it on my future project which will be detailed on the future plain section. I faced many problems during the implementation. For example, all the modules had to be compatible with each other and its very time consuming to find the current versions.

5. Future plans

The focus of further developments would include more traffic signs, increasing the efficiency of the model using a loss function, and implementing the model onto a line following robot for a real-world demonstration. There would be miniature traffic signs and traffic lights around the track. While the robot follows the line, model will check the traffic signs then makes decisions. For instance, if algorithm recognizes “max speed 50” sign then robot is going to limit the speed to “50” percent, if recognizes “max speed 30” sign then robot’s speed will be “30”. When model recognizes stop sign or red traffic light will stop immediately until traffic light turns to green.

References

[1] "iotforall.com," Autonomous driving levels, [Online]. Available: https://www.iotforall.com/5-autonomous-driving-levels-explained/.
[2] "YOLOv3 Official Website," [Online]. Available: https://pjreddie.com/darknet/yolo/.
[3] "COCO (Common Object in Context," [Online]. Available: http://cocodataset.org/. [Accessed 03 03 2020].
[4] "Yolo vs others," [Online]. Available: https://medium.com/analytics-vidhya/everything-you-need-to-know-to-train-your-custom-object-detector-model-using-yolov3-1bf0640b0905.
[5] "CNN Convolutional Neural Network," [Online]. Available: http://cs231n.github.io/convolutional-networks/.
[6] G. S. Tran, "ResearchGate CNN Architecture," [Online]. Available: https://www.researchgate.net/publication/324549019_Towards_Real-Time_Smile_Detection_Based_on_Faster_Region_Convolutional_Neural_Network.
[7] S. Djahel, "Researchgate," [Online]. Available: https://www.researchgate.net/publication/320592354_A_Novel_YOLO-based_Real-time_People_Counting_Approach#pf1
[8] "Evolution of Object Detection and Localization Algorithms," [Online]. Available: https://mc.ai/evolution-of-object-detection-and-localization-algorithms/.
[9] J. Du, "ResearchGate," [Online]. Available: https://www.researchgate.net/publication/324754316_Understanding_of_Object_Detection_Based_on_CNN_Family_and_YOLO.
[10] A. Ng, "Anchor boxes," [Online]. Available: https://www.coursera.org/lecture/convolutional-neural-networks/anchor-boxes-yNwO0.
[11] Mathworks.com, "Anchor Boxes explained," [Online]. Available: https://www.mathworks.com/help/vision/ug/anchor-boxes-for-object-detection.html.
[12] J. Hui, "medium.com," [Online]. Available: https://medium.com/@jonathan_hui/real-time-object-detection-with-yolo-yolov2-28b1b93e2088.

Acknowledgements

I cannot express enough thanks to my supervisor Mr. Timotei István Erdei for his continued support and encouragement. I am also very grateful to all my teachers who helped me in this project. The research was supported by the University of Debrecen, Doctoral School of Informatics.

IOP Conf. Series: Materials Science and Engineering **898** (2020) 012036 doi:10.1088/1757-899X/898/1/012036

Researches regarding the inner turning of the Polytetrafluoroethylene at small and medium feeds

I Pascu[1,2], I Popescu[1] and A Didu[1]

[1] Automotive, Transport and Industrial Engineering Department, Faculty of Mechanics, University of Craiova, Calea Bucuresti Str., no.107, Craiova, 200512, Romania

i_pascu@yahoo.com

Abstract. Experiments were made of the inner cylindrical turning of the Teflon with small and medium feeds, collecting the resulting chips. The shapes and dimensions of these chips are analyzed, explaining their correlation with the advance and with the deformation of the chips sections. The shapes of the chips resulting from considerations of the cutting process of this material are explained. From the analysis of the shapes of the chips, conclusions can be drawn regarding the deformation of the part material in the cutting area, meaning an appreciation of the machinability of the material. The quality of the chip surface indicates the marks left on the chip by the fine chip elements, some of them sticking to the resulting chip. The visualization of the aspect of the machined surface on the piece shows the lack of deposits on the edge, with small exceptions, the uniformity of the channels generated by the tip of the turning tool on the workpiece. All these considerations, as well as the final table regarding the deformation of the chip during processing, show how the Teflon behaves when turning inside with small feeds.

1. Introduction

Research on the machinability of Teflon is rare in the literature. In [1] the Ra parameter of the roughness is studied at the external cylindrical turning of a Teflon part, establishing the parameters of the cutting regime that ensures an imposed roughness. In [2], the temperature in the cutting area at the Teflon drill is measured, finding that the maximum temperature of 40 degrees appears at the cutting speed of 200 m / min. Questions and details about cutting Teflon appear on various Internet forums, and on YouTube there are videos, without comments, showing how a piece of Teflon is turned [3-6].

The shapes of the chips, as indicators of the phenomena in the cutting area, have been studied in some works, but for turning steel. Thus, in [7] the chips resulting from turning a steel are analyzed, depending on the geometric parameters of the knife, looking for the conditions in which convenient chips appear. Similar studies, but when turning an austenitic steel with high cutting speeds, are given in [8]. In [9] the 1045 steel was turned and the shapes of the chips correlated with the noises from the cutting process were studied, evaluating the chips into convenient and inconvenient.

Some approaches about plastics machining are investigated in [10].

In [11] a research about Teflon composites turning using a polycrystalline diamond tool is presented. The effect of the cutting parameters and insert radius on the cutting force and surface roughness was studied and a predictive model have been derived. Also, in [12] a design optimization of turning parameters of Teflon cylindrical rods by using ANOVA methodology is presented.

2. Experimental data

The Teflon is a synthetic plastic, composed of carbon and fluorine, with the chemical name of Polytetrafluoroethylene; it has a low coefficient of friction, is resistant to corrosion and wear, being used in the construction of machines for bearings, gears, various bushings, gaskets, etc. For the experiment, a Teflon bush with an outer diameter of 58 mm and an inner diameter of 32 mm was turned inside. The Teflon used have hardness 6.2 HV. The cutting regime used was: v = 50.24 m / min, the cutting depth ap = 0.5 mm, and the feed steps were: f = 0.096; f = 0.208; f = 0.302 mm / rot. No cutting fluid was used.

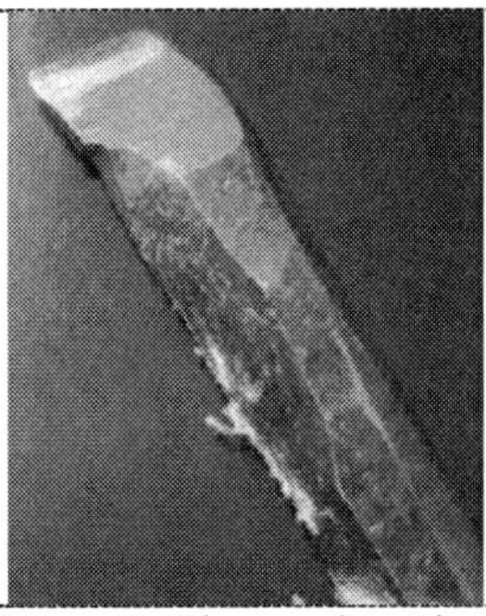

Figure 1. The used cutting tool

3. The shapes of the resulting chips

The resulting chips were studied in order to draw conclusions regarding the development of the cutting process when turning this material. All the chips are flowing, sliding slightly on the clearance face and on the channel on the cutting tool, the flow being also due to the low coefficient of friction between Teflon and the high tool steel. The chips were initially ribbon-shaped, then coiled in a ball, due to the low strength. In length the chips are not smooth, but have sinusoidal succession shapes, in different planes with variable steps and amplitudes. The color of the chips is white.

3.1 Feed f = 0.096 mm / rot

In figure 2 the aspect of the chips obtained at f = 0.096 mm / rot is presented.

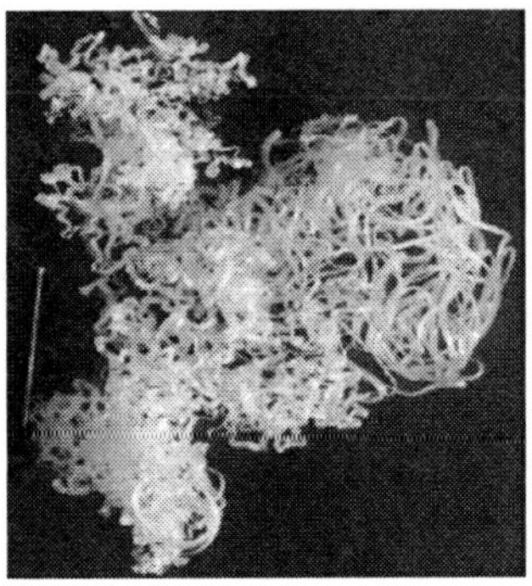

Figure 2. The chips at f = 0.096

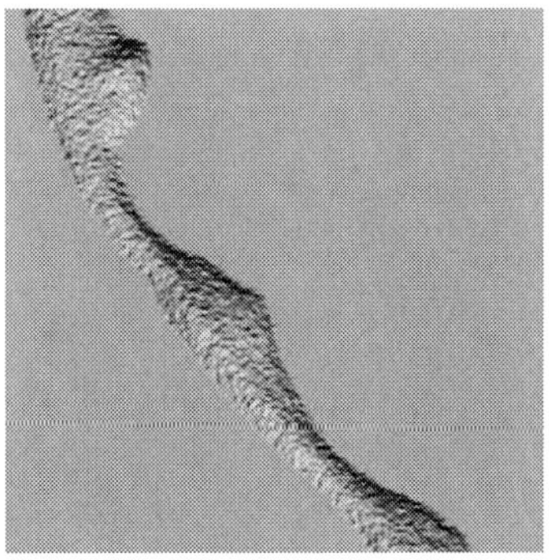

Figure 3. The thickness of the chip

Figure 4. Processed surface

The image also shows a bold, considered a scale standard (images change their size when typing); the bold has a length of 28 mm and a diameter of 0.76 mm.

The thickness of the chip can be seen in figure 3.

A variable thickness is observed, but here also intervenes the twisted shape of the chip.

In figure 4 the aspect of the processed surface with this feed is shown.

There are quite clear the traces of the tip of the cutting tool, but also some agglomerations of chips, which shows that deformations of the initial chip appear in the cutting area. Because the image scale changes when editing, an approximation of the sizes can be done by evaluating the distance between the channels, equal to the size of the feed.

3.2 Feed f = 0.209 mm / rot

The resulting chips are given in figure 5. Here, too, the chips are coiled because of their low resistance to bending, not remaining rigid.

Figure 6 shows the variation of the chip width and figure 7 shows the chipped surface. Longitudinal channels are also observed, caused by the non-uniformity of the main edge. The size of the image can be assessed by the size of the chip width, measured and given in Table 1 below.

Here, too, small agglomerations of chips can be seen on the surface of the piece (figure 7).

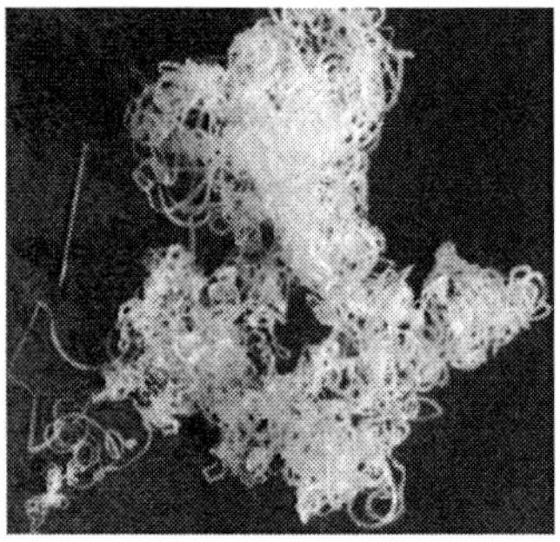

Figure 5. The shapes of chips

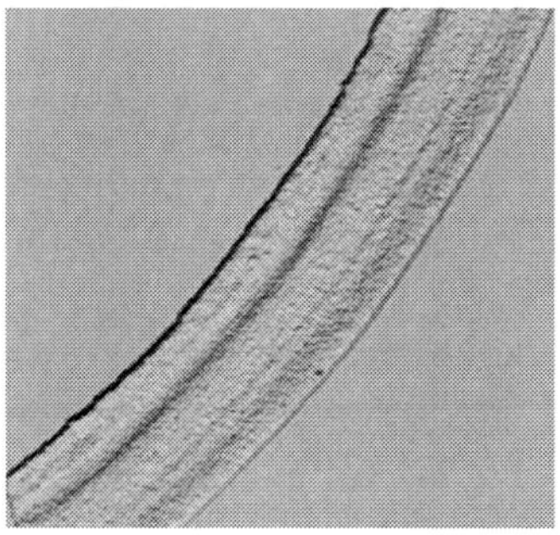

Figure 6. The chip's width

Figure 7. Processed surface

3.3 Feed f = 0.302 mm / rot

For this sample, the images of figures 8, 9, 10 were obtained. In this case the chips are thicker than in the previous samples, so that less deformed shapes appear.

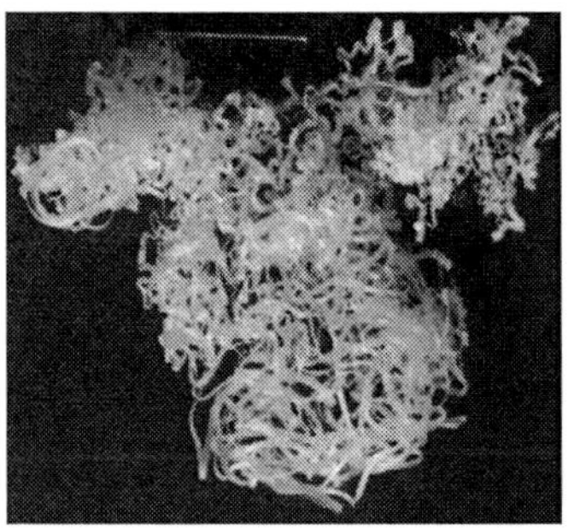

Figure 8. The shapes of chips

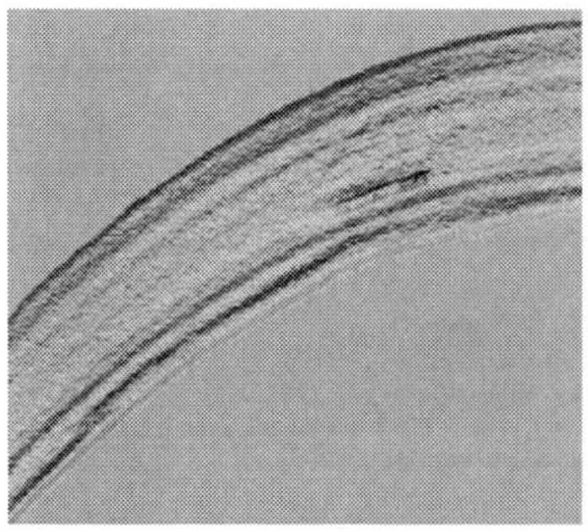

Figure 9. The chip's width

Figure 10. The surface of the part

Here also channels can be observed on the width of the chip, but they are quite uniform.

No chip agglomerations appear on the surface of the part, but the depths of the channels are not uniform, of the chip from the part breaking.

The images in figure 2, 5, 8 seem similar, but they are different not only in the shape in which they coiled, but also in size, the lengths of some bent areas being different from one case to another, so that on a large scale, compared to the bold used as a coefficient of scale, have significant differences.

4. The dimensions of the chips

The theoretical section of the chip in this operation is given in figure 11, where the theoretical section of the chip appears, the plan approach angle k_r and the corner angle of the chip α.

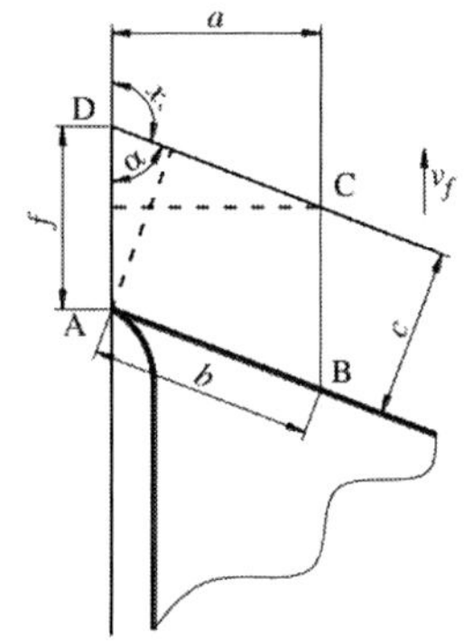

Figure 11. The theoretical chip section

The following relationships are written:

$$\alpha = \pi - k_r \tag{1}$$

$$a = b \sin\alpha \tag{2}$$

$$c = f \sin\alpha \tag{3}$$

Having a = 0.5 mm and k_r = 110°, the result is b = 0.532 mm = constant. The thicknesses and widths of the obtained chips were measured (c and b1), the results being given in table 1. In the table also appear the coefficients of chip thickness, ka and chip width, kb:

$$k_a = a1/c;\ k_b = b1/b \tag{4}$$

Table 1. Chip section results

f [mm]	c [mm]	a1 [mm]	b1 [mm]	k_a [mm]	k_b [mm]
0.096	0.09021	0.08	0 .78	0..8868151	1.466165
0 .209	0.1963957	0 .15	0 .89	0 .7637642	1.672932
0 .302	0 .2837871	0 .19	0 .91	0.6695161	1.710526

The followings are noted:

- the thickness c of the theoretical chip increases linearly with the feed;
- the thickness of the chip a_1 and its width b_1 increase as the feed increases; this is explained by shortening the length of the chip and thickening its section; therefore, the real chip is plastically deformed, differing from the theoretical one;
- the coefficient k_a decreases at the beginning, then, increasing the feed, it increases.

5. Conclusions

The shapes and dimensions of the chips resulted after Teflon inner turned were studied.

Flowing chips were obtained, long, coiled, and by increasing the feed the sections are enlarged, becoming more resistant, they no longer formed straight ribbons, but they coiled in the form of circular arches.

The thickness of the chips is uneven, but the widths are quite uniform, with some irregularities on one edge. Traces of the tip of the knife can be clearly seen on the processed surfaces, but also rare piles of chip elements.

Measurements have shown that the chips become thicker and wider as the feed rate increases. All this shows that in the cutting process of this material there are remarkable plastic deformations in the cutting area as when turning steels.

References

[1] Sanjeev Kumar M , Kaviarasan V and Venkatesan V *Int. J. Mod. Eng. Res. Technol.* 2018 **2** 143
[2] Quran Al. F *J Appl Sci,* 2007 **7** 141
[3] http://forums.bit-tech.net/showthread.php?t=163306&page=5
[4] http://www.lasercuttingshop.com/?projects=teflon
[5] https://www.youtube.com/watch?v=ioQrVA-vSs0
[6] https://www.youtube.com/watch?v=pUOO28H6T0o
[7] Mashasar J and Murugan M Int.J.Machining and Machinability of Materials 2009 **5** 452
[8] Lin W.S, Achiev J Mater. Manuf. Eng. 2008 **27** 191
[9] Karam S and Teti R Procedia CIRP Conference on Intelligent Computation in Manufacturing Engineering 2013 **12** 7
[10] Quadrini F, *Polym. Eng. Sci*, 2007 **48** 434
[11] Fetecau F and Stan F *Measurement* 2012 **45** 1367
[12] Raj I.J.A, Vijayakumar P, Mahadevan K.T.T.M, Kumar P, Ragavan R.V *Int. J. Appl. Eng. Res.* 2019 **11** 518

IOP Conf. Series: Materials Science and Engineering **898** (2020) 012037 doi:10.1088/1757-899X/898/1/012037

Research about using the Failure Mode and Effects Analysis method for improving the quality process performance

I Pascu[1,3], D Paraschiv[2] and A Didu[1]

[1] Automotive, Transport and Industrial Engineering Department, Faculty of Mechanics, University of Craiova, Calea Bucuresti Str., no.107, Craiova, 200512, Romania
[2] Cummings Generator Technologies Romania, Decebal str., no. 116A, Craiova, 200781, Romania

i_pascu@yahoo.com

Abstract. Failure Mode and Effects Analysis (FMEA) is a method of analysing potential failures of a product or process, in order to plan the necessary measures to prevent their occurrence. Being a critical analysis method, FMEA has extremely precise objectives: determining the weak points of a technical system; searching for the initiating causes of component dysfunction; determining the needs of technology upgrading and modernization of production; increasing the level of communication between work compartments, people, hierarchical levels; analysis of the consequences on the environment, the safety of operation, the value of the product; providing corrective actions to remove the causes of defects; providing a plan to improve product quality and maintenance. The FMEA must be used from the design phase before the product is made.This paper describes a study about the improving the quality process for an axis for packaging using FMEA method. The potential causes of the defects have been studied and improvement measures have been proposed. Among these we mention: implementation and monitoring of the preventive maintenance program; providing specific compliance; self-control, flow or sample control; periodic training and retraining; upgrading existing equipment, acquisition of more advanced software.

1. Introduction

Worldwide due to customer expectations coupled with the continuing increase in the complexity of products, declining design and launch periods have necessitated systematic quality planning [1].

The premise of traditional quality assurance, based on the detection and elimination of defective products, is no longer relevant, the argument being extremely simple and intuitive: defects that can be avoided before launching the product, do not need to be corrected later [2]. Modern methods of systemic quality design are answers to new quality requirements, they must allow the analysis and elimination of potential defects from the design and implementation stage [3], so that more and more often the notion of "quality design" is encountered [4]. One of these methods that has become increasingly used in recent decades is the FMEA [5,6]. Failure Modes, Effects & Analysis (FMEA) is a systematic method of determining and preventing errors, defects and risks that may occur, applicable to a process, product or machine used in the process [7,8]. This method consists in detecting possible defects, inventorying the causes that could cause these failures, demerits effects on users in order to plan the necessary measures to prevent their occurrence [9]. This method was originally developed in 1949 by the US military, thus the military procedure MIL-P 1629 entitled "Procedures for Performing a Failure Mode, Effects and

Criticality Analysis" having applicability to projects aimed at ensuring maximum availability of strategic military equipment [10]. The first notable applications of FMEA techniques (acronym AMDEC in French) are related to NASA (60's) and later in the 90's by the 3 major US car manufacturers: GM, Ford and Chrysler by including them in the prescriptions of the quality standard QS 9000, figure 1 [11-13].

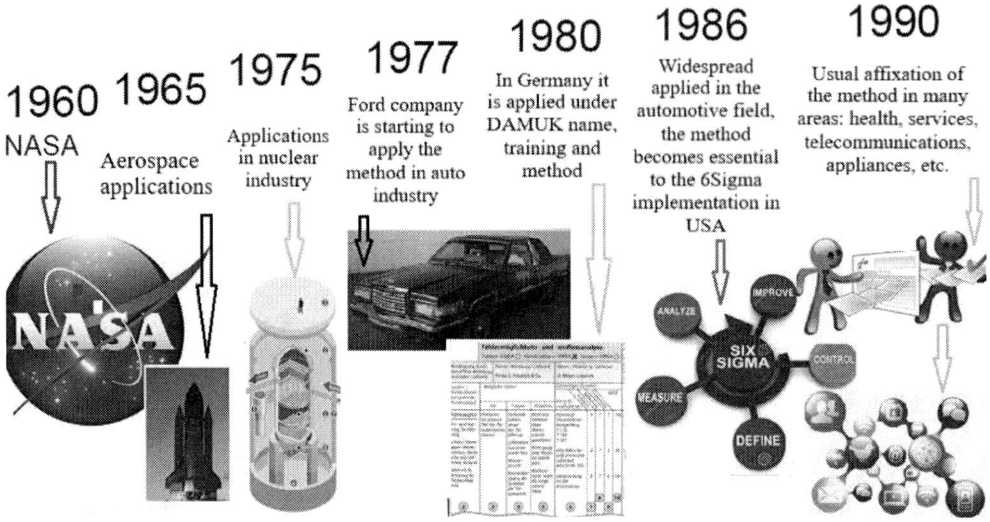

Figure 1. Time evolution of the FMEA workability.

There are two main types of FMEA [14]: product development (design), DFMEA and process development, PFMEA. In addition to the two types of FMEA listed above (DFMEA and PFMEA), the following are also known and used: FMEA system (system) focused on the study of the functions of the components of the subsystem, machine–or technological equipment; FMEA service that is used to analyze the functions of the service; FMEA software focused on the study of software functions and computer components [15].

2. Methodical Basics for FMEA Analysis

The methodology for applying the FMEA method was presented in [16, 17]. The application of the process FMEA involves, in a first stage, the description of the process functions. Starting from these functions, the potential faults are identified and the critical stages of the process are highlighted. The necessary corrective measures shall be taken to prevent the occurrence of defects.

The designated team of specialists evaluates the defects in terms of two criteria: probability of occurrence (Occ) and probability of detection (Det), which are expressed using the same notation scale with marks from 1 to 10. The quantification of these probabilities depends on the type of product or the analyzed process. Severity (Sev) means the value that characterizes how serious is the effect established for the failure mode and how it affects the customer, in terms of product / process failure / failure, their effects and grading being achieved by grades given from 1 at 10.

The assessment of the importance of defects is done using the notation scale based on probabilities of occurrence (Occ) and detection (Det), and occurrence (Occ), the RPN risk index coefficient is determined by the relation:

$$RPN = Sev\ x\ Det\ x\ Occ \quad (1)$$

The estimations of the analysis are written in tabular form, similar to the one described in figure 2.

Potential Failure Mode and Effects Analysis (Process FMEA)

Item:	Process Responsibility:		
Type of component:	Key Date:		Rev.
Core team:			

Process Step / Function	Requirement	Potential failure mode	Potential Effect(s) of Failure	SEV	CLASS	Potential Cause(s) of Failure	Current Process: Curent Process Controls Prevention	Current Process: OCC	Current Process: Current Process Controls Detection	Current Process: DET	RPN	Responsibility & Target Completion Date	Action Results: Action taken	Action Results: SEV	Action Results: OCC	Action Results: DET	Action Results: RPN

Figure 2. The table header used for FMEA analysis [18]

The RPN coefficient has values between 0 and 1000. It is generally considered that measures are needed to prevent potential failures when the CR risk coefficient is greater than 100.

3. FMEA process Analysis for a Packing Shaft

Figure 3 shows the packing axis for which the process FMEA analysis was performed.

Figure 3. Packing shaft

The operations and stages necessary for the obtaining of the component in the chronological order of their development, are: qualitative inspection - reception of materials; semi-finished cutting; milling and centering; turning 1; milling and drilling; turning 2; grinding; final adjustment.

After the application of the FMEA, it was found that there are operations that lead to serious disturbances of the technological process, having a high RPN, of over 100 points, these being the following: when cutting the semi-finished product, when the semi-finished product shows arrow curvature due to deviation from misalignment due to large successive temperature variations (severity - score 5, occurrence - score 3, detection - score 10), RPN value of 150, certificate of material from the supplier; when cutting the semi-finished product, when the documents are incorrectly completed or incomplete, having as affect the mounting of the shaft with material of different specification due to the filling error at the material suppliers (severity - point 9, occurrence - point 4, detection - point 3), RPN with a value of 108, certification and control of material from the supplier is recommended; when cutting the semi-finished product, when the cut semi-finished product is longer than necessary, it may have the effect of affecting the on time delivery (OTD) due to non-compliance with the working instructions (severity - score 10, occurrence - score 4, detection - score 3), RPN value of 120, training, periodic retraining and flow self-monitoring are recommended; at turning operation 1 when the milled and drilled shaft is not in accordance with the documentation may cause rejection of the shaft or its reclassification due to human error, respectively operator is trained but does not comply with the manufacturing process (severity - point 10, occurrence - point 5, detection - point 4), RPN with a value of 200, standard working instructions describing the process steps are recommended. At the turning operation 1 when the overhaul is exceeded, the shaft may be reshaped, reclassified or rejected due to the delayed launch by the design, respectively the operator is trained but does not respect the manufacturing process (severity - score 10, occurrence - score 3, detection - score 4), RPN with a value of 120, the following modification sheet is recommended, audit.

The RPN values in order of their appearance are presented in the diagram in the figure 4.

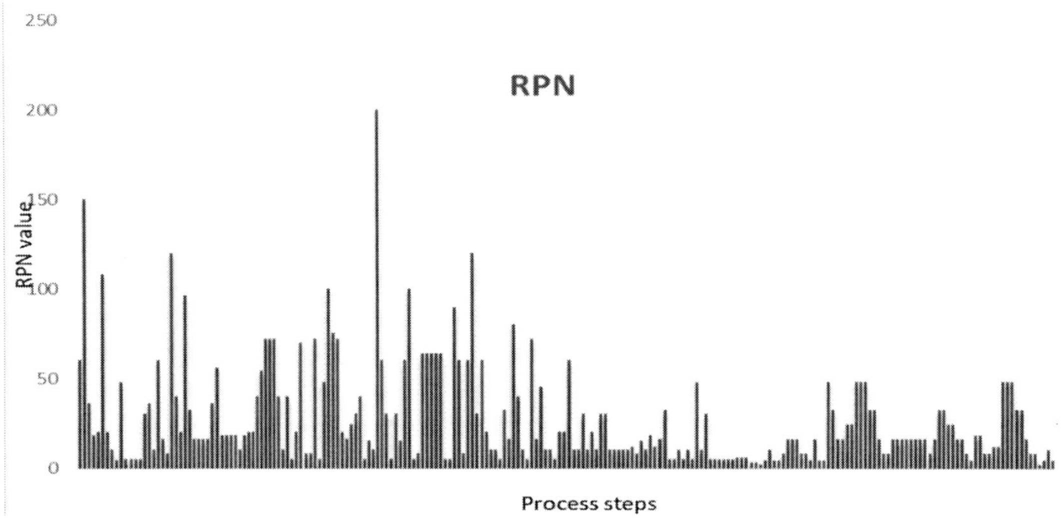

Figure 4. The RPN values in order of their appearance.

In figure 5 the RPN values in descending order of parts subjected to FMEA analysis is represented. After taking the necessary measures to implement the recommendations for each operation with a high RPN, it decreased to zero, leading to finished products with "zero defects".

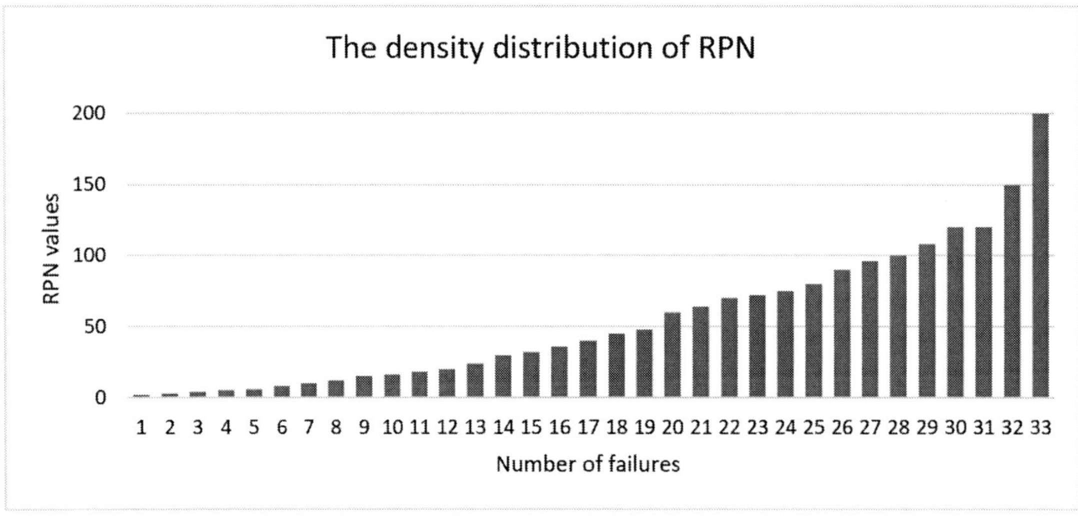

Figure 5. Diagram of RPN distribution

IOP Conf. Series: Materials Science and Engineering **898** (2020) 012037 doi:10.1088/1757-899X/898/1/012037

4. Conclusions

Timely use of FMEA - Process analysis can avoid costly changes to the technological process of making the product "packaging tree" by identifying potential defects, avoiding them and assessing the risks and potential consequences of defects and obtaining "zero defects" as target products.

The main potential defects analysed are: misalignment due to large successive temperature variations, shaft mounting with different specification material, milled and drilled shaft is not in accordance with the documentation, reshaping, reclassifying or rejecting the shaft, damage affecting the dimensions on the part. The potential causes of the defects have been studied and improvement actions have been proposed. Among these we mention:

- implementation and monitoring of the preventive maintenance program;
- modernization of the existing milling machine;
- revamping of the numerically controlled lathe on which the turning operations are performed;
- update for the software of the processing program - introduced in the program, with the processing and revision of the drawings;
- acquisition of a program for simulating the turning operation, periodic testing of the program in production;
- standard work instructions that describe the process steps;
- monitoring of operators through actions to observe operators' non-conformities and their regular training;
- periodical trainings and annual audits.

References

[1] Swamidass P.M 2000 *Encyclopedia of Production and Manufacturing Management* (Boston: Springer MA)
[2] Pekuri A, Haapasalo H, Herrala M 2011 *Int. J. Product. Perform. Manag.* **1** 39.
[3] Ott E.R, Schilling E.G, Neubauer D.V 2005 *Process Quality Control: Troubleshooting and Interpretation of Data*, Fourth Edition (Milwaukee: ASQ Quality Press).
[4] Rawlins, Ashley R 2008 *Total Quality Management (TQM)* (Bloomington: AuthorHouse,).
[5] Mhetre RS, Dhake RJ 2012 *Int. J. Adv. Res. Eng. Appl. Sci.* **2** 302
[6] Segismundo P, Cauchick M 2008 *Int. J. Qual. Reliab. Manag.* **25** 899
[7] Carlson C.S 2012 *Effective FMEAs: Achieving Safe, Reliable, and Economical Products and Processes using Failure Mode and Effects Analysis* (New Jersey: John Wiley & Sons)
[8] Paciarotti C *et al* 2014 Int. J. Qual. Reliab. Manag. **7** 788
[9] Jain K 2017 *Int. J. Health Care Qual. Assur.* **2** 175
[10] Carlson C.S, 2014 *Proc. IEEE 2014 Annual Reliability and Maintainability Symposium (RAMS) (Colorado Springs)* vol.1 (New York: Institute of Electrical and Electronics Engineers Publ.) p 1
[11] Ványi G, Loránd E 2016 *Acta Univ. Sapientiae Inform.* **8** 82
[12] Pries K.H 1998 *Automotive Electronics Reliability* vol.2, ed. R K Jurgen (Warrendale: SAE International) p 351
[13] Lakhmi C.J 2014 *Handbook on Decision Making: Vol 1: Techniques and Applications,* ed. C P Lim (Berlin: Springer)
[14] Zheng L.Y, Liu Q, McMahon C.A 2010 *Proc. 6th CIRP-Sponsored International Conference on Digital Enterprise Technology, Advances Intelligent and Soft Computing Integration of Process* eds G Huang, K L Mak, G Maropoulos Berlin:Springer-Verlag Inc) p. 1673
[15] Chrysler Corporation, Ford Motor Company, General Motors Corporation 2004 *Potential Failure Mode and Effects Analysis (FMEA),* Reference Manual.
[16] Scutti J, McBrine W 2002 *Introduction to Failure Analysis and Prevention*,(Cleveland: ASM International Materials Park, 2008, pp. 146–151.
[17] Pascu I, Paraschiv D, *Appl. Mech. Mater.* **822** 429

IOP Conf. Series: Materials Science and Engineering **898** (2020) 012038 doi:10.1088/1757-899X/898/1/012038

Static analysis of a vehicle suspension with leaf springs and reaction bars

C Alexandru

Transilvania University of Braşov, Romania

E-mail: calex@unitbv.ro

Abstract. The work approaches the static analysis of a suspension mechanism with leaf springs and reaction bars used for motor vehicles. The reaction bars are used for compensating the twist of the springs into an 'S' shape, a phenomenon that occurs in the classic suspension system with soft leaf springs during hard acceleration or braking. The virtual prototyping software package ADAMS is used for modelling and simulating the suspension system. The static analysis results reveal the improved behaviour of the suspension system with reaction bars relative to the classical leaf spring suspension system.

1. Problem statement

The automotive industry is one of the most important and performing economic branches, the technical-material resources oriented to this field constituting peaks of current technology. If in the first vehicles, the main problem was self-propulsion, with the emergence and development of internal combustion engines, the car manufacturers' researches were oriented both towards optimizing the form (design) and improving the dynamic performance of the vehicle, in terms of stability, comfort, manoeuvrability and reliability.

In the relative motion to car body (chassis), the vehicle wheels can be guided independently - by means of a guiding mechanism for each wheel (independent suspension), or dependent - by a guiding mechanism of the rigid axle (dependent suspension). The first solution is frequently used for passenger cars (for both front and rear wheels), while the second solution is mainly used for the rear axles of the larger gauge vehicles (e.g. commercial vehicles) [1-4].

In the classic suspension solution of the rear axle with leaf springs, the spring ensures (besides the main role of elastic element) the axle guiding relative to chassis. A solution to improve the vehicle comfort in the case of suspension with leaf springs consists of the use of soft springs (with a small number of leaves/foils, one or two usually). In the case of these soft spring suspensions, the twisting of the leaf springs into an 'S' shape occurs during hard acceleration or braking, with negative effects on the vehicle dynamics. The 'S'-the twisting phenomenon is produced/determined by the longitudinal contact forces between wheels and road, which causes the axle to twist around its own axis, thus resulting in high bending stress of the leaf spring in the recessed area, and so its deformation into an 'S' shape (Figure 1).

A possible, and also used/implemented, the solution to compensate for the 'S'-twisting consists of the asymmetrical arrangement of the shock absorbers (dampers) relative to the axle longitudinal plane (Figure 2). A semi-constructive variant of such a suspension system is presented in Figure 3. However, this solution is not very effective (efficient) when the wheel contact forces are variable in size and direction.

IOP Conf. Series: Materials Science and Engineering **898** (2020) 012038 doi:10.1088/1757-899X/898/1/012038

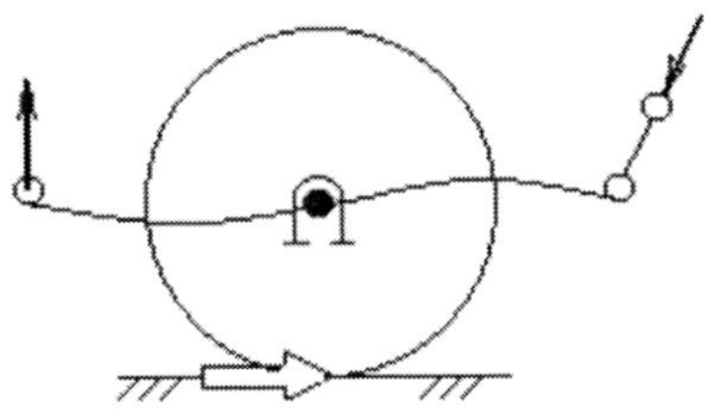

Figure 1. The 'S'-twisting of the leaf spring.

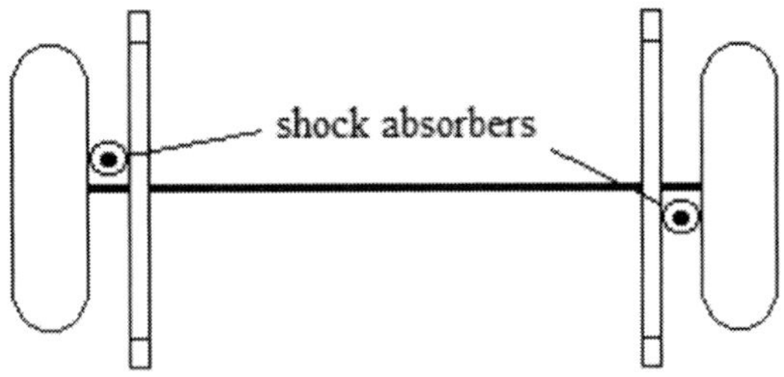

Figure 2. Suspension system with shock absorbers arranged asymmetrically.

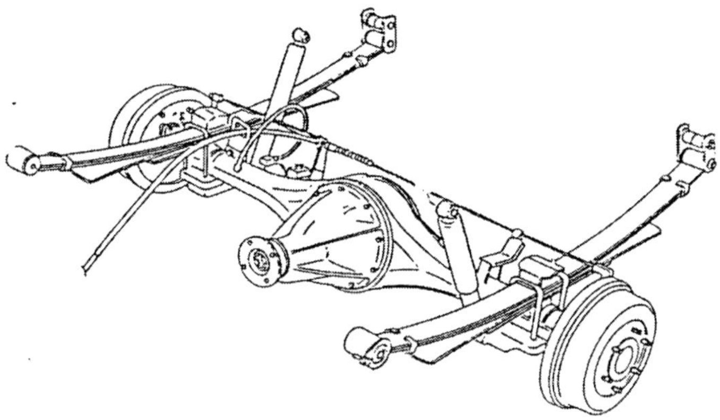

Figure 3. Semi-constructive solution with shock absorbers arranged asymmetrically.

The best solution for minimizing/avoiding the 'S'-twisting of the leaf springs consists in equipping the vehicle suspension with two longitudinal reaction bars (also called traction or slapper bars), as shown in Figure 4.a, with a corresponding semi-constructive model presented in Figure 4.b. The reaction bars can be disposed (arranged) in upper or lower plan in relation to the axle, and they are hinged to axle and chassis by bushings (also called flexiblocks). These compliant joints of rubber have 6 elastic restricted degrees of mobility (DOM), in other words they generate elastic reaction forces and torques on all axes [5], as shown in the theoretical model from Figure 5.

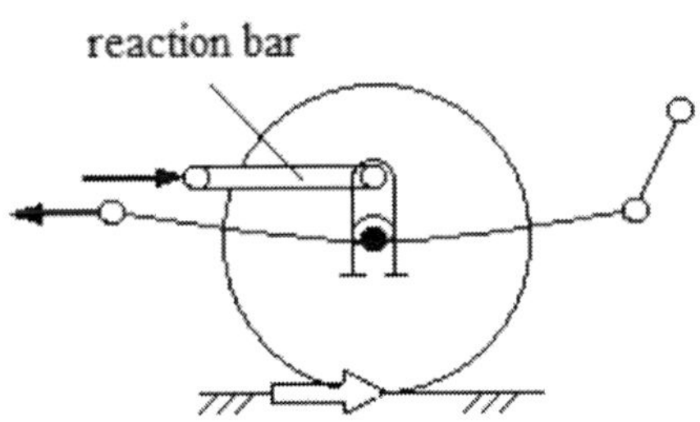

a.

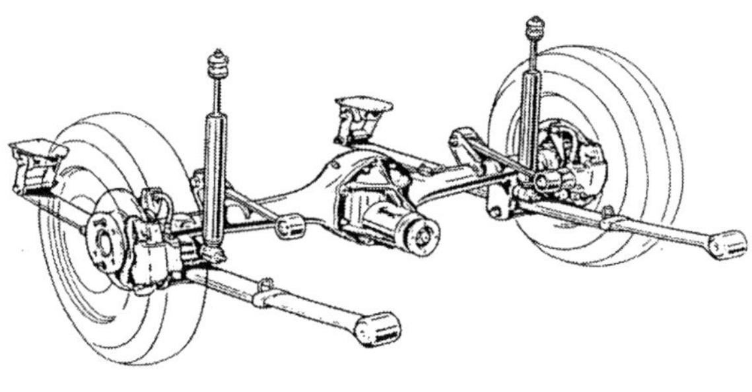

b.

Figure 4. The rear axle suspension system with leaf springs and reaction bars.

The role of reaction bars is to take over the longitudinal contact forces from the wheels so that the springs are only loaded with traction and compression, without being subjected to bending. The so obtained suspension system combines the features of the multi-link axle guiding mechanisms with coil springs (as they are defined and systematized in [1]) and the classical suspension with hard leaf springs.

In this paper, the effectiveness of the vehicle suspension with leaf springs and reaction bars is verified through a comparative analysis with the corresponding classical suspension with leaf springs. In this regard, the static models of the two suspension systems have been conceived and analyzed through the use of the facilities achieved by the virtual prototyping package ADAMS, which is currently developed and maintained by MSC Software, a global leader in CAE software development and services. The benefits of using such advanced software solutions in the design process of mechanical & mechatronic systems are reflected in more competitive products, for various types of applications [6-9].

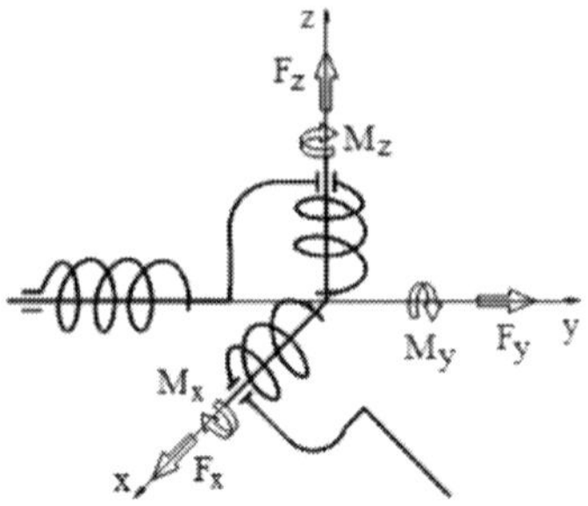

Figure 5. The modelling of the bushing.

2. Defining the static model

It is well-known that the statics is an out-of-time analysis, which is concerned with the analysis of loads acting on systems that do not experience an acceleration. There are the following input parameters for the static analysis: the assembled configuration of the mechanical system (including the bodies and the connections between bodies), the loads through forces and/or torques (excepting the forces that depend on velocity and acceleration, such as damping and inertia forces/torques). The output of the static analysis is represented by the equilibrium/balance configuration (position), when the resultant of all the acting and reacting forces/torques is null [10, 11]. For this work, the static equilibrium configuration of the suspension system will be reported by the twisting angle of the rear axle, corresponding to the rotational movement of the axle around its own axis.

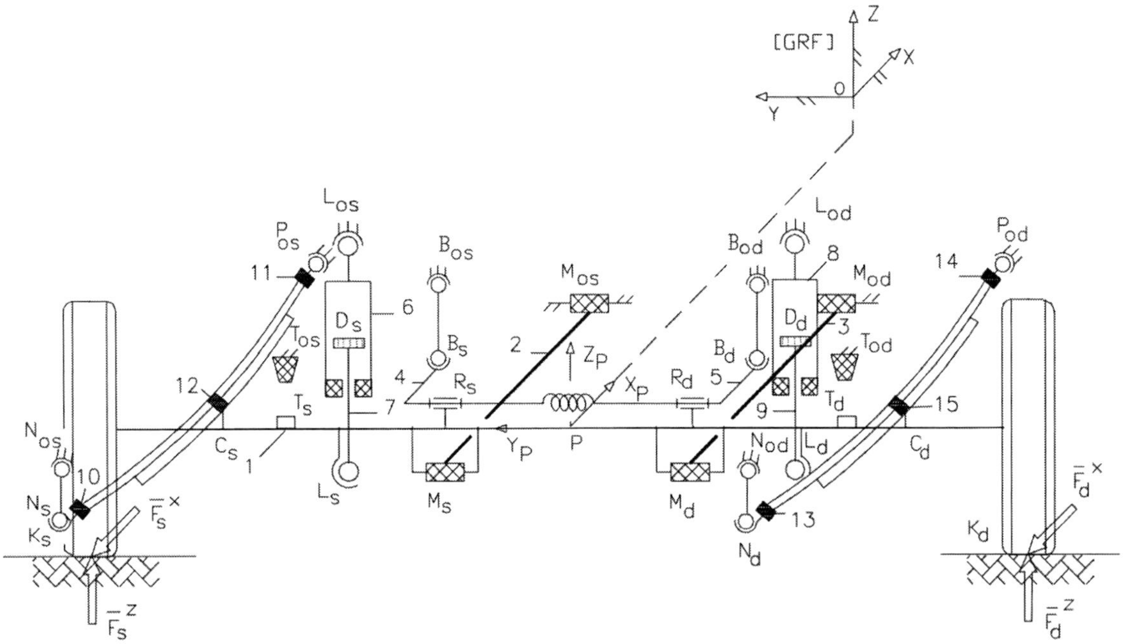

Figure 6. The multi-body model of the axle suspension system with leaf springs and reaction bars.

In this regard, the multi-body model of the suspension with leaf springs and reaction bars is presented in Figure 6. The spatial positioning of the axle coordinate system $PX_PY_PZ_P$ with respect to the global coordinate system OXYZ determines the relative movement in the suspension system. The global frame is attached to chassis (car body), which is considered to be fixed in the static analysis. When the vehicle, therefore the suspension system, is at rest (this being the initial reference position), the local axes X_P, Y_P and Z_P are arranged parallel to the corresponding ones from the global frame. The suspension system is externally loaded by the vertical (F^Z) and longitudinal (F^X) contact forces between wheels and road (ground), corresponding to the traction-braking regime.

Because the leaf spring assures, in addition to the main role of elastic suspension element, the guiding function of the axle (denoted by '1' in Figure 6) relative to chassis, three parts were used to materialize a leaf spring in the multi-body model conceived in ADAMS. Thus, the leaf spring was modeled by two point masses placed at the ends of the spring ('10-11' for the left spring, and '13-14' for the right spring), which are connected to the car body (at the front end) and respectively to the spring ring (at the rear end) by spherical joints, while the third part of the spring ('12', and respectively '15'), which concentrates its mass, it is located in the middle area, being rigidly connected to the axle (in fact, this fixed joint materializes the spring flange/clamp). The three bodies that model a leaf spring are interconnected by elastic elements with uniform cross-section ("massless beam" in ADAMS). The beam (whose modelling is based on the material properties of the leaf spring) transmits forces according to the Timoshenko beam theory [12].

The spring ring, which has a reduced mass compared to the other bodies from the suspension system, was modeled as a constant distance constraint (spherical - spherical composed restriction, namely N_{0s}-N_s and N_{0d}-N_d, by case) between the adjacent bodies (the car body and the spring rear part '10/13'). The same type of constraint was used for the modeling of the anti-roll bar tierods (B_{0s}-B_s and B_{0d}-B_d), the anti-roll bar itself being modeled by two parts (denoted by '4' and '5' in Figure 6) between which a torsion spring is disposed.

The shock absorbers were modeled not only as internal force generating elements (which, however, for the static analysis have no influence, because the damping forces depend on velocity, so time), but also as kinematic elements, through two bodies/parts per damper, materializing its cylinder and piston ('6-7' for the left damper, and '8-9' for the right damper), which are connected to chassis and axle by spherical joints, and coupled together by a cylindrical joint.

In these terms, the static model of the axle suspension system with leaf springs and reaction bars (denoted by '2' and '3' in Figure 6) is defined by the following:

- generalized coordinates for 15 mobile bodies: $15 \times 6 = 90$;
- geometrical constraints generated by the mechanical connections:
  - spherical - spherical composed restrictions (B_{0s} - B_s, B_{0d} - B_d, N_{0s} - N_s, N_{0d} - N_d): $4 \times 1 = 4$,
  - revolute joints (R_s, R_d): $2 \times 5 = 10$,
  - spherical joints (L_{0s}, L_{0d}, L_s, L_d, P_{0s}, P_{0d}): $6 \times 3 = 18$,
  - cylindrical joints (D_s, D_d): $2 \times 4 = 8$,
  - fixed joints (C_s, C_d): $2 \times 6 = 12$.

The degree of freedom (DOF) of the suspension mechanism, which expresses the number of uncontrolled (independent) movements, which take place under forces action, can be computed by using the Gruebler's count, $DOF = 6n - \Sigma r$, where n - the number of mobile bodies, and Σr - the sum of geometric constraints, resulting: $DOF = 90 - 52 = 38$. In a similar way, for the classical suspension system with leaf springs (without reaction bars, so with 13 mobile bodies), there is obtained: $DOF = 78 - 52 = 26$. It should be mentioned that in the classical suspension there is the same number of geometric constraints as for the suspension with reaction bars, although the reaction bars are missing (so are their connections) because these connections are made by bushings, which - as mentioned - do not introduce constraints, but they are force generating elements (as shown in Figure 5). Therefore, the difference in number of degrees of freedom between the two suspension systems is only given by the number of generalized coordinates of the reaction bars.

3. Results and conclusions

The above presented suspension models (with and without reactions bars) were modeled in ADAMS/View (which is the general pre-processing interface of the MSC.ADAMS suite of software), the static analysis being performed by using ADAMS/Solver (a powerful numerical analysis application that automatically formulates and solves the motion equations).

The comparative analysis was performed by considering the following values for the contact forces on wheels: normal (vertical) reaction forces: $F^{Z}_{s,d} = 292$ daN, longitudinal reaction forces: $F^{X}_{s,d} \in [-400, 400]$ daN.

As mentioned, the parameter of interest is the twisting angle of the axle (denoted η_y in Figure 7), which actually determines the 'S'-twisting of the leaf springs. In these terms, in accordance with the diagrams shown in Figure 7, in case of the suspension system with reaction bars (curve 'a') the 'S'-twisting of the leaf springs is very small compared with that corresponding to the classical suspension system (curve 'b'). The slight variation of the axle twisting angle that occurs in the case of the reaction bar based suspension is due to the elasticity/deformability of the bushings by which the bars are connected to the adjacent elements (mainly those from the axle).

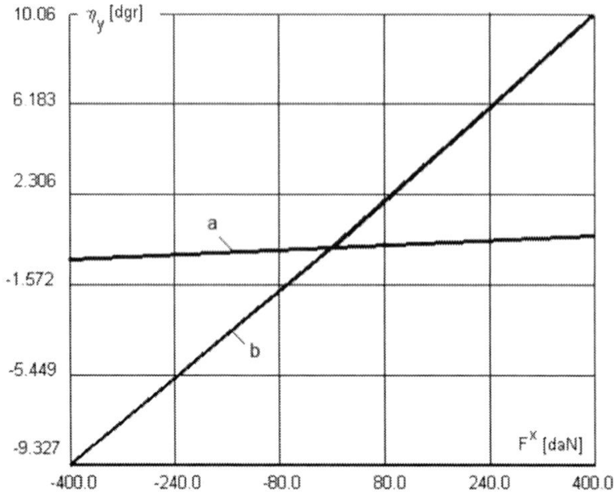

Figure 7. The twisting angle of the rear axle.

In conclusion, the use of the reaction bars makes it possible to use soft leaf springs in the axle suspension system, with the purpose to improve the comfort performance of the vehicle, but without affecting its dynamic behavior.

References

[1] Alexandru C 2009 The kinematic optimization of the multi-link suspension mechanism used for rear axle of the motor vehicle *Proceedings of the Romanian Academy - Series A* **10**(3) pp 244-253

[2] Alexandru P, Macaveiu D and Alexandru C 2012 Design and simulation of a steering gearbox with variable transmission ratio *Proceedings of the Institution of Mechanical Engineers, Part C: Journal of Mechanical Engineering Science* **226** pp 2538-2548

[3] Knapczyk J and Maniowski M 2006 Elastokinematic modeling and study of five-rod suspension with subframe *Mechanism and Machine Theory* **41**(9) pp 1031-1047

[4] Knapczyk J and Maniowski M 2010 Optimization of 5-rod car suspension for elastokinematic and dynamic characteristics *The Archive of Mechanical Engineering* **52**(2) pp 133-147

[5] Alexandru C 2019 Method for the quasi-static analysis of beam axle suspension systems used for road vehicles *Proceedings of the Institution of Mechanical Engineers, Part D: Journal of Automobile Engineering* **233**(7) pp 1818–1833

[6] Berceanu C and Tarniță D 2010 Aspects regarding the fabrication process of a new fully sensorized artificial hand *Proceedings of the International Conference ModTech* pp 123-126

[7] Geonea ID, Alexandru C, Margine A and Ungureanu A 2013 Design and simulation of a single DOF human-like leg mechanism *Applied Mechanics and Materials* **332** pp 491-496

[8] Ioniță M and Alexandru C 2012 Dynamic optimization of the tracking system for a pseudo-azimuthal photovoltaic platform *Journal of Renewable and Sustainable Energy* **4** pp 053117

[9] Tarniță D, Catană M and Tarniță DN 2016 Design and simulation of an orthotic device for patients with osteoarthritis *Mechanisms and Machine Science* **38** pp 61-77

[10] Alexandru C 2016 Analytical method for determining the static equilibrium position of the rear axles guiding mechanisms of the motor vehicles *Applied Mechanics and Materials* **841** pp 59-64

[11] Alexandru C 2018 Method for the kinetostatic analysis of the road vehicles axle suspensions *Mechanisms and Machine Science* **57** pp 57-65

[12] Timoshenko SP, 1921 On the correction factor for shear of the differential equation for transverse vibrations of bars of uniform cross-section *Philosophical Magazine* pp 744

IOP Conf. Series: Materials Science and Engineering **898** (2020) 012039 doi:10.1088/1757-899X/898/1/012039

Research in the field of vibration of automotive systems

M Stanescu[1], V Ionica[1], I Geonea[1], C Miritoiu[1] and A Bolcu[1]
[1]University of Craiova, Faculty of Mechanics

igeonea@yahoo.com

Abstract. The aim of the research is to improve knowledge about the behaviour of materials and structures, to develop models and tools useful in the process of design of structures and machines and to capitalize on a technical culture in terms of analysis, design and manufacturing methodologies. We also set out to show the influence of which types of vibrations have an effect on the operation of the crankshaft mechanism of a car engine, as well as their influence on the operation of the camshaft system, all of which are taken into account in any design process in this field. The generalization of the theory of elasticity and the dynamics of viscous fluids led to the creation of the linear theory of viscos-elasticity, the bodies with such a behaviour obeying the principle of overlapping Boltzmann. Thus, mechanical models were built consisting of spring and damping systems, the springs describing the elastic properties of the body, and the dampers on the viscous ones.

1. Introduction

A high level of safety calculations can ensure the use of multiple systems of computational mathematics, what we see in this work by comparing graphs that we obtained using two computers where cam follower system. The trend of replacing metallic materials with materials with rheological behaviour also belongs to the current field, the latter having a series of advantages, one of them being very important, being that of the existence of reduced forces and inertia torques due to the lower specific mass, compared with metallic materials under comparable stiffness, which appears in this work. Currently, research activities focus on the development of multidisciplinary experimental, theoretical and numerical skills put into play when designing structures, machine elements or, in general, mechanical systems [1, 2, 3]. The aim of the research is to improve knowledge about the behaviour of materials and structures, to develop models and tools useful in the process of design of structures and machines and to capitalize on a technical culture in terms of analysis, design and manufacturing methodologies [4, 5, 6]. This research is supported by the fields of materials science, nonlinear mechanics of solids, fluids and coupled systems, acoustics, forming and processing techniques, experimental measurement methods and numerical modelling. Research in the field of vibrations is the subject of research of numerous articles, focused especially on vibrations in gears, suspension systems for vehicles [7-15].

2. Vibrations of the crankshaft mechanism

Figure 1a show the mechanical model of the crank connecting rod mechanism of a motor subjected to transverse and longitudinal vibrations. Figure 1b shows the coupling of an engine with a layout, which consists of the crankshaft 1, the connecting rods 2 and 3, the two pistons and with the bolts 4, the cylinder head 7 (8), the cylinder liner 11 with the cooling fins 6 and inlet and outlet valves 9 respectively 10. The device occupies the position in figure 1c.

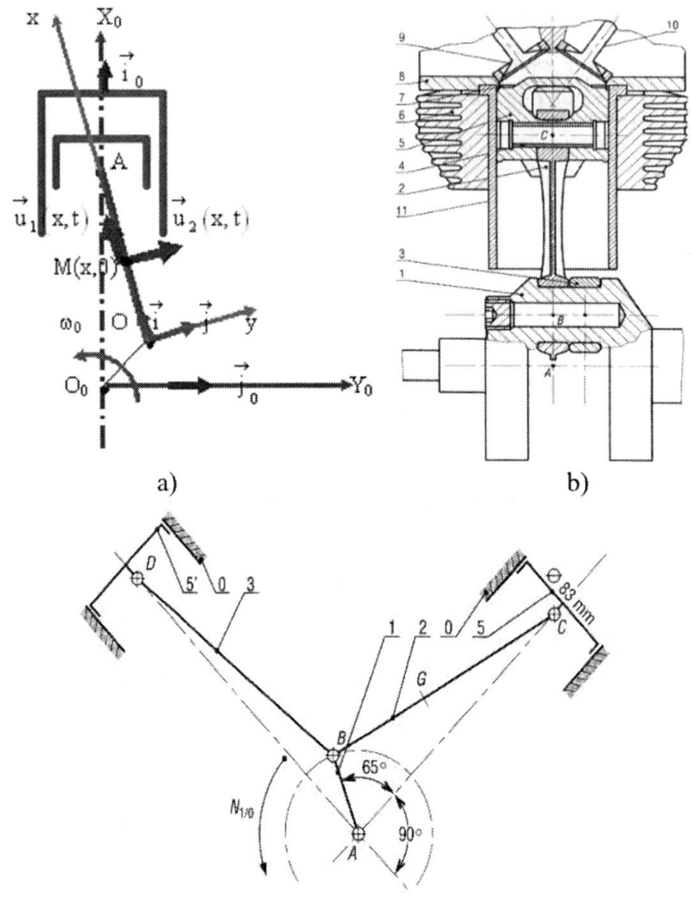

Figure 1. The crankshaft mechanism of an engine.

Let be the laws of definition of the longitudinal and transverse displacements of the viscoelastic connecting rod, respectively:

$$u_1^{(1)}(x,t) = a_1 + a_2 \cdot cos^2(\omega_0 t) + a_3 \cdot cos(\omega_0 t) +$$
$$\frac{2}{L} \cdot \sum_{n=1}^{\infty} \left[a_{1,n} + a_{2,n} \cdot cos^2(\omega_0 t) + a_{3,n} \cdot cos(\omega_0 t) + a_{4,n} \cdot cos\left(\omega_{n,1} t\right)\right] \cdot cos(\alpha_n x), \quad (1)$$

$$u_2^{(1)}(x,t) = \frac{2b}{L}\sum_{n=1}^{\infty}\frac{1}{b_n}\left[\left(\sum_{j=1}^{4} b_{j,n}\right)sin(\omega_0 t) + \left(b_{5,n}+b_{6,n}\right)sin(2\omega_0 t) + \left(\sum_{j=1}^{4} c_{j,n}\right)sin\left(\omega_{n,2}t\right)\right]\cdot sin(\alpha_n x) \quad (2)$$

Consider the specific case of an internal ignition engine in which: L=L$_b$=0,164 m, is the connecting rod length; r=0,043 m, is the crankshaft radius; ω_0 =680,333 s^{-1}, is the crankshaft angular velocity; I$_{zz}$=7376,28 mm^4, is the connecting rod moment of inertia ; ρ =7800 Kg/m^3, is the rod specific mass; E=2,1$\cdot$10^{11} N/m^2, represents the Young's modulus; A=283,68$\cdot$10^{-6} m^2, is the connecting rod cross section. With the laws of definition (1), (2) and with the above concrete data results, with the Mathematica program, the graphical representations from figures 2, 3, 4, 5, of the functions $u_1 = u_1^{(1)}(x,t)$, the longitudinal displacements, $u_2 = u_2^{(1)}(x,t)$, the transversal displacement, $u_1 = u_1^{(1)}\left(\frac{L}{3},t\right)$, respective $u_2 = u_2^{(1)}\left(\frac{L}{3},t\right)$, in a first approximation.

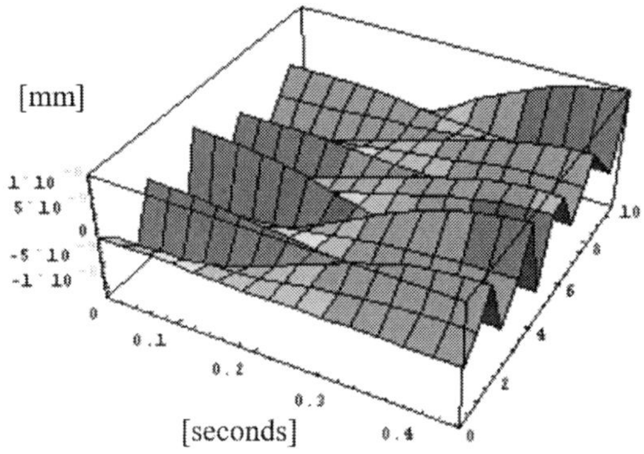

Figure 2. Graphical variation of displacement u_1, plotted in Mathematica software.

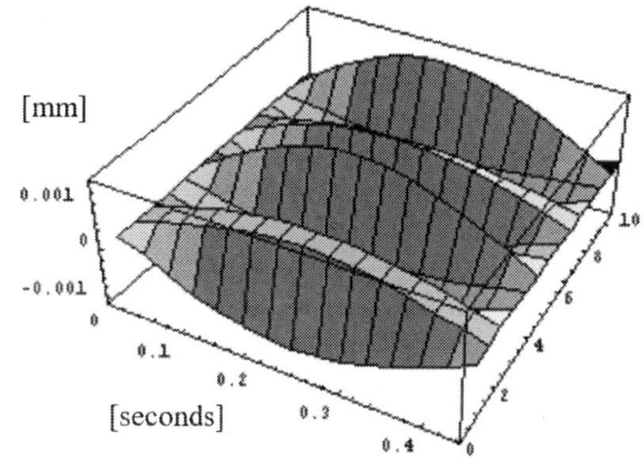

Figure 3. Graphical variation of displacement u_2, plotted in Mathematica software.

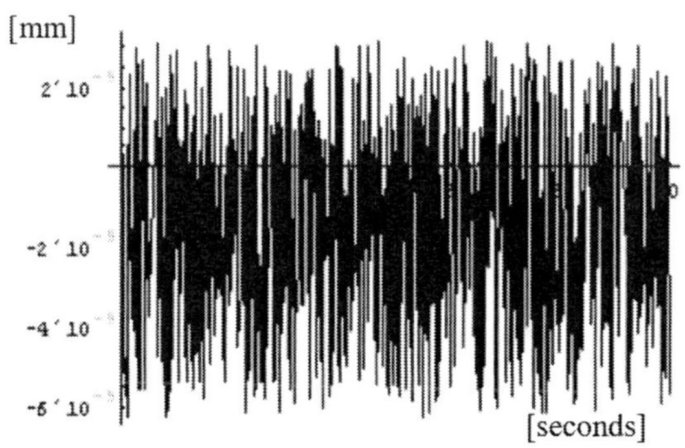

Figure 4. Graphical variation of displacement u_1, plotted in Maple software.

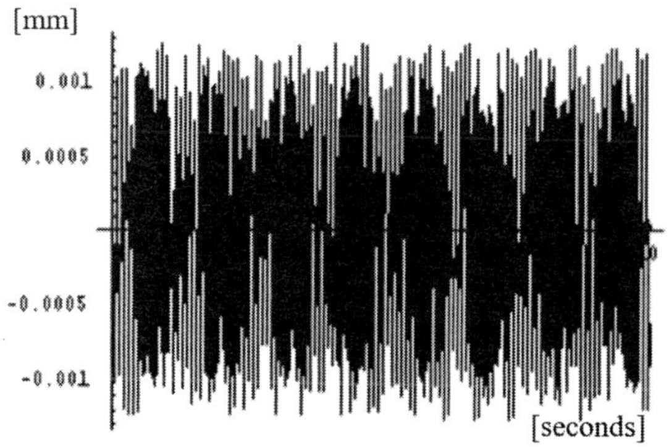

Figure 5. Graphical variation of displacement u_2, plotted in Maple software.

IOP Conf. Series: Materials Science and Engineering **898** (2020) 012039 doi:10.1088/1757-899X/898/1/012039

Stopping the iterative process at the third approximation, the graphical representation of the transverse displacement function at the abscissa point results $x = L/3$, from figure 6.

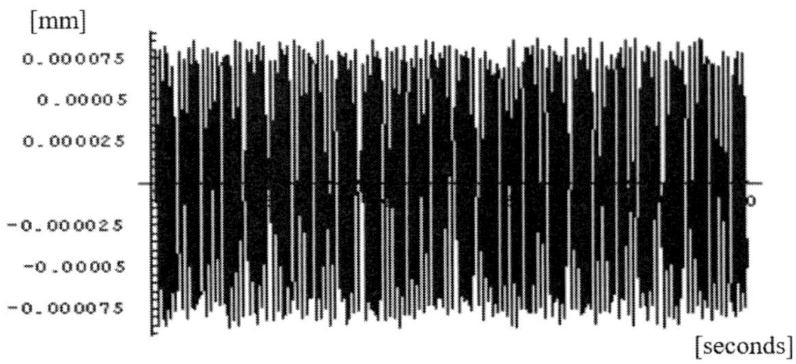

Figure 6. Graphical variation of displacement u_2, for $x = L/3$, plotted in Maple software.

As can be seen, the field of longitudinal displacements has a negligible influence, which is why we continued the iterative process only for transverse displacements. As the displacements are very small, it is convenient to give up ferrous materials, the materials with such behaviour being much cheaper.

3. Cam-follower system vibration

The vibrating system actuated by the cam in figure 7a is considered.

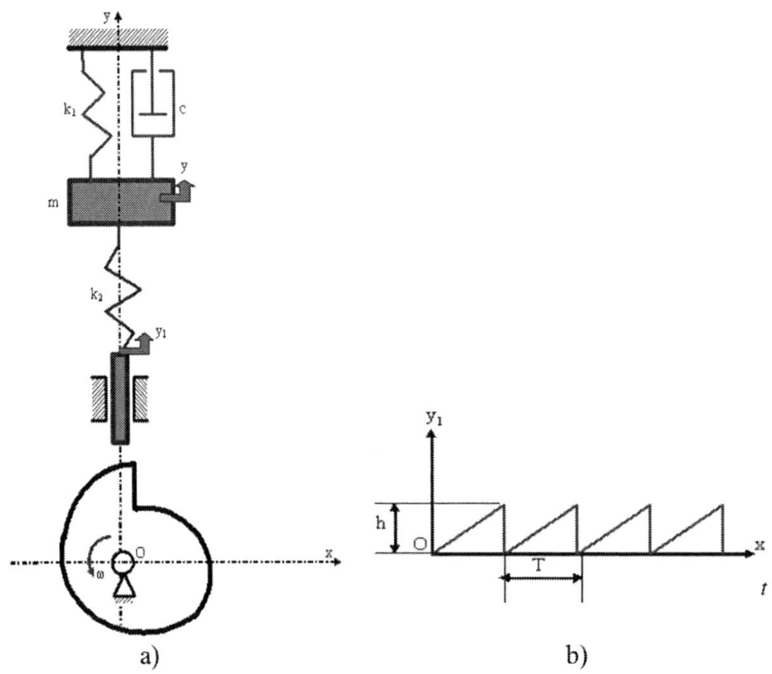

Figure 7. The mechanical model of the cam-driven vibrating system.

The lower end of the spring of elastic constant k_2 has the function of follower, being driven by a cam rotating at a constant angular velocity ω and having a sawtooth-type law of motion, with the maximum displacement h, as in figure 7 b. In the figure 7b the time variable t was denoted by x, notation that will use further. In Lagrange's or Newton's formalism we obtain the mathematical model of the motion of the mass body m (valve):

$$m\ddot{y}(t) + c\dot{y}(t) + k_1 y(t) + k_2[y(t) - y_1(t)] = 0 \tag{3}$$

or, in other words:

$$\ddot{y}(t) + 2n\dot{y}(t) + \omega_n^2 y(t) = \varpi_n^2 y_1(t) \tag{4}$$

where: $k_e = k_1 + k_2, \frac{c}{m} = 2n, \frac{k_e}{m} = \omega_n^2, \frac{k_2}{m} = \varpi_n^2$, c being the damping constant.

The law of motion of the cam for a period T is:

$$y_1(t) = \frac{h}{T}t, 0 < t < T \tag{5}$$

so that the mathematical model (4) takes shape:

$$\ddot{y}(t) + 2n\dot{y}(t) + \omega_n^2 y(t) = \varpi_n^2 \frac{h}{T}t. \tag{6}$$

Consider a numerical application in which: $m = 1Kg, h = 0{,}05\, m, T = 1\, s, k_1 = 300\frac{N}{m}, k_2 = 20\frac{N}{m}$, $c = 15Nsm^{-1}$, so that the equation (6) become:

$$\ddot{y}(t) + 15\dot{y}(t) + 320y(t) = t. \tag{7}$$

Applying to equation (7) the unilateral Laplace transform with respect to time, under the initial conditions $y(0) = 0, \dot{y}(0) = 1\left[\frac{m}{s}\right]$, the algebraic equation results:

$$s^4\tilde{y}(s) - s^2 + 15s^3\tilde{y}(s) + 320s^2\tilde{y}(s) = 1 \tag{8}$$

with the solution:

$$\tilde{y}(s) = \frac{s^2 + 1}{s^2(s^2 + 15s + 320)}. \tag{9}$$

Inverting in (9), in the *Mathematica* calculation system, results the solution of equation (7) in the form of the time function:

$$y(t) = \frac{1}{21606400}\cdot \tag{10}$$

$$\cdot\left\{-3165 + 67520t + e^{-\frac{15}{2}t}\left[3165\cos\left(\frac{\sqrt{1055}}{2}t\right) + 40877\sqrt{1055}\sin\left(\frac{\sqrt{1055}}{2}t\right)\right]\right\}$$

whose representation is given, in the same calculation system, in figure 8:

IOP Conf. Series: Materials Science and Engineering **898** (2020) 012039 doi:10.1088/1757-899X/898/1/012039

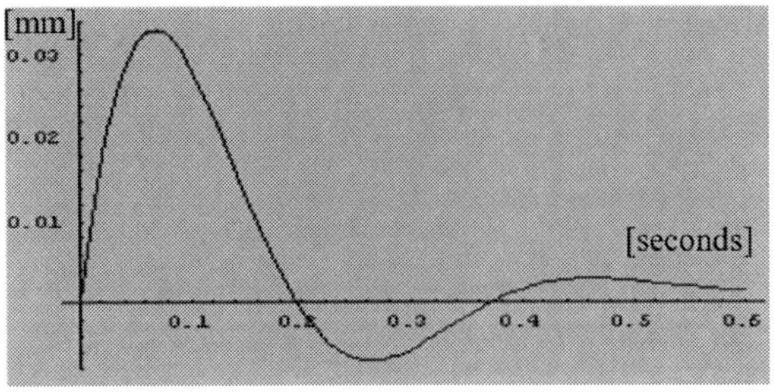

Figure 8. Mathematica representation of y function.

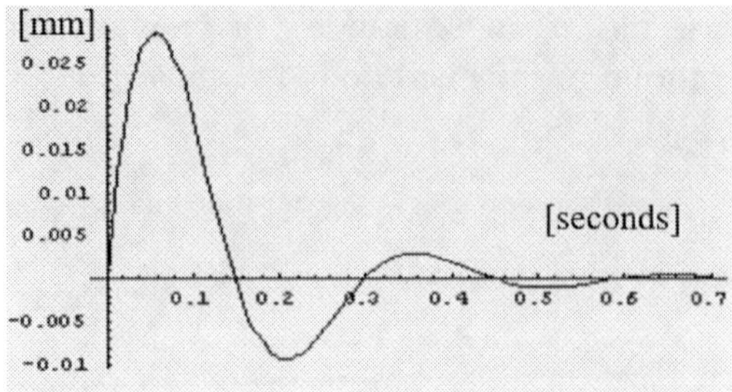

Figure 9. Maple representation of y function.

In the *Maple* software, the time function is obtained as in equation (11), and plotted in figure 9.

$$y(t) = e^{-7,5t}[0,6186 \cdot 10^{-5} cos(21,0653t) + 0,0047 sin(21,0653t)] + \\ e^{-0,3921 \cdot 10^{-107}}[0,000064 sin(3,1415t) - 0,6186 \cdot 10^{-5} cos(3.1415t)] \quad (11)$$

We presented only the first harmonic because, for the chosen numerical data, the variation of the function $y = y(t)$ according to the number of harmonics used is not significant. As we have shown above, a high level of computational security can be ensured by the use of several systems of computational mathematics, which we observe in this example by comparing the two graphs we obtained using two computational systems. The periodic function $y_1 = y_1(t)$, given by the law of definition (5), is decomposed into the Fourier series:

$$y_1(t) = \sum_{i=1}^{n} a_i \, sin(i\pi t) \quad (12)$$

where: a_1=0,03183, a_2=-0,01591, a_3=0,01061, a_4=-0,00795, a_5=0,00636, a_6= - 0,005305.

The considered decomposition presenting terms only in the sine function, it being obtained by using the *Maple* calculation system. The following figures show this decomposition, observing the effect that the number of terms, taken into account, has on this dependence. Consideration of several terms makes the representation of the function $y_1 = y_1(t)$ as relevant as possible. In our application it is observed that, starting from the representation of even the first 15 harmonics, the variation of this dependence depending on the number of harmonics used is no longer significant. However, if we take a concrete example, such as the valves of an internal ignition engine, it requires the consideration of over 150 harmonics.

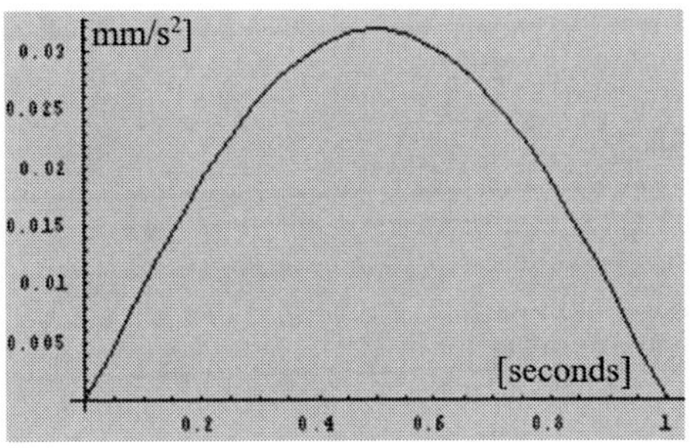

Figure 10. Representation of the first harmonic.

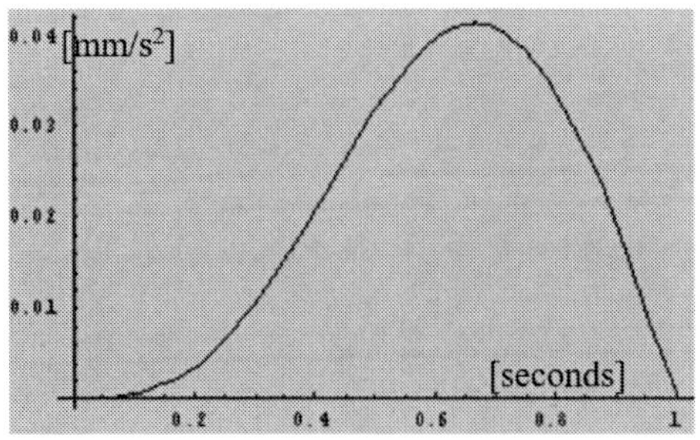

Figure 11. View of the first two harmonics.

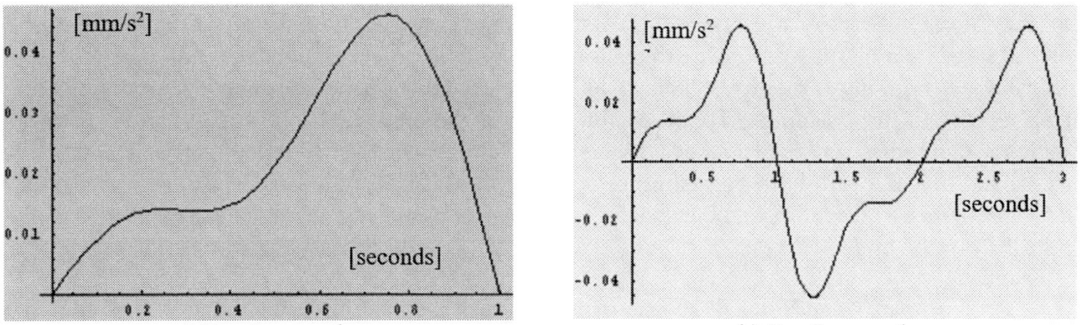

a) For 1 second b) For 3 seconds

Figure 12. Representation of the first three harmonics $y_1(t)$.

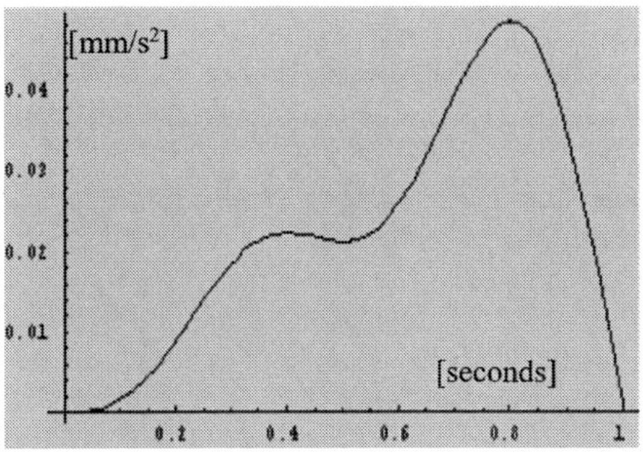

Figure 13. Representation of the first four harmonics y_1.

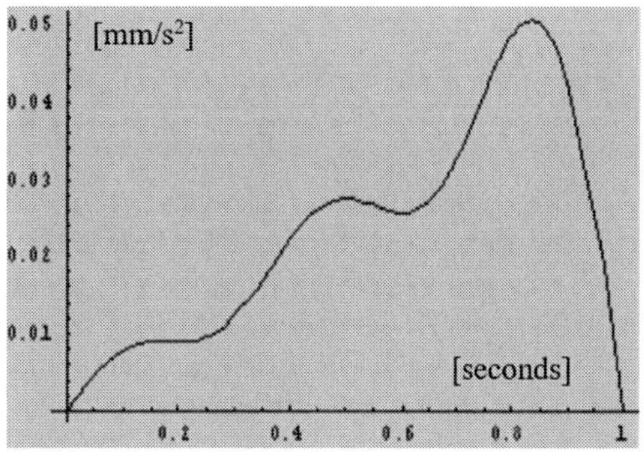

Figure 14. Representation of the first five harmonics y_1.

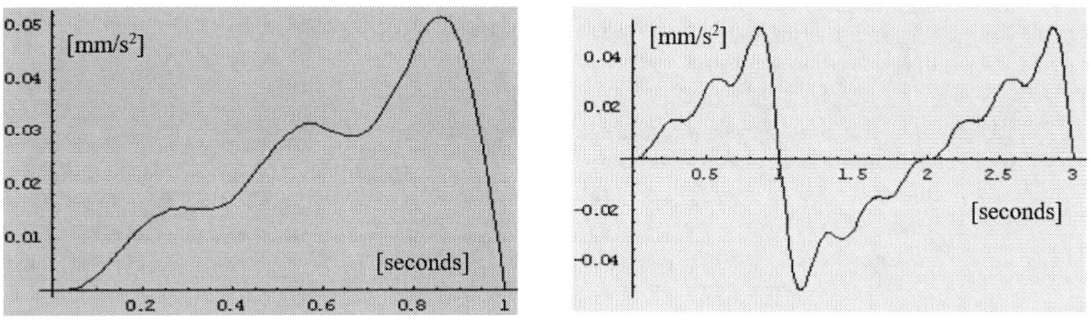

a) For 1 second. b) For 3 seconds.

Figure 15. Representation of the first six harmonics $y_1(t)$.

4. Conclusions

A high-level of computational security can be ensured by the use of several systems of computational mathematics, which we observe in this paper by comparing the graphs we obtained using two computational systems in the case of the dashboard system. Currently, research activities focus on the development of multidisciplinary experimental, theoretical and numerical skills put into play when designing structures, machine elements or, in general, mechanical systems. The aim of the research is to improve knowledge about the behaviour of materials and structures, to develop models and tools useful in the process of design of structures and machines and to capitalize on a technical culture in terms of analysis, design and manufacturing methodologies. This research is based on the fields of materials science, nonlinear mechanics of solids, fluids and coupled systems, acoustics, forming and processing techniques, experimental measurement methods and numerical modelling. The trend of replacing metallic materials with materials with rheological behaviour also belongs to the current field, the latter having a series of advantages, one of them being very important, being that of the existence of reduced forces and inertia torques due to the lower specific mass. in relation to metallic materials under comparable conditions of rigidity, which should be noted in this paper. The generalization of the theory of elasticity and the dynamics of viscous fluids led to the creation of the linear theory of viscous-elasticity, the bodies with such a behaviour obeying the principle of overlapping Boltzmann. Thus, mechanical models were built consisting of spring and damping systems, the springs describing the elastic properties of the body, and the dampers on the viscous ones.

References

[1] Abousleiman V and Velex, P 2006 A hybrid 3d finite element/lumped parameter model for quasi-static and dynamic analyses of planetary/epicyclic gear sets *Mechanism and Machine Theory* **41(6)** pp 725-748

[2] Băgnaru D G Stănescu M M and Cuţă P 2008 Vibrations of the tool and tool holder unit for a shaper device, *8th International Conference, Research and Development in Mechanical Industry RaDMI* 2008 pp 202-205

[3] Băgnaru D G 1997 On the analysis of the vibrations of viscoelastic bars, the constituent elements of the plane mechanisms *SYROM 97*, pp 25-30

[4] Băgnaru D G Hadăr A Bolcu D Stănescu M M and Cuţă P 2009 The vibrations influence on the field of accelerations of the linear-elastic connecting rod of a mechanism connecting rod lug *Annals of DAAAM for 2009 & Proceedings* **20(1)** pp 833-835

[5] Băgnaru D G, Stănescu M M, Bolcu D and Cuţă P 2008 Approximate method in order to determine the displacement fields for the kinematic elements standing in vibration *Second international congres automotive, safety and environment SMAT* pp 9-12

[6] Blankenship G W and Singh R 1992 A comparative study of selected gear mesh interface dynamic models *International Power Transmission and Gearing Conference ASME* **1** pp 137-146

[7] Bard C 1995 Modeling of the dynamic behavior of gear transmissions *L'institut National des Sciences Appliquees de Lyon*

[8] Craig R R and Bampton M C C 1968 Coupling of substructures for dynamics analysis *AIAA Journal* **6(7)** pp 1313–1319

[9] Geonea I D Rosca A S Racila L Rinderu P and Ionica V 2020 About the Vibration and Dynamics of a Bus Chassis *Applied Mechanics and Materials* **896** pp 97-104

[10] Ionica V Bagnaru D Geonea I and Ciurezu L 2019 Vibrations Influence on Automotive Drive *SIAR International Congress of Automotive and Transport Engineering: Science and Management of Automotive and Transportation Engineering* pp 565-572

[11] Ionica V Stanescu M Miritoiu C and Geonea I 2019 Vibrations influence on the operation of gears *In IOP Conference Series: Materials Science and Engineering* **568**

[12] Kahraman A 1994 Planetary gear train dynamics *Journal of mechanical design* **116(3)** pp 713-720

[13] Lin J and Parker R G 1999 Analytical characterization of the unique properties of planetary gear free vibration *Journal of vibration and acoustics* **121(3)** pp 316–321

[14] Ohji K Ogura K Kubo S and Katada Y 1978 A study on governing mechanical parameters of creep crack growth rate by using notched round bar specimens with proportional geometry *Proc. 21st Japan Congr. Materials Res. Soc. Materials Sci.* pp 99-104

[15] Özguven H N and Houser D R 1988 Mathematical models used in gear dynamics -A review *Journal of Sound and Vibrations* **121** pp 383-411

IOP Conf. Series: Materials Science and Engineering **898** (2020) 012040 doi:10.1088/1757-899X/898/1/012040

3D Modelling and FEM Analysis on Die Clash Mint Error

C C Gavrila[1] and M T Lates[1]

[1]Product Design, Mechatronics and Environment Department, "Transilvania" University of Brasov, Brasov, Romania

E-mail: cgavrila@unitbv.ro

Abstract. During the last century, a large number of modern coins which wearing different error types were minted and released. In literature, there are known various error types, each is provided by some causes, as the coin metal, the coin dimension and also the manufacturing step where it could appear. The die clash error appears when the striking tools – the obverse and reverse negative dies - heat each other without the coin blank inside them. The paper presents the influence of some initial condition on the amplitude of this error, as the die size, the model flat field size or the model shape. In the first part of the paper, some introductive aspects of the die clash error are presented with some particular examples. Then, the obverse and reverse die 3D model is presented, followed by the finite element analysis, which is realized and achieved for considered initial conditions. In the final paper part, the analysis results and conclusions are presented.

1. Introduction

Striking metal coin currency is followed in most of the situations by a various amount of error types. These error types could appear on every coin's manufacturing step: at the coin metal alloy manufacture, on the blank coin manufacture, at the striking tools manufacturing, or at the properly coin striking. Usually, when the errors appear on dies, the model is slightly changed, and the pressed coins will result in the changed model with an error. As for collecting value, for increased spectacular error on coin or medal, the value for the subjected piece is increased [1, 2]. We consider that, from this point of view, it is important to know which of the initial coin strike conditions could influence some error to become more valuable.

The metal coins are manufactured by pressing at high loads the coin blank with hardened steel negative obverse and reverse dies. The both negative dies relief (composed by properly incused figures surrounded by flat field) forms a closed space to be filled by the coin material [3, 4]. These negative dies are obtained before hardening, by pressing with a hardened steel positive die with the relief model. Since the model on the negative die is incused, after the striking, result of the relief model on the coin.

If the coin blank fails to get into the striking space between negative dies, die clash occurs; then, the dies could hit each other. Generally, the dies are adjusted so that, at their closest approach, there is still a minimum clearance between them, even in the absence of a coin blank. There are a lot of situations when the dies are out of adjustment, and direct contact between them appears. In this situation, each die leaves a light or deep impression of some portions of its design on the opposing die. The die negative most vulnerable model's areas to the die clash are the closest areas from the die face. The flat face field around the model and the shallowest recessed on the die face are included; at striking, these led to the areas of lowest relief on the coin. The basic model, together with these impressions is pressed, as a mirrored image, to strike coins. These coins, which wear the incuse design elements are known as clash

IOP Conf. Series: Materials Science and Engineering **898** (2020) 012040 doi:10.1088/1757-899X/898/1/012040

die error coins [2]. The most types of die clash errors coins have reduced diameter and especially thin blank; rarely, the die clash error could appear also at some increased size coins. Also, the die clash could appear once, when the traces impressed are single, or multiple, when the traces are also multiple. For multiple traces, these could be offset and close each other, when the dies maintain the same position during strike, or far each other, rotated, when major misalignments appear.

The dies clash error is known at various Romanian and foreign modern coins [2, 5]. As is presented in examples from figure 1, the negative impressions appear on the coin face flat field, which is the lowest relief on the coin. Most of these impressions could be observed near exterior coin rim, around the effigy or country arms, or inside the effigy or country arms, if there are some lowest than field relief portions as ears, eyes or shield hatches. It has to be mentioned that, these negative impressions are rarely on the entire coin face field, as is presented in figure 1, a, on a 2 bani coin issued in 1900. Usually appears on both coin faces, on a portion which is combined with the other face corresponding portion, as in figure 1, c, on a 500 lei coin, issued in 1944. Other found examples, where one face wears on a portion these impressions, are presented in figure 1, b, for 1 leu coin, issued in 1941, and, respectively, in figure 1, d, for a 3 lei coin, issued in 1966.

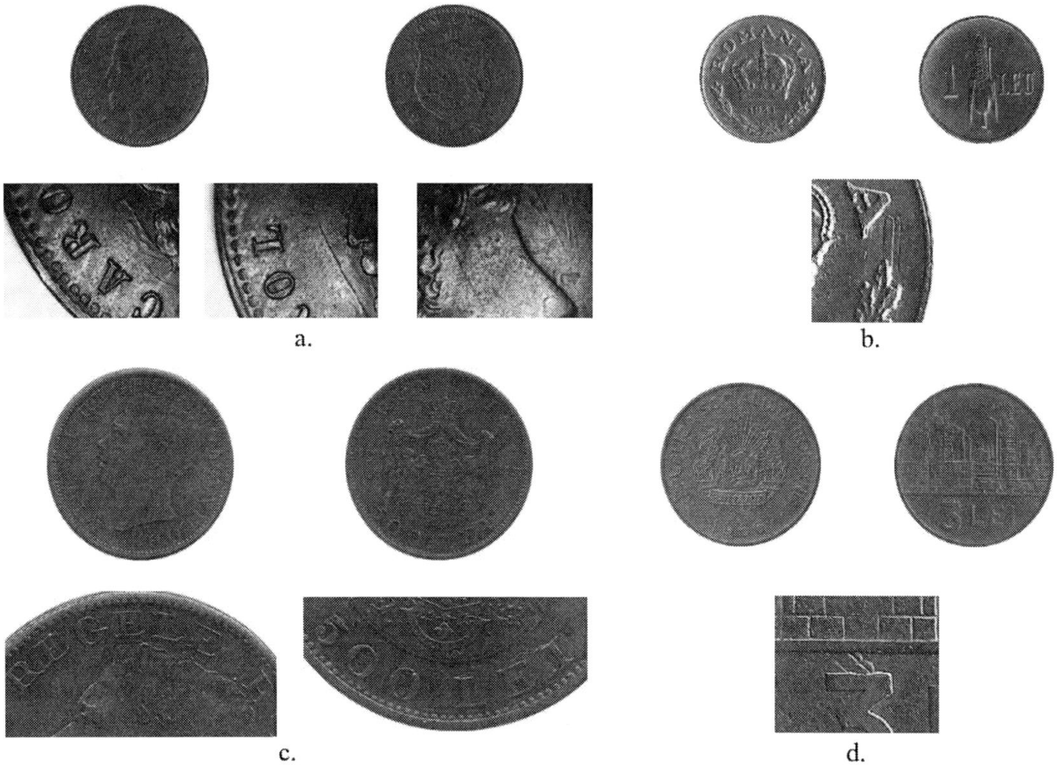

Figure 1. Various types of die clash error on coins

Some of the position of these negative impressions can be combined with the coin face basic design, resulting some interesting representation as effigy with wrinkles, tears, glasses and more others, desirable for private collections [6].

The traces left on the dies are given by the contour edges portion which touches each other. The combination of this traces are given by some design details, as the figures to be impressed on coin, but also by the dies position: medal position - when the obverse-reverse angle is 0^0 (example presented in figure 1, d), or coin position - when the obverse-reverse angle is 180^0 (the other examples from figure 1). In the literature, the side wearing the effigy face is considered the obverse and the side wearing the country arms, reverse [5, 6]. Another important detail consists in the flatness of the face field around relief contour edges. The reference case is to be considered as a plane surface; in the real case, due to some dies constructive reasons, could be slightly concave or convex [4]. In figure 2, there are presented the contours and traces defined by the dies model, for obverse-reverse medal position. In this example, the contact area between dies covers the full field common area, situation possible in the presumption when, there are no the misalignments inside of the pressing machine and the face field is plan [7]. Otherwise, the contact area is decreased. It can be observed that, depending by the coin design, the obverse flat field area (composed by red and green portions) is increased than the reverse (composed by red and blue portions).

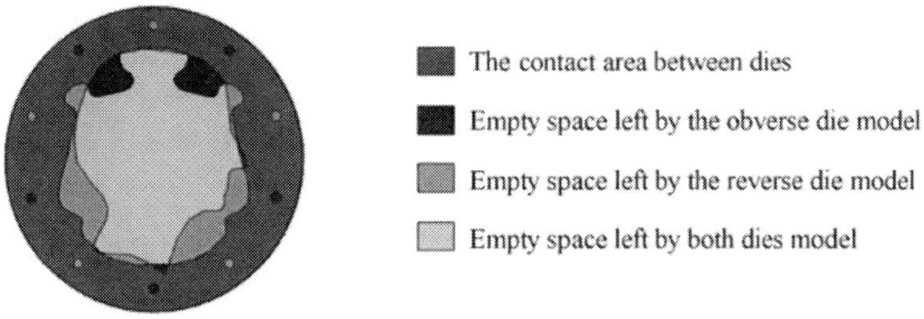

Figure 2. The clash dies contours and traces

2. The dies virtual model

As was presented, on a coin face, there are multiple complex details, as figures and also letter and number inscription. On a virtual model, these details cannot be faithfully reproduced. So, simplified models should be computed for study, using the CATIA software [8, 9].

a. b.

Figure 3. The dies 3D model

The each die simplified model consist in a cylinder with engraved model on top face. For obverse die model, presented in figure 3, a, an incused effigy contour is represented and some exterior ornaments; for reverse, presented in figure 3, b, an incused arms contour and also exterior ornaments. The considered exterior diameter for both dies is 30 mm. After the sketching of the parts, it follows the each part's virtual model computing, using the Part Design module. Using the Assembly Design module, the assembly is computed, following the corresponding constraints. As load, the dies will be pressed each other on the incused relief face. The dies model are presented in figure 3.

3. Finite element analysis, simulation and results.

IOP Conf. Series: Materials Science and Engineering **898** (2020) 012040 doi:10.1088/1757-899X/898/1/012040

For the analysis, the ANSYS software is used. The analysis objective is to determine the pressed dies ensemble behaviour under the load. For the analysis, the simplified version of the virtual model is used, as is presented in figure 4, a, b. The considered material for dies is steel; the properties as Young modulus, Poisson coefficient, Tensile Yield Strength, Ultimate Strength should be defined [10, 11].

In the contact area it is chosen a smooth mesh with the minimum edge length equal with 0.001 mm. The applied normal force is equal with 650 KN, in order to obtain high contact pressures, over the allowable stress – 1600 MPa [3, 11].

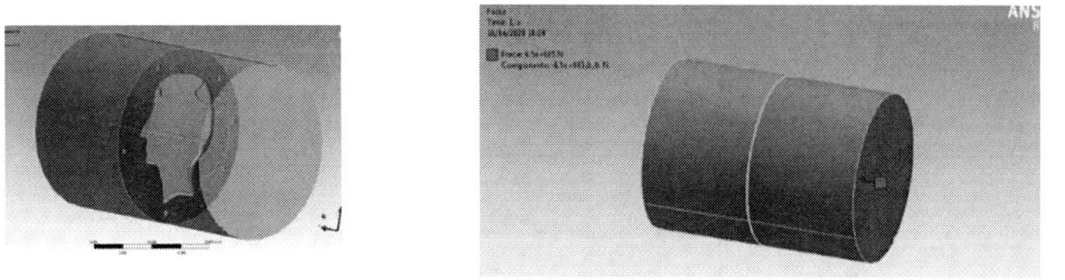

a. b.

Figure 4. The finite element model

The first studied case is when the negative dies contact is defined by entire plan face field common area. The results, presented in figure 5, a, b, and also in table 1, consist in the contact pressure maximum values and also the maximum values of the penetration in material.

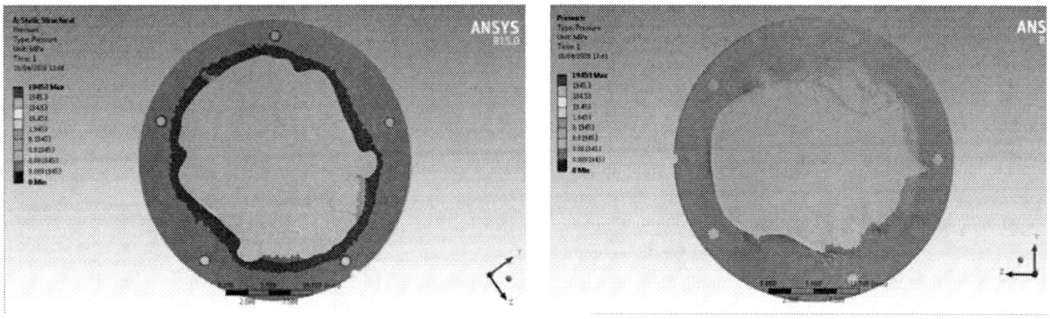

a.

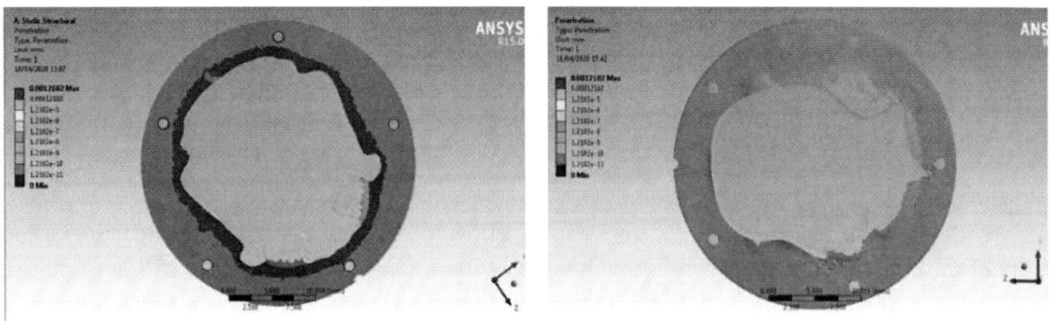

b.

Figure 5. Contact pressure and penetration in material, first dies contact case

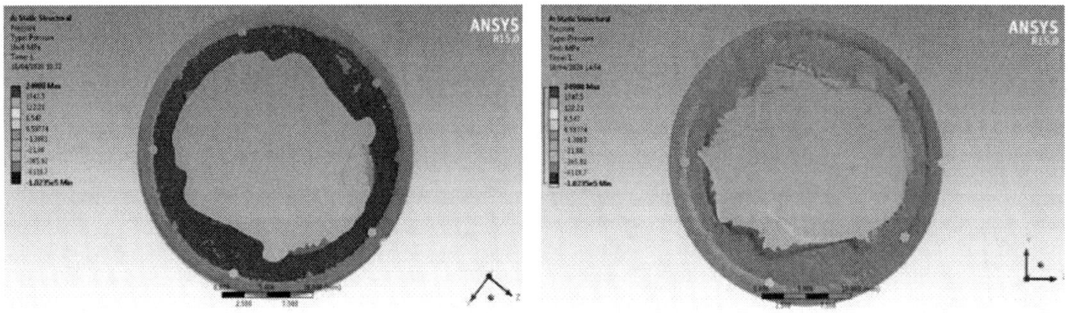

a.

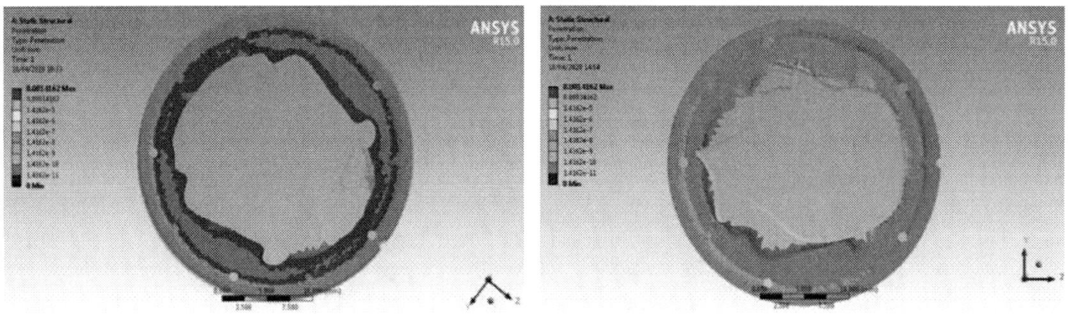

b.

Figure 6. Contact pressure and penetration in material, second dies contact case

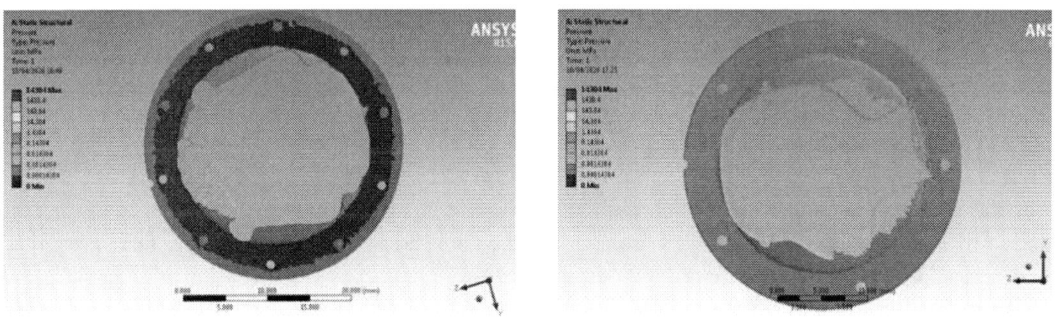

a.

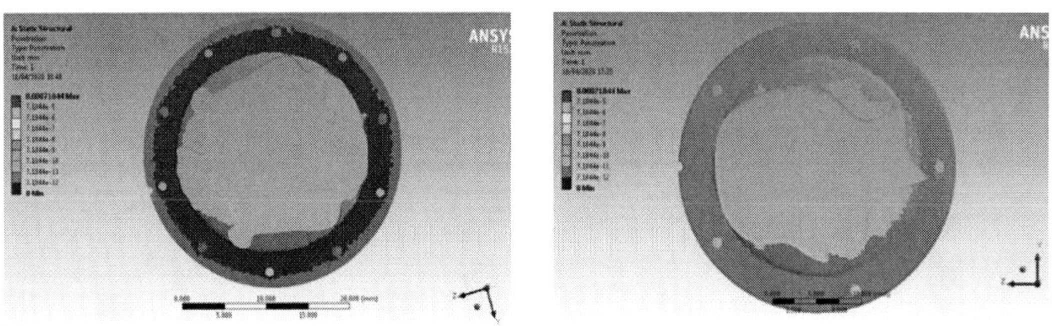

b.

Figure 7. Contact pressure and penetration in material, third dies contact case

The second studied case is when the face field is convex for both dies. The contact area is considered defined by common plan surface surrounding the relief contour. The results, the contact pressure maximum values and also the maximum values of the penetration in material are presented in figure 6 a, b, and table 1.

The third studied case is when the face field is concave for both dies. The contact area is considered defined by common plan crown circle surface, exterior bordering the die field surface. The results, the contact pressure maximum values and also the maximum values of the penetration in material are presented in figure 7, a, b, and table 1.

4. Conclusions

In the plane field case, when the dies are pressed each other, the first contact appears on the full field common area. The negative contours are impressed each other and the maximum contact pressure and penetration are on this contour edge. Due to the material rigidity, deformation and also the reverse flat field decreased area, the reverse side is more loaded than the obverse.

In the convex field case, the first contact appears on the incused model contour and surrounding plane area. The maximum contact pressure and penetration are on this contour. The each other impression begins on interior incused relief contour; the margin relief contours, close to the die exterior edge are not impressed. As in previous situation, the reverse side is more loaded than the obverse. This case has the worse condition on clash die, but for its behaviour at striking coins, in literature is considered the most desirable [4].

In the concave field case, when the first contact appears on the field exterior plan area, the maximum contact pressure and penetration are on this contour. The each other impression begins on this exterior contour, close to the die exterior edge; the interior relief contours are not impressed. Also here, as in previous situations, the reverse side is more loaded than the obverse. The maximum values of the contact pressure and penetration are lowest than to the previous cases; in literature is considered to be avoided for striking coins [4].

From the pressure and penetration variation it can be observed that, besides the traced contour on the die field, appear a deformation on the flat surface in the contact area. In all studied cases, due to the material rigidity, deformation and also to the flat field size, the effigy contact area resist better than the coat of arms contact area. This could explain the different die clash impressed traces: deep on one coin face and light on the other, in some coin cases and, respectively, both coin faces similar impressed, deep or light, on other coin cases.

If the appearance of this error is combined with some misalignments inside the pressing machine, this led to decreased die clash contact area and worse conditions. So, the obtained values presented in table 1 should be considered as relative values, to be used to compare the different studied cases.

Despite of the more or less spectacular traces impressed by dies on the coins at striking, repeated die clash could lead to the dies destruction.

Table 1. The contact pressure and the penetration in dies material, maximum values

The considered contact between dies	Contact pressure, MPa	Penetration in the material, mm
Full field plan contact area	19453	0.0012102
Convex face field, contact surrounding the relief contour	24988	0.0014162
Concave face field, contact on the exterior crown circle	14304	0.00071844

IOP Conf. Series: Materials Science and Engineering **898** (2020) 012040 doi:10.1088/1757-899X/898/1/012040

References

[1] *** 2007 Euro 4, Coins and Banknotes (Euro4, Monnaies et Billets), *Edition Les Chevau-Lègers* Paris France (in french) pp. 193 – 204.

[2] Diamond M 2007 Radically Misaligned Die Clashes in Recent Lincoln Cents - The Roster Grows *Mint Errors News **17*** Mike Byers, USA pp 41 – 45.

[3] Iliescu C and Tureac O 1987 Cold Pressing Technology (Tehnologia presării la rece) *Editura Universității din Brașov* (in romanian) pp. 266 - 267.

[4] *** 1945 Finance Ministry, National Mint, Ten Years of Activity (Ministerul Finanţelor, Monetăria naţională. Zece ani de activitate) *Monitorul oficial şi Imprimeriile statului* Bucureşti (in romanian) p. 75.

[5] Gavrilă C C 2019 Few Error Types on Romanian Modern and Contemporan Coins (Câteva tipuri de erori identificate la monedele româneşti moderne şi contemporane) *Oltenia. Studii şi comunicări. Arheologie-istorie **XXVI*** Muzeul Olteniei Craiova (in romanian) pp 174 - 184.

[6] Buzdugan G Luchian O and Oprescu C C 1977 Romanian Coins and Banknotes (Monede şi bancnote româneşti) *Editura Sport Turism* Bucureşti (in romanian) pp 256 - 277.

[7] Hilbert H 1938 Punching technique (Stanzereitechnik) Part I *Carl Hanser Verlag* Munchen (in german) p 14, 196.

[8] Gavrilă C C and Velicu R 2016 Virtual Modeling, Detail Design and FEM Analysis for a Testing Device *Current Solutions in Mechanical Engineering* Trans Tech Publications pp. 3 - 6.

[9] Ghionea I G 2009 CATIA v5. Application in Mechanical Engineering (CATIA v5. Aplicaţii în ingineria mecanică) *Editura Bren* Bucureşti, (in romanian).

[10] Lateş M T 2008 Metoda elementelor finite. Aplicaţii *Editura Universității Transilvania Brașov* (in romanian).

[11] Lee H H 2012 Finite Element Simulations with ANSYS Workbench 14. Theory, Applications, Case Studies. *Schroff Development Corporation* Kansas USA.

IOP Conf. Series: Materials Science and Engineering **898** (2020) 012041 doi:10.1088/1757-899X/898/1/012041

A modular approach to the kinematics of the vehicle axle suspension linkages

C Alexandru

Transilvania University of Braşov, Romania

E-mail: calex@unitbv.ro

Abstract. The work deals with a modular approach on the kinematics of the suspension linkages used for the rear axle of motor vehicles. The multi-body systems theory/method was used to formulate the motion equations for the kinematic analysis, by individually considering the basic binary connections (which are defined by considering the types geometric constraints between the guiding arms/links and the adjacent bodies - car body and axle, respectively) that constitutes the suspension mechanism. Subsequently, by combining these binary connections in various combinations, the kinematic analysis of any type of axle suspension linkage (at least those commonly used) can be easily performed.

1. Introduction

Relative to car body (chassis), the vehicle wheels can be guided independently - by means of a guiding mechanism for each wheel (independent suspension), or dependent - by a guiding mechanism of the rigid axle (dependent suspension). The first solution is frequently used for passenger cars (for both front and rear wheel suspensions), while the second one (which is addressed in this paper) is mainly used for the rear axles of larger gauge cars (e.g. commercial or off-road vehicles).

This paper deals with a study on the kinematics of the multi-link axle suspension linkages, which is a continuous research concern and challenge, the literature revealing various methods, more or less complex, and with a wider or narrower applicability [1-5]. A modular approach based on the multi-body systems (MBS) theory is proposed by decomposing the guiding mechanisms in the basic binary links by which is axle is guided in the relative movement to the vehicle chassis. Subsequently, by combining the binary links, the kinematics can be carried out for most types of axle guiding linkages.

2. The kinematics of the axle suspension linkages

The guiding links from the axle suspension linkages are hinged to axle and chassis by bushings (flexiblocks), which are compliant joints of rubber with 6 elastic restricted degrees of mobility. For the kinematic study, the bushings are frequently modeled by spherical joints, thus ignoring the linear deformations, which are generally insignificant [1, 3, 5-8]. In the case of the triangular guiding arms, which are double hinged to the vehicle chassis, the two corresponding spherical joints determine a rotational (revolute) joint, whose axis is defined by the centers of the spherical joints. A comprehensive systematization of the vehicle axle suspension linkages was carried out in [1].

From multiple possible variants of beam axle guiding, the guidance on circle (spherical - revolute binary connection between axle and chassis - Figure 1,a) and respectively the guidance on sphere (spherical - spherical binary connection between axle and chassis - Figure 1,b) are frequently used/implemented.

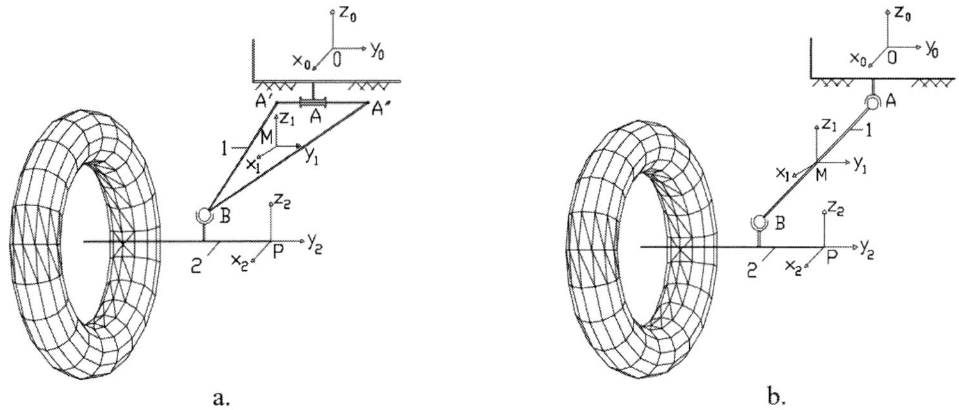

Figure 1. The basic binary links/connections for guiding the vehicle rear axle.

One of the most used and performing methods for the kinematic, static and dynamic analysis of the mechanical systems is the one called in short MBS (the acronym stands from Multi-Body Systems), which treats the mechanical system as a set of bodies, connected by geometrical constraints (joints), elastic and damping elements, on which a varied system of external and reaction forces/torques can act. The MBS method underlies several powerful commercial software solutions, such as ADAMS or DYMES, which are used in a large variety of applications [9-14]. In the MBS concept, the spatial movement of a part is defined by 6 generalized coordinates: the coordinates of the origin of the local coordinate system attached to the part, and the orientation of the axes of this frame relative to the ones of the global (inertial) coordinate system.

The kinematic model of the axle suspension linkages contains several moving bodies (the axle and the guiding links/arms). For each moving body, there is defined a local coordinate system, namely $MX_1Y_1Z_1$ for the guiding arm "1" and "$PX_2Y_2Z_2$" for the rear axle "2", in the case of the basic binary connections shown in Figure 1. In the kinematic study, the car body (chassis) is considered to be rigidly connected to ground, so it is the reference part to which the global coordinate system $OX_0Y_0Z_0$ is attached, where X_0, Y_0 and Z_0 are the longitudinal, transversal and vertical axes of the vehicle.

Each geometric restriction (joint) in the axle suspension linkage replaces a number of degrees of freedom, by adding algebraic constraint equations of the following form (the first three equations constrain translational movements, while the last three restrict rotational movements):

- $X_i - X_j = 0$ - the global coordinate X (which is the longitudinal axis in Figure 1) of the "i"-body reference frame equals the corresponding coordinate of the "j"-body reference frame,
- $Y_i - Y_j = 0$ - the global coordinate Y (the transversal axis in Figure 1) of the "i"-body reference frame equals the corresponding coordinate of the "j"-body reference frame,
- $Z_i - Z_j = 0$ - the global coordinate Z (the vertical axis in Figure 1) of the "i"-body reference frame equals the corresponding coordinate of the "j"-body reference frame,
- $Z_i \cdot X_j = 0$ - the global coordinate Z of the "i"-body reference frame remains normal (perpendicular) on the X axis of the "j"-body reference frame,
- $Z_i \cdot Y_j = 0$ - the global coordinate Z of the "i"-body reference frame remains normal (perpendicular) on the Y axis of the "j"-body reference frame,
- $X_i \cdot Y_j = 0$ - the global coordinate X of the "i"-body reference frame remains normal (perpendicular) on the Y axis of the "j"-body reference frame.

In the following, for the main types of joints (connections) from the axle suspension linkages, the geometric constraint equations will be defined. For the spherical - revolute binary connection (see Figure 1.a) the corresponding equations are the ones shown in Table 1, while Table 2 corresponds to the spherical - spherical binary connection (see Figure 1.b). In these tables, in the constraint equations area, T stands from translational and R from rotational (the movements canceled by the binary links/connections).

IOP Conf. Series: Materials Science and Engineering **898** (2020) 012041 doi:10.1088/1757-899X/898/1/012041

Table 1. The constraint equations for the spherical - revolute binary connection.

Element / part			Car body (0)	Guiding bar (1)		Axle (2)
Joint			A - A'	A - A''	B	B
Coordinates	Local	X	$X_A, X_{A'}$	$X_{A(M)}, X_{A''(M)}$	$X_{B(M)}$	$X_{B(P)}$
		Y	$Y_A, Y_{A'}$	$Y_{A(M)}, Y_{A''(M)}$	$Y_{B(M)}$	$Y_{B(P)}$
		Z	$Z_A, Z_{A'}$	$Z_{A(M)}, Z_{A''(M)}$	$Z_{B(M)}$	$Z_{B(P)}$
	Global	X	$X_A, X_{A'}$	$X_M + m_{11}\cdot X_{A(M)} + m_{12}\cdot Y_{A(M)} + m_{13}\cdot Z_{A(M)}$ $X_M + m_{11}\cdot X_{A''(M)} + m_{12}\cdot Y_{A''(M)} + m_{13}\cdot Z_{A''(M)}$	$X_M + m_{11}\cdot X_{B(M)} + m_{12}\cdot Y_{B(M)} + m_{13}\cdot Z_{B(M)}$	$X_P + w_{11}\cdot X_{B(P)} + w_{12}\cdot Y_{B(P)} + w_{13}\cdot Z_{B(P)}$
		Y	$Y_A, Y_{A'}$	$Y_M + m_{21}\cdot X_{A(M)} + m_{22}\cdot Y_{A(M)} + m_{23}\cdot Z_{A(M)}$ $Y_M + m_{21}\cdot X_{A''(M)} + m_{22}\cdot Y_{A''(M)} + m_{23}\cdot Z_{A''(M)}$	$Y_M + m_{21}\cdot X_{B(M)} + m_{22}\cdot Y_{B(M)} + m_{23}\cdot Z_{B(M)}$	$Y_P + w_{21}\cdot X_{B(P)} + w_{22}\cdot Y_{B(P)} + w_{23}\cdot Z_{B(P)}$
		Z	$Z_A, Z_{A'}$	$Z_M + m_{31}\cdot X_{A(M)} + m_{32}\cdot Y_{A(M)} + m_{33}\cdot Z_{A(M)}$ $Z_M + m_{31}\cdot X_{A''(M)} + m_{32}\cdot Y_{A''(M)} + m_{33}\cdot Z_{A''(M)}$	$Z_M + m_{31}\cdot X_{B(M)} + m_{32}\cdot Y_{B(M)} + m_{33}\cdot Z_{B(M)}$	$Z_P + w_{31}\cdot X_{B(P)} + w_{32}\cdot Y_{B(P)} + w_{33}\cdot Z_{B(P)}$
Constraint equations		T	$X_A = X_M + m_{11}\cdot X_{A(M)} + m_{12}\cdot Y_{A(M)} + m_{13}\cdot Z_{A(M)}$		$X_M + m_{11}\cdot X_{B(M)} + m_{12}\cdot Y_{B(M)} + m_{13}\cdot Z_{B(M)} = X_P + w_{11}\cdot X_{B(P)} + w_{12}\cdot Y_{B(P)} + w_{13}\cdot Z_{B(P)}$	
			$Y_A = Y_M + m_{21}\cdot X_{A(M)} + m_{22}\cdot Y_{A(M)} + m_{23}\cdot Z_{A(M)}$		$Y_M + m_{21}\cdot X_{B(M)} + m_{22}\cdot Y_{B(M)} + m_{23}\cdot Z_{B(M)} = Y_P + w_{21}\cdot X_{B(P)} + w_{22}\cdot Y_{B(P)} + w_{23}\cdot Z_{B(P)}$	
			$Z_A = Z_M + m_{31}\cdot X_{A(M)} + m_{32}\cdot Y_{A(M)} + m_{33}\cdot Z_{A(M)}$		$Z_M + m_{31}\cdot X_{B(M)} + m_{32}\cdot Y_{B(M)} + m_{33}\cdot Z_{B(M)} = Z_P + w_{31}\cdot X_{B(P)} + w_{32}\cdot Y_{B(P)} + w_{33}\cdot Z_{B(P)}$	
		R	$(Y_{A'}-Y_A)\cdot \sum_{i=1,K=X}^{3,Z} m_{3i}\cdot(K_{A''(M)}-K_{A(M)}) = (Z_{A'}-Z_A)\cdot \sum_{i=1,K=X}^{3,Z} m_{2i}\cdot(K_{A''(M)}-K_{A(M)})$ $(Y_{A'}-Y_A)\cdot \sum_{i=1,K=X}^{3,Z} m_{1i}\cdot(K_{A''(M)}-K_{A(M)}) = (X_{A'}-X_A)\cdot \sum_{i=1,K=X}^{3,Z} m_{2i}\cdot(K_{A''(M)}-K_{A(M)})$		–	
Kinematic parameters			–	X_M, Y_M, Z_M $\varphi_{1X}, \varphi_{1Y}, \varphi_{1Z}$		X_P, Y_P, Z_P $\varphi_{2X}, \varphi_{2Y}, \varphi_{2Z}$
Geometrical parameters (input data)			X_A, Y_A, Z_A $X_{A'}, Y_{A'}, Z_{A'}$ $\varphi_{0X}, \varphi_{0Y}, \varphi_{0Z}$	$X_{A(M)}, Y_{A(M)}, Z_{A(M)}; X_{A''(M)}, Y_{A''(M)}, Z_{A''(M)}$ $X_{B(M)}, Y_{B(M)}, Z_{B(M)}$		$X_{B(P)}, Y_{B(P)}, Z_{B(P)}$

Table 2. The constraint equations for the spherical - spherical binary connection.

Element / part			Car body (0)	Guiding bar (1)		Axle (2)
Joint			A	A	B	B
Coordinates	Local	X	X_A	$X_{A(M)}$	$X_{B(M)}$	$X_{B(P)}$
		Y	Y_A	$Y_{A(M)}$	$Y_{B(M)}$	$Y_{B(P)}$
		Z	Z_A	$Z_{A(M)}$	$Z_{B(M)}$	$Z_{B(P)}$
	Global	X	X_A	$X_M + m_{11} \cdot X_{A(M)} + m_{12} \cdot Y_{A(M)} + m_{13} \cdot Z_{A(M)}$	$X_M + m_{11} \cdot X_{B(M)} + m_{12} \cdot Y_{B(M)} + m_{13} \cdot Z_{B(M)}$	$X_P + w_{11} \cdot X_{B(P)} + w_{12} \cdot Y_{B(P)} + w_{13} \cdot Z_{B(P)}$
		Y	Y_A	$Y_M + m_{21} \cdot X_{A(M)} + m_{22} \cdot Y_{A(M)} + m_{23} \cdot Z_{A(M)}$	$Y_M + m_{21} \cdot X_{B(M)} + m_{22} \cdot Y_{B(M)} + m_{23} \cdot Z_{B(M)}$	$Y_P + w_{21} \cdot X_{B(P)} + w_{22} \cdot Y_{B(P)} + w_{23} \cdot Z_{B(P)}$
		Z	Z_A	$Z_M + m_{31} \cdot X_{A(M)} + m_{32} \cdot Y_{A(M)} + m_{33} \cdot Z_{A(M)}$	$Z_M + m_{31} \cdot X_{B(M)} + m_{32} \cdot Y_{B(M)} + m_{33} \cdot Z_{B(M)}$	$Z_P + w_{31} \cdot X_{B(P)} + w_{32} \cdot Y_{B(P)} + w_{33} \cdot Z_{B(P)}$
Constraint equations		T	$X_A = X_M + m_{11} \cdot X_{A(M)} + m_{12} \cdot Y_{A(M)} + m_{13} \cdot Z_{A(M)}$		$X_M + m_{11} \cdot X_{B(M)} + m_{12} \cdot Y_{B(M)} + m_{13} \cdot Z_{B(M)} = X_P + w_{11} \cdot X_{B(P)} + w_{12} \cdot Y_{B(P)} + w_{13} \cdot Z_{B(P)}$	
			$Y_A = Y_M + m_{21} \cdot X_{A(M)} + m_{22} \cdot Y_{A(M)} + m_{23} \cdot Z_{A(M)}$		$Y_M + m_{21} \cdot X_{B(M)} + m_{22} \cdot Y_{B(M)} + m_{23} \cdot Z_{B(M)} = Y_P + w_{21} \cdot X_{B(P)} + w_{22} \cdot Y_{B(P)} + w_{23} \cdot Z_{B(P)}$	
			$Z_A = Z_M + m_{31} \cdot X_{A(M)} + m_{32} \cdot Y_{A(M)} + m_{33} \cdot Z_{A(M)}$		$Z_M + m_{31} \cdot X_{B(M)} + m_{32} \cdot Y_{B(M)} + m_{33} \cdot Z_{B(M)} = Z_P + w_{31} \cdot X_{B(P)} + w_{32} \cdot Y_{B(P)} + w_{33} \cdot Z_{B(P)}$	
		R	–		–	
Kinematic parameters			–	X_M, Y_M, Z_M $\varphi_{1X}, \varphi_{1Y}, \varphi_{1Z}$		X_P, Y_P, Z_P $\varphi_{2X}, \varphi_{2Y}, \varphi_{2Z}$
Geometrical parameters (input data)			X_A, Y_A, Z_A $\varphi_{0X}, \varphi_{0Y}, \varphi_{0Z}$	$X_{A(M)}, Y_{A(M)}, Z_{A(M)}$ $X_{B(M)}, Y_{B(M)}, Z_{B(M)}$		$X_{B(P)}, Y_{B(P)}, Z_{B(P)}$

The connection matrix M_{10} between the local ($X_1Y_1Z_1$) and global (XYZ) coordinate systems has the following form:

$$M_{10} = \begin{bmatrix} m_{11} & m_{12} & m_{13} \\ m_{21} & m_{22} & m_{23} \\ m_{31} & m_{32} & m_{33} \end{bmatrix} = \begin{bmatrix} \bar{i} \cdot \bar{i}_1 & \bar{i} \cdot \bar{j}_1 & \bar{i} \cdot \bar{k}_1 \\ \bar{j} \cdot \bar{i}_1 & \bar{j} \cdot \bar{j}_1 & \bar{j} \cdot \bar{k}_1 \\ \bar{k} \cdot \bar{i}_1 & \bar{k} \cdot \bar{j}_1 & \bar{k} \cdot \bar{k}_1 \end{bmatrix} = \begin{bmatrix} \cos(X, X_1) & \cos(X, Y_1) & \cos(X, Z_1) \\ \cos(Y, X_1) & \cos(Y, Y_1) & \cos(Y, Z_1) \\ \cos(Z, X_1) & \cos(Z, Y_1) & \cos(Z, Z_1) \end{bmatrix}. \quad (1)$$

3. Results and conclusions

By linking in parallel the above presented guidance cases, a large variety of axle suspension linkages can be obtained, as they are systematized in [1], in which the geometric constraints equations have the form of those shown in Tables 1 and 2. For example, by interposing between axle and chassis two spherical - spherical connections and one spherical - revolute connection, the so-called 2S1C (where S means guidance on sphere , while C - guidance on circle) suspension mechanism is obtained, which is shown in Figure 2 (as structural model and semi-constructive solution).

This mechanism has two mobilities, which correspond to the left and right wheels vertical movements, which can be transposed into the vertical displacement of the axle center (Z_P) and the axle roll rotation around the longitudinal axis (φ_{4X}).

IOP Conf. Series: Materials Science and Engineering **898** (2020) 012041 doi:10.1088/1757-899X/898/1/012041

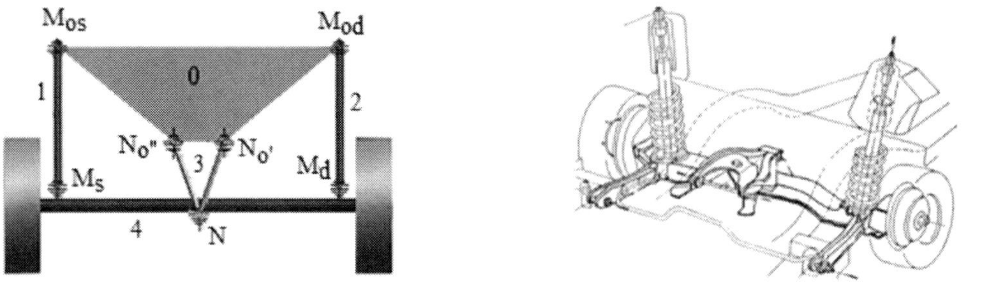

a. b.

Figure 2. Structural model (a) and semi-constructive solution (b) for the 2S1C axle guiding linkage.

By transposing the general forms of the constraint equations from Tables 1 and 2 for this particular type of axle suspension linkage, the following equations are obtained:

- for the joints of the guiding bars "1-3" to axle "4":

$$F_{14}=\begin{bmatrix}X_P\\Y_P\\Y_P\end{bmatrix}+M_{40}\cdot\begin{bmatrix}X_{Ms}\\Y_{Ms}\\Z_{Ms}\end{bmatrix}_4-\begin{bmatrix}X_{Mos}\\Y_{Mos}\\Z_{Mos}\end{bmatrix}-M_{10}\cdot\begin{bmatrix}X_{Ms}\\Y_{Ms}\\Z_{Ms}\end{bmatrix}_1=0,F_{24}=\begin{bmatrix}X_P\\Y_P\\Y_P\end{bmatrix}+M_{40}\cdot\begin{bmatrix}X_{Md}\\Y_{Md}\\Z_{Ms}\end{bmatrix}_4-\begin{bmatrix}X_{Mod}\\Y_{Mod}\\Z_{Mod}\end{bmatrix}-M_{20}\cdot\begin{bmatrix}X_{Md}\\Y_{Md}\\Z_{Md}\end{bmatrix}_2=0,$$
$$F_{34}=\begin{bmatrix}X_P\\Y_P\\Y_P\end{bmatrix}+M_{40}\cdot\begin{bmatrix}X_N\\Y_N\\Z_N\end{bmatrix}_4-\begin{bmatrix}X_{No}\\Y_{No}\\Z_{No}\end{bmatrix}-M_{30}\cdot\begin{bmatrix}X_N\\Y_N\\Z_N\end{bmatrix}_3=0; \quad (2)$$

- for the joints of the guiding bars "1-3" to car body "0":

$$F_{10}=\begin{bmatrix}X_{Mos}\\Y_{Mos}\\Z_{Mos}\end{bmatrix}-\begin{bmatrix}X_O\\Y_O\\Z_O\end{bmatrix}-\begin{bmatrix}a_1\\a_2\\a_3\end{bmatrix}=0,F_{20}=\begin{bmatrix}X_{Mod}\\Y_{Mod}\\Z_{Mod}\end{bmatrix}-\begin{bmatrix}X_O\\Y_O\\Z_O\end{bmatrix}-\begin{bmatrix}b_1\\b_2\\b_3\end{bmatrix}=0,F_{30}=\begin{bmatrix}X_{No}\\Y_{No}\\Z_{No}\\\delta_{3x}\\\delta_{3z}\end{bmatrix}-\begin{bmatrix}X_O\\Y_O\\Z_O\\\delta_{ox}\\\delta_{oz}\end{bmatrix}-\begin{bmatrix}c_1\\c_2\\c_3\\c_4\\c_5\end{bmatrix}=0. \quad (3)$$

The matrices M_{i0} (i=1…4) have similar forms with the one in equation (1). The geometric constants a_i, b_i, c_j (i=1…3, j=1…5) can be obtained from the initial configuration of the suspension system. The equations (2) and (3) define a system of 20 scalar equations, with 24 generalized coordinates (6 per moving body). As mentioned, the independent parameters are Z_P and φ_{4x}, while the rotations of the guiding bars "1" (φ_{1x}) and "2" (φ_{2x}) around their own axes are kinematically passive. Therefore, 20 unknown kinematic parameters will be established by processing the system of equations (2) and (3), thus determining the kinematic behavior of the axle suspension linkage, as it is defined in [1].

The global coordinates (X, Y, Z) of the joints in the initial position of the guiding mechanism/vehicle are input parameters for the kinematic study, as follows (in correlation with the notations in Figure 2): $M_{0s/d}$ (2014.5, ±536.0, 40.0), $M_{s/d}$ (2523.5, ±536.0, 40.0), $N_{0'/''}$ (2362.0, ±100.0, 168.0), N (2629.0, 0.0, 168.0). In the same position, the axle center has the global coordinates P (2596.0, 0, 111.0).

To exemplify, it was considered the case where the wheels move vertically from the initial position with the same rate, but in opposite directions (up/down to the altitude of ±80mm), and then in reverse, until the wheels perform a full up-down / down-up race, with a return to the starting (initial) position.

These vertical displacements of the wheels are equivalent with the following variations of the two independent kinematic parameters: Z_P = 111.0 mm (it remains constant during the analysis, so $\Delta Z_P = 0$), and $\varphi_{4X} \in [-6.676°, 6.676°]$. From the results of the analysis thus carried out, Figure 3 shows the variation diagrams for the displacements of the axle centre along the longitudinal (ΔX_P) and transversal (ΔY_P) axes, as well as the twisting angle of the axle (φ_{4Y}) and the pivoting angle around the vertical axis (φ_{4Z}), depending on the roll angle of the axle (φ_{4X}).

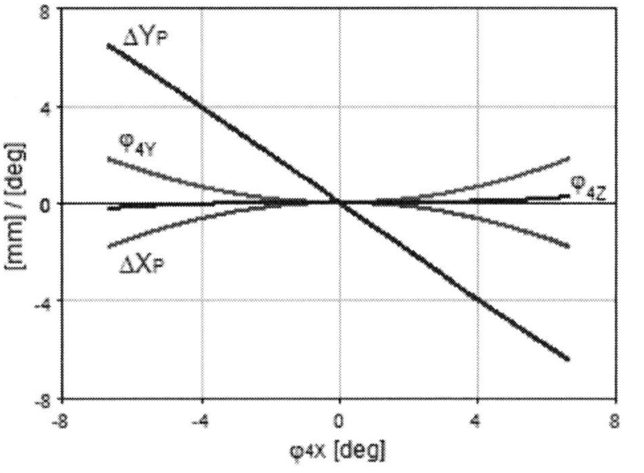

Figure 3. Kinematic analysis results.

The kinematic analysis method can be implemented for most types of axle guiding linkages, and it avoids the disadvantages of the numerical methods based on nonlinear equation systems. At the same time, the method is easy to implement for the independent wheel suspension mechanisms.

References

[1] Alexandru C 2009 The kinematic optimization of the multi-link suspension mechanism used for rear axle of the motor vehicle *Proceedings of the Romanian Academy - A* **10**(3) pp 244-253
[2] Attia HA 2003 Kinematic analysis of the multi-link five-point suspension system in point coordinates *Journal of Mechanical Science and Technology* **17**(8) pp 1133-1139
[3] Balike KP, Rakheja S and Stiharu I 2008 Kinematic analysis and parameter sensitivity to hard points of five-link rear suspension mechanism of passenger car *Proceedings of the Design Engineering Technical Conference* pp 755-764
[4] Hiller M and Woernle C 1985 Kinematical analysis of a five point wheel suspension *ATZ* **87** pp 59-64
[5] Simionescu PA and Beale D 2002 Synthesis and analysis of the five-link rear suspension system used in automobile *Mechanism and Machine Theory* **37**(9) pp 815-832
[6] Knapczyk J and Maniowski M 2002 Selected effects of bushings characteristics on five-link suspension elastokinematics *Mobility and Vehicle Mechanics* **3**(2) pp 107-121
[7] Knapczyk J and Maniowski M 2006 Elastokinematic modeling and study of five-rod suspension with subframe *Mechanism and Machine Theory* **41**(9) pp 1031-1047
[8] Tică M, Dobre G and Mateescu V 2014 Influence of compliance for an elastokinematic model of a proposed rear suspension *International Journal of Automotive Technology* **15**(6) pp 885-891
[9] Alexandru C and Comşiţ M 2007 Virtual prototyping of the solar tracking systems *Renewable Energy and Power Quality Journal* **1**(5) pp 105-110
[10] Alexandru C and Pozna C 2008 Virtual prototype of a dual-axis tracking system used for photovoltaic panels *Proceedings of the IEEE International Symposium on Industrial Electronics - ISIE* pp 1598-1603
[11] Alexandru P, Macaveiu D and Alexandru C 2012 Design and simulation of a steering gearbox with variable transmission ratio *Proceedings of the Institution of Mechanical Engineers, Part C: Journal of Mechanical Engineering Science* **226** pp 2538-2548
[12] Geonea ID, Alexandru C, Margine A and Ungureanu A 2013 Design and simulation of a single DOF human-like leg mechanism *Applied Mechanics and Materials* **332** pp 491-496
[13] Ioniţă M and Alexandru C 2012 Dynamic optimization of the tracking system for a pseudo-azimuthal photovoltaic platform *Journal of Renewable and Sustainable Energy* **4** pp 053117
[14] Tarniţă D, Catană M and Tarniţă DN 2016 Design and simulation of an orthotic device for patients with osteoarthritis *Mechanisms and Machine Science* **38** pp 61-77

IOP Conf. Series: Materials Science and Engineering **898** (2020) 012042 doi:10.1088/1757-899X/898/1/012042

Studies related the measurement errors determination in case of inclined surfaces measurement on Coordinate Measurement Machines

P E Serban[1], **F Peti**[2]

[1] Metrology Research, Quality Department, CIEMatricon SA, GhDoja street, no.155, TarguMures, Romania
[2]Industrial Engineering and Management Department, Faculty of Engineering and Information Technology, University of Medicine, Pharmacy, Sciences and Technology "George Emil Palade"of TarguMures, N. Iorga street, no. 1, TarguMures, Romania

serban.petru@yahoo.com

Abstract. Coordinate measurement machines (CMM) are an important part of production process technology that help and give a fast response in complex measurement situations. Although the technologies advance in high speed, old coordinate measurement machines are capable to give accurate results. The present paper will give an alternative to use this type of machines, machines that are operated manual or with rudimental software that are not upgraded with actual technologies. In order to be possible to measure an inclined surface with a coordinate measurement machine we need to consider that the contact approach between the coordinate measurement machine tool and the measured part must respect some elementary steps: the movement of the coordinate measurement machine probe must be on same direction as the surface vector that we want to measure, the contact force must be the same as the force used for calibration of the probe. All this are very difficult to be respected with a manual or oldcoordinate measurement machine. During measurement with machines that doesn't allow probe angle control an error can appear for surfaces that are positioned at an angle to the reference surface. This article presents one way to can eliminate measurement errors on Coordinate measurementmachines for inclined surfaces.

1. Introduction

The appeared error for surfaces that are positioned at an angle to the reference surface is due to the fact that the stylus enters in contact earlier with the work piece. It is supposed that calculated dimension will take into account Point 1 from the surface of the stylus, but in fact the stylus is coming in contact with the work piece in Point 2, phenomenon graphically illustrated on Figure 1.
This difference will generate a measurement error, and the measured value can be lower or higher than the real distance, in function of inner or outer surface is measured.
This error was observed by measuring control points on complex surfaces of the electrodes used for spark erosion in the manufacturing of the cavities of the high pressure diecasting tools.

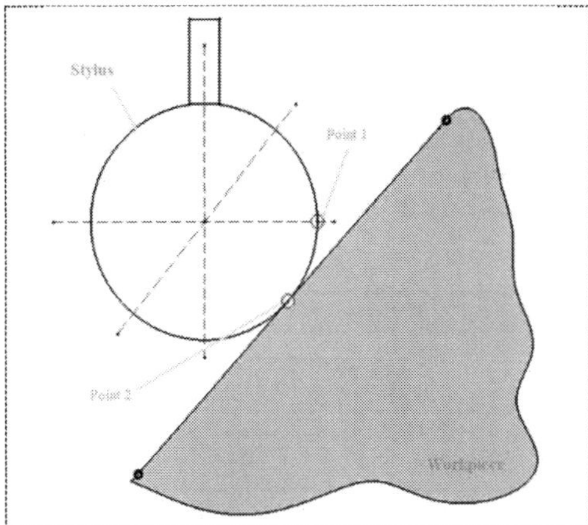

Figure 1. Theoretical and real contact points between the stylus the work piece

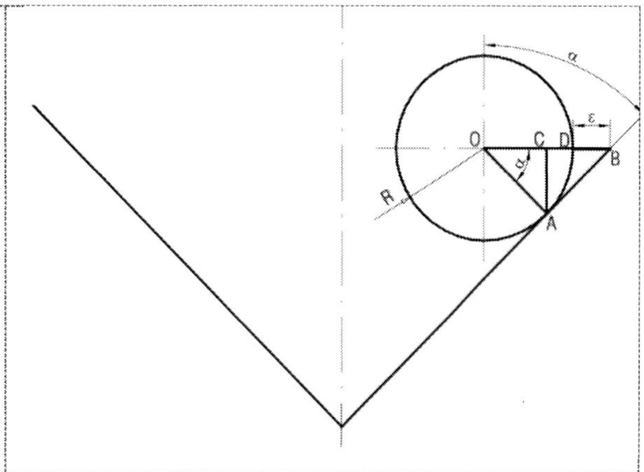

Figure 2. Schematical representation of the stylus in contact with the work piece

2. Theoretical consideration

In the Figure 2 are represented the main parameters that have influence on the error determination: ε – measurement error.

These parameters are:

R – Radius of the stylus used during the measurement
α – draft angle of the surface on which the asked length is measured

3. Methodology and material .Working procedure

For the determination of the measurement error ε was used the following algorithm:

$$\Delta OAB\colon \cos \propto = \frac{OA}{OB} = \frac{OA}{OD + DB} = \frac{R}{R + \varepsilon} \Rightarrow \quad (1)$$

$$\cos \propto \cdot (R + \varepsilon) = R \Rightarrow \quad (2)$$

$$\varepsilon = \frac{R - R\cos\alpha}{\cos\alpha} = \frac{R \cdot (1 - \cos\alpha)}{\cos\alpha} \quad (3)$$

The relation has been verified by measuring a prism with $2 \cdot \alpha = 90^{\circ}$, respectively a prism with $2 \cdot \alpha = 130^{\circ}$ on a Zeiss Vista CMM machine.
In the tables 1 and 2 are summarized the values of the errors for exact draft angles in case of stylus with a sphere diameter of 1 mm, respectively 2 mm.

IOP Conf. Series: Materials Science and Engineering **898** (2020) 012042 doi:10.1088/1757-899X/898/1/012042

Table 1. Measurement errors in function of draft angle with stylus diameter of 1 mm

D	R	α	ε	D	R	α	ε
1	0.5	1	0.0001	1	0.5	46	0.2198
1	0.5	2	0.0003	1	0.5	47	0.2331
1	0.5	3	0.0007	1	0.5	48	0.2472
1	0.5	4	0.0012	1	0.5	49	0.2621
1	0.5	5	0.0019	1	0.5	50	0.2779
1	0.5	6	0.0028	1	0.5	51	0.2945
1	0.5	7	0.0038	1	0.5	52	0.3121
1	0.5	8	0.0049	1	0.5	53	0.3308
1	0.5	9	0.0062	1	0.5	54	0.3507
1	0.5	10	0.0077	1	0.5	55	0.3717
1	0.5	11	0.0094	1	0.5	56	0.3941
1	0.5	12	0.0112	1	0.5	57	0.4180
1	0.5	13	0.0132	1	0.5	58	0.4435
1	0.5	14	0.0153	1	0.5	59	0.4708
1	0.5	15	0.0176	1	0.5	60	0.5000
1	0.5	16	0.0201	1	0.5	61	0.5313
1	0.5	17	0.0228	1	0.5	62	0.5650
1	0.5	18	0.0257	1	0.5	63	0.6013
1	0.5	19	0.0288	1	0.5	64	0.6406
1	0.5	20	0.0321	1	0.5	65	0.6831
1	0.5	21	0.0356	1	0.5	66	0.7293
1	0.5	22	0.0393	1	0.5	67	0.7797
1	0.5	23	0.0432	1	0.5	68	0.8347
1	0.5	24	0.0473	1	0.5	69	0.8952
1	0.5	25	0.0517	1	0.5	70	0.9619
1	0.5	26	0.0563	1	0.5	71	1.0358
1	0.5	27	0.0612	1	0.5	72	1.1180
1	0.5	28	0.0663	1	0.5	73	1.2102
1	0.5	29	0.0717	1	0.5	74	1.3140
1	0.5	30	0.0774	1	0.5	75	1.4319
1	0.5	31	0.0833	1	0.5	76	1.5668
1	0.5	32	0.0896	1	0.5	77	1.7227
1	0.5	33	0.0962	1	0.5	78	1.9049
1	0.5	34	0.1031	1	0.5	79	2.1204
1	0.5	35	0.1104	1	0.5	80	2.3794
1	0.5	36	0.1180	1	0.5	81	2.6962
1	0.5	37	0.1261	1	0.5	82	3.0926
1	0.5	38	0.1345	1	0.5	83	3.6028
1	0.5	39	0.1434	1	0.5	84	4.2834
1	0.5	40	0.1527	1	0.5	85	5.2369
1	0.5	41	0.1625	1	0.5	86	6.6678
1	0.5	42	0.1728	1	0.5	87	9.0537
1	0.5	43	0.1837	1	0.5	88	13.8269
1	0.5	44	0.1951	1	0.5	89	28.1493
1	0.5	45	0.2071				

Table 2. Measurement errors in function of draft angle with stylus diameter of 2 mm

D	R	α	ε	D	R	α	ε
2	1	1	0.0002	2	1	46	0.4396
2	1	2	0.0006	2	1	47	0.4663
2	1	3	0.0014	2	1	48	0.4945
2	1	4	0.0024	2	1	49	0.5243
2	1	5	0.0038	2	1	50	0.5557
2	1	6	0.0055	2	1	51	0.5890
2	1	7	0.0075	2	1	52	0.6243
2	1	8	0.0098	2	1	53	0.6616
2	1	9	0.0125	2	1	54	0.7013
2	1	10	0.0154	2	1	55	0.7434
2	1	11	0.0187	2	1	56	0.7883
2	1	12	0.0223	2	1	57	0.8361
2	1	13	0.0263	2	1	58	0.8871
2	1	14	0.0306	2	1	59	0.9416
2	1	15	0.0353	2	1	60	1.0000
2	1	16	0.0403	2	1	61	1.0627
2	1	17	0.0457	2	1	62	1.1301
2	1	18	0.0515	2	1	63	1.2027
2	1	19	0.0576	2	1	64	1.2812
2	1	20	0.0642	2	1	65	1.3662
2	1	21	0.0711	2	1	66	1.4586
2	1	22	0.0785	2	1	67	1.5593
2	1	23	0.0864	2	1	68	1.6695
2	1	24	0.0946	2	1	69	1.7904
2	1	25	0.1034	2	1	70	1.9238
2	1	26	0.1126	2	1	71	2.0716
2	1	27	0.1223	2	1	72	2.2361
2	1	28	0.1326	2	1	73	2.4203
2	1	29	0.1434	2	1	74	2.6280
2	1	30	0.1547	2	1	75	2.8637
2	1	31	0.1666	2	1	76	3.1336
2	1	32	0.1792	2	1	77	3.4454
2	1	33	0.1924	2	1	78	3.8097
2	1	34	0.2062	2	1	79	4.2408
2	1	35	0.2208	2	1	80	4.7588
2	1	36	0.2361	2	1	81	5.3925
2	1	37	0.2521	2	1	82	6.1853
2	1	38	0.2690	2	1	83	7.2055
2	1	39	0.2868	2	1	84	8.5668
2	1	40	0.3054	2	1	85	10.4737
2	1	41	0.3250	2	1	86	13.3356
2	1	42	0.3456	2	1	87	18.1073
2	1	43	0.3673	2	1	88	27.6537
2	1	44	0.3902	2	1	89	56.2987
2	1	45	0.4142				

4. Conclusions

It is recommended that during the measurements of inclined surfaces on CMM machines that doesn't allow probe angle controlthe value of this determinate error to be taken into account, respectively the calculated correction to be applied. Under the correction appliance is understood the adding or subtracting of the value of the correction.

Under correction is understood the value which has to be added or subtracted to the/ from the measured value.

This value is determinate as follows:

- For internal conical or inclined surfaces when the inner diameter or dimension is determinate, to the measured value will be added 2 ε
- For external conical or inclined surfaces when the external diameter is determinate ,from the measured value will be subtracted 2 ε
- For the internal inclined surfaces when the distance of a point or axis is determinate to a reference, to the measured value is added ε;
- For the external inclined surfaces when the distance of a point or axis is determinate to a reference, from the measured value is subtracted ε;

References

[1] David Flack - *Measurement Good Practice Guide No. 42 CMM Verification*, Engineering Measurement Division National Physical Laboratory,

[2] Fernando A. M. Ferreiraa,*, Jesus de Vicente y Olivab, Angel M. Sanchez Perezc, *Evaluation of the performance of coordinate measuring machines in the industry, using calibrated artefacts*, Procedia Engineering 63 (2013) p 659 – 668

[3] I. Puertasa,*, C.J. Luis Péreza, D. Salcedoa, J. Leóna, R. Luria, J.P. Fuertesa - *Precision study of a coordinate measuring machine using several contact probes*, Procedia Engineering 63 (2013) p547 – 555

[4] ISO 10360-2:2009 *Geometrical product specifications (GPS) — Acceptance and reverification tests for coordinate measuring machines (CMM) — Part* 2: CMMs used for measuring linear dimensions

[5] ISO 5459: *Geometrical product specifications – Geometrical tolerancing – Datums and datum systems*

[6] www.hexagon.com

[7] www.mitutoyo.com

[8] www.renishaw.com

[9] www.zeiss.com

Acknowledgments

This work was partially supported by the UMFST "GE Palade" Quality Engineering and Digital Manufacturing Research Center.

IOP Conf. Series: Materials Science and Engineering **898** (2020) 012043 doi:10.1088/1757-899X/898/1/012043

Study of modern perforation processes specific to the automotive industry

D I Poiana [1], I Ionel [2]

[1] Politehnica University Timişoara, Piata Victoriei nr. 2, 300006, Timis, Romania

dinupoiana@yahoo.com

Abstract. The purpose of this paper is to critically present the trend in recent years in terms to use of perforation processes for perforation parts of the cars, operations that have real advantages, including reducing the environmental pollution in general. A study and a critical and comparative analysis are performed, referring to three perforation processes: hydroperforation, laser perforation and thermal erosion perforation. Hydroperforation is a process that is applied in a liquid environment and usually at the end of plastic deformation operations of some tubes by hydroforming in installations specially designed for this purpose. The paper presents some defining characteristics of these processes used in the industry being exemplified for the automotive industry. In general, all three perforation processes are performed at the end of the part formation by plastic deformation by hydroforming.

1. Introduction.

The current global trend regarding the accelerated use of efficient, economic and ecological processing processes imposes the need to study in-depth, in the future, the processing processes of current technical and economic global interest, as well as the need to approach theoretical research in the field. Last but not least, the current requirements regarding the reduction of specific energy consumption per production unit must be implemented. Even if it is relatively small, considering a large number of subansamble / products made by this technology, it will be possible, as a whole, to achieve a satisfactory overall result.

At present, depending on the chosen technology, different perforation/drilling processes are used in the case of tubular parts, namely:

1) in the air,
   - with associated edges, by plastic deformation
   - by using a single blade, that is without associated cutting edges, typically a punch, by plastic deformation
   - with laser
   - by thermal erosion
2) in a liquid medium
   - by hydroperforation
   - by impulse erosion

In recent years, automotive manufacturing and aeronautics have been the main drivers of the ever-increasing application of various processing processes to meet certain production requirements.

The automotive industry has developed, since the 1960s, the deformation of pipes by hydroforming and at the same time various perforation processes to make holes in various tubular parts of the automobile.

These perforation processes propelled and developed by the automotive industry, are:

1) hydroperforation
2) laser perforation
3) perforation by thermal erosion

2. Specific types of tubular parts, with perforated holes, chassis and car body components

In the case of parts of the long pipe type, a series of technological problems appear, such as:

- the impossibility, in most of these cases, to use two associated edges, a punch and a cutting plate, because a cutting plate cannot be inserted inside the tubular part;
- occurrence of deformations of the part, in the perforated area, if no cutting plate is used, due to the action of the punch on the tubular wall of the part;
- in most cases, the need to use the special devices, fasteners and tighteners in the holes area, for executing the operation of punching through plastic deformation technology.

The tubular parts are presented in a wide range of constructive forms, having surfaces of revolution / curves or even in the form of pipes, of small, medium or large dimensions. In recent decades, the need for such tubular parts has increased due to the growing demand in the automotive market and in the aerospace industry.

Some characteristic types of such long tubular parts used in car construction are shown in the figures 1- 6 [1].

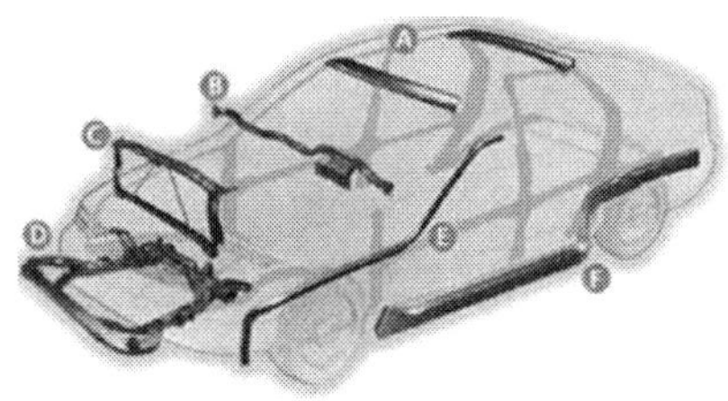

Figure 1. Tubular parts positioned on the car body [1].
A- roof headers; B- instrument panel supports; C- radiator supports;
D- engine cradles; E- roof rails; F- frame rails.

Figure 1 [1] shows perforated tubular parts, as they are positioned to the car. There is a wide range of parts required in most car installations and in its safety structure: rood headers, instrument panel supports, radiator supports, engine cradles, rood rails, frame rails. These parts have certain holes, having a functional or constructive role.

The characteristics of these tubular parts are closely linked to the functional role they must play in the product to which they belong.

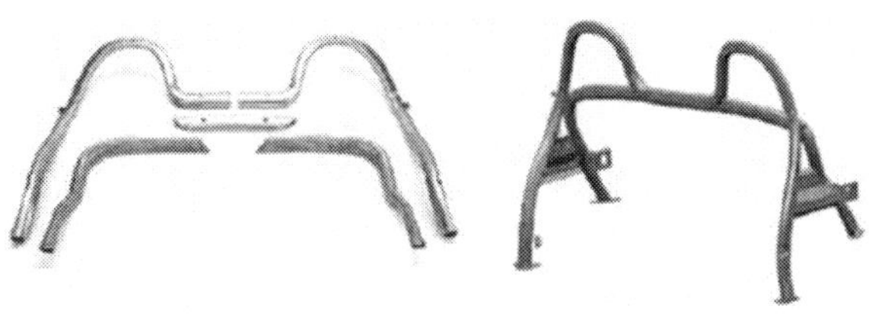

a *b*

Figure 2. Rollover bar protection system in Porsche Boxter [1].

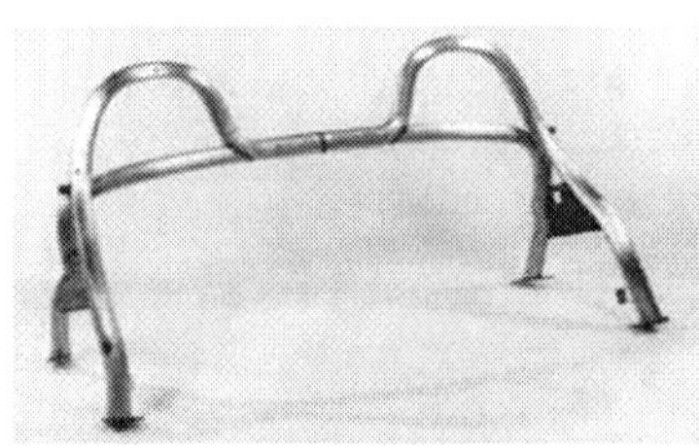

Figure 3. Rollover bar protection [1].

IOP Conf. Series: Materials Science and Engineering **898** (2020) 012043 doi:10.1088/1757-899X/898/1/012043

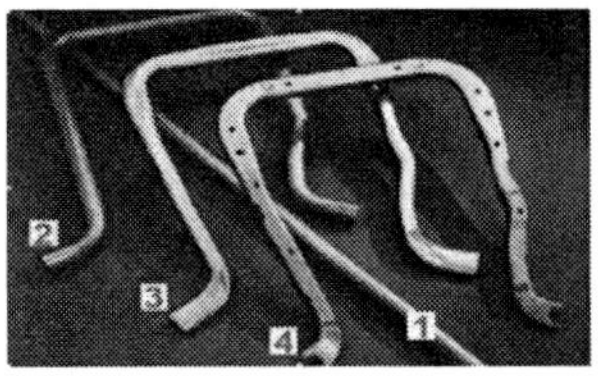

Figure 4. Hydroformed engine cradle: (1)the initial straight tubular blank, (2)bent, (3)pre-formed, (4) hydroformed and pierced (Courtesy of Schuler Inc) [1].

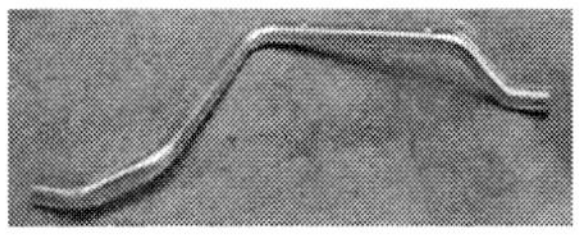

Figure 5. Safety frame **[1]**.

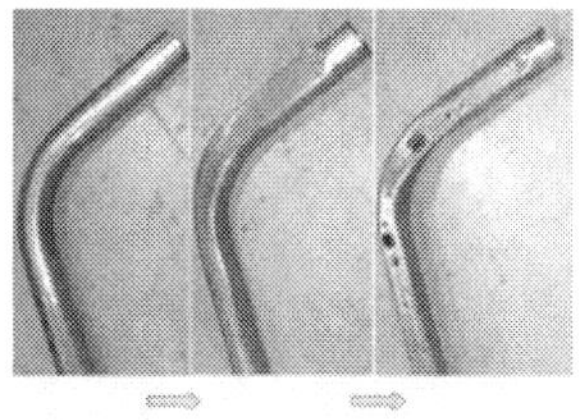

Figure 6. Steps of formation and perforation [1].

Some of these parts have a role in the resistance structure of the product, others have a role of protection against external agents, others have the role of sealing or the role of transport and/or control of fluids in certain installations, or the role of selection/filtering/sorting in the composition of some installations.

3. The hydroperforation process.

3.1. Defining the process

Hydroperforation_is a processing process that uses a special type die, which has a single edge to perform the processing, using a hydraulic pressure.

There are two variants of hydroperforation:

a) The first option of hydroperforation is to exert a punching force from the outside of the pipe, by rounding the punch in the piece, the punch penetrates the piece and loosens the portion of material to the inside of the pipe, as shown in figure 7 [4]. Because the fluid pressure in the inside of the deformable piece wall, in the area around the hole, the deformation is minimal.În figure 7 is denoted by D_p, the diameter of the punch, with D_d, the diameter of zone and a slight deformation of the piece, around the hole made, and with X was noted the distance of deformation of the piece wall, around the hole made.

b) The second option of hydroperforation is with the exercise of a hydraulic perforating force inside the pipe, by superficially pressing the punch on the piece and then withdrawing it, to make the cut the portion of the pipe wall. It is represented schematically in figure 8 [4], it consists in the perforation due to the pressure of the liquid inside the part, which determines the cutting of the wall of the part near the hole in which the plunger punch acts, meaning when the controlled retraction of the punch, the perforation is perfomed.After perforation, by the reverse controlled movement of the punch in the inside of the tubular part, the cut piece from the wall of the piece is removed from the bore of the punch .În this second option, the piece wall not appear deformed in the hole area, as in the case of the first option.

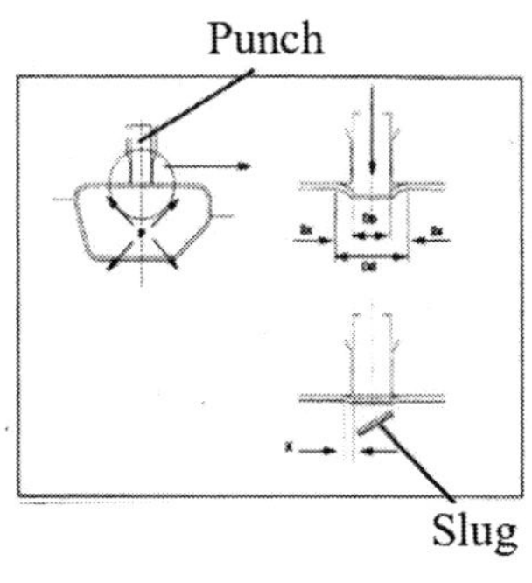

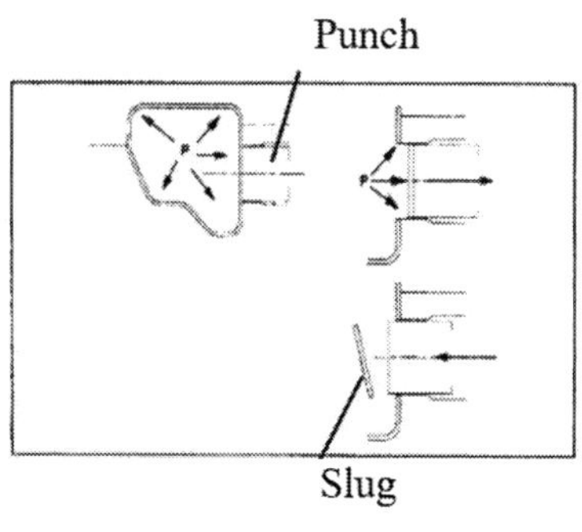

Figure 7. The punch penetrates outside the piece and detach the slug of the hole [4].

Figure 8. The punch partially penetrates the piece and detachment is done by hydraulics pressure inside the piece [4].

Special installations are used for hydroperforation, they are composed of certain elements that allow drilling operations for various long tubular parts. These elements have been called modules and have a specific structure. Figure 9 [4] shows schematically such an element used in the hydroperforation installation, an element called "module" and comprising:
- the upper part of the die
- the lower part of the die
- the punch that moves in a special and tight bore in the body of the die
- the hydraulic cylinder, which actuates the punch.

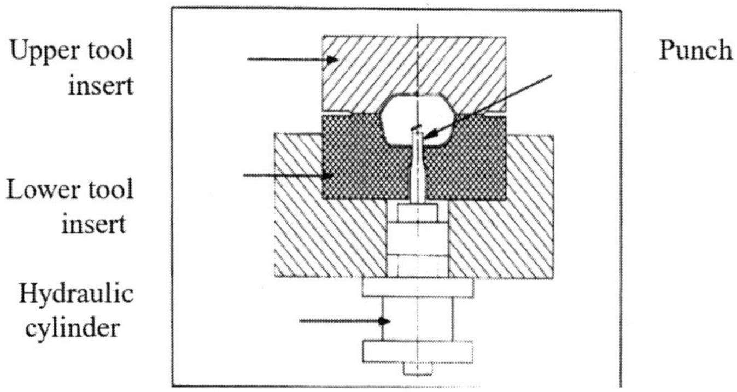

Figure 9. The scheme of a module used in installations of hydroforming and hydroperforation [4].

The module presented is a unit with a simple structure, because it provides only one hydraulic cylinder, but it can act more punches. This structure allows easy use for various cases of perforation, by relatively simple mounting on the positioning of the punches in relation to the given part.

A problem with hydroperforation is the removal of the slug, resulting from perforation.

Figure 10 [4] shows the case where the slug it remains partially on the wall of the piece. This case is characteristic of the first perforation option presented. If this slug if this residual material interferes with the operation of the part then it must be removed by secondary operations, subsequent to the perforation and the part must remain clean.

Figure 11 [4] shows the case where, after perforation, the slug is attached to inside edge of the hole.

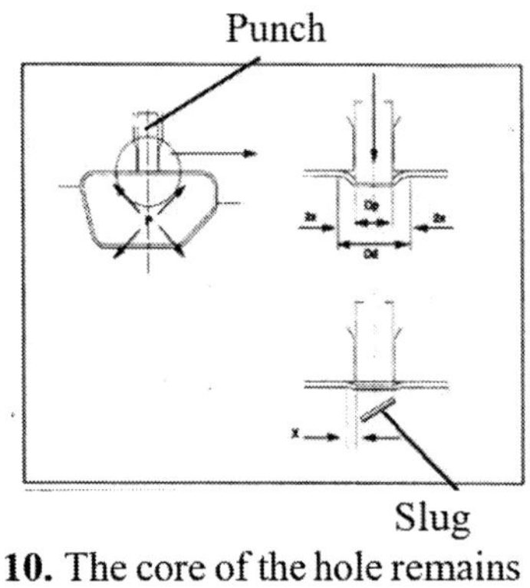

Figure 10. The core of the hole remains partially attached attached to inside edge of the hole [4].

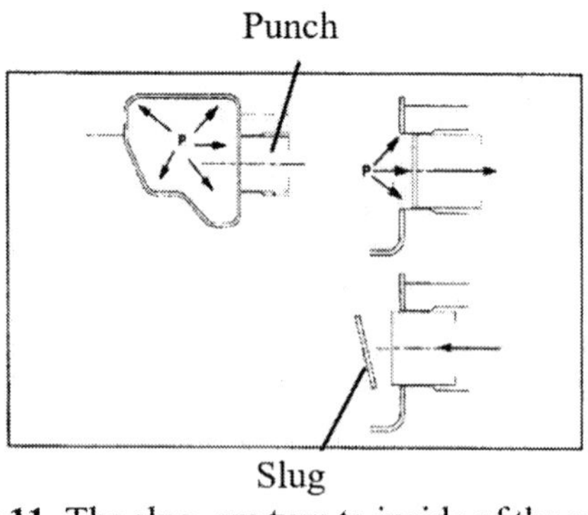

Figure 11. The slug are turn to inside of the part [4].

Figure 12 [4] shows the schemes of the different cases of detachment of the slug at hydroperforation. These presents for cases:

- total detachment and movement inside the part (top A, in figure 12);
- partial detachment and remains trapped inside by the edge of the hold (left B, in figure 12);
- total detachment and movement to the outside of the part (bottom D, in figure 12);
- detachment and movement towards the inside of the part and the remaining hole edges, turned inwards (right C, in figure 12).

The holes made by hydroperforation have small positioning tolerances and a good dimensional accuracy .

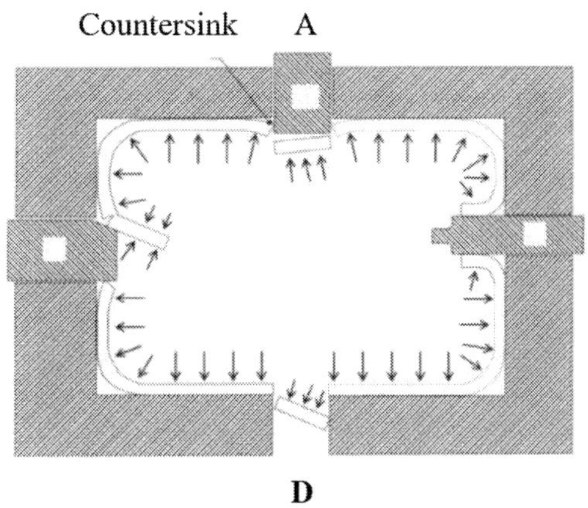

Figure 12. Cases of detachment of the core of the holes when hydroperforating parts [4].

3 .2. Technological constraints on hydroperforation

The diameters of the holes are limited to a minimum of 2 g, where „g" is the wall thickness of the tubular part.

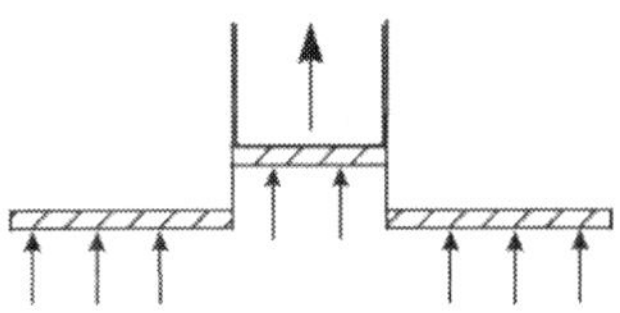

Figure 13. The forces that appear when the punch detaches the material [5].

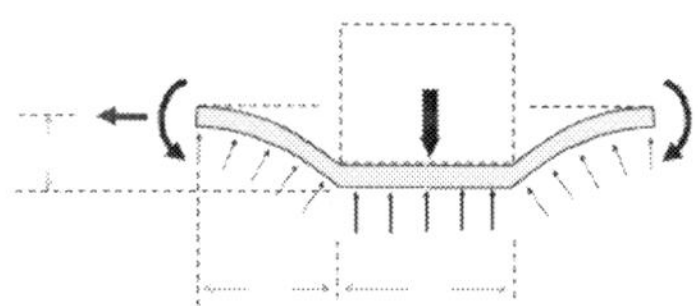

Figure 14. The forces that appear when the punch is withdrawn and the detachment is made by hydraulic pressure [1].

Figure 13 [5] shows the forces that appear during the perforation process, if the punch is the one that penetrates the material, when the force acts from the outside of the tube to the inside of it.

Figure 14 [1] shows the forces that appear in the perforation process, when the hydraulic force acts on the perforation and the punch retracts, the pressing force is from the inside to the outside of the tube.

3 .3. Advantages of hydroperforation

The use of hydroforming has been able to expand due to research in the field of materials, with the current accentuated tendency to obtain light materials but with very high mechanical resistance , materials increasingly required for the manufacture of cars and airplanes. Such materials can positively influence the Weight/Power characteristic, as it must be as small as possible, in the case of cars and airplanes, in order to be able to develop higher speeds and accelerations, with maximum economic efficiency.

Figure 15 [2] shows a part where a minimum and a maximum hole they are made by hydroperforation, for the wall thickness of the tubular part, between 2 and 6 mm.

Figure 15. Maximum and minimum hole obtained by hydroperforation [2].

Figure 16 [2] shows a picture of a hydroforming installation used to make automotive components.
Figure 17 [2] shows an automated hydroforming and hydroperforation manufacturing line from Opel, a car manufacturer.

Figure 16. Hydroforming installation [2].

Figure 17. Automated hydroforming and hydroperforation manufacturing line [2].

Advantages of the hydroforming and hydroperforation, compared to classical processes:

- the holes obtained by hydroperforation have small positioning tolerances and a good dimensional accuracy;
- reducing the weight of semi-finished products;
- reduction the number of part components obtained by assembly;
- lower costs of the components and the assembly;
- lower equipment costs;
- increased strength and rigidity of the finished parts obtained;
- a high dimensional stability (geometric tolerance and positioning of the hydroperforated holes) of the finished parts obtained.

4. The laser perforation process

Laser perforation has seen a spectacular development through the recent use of robotics in drilling tubular parts such as long pipes. Laser beams are monochrome and coherent electromagnetic radiation in the optical field, characterized by exceptional directionality and brightness.
Elements for carrying out the process, are:

- collimated laser beam
- optical focusing system
- environment work

- the object to be processed

When laser perforation, a laser beam makes non-contact holes with a range of very small diameters to very large diameters in various materials.When laser drilling, a short laser pulse, introduces energy into the part in a very short time, so that the material is melted in that area and vaporizes, and due to the resulting vapor pressure the molten material is evacuated from the hole. When processing with the help of lasers with ultrashort pulses in a regime of picoseconds, the solid state material evaporates directly, without liquefaction, avoiding the heating of the part.

There are several laser perforation procedures, as follows:

➢ Perforation with individual impulses

In this case, an individual laser pulse with a comparatively high energy performs the perforation. Numerous holes can be made very quickly.

Figure 18 [8] shows the simplified scheme of the perforation with individual impulses process.

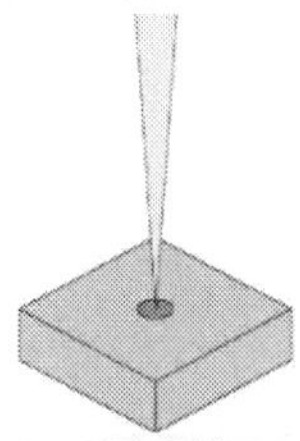

Figure 18. Scheme of the perforation with individual impulses [8]

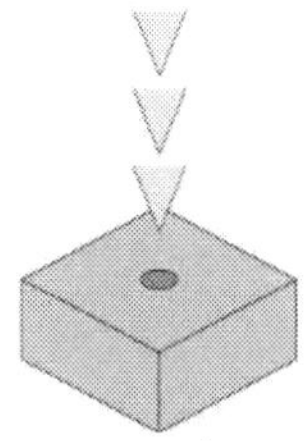

Figure 19. Scheme of the percussion perforation [8]

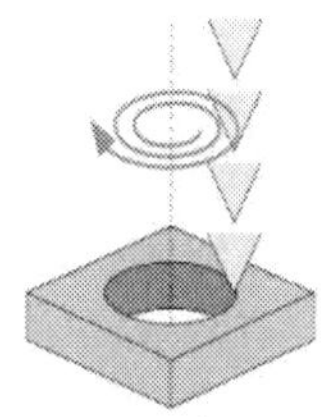

Figure 20. Scheme of the perforation by drilling [8]

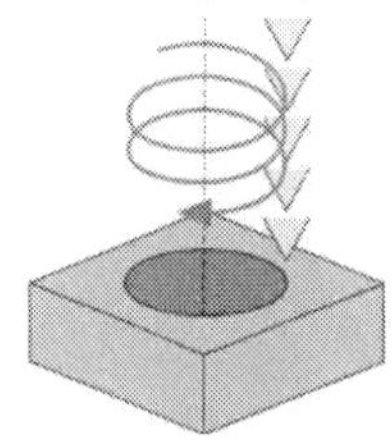

Figure 21. Scheme of the helical perforation [8]

➢ Percussion perforation

When percussion drilling, the punching is done through several pulses with reduced durations and energies. This perforation process allows to obtain deeper and more precise holes, compared to the perforation with individual pulses. Also, by percussion perforation smaller diameters of the holes are possible.

Figure 19 [8] shows the simplified scheme of the percussion drilling process.

➢ Perforation by drilling

Drilling involves perforating by means of several laser pulses. First, an initial hole is made by percussion perforation. Then the original hole is enlarged, moving over the piece on several circular paths, getting bigger and bigger. Thus, most of the molten material is discharged down from the orifice.

Figure 20 [8] shows the simplified scheme of the perforation by drilling process.

➢ Helical perforation

No initial hole is made in the helical perforation. The laser moves from the first pulses on a circular path above the material. In this way, much of the material is discharged upwards. The laser processes in depth, in the form of a helical ladder. Thus, it is possible to monitor the focal point, so that it is permanently at the bottom of the perforation. After the laser penetrates the material, several more trajectories can be added. They serve to extend the lower part of the perforation and to smooth the edges. The helical perforation allows the realization of very large and deep holes, of high quality.

Figure 21 [8] shows the simplified scheme of the helical perforation process.

Making different diameters can be obtained by the size of the holes of the laser nozzle.

Figure 22 [8] shows the laser nozzle.

Figure 22. The laser nozzle [8]

Materials processed currently by laser are:

- ceramic materials, such aluminium, ruby, and synthetic sapphire, natural or synthetic diamond;
- macromolecular materials such polyethylene, polycarbonates, polystyrene, polyacetates;
- high alloyed steels with Ni, Cr, Co (inclusively stainless and refractory steels, W, Mo, Ti, Cu);
- brass;
- metal-ceramics materials (metal carbides).

Types of parts subjected to laser perforation are bore parts, tubular parts, pipes.

Figure 23 [3] shows the laser perforation operations:

- perforation by copying
- perforation
- enlargement

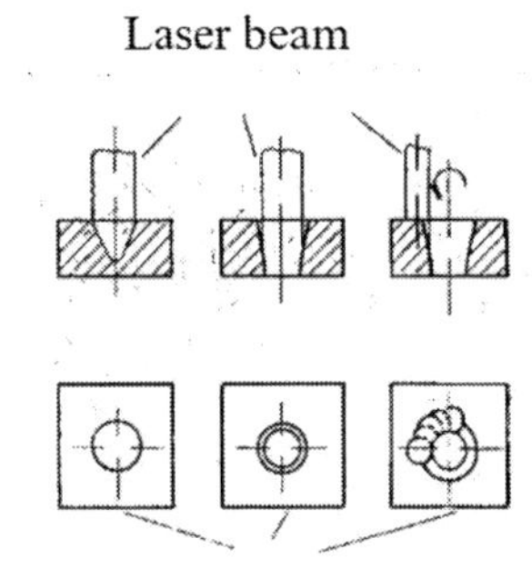

Figure 23. Various operations of perforation carried out using a laser [3].

Laser applications in the automotive industry are to reduce the weight of vehicles, massive components are increasingly being replaced by pipe structures. Thus, pipes are used in the manufacture of head restraints, transverse beams and seat structures [8].

The advantages of using laser perforation are:

- a very good machining accuracy;
- a high productivity and efficiency, the process can be automated or even robotic today.

5. The thermal erosion perforation process

Thermal erosion perforation of long tubular parts is an innovative process, which uses a rotary tool with sharp edges, which penetrates the part and causes a strong heating of the part in the work area and the material is removed, resulting the holes in the parts. Schematically, this processing process is presented in figure 24 [4]. This process has a high productivity, the processing time of an orifice being 5-6 seconds.

IOP Conf. Series: Materials Science and Engineering **898** (2020) 012043 doi:10.1088/1757-899X/898/1/012043

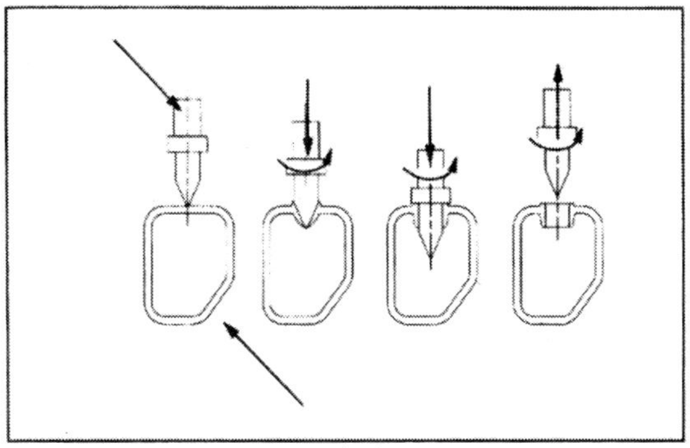

Figure 24. The scheme of the thermal erosion perforation process [4].

According to [7] and in the last decade, thermal perforation has been widely used in the aerospace and automotive industries, due to its unique advantages over the conventional perforation process: finishing the surface of the thermally perforation hole.

The surface roughness of the thermally perforated hole on galvanized steel is very good and significant variable parameters are made, such as:

- axial speed;
- tool angle;
- the thickness of the workpiece.

Figure 25 [6] shows examples of tubular parts in which holes were made by thermal perforation, some with thread.

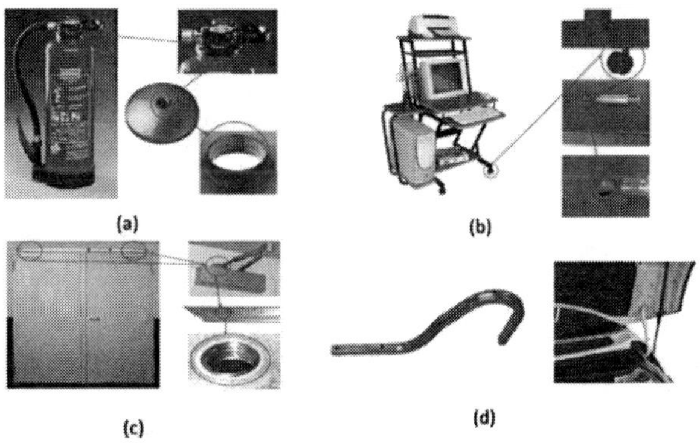

Figure 25. Various tubular parts with holes made by thermal erosion [6].

According to [7] the conventional perforation process produce a massive environmental pollution, because they produce a waste of metal chips, span.

A thermal perforation tool is shown in figure 26 [6].

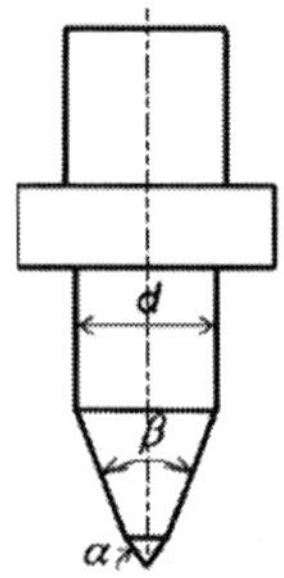

Figure 26. Representation of a tool used for thermal erosion perforation [6].

The various tools used for thermal perforation are shown in figure 27 [6].

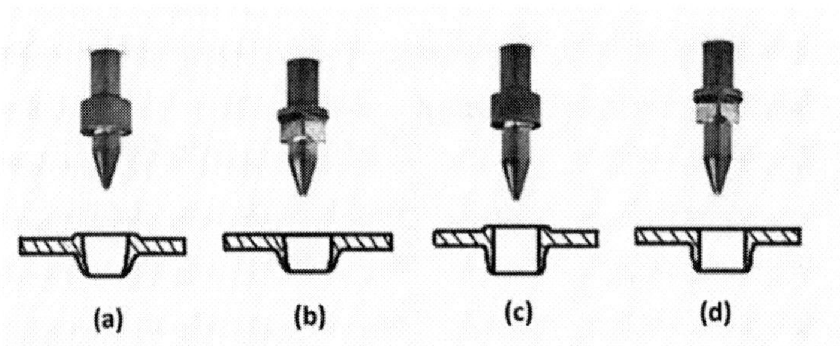

Figure 27. Various tools used for thermal erosion perforation [6].

Figure 28 [6] shows different thermal perforation tools, with two types of contact friction surfaces 100% (a) and 50% (b). It is observed that the conical portions can have angles of 30 ^{0}, 45 ^{0} or 60 ^{0}.

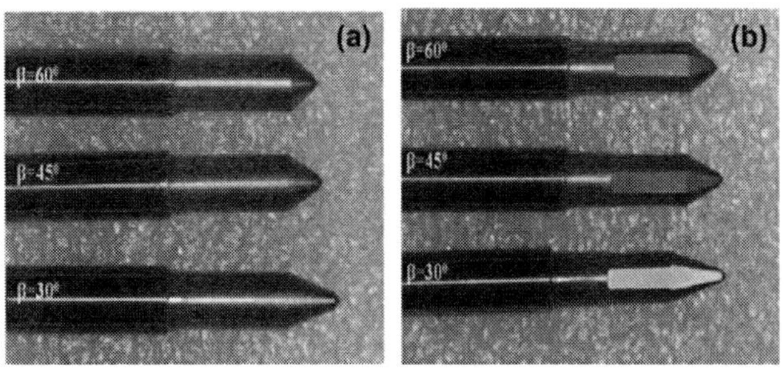

- with an angle of 60 °

- with an angle of 45 °

- with an angle of 30 °

(a) (b)

Figure 28. Thermal perforation tools:
(a) with 100% contact friction surface (b) with a contact friction surface of 50% [6].

The thermal perforation tools practically creat a reflection of the material inside the tubular part, and at the end of the perforation, the tools also make a compaction of the upper part of the work area, and make a flat surface on the part, only that no chips are produced.

Pipe-type tubular parts are subjected to this process, having the holes made for self-tapping screws.

Figure 29 [4] shows on the same parts, the holes made by thermal erosion perforation (a) and holes made by laser perforation (b).

IOP Conf. Series: Materials Science and Engineering **898** (2020) 012043 doi:10.1088/1757-899X/898/1/012043

a) Flow perforation holes, suitable for 6 mm self-tapping screws

b) Laser cut slots

Figure 29. a) Flow perforation holes, suitable for 6 mm self-tapping screws; b) Laser cut slots [4].

Advantages of the thermal erosion perforation process are indicate in the following:

- in order to reduce emissions pollution, it should replace the conventional drilling process;
- is an energy efficient and clean perforation process, which has attracted several car manufacturers and the aerospace industry in recent years;
- the processing time and production cost of perforation and the risk of rejection are reduced;
- due to the high speed of the tool, a large amount of thermal energy is produced in the perforation area and thus a movement of the material is achieved, so that no metal chips result.

Conclusions

In the present paper, three modern perforation processes have been studied and critically analyzed: hydroperforation, laser perforation and thermal erosion perforation. They are important processes as they were have been successfully applied in recent years, especially in the automotive industry.

These perforation processes are a consequence of the requirements in certain sectors of activity their results being demanded by novel engineering studies and technology. The results were spectacular leading to the development of technologies to produce parts and subassemblies that are as light as possible, as durable as possible, as cheap as possible, and as environmentally friendly as possible.

Compared to the classic perforation/drilling processes, these three innovative processes have been intensivelly applied also do to the fact that light materials with very good mechanical and technological characteristics have been obtained, such as alloy steels, magnesium alloys, aluminum alloys.

In parallel with the hydroperforation, the laser perforation process and more recently the thermal erosion perforation process were applied for certain components.

One proved that these three processes are very adequate, because they have a good productivity, a good dimensional accuracy and positioning of the holes, they achieve a better protection of the environment by reducing polluted emissions. Finally the advantages of these processes were presented, compared to the classic perforation/drilling procedures, namely:

➢ Hydroperforation

- holes with low positioning tolerances and good dimensional accuracy are obtained by hydroperforation;
- allows to reduce the weight of semi-finished products;
- implicitly leads to the reduction of the number of components of a part obtained by assembly;
- achieves lower costs of components and assembly;
- leads to lower costs of car equipment;

- finished parts with increased strength and rigidity are obtained;
- is characterized by a high dimensional stability of the finished parts obtained (geometric tolerance and positioning of the hydroperforated holes).

➢ Laser perforation

- parts with holes with a very good processing precision are obtained;
- the process has a high productivity and efficiency, the process can be automated and robotized;
- allows the realization of holes in parts made of different materials (high steels alloyed with Ni, Cr, Co, brass, metal-ceramic materials, macromolecular materials).

➢ Thermal erosion perforation

- when making the holes, there are no metal chips, due to the high speed of the tool, a large amount of thermal energy is produced in the perforation area and thus a flow of the material is made, without steel chips;
- processing time and production cost of perforation are reduced;
- the risk of rejecting the parts is reduced;
- it is an energy efficient and clean perforation process, and that is why in recent years it has started to be used more and more in the production of cars and in the aerospace industry;
- is part of technological processes that reduce polluted emissions.

One considers that the purpose of this paper has been fully achieved, meaning to make a study and a critical analysis of these three innovative perforation processes, used successfully in the automotive industry.

References

[1] Koç M 2008 *Hydroforming for advanced manufacturing*, Woodhe Publishin and Maney Publishing on behalf of The Institute of Materials, Minerals & Mining, CRC PressBoca Raton Boston New York Washington, DC.

[2] Neugebauer R 2007 *Hydro-Umformung*, ISBN-13978-3-540-21171-6, Springer Berlin,Heidelberg, New York, http:/dnb.ddb.de.@Springer-Verlag Berlin Heidelberg

[3] Nichici Al a.o. 1983 *Erosion processing in machine building (Prelucrarea prin eroziune în construcţia de maşini)*, Ed. Facla, Timişoara.

[4] Singh H 2003 *Fundamentals of Hydroforming*, Copyright © 2003 by the Society of Manufacturing Engineers 987654321.

[5] *** *Cold pressing technologies (Tehnologii de presare la rece)* 2015 http://ro.scribd.com/doc/36881830/Tehnologie-Presare-La-Rece-2#scribd, accessed on 10.11.2019

[6] *** Kumar R, Hynes R J 2019 Thermal drilling processing on sheet metals *A review, International Journal of Lightweight Materials and Manufacture,* journal homepage: https://www.sciencedirect.com/journal/international-journal-of-lightweight-materials-and-manufacture, pp193 – 197, Journal homepage: www.elsevier.com/ locate/jestch, accessed on 05.02.2020

[7] *** Kumar R, Hynes R J 2019 Prediction and optimization of surface roughness in thermal drilling using integrated ANFIS and GA approach, *Engineering Science and Technology an International Journal*, pp 1–12, Journal homepage: www.elsevier.com/ locate/jestch, http://creativecommons.org/licenses/by-nc-nd/4.0/, accessed on 05.02.2020

[8] *** Trumpf GmbH + Co. KG, *TRUMPF at LASER WORLD of PHOTONICS*, May 23 to 26, 2011, Munich Exhibition Grounds, Hall C2, Booth 341, www.trumpf.com, accessed on 05.02.2020

IOP Conf. Series: Materials Science and Engineering **898** (2020) 012044 doi:10.1088/1757-899X/898/1/012044

Digitalization concept: Cyber-risks and damages for companies in adhered industries

C Vartolomei[1] and S Avasilcăi[1]

[1]"Gheorghe Asachi" Technical University of Iasi, Department of Engineering and Management, Mangeron Blvd. 28, TEX1, 700050, Romania

cristian.vartolomei@tuiasi.ro

Abstract. Starting with the modern binary number system in 1679, followed by the universal computation as a result of the Boolean algebra in 1847, then continued in 1948 with "Mathematical Theory of Communication", capacitors and microchips between 1950s and 1970s, Apple I desktop computer in 1976 and until 2020 when we have devices as smartphones, computers, laptops, smart houses, cities and cars, all these inventions are small but realistic steps that built the current concept adopted by the humankind, called digitalization. Considering this impactful concept during our research, we propose to perform a literature review regarding its implementation in industries as medical, banking, automotive, along with generic examples from undefined ones, followed by the risks and disasters as a result of cyber-attacks, which can be seen as disadvantages of implementing this concept. In the end, we propose to discover the effects on the virtual business environment by exemplifying the cyber-attacks and analyzing them.

1. Introduction

Digitalization has begun to make its presence known with the transition from physical paper documents to digital ones. In this way, the beneficiaries of joining the technologies that made this possible, have observed that these types of digital documents offer them a more accurate way in manipulating the data, finding and reviewing the mistakes, along with a much faster writing speed. From this point on, the mankind began to work on the development of these technologies, to the point where it ensured the interconnection of a variety of systems such as smartphones, laptops, tablets, autonomous cars and smart homes and cities through communication channels as satellites, optical fiber and wireless networks. But even if the benefits of joining this concept continue to evolve, at the same time it joins the disadvantages of accessing data from a system connected to the Internet from anywhere in the world by different cyber-attackers, having an increased level of knowledge regarding the information systems technical side, being able to penetrate the host system and stop its functionality, tamper with its data or steal the information.

The main purpose of this paper is to perform a literature review regarding the impact of the digitalization in different industries, by highlighting in the beginning the way that cardiology medical field, along with the banking and automotive industries have evolved adhering to and implementing different technologies. Along with the aforementioned aspects, this paper also aims to classify the cyber-attacks that can be performed on companies which implemented the digitalization, in order to not being displaced by competition and ensuring benefits as, fast access to information, low handling and storing costs, a better communication between the employees and with the clients, a better organization within

IOP Conf. Series: Materials Science and Engineering **898** (2020) 012044 doi:10.1088/1757-899X/898/1/012044

the company, increasing the profit and delivering a qualitative service/product. In this way, by performing this research, we look forward to obtain and enumerate the effects on the virtual business environment for the companies in different industries, as results of cyber-attacks.

2. Digitalization in various industries at a glance

Digitalization has always been important for the humankind, because, for example, considering the medical technologies used for establishing diagnoses, it offered a new and more accurate perspective for the doctors, starting from the arrival of the patient to the hospital until he leaves at home, in some cases without any disease. In this way, considering the cardiology field in medicine, the fact that it has so far occupied the world rank as a mortality rate, cardiovascular diseases are becoming a field of exploration in creation the concept of artificial intelligence, that is an important aspect of the digitalization process. Early detection, accuracy and fast diagnosis are the three most important pillars in active research. Detection of electrical changes at the cardiomyocytes level, before these will become electrocardiographically visible, as well as the detection of episodic atrial fibrillation in a heart with apparently normal rhythm are two useful medicinal tools in a more accurate assessment of myocardial infarction risk, heart failure, as well as the risk of being hospitalized again in the coming years [1].

Along with this example, the banking field has also took advantages by offering the availability of the services 24/7 to the clients, most used one at the moment being the internet banking (available on a computer, a smartphone or a tablet), which ensure national and international money transfers, payment of the invoices, opening of savings accounts and last but not least, payments to the online stores.

Another important field which due to the adoption of the digitalization had a significant impact regarding the product offered to the customer is automotive, by implementing the following concepts:

- Autonomous Driving (AD) – vehicle that is able of sensing its environment and proceeding in a safe way with almost no human interaction [2];
- Connected Car (CC) – vehicles are connected through different communication channels, as Vehicle to Vehicle (V2V) (intercommunication between vehicles in a defined range), Vehicle to Infrastructure (V2I) (communication with a network of vehicles or road infrastructures), Vehicle to Pedestrians (V2P) (a pedestrian can use a computer or mobile application in order to locate nearby taxis along with the estimation of the arrival time – e.g. Uber) and Vehicle to Network (V2N) (different organizations are able to connect with the connected car in order to alert the drivers regarding different changes as weather conditions or traffic jams) [3];
- Internet of Things (IoT) - vehicle that got implemented in its architecture internet connectivity through LAN and WLAN, giving the possibility to access/send/download data and communicate with Internet of Things devices [4];

But even if the adoption of this concept has so many advantages for our actual level of technology progress in various industries, the rise of disadvantages is directly proportional considering the cyber-attacks which can occur.

3. Risks and disasters as a result of cyber-attacks

In order to classify the cyber-attacks which can occur over a company adhered to the digital transformations, during this research we will use the STRIDE model, developed by Microsoft in order to identify computer threats, in six categories as Spoofing, Tampering, Repudiation, Information Disclosure, Denial of Service and Elevation of Privilege [5].

In this way, Figure 1 classifies the attacks from the second row (excepting the "Attack Class") in subcategories of the STRIDE model (first row), all of them being defined as attack classes, by the followings:

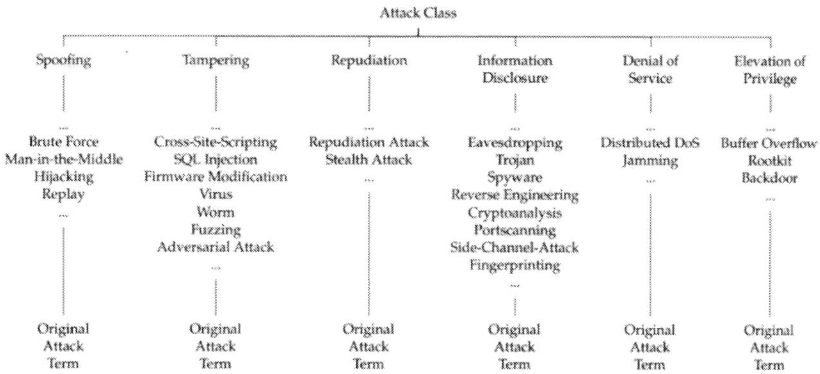

Figure 1. Classification of the attacks based on STRIDE model [6]

3.1 Spoofing

Spoofing element is defined by the brute force attack, in which an attacker can perform authentication attempts on an organization's system in order to obtain access on it, followed by the Man-in-the-Middle (MiM) used to capture and (depending on the attack) to tamper, (e.g. with the data traffic between an employee and the customer), hijacking that results in taking over the control of an established Internet connection between two or more parties [7] and replay attacks where a client-server communication is maliciously or fraudulently repeated or delayed [8], in order to make it unavailable.

3.2 Tampering

In the Tampering element, Cross-Site-Scripting (or XSS) is a type of attack in which the attacker inserts a malicious script in an organization's web-page's field, in order to behave in ways as, e.g. stealing user's credentials. A successful SQL Injection will steal all the info from an organization's database and a firmware modification will change the behavior of a hardware system (e.g. router, switch, smart TV, etc.), while a virus or a worm will replicate itself in order to spread to other systems in the organization's internal network [9]. In the end, a fuzzing attack will send crafted input in order to obtain some traces that will make the attacker to further understand the functionality of a system (being it hardware or software). Also, an adversarial attack is performed in a structured way, considering the level of information gathered, in this concept artificial intelligence being involved.

3.3 Repudiation

Repudiation element, containing also the stealth attack, happens when an organization does not adopt controls to properly track users' actions, thus permitting malicious manipulation of further actions [10], in this way giving the possibility to the attacker to steal data from an organization without being detected.

3.4 Information Disclosure

In Information Disclosure we find some various attacks, as eavesdropping which is performed by capturing data from unencrypted communication channels, followed by the trojan malware which is used to delete, block, modify, copy or corrupting organization's or customer's Intellectual Property (IP) [11]. Forwards, spyware can be defined as being a software used for stealing the internet browsing history (and not limited to) [12], reverse engineering used in decompiling (e.g. an organization's software, in order to obtain its source code for various further attacks), cryptoanalysis that is the study of analyzing an encrypted connection between two parties in order to access the information exchanged [13], port scanning which is a technique used in finding open ports and related running services for an operating system [14], Side-Channel-Attack that breaks encrypted data by monitoring the electromagnetic field radiation emitted by a computer screen to view the data before the encryption process [15] and traffic fingerprinting, that sniffs an encrypted communication in order to analyze the packets' flow pattern [16].

3.5 Denial of Service
The difference between Denial-of-Service and Distributed Denial-of-Service is that the first one uses only a source to flood until unavailability a victim's system (e.g. an entire organization or only an employee), while the second one uses multiple sources. Also, jamming attack is used in the field of wireless sensor networks, by redirecting electromagnetic signals in order to disturb or interrupt the signal transfer.

3.6 Elevation of Privileges
Elevation of Privileges contains Buffer Overflow, Rootkit and Backdoor. The first one is used in leading a program to put much more data in a buffer that it can hold, while the difference between the followed attacks is that a Rootkit will be applied as a set of software that are used by an attacker when finally gained access to the victim's machine, in order to establish an open-way for anytime he wants to turn back without being detected. In the end, a Backdoor is used to access the victim's machine by installing, e.g. a virus or even by legitimate programs [17].

4. Effects on virtual business environment
Considering all the types of attacks that have been mentioned in the last chapter, during our research we discovered that the effects on the virtual business environment can be defined as:

- Loss or damage of the hardware and software systems that are leading in generating extra expenses for an organization (e.g. in 2014, financial fraud actions in e-commerce in UK lead to costs of £217.4 million, and £60.4 in online banking frauds [18]);
- Loss of income as a result of various attacks that stopped the activity for a period of time or by not having access to the IP anymore (e.g. Kaspersky reports that the average cost for a company as a result of a DDoS attack was $2 million, in 2017 [19]);
- If an attacker steals the IP of a victim's customer, then the victim can be sued for information leakage (e.g. Company Zynga Inc. has been sued in the district of California because a hacker claimed to have accessed the company's database and stolen 218 million user accounts [20]);
- The extorsion losses, as being blackmailed by an attacker to pay a big amount of money in order to not make public some information about own or customers' IP (e.g. In 2019, hackers blackmailed a German IT company with stolen financial and private information whose clients include Oracle, Airbus and Porsche [21]);
- Last but not least, a cyber-attack can damage an organization's reputation in a very serious way, leading in being very hard to find a new customer to deliver a service or a product (e.g. Company TalkTalk received complaints from customers who were waiting to be compensated as a result of a cyber-security breach in October 22, 2015 [22].

5. Conclusions
In the end, along with the literature review regarding the digitalization transformations in the cardiology medical field and industries as banking and automotive, during this research we also described different types of attacks on companies from any industry that implemented the technologies as a part of the aforementioned concept, based on categories provided by the STRIDE model.

Considering all these aspects, we can conclude that the cyber-attacks are a very serious risk for all organizations which adhered to the digitalization, because depending on the attacker's knowledge and capabilities, all the software and hardware used by an employee can be attacked in different ways and with different purposes. In this way, the impact in the virtual business environment can wear a significant risk for the attacked company, as losing the trust on the market regarding the security of the clients'

Intellectual Property, brand's reputation, extra-expenses which can occur, unavailability of the company's both hardware and software systems (which are used by the employees in the development of the product/service or in the communication with the market or with the customers), along with the blackmail situations in which a company can be pushed.

References

[1] Mayo Clinic Artificial Intelligence in Cardiology: Introduction to A.I. https://www.youtube.com/watch?v=mGODKIbiXpE
[2] Taeihagh A and Lim H S M 2019 Governing autonomous vehicles: emerging responses for safety, liability, privacy, cybersecurity, and industry risks Transport Reviews 39 (1):103-128. arXiv:1807.05720. doi: 10.1080/01441647.2018.1494640. ISSN 0144-1647
[3] Application of IoT in Automotive Industry. Future of Automobiles https://www.biz4intellia.com/blog/iot-applications-in-automotive-industry/
[4] Kaya I 2018 Connected Car Experiences in 2019: Exploring the Possibilities https://www.cmswire.com/digital-experience/connected-car-experiences-in-2019-exploring-the-possibilities/
[5] Wikipedia https://en.wikipedia.org/wiki/STRIDE_(security)
[6] Sommer F, Dürrwang J and Kriesten R 2019 Survey and Classification of Automotive Security Attacks MDPI 10(4): doi.org/10.3390/info10040148
[7] Rouse M Email spoofing https://searchsecurity.techtarget.com/definition/email-spoofing
[8] Wikipedia https://en.wikipedia.org/wiki/Replay_attack
[9] Wikipedia https://en.wikipedia.org/wiki/Computer_worm
[10] OWASP https://owasp.org/www-community/attacks/Repudiation_Attack
[11] Kaspersky https://www.kaspersky.com/resource-center/threats/trojans
[12] Norton https://us.norton.com/internetsecurity-how-to-catch-spyware-before-it-snags-you.html
[13] Wikipedia https://en.wikipedia.org/wiki/Cryptanalysis
[14] Whatismyip from https://www.whatismyip.com/port-scanner/
[15] Porup J M 2019 What is a side channel attack?How these end-runs around encryption put everyone at risk https://www.csoonline.com/article/3388647/what-is-a-side-channel-attack-how-these-end-runs-around-encryption-put-everyone-at-risk.html
[16] Gu X, Yang M, Zhan Y, Pan P and Ling Z 2018 Fingerprinting Network Entities Based on Traffic Analysis in High-Speed Network Environment Intrussion Detection and Prevention in Cloud, Fog, and Internet of Things doi.org/10.1155/2018/6124160
[17] Sqasolar https://www.sqasolar.org.uk/solar/material/IS01CGCD/page_19.htm
[18] Understanding the costs of cyber-crime. A report of key findings from the Costs of Cyber Crime Working Group 2018 Home Office Science Advisory Council https://assets.publishing.service.gov.uk/government/uploads/system/uploads/attachment_data/file/674046/understanding-costs-of-cyber-crime-horr96.pdf
[19] Kobialka D 2018 Kaspersky Lab Study: Average Cost of Enterprise DDoS Attack Total $2M https://www.msspalert.com/cybersecurity-research/kaspersky-lab-study-average-cost-of-enterprise-ddos-attack-totals-2m/
[20] Coble S 2020 Zynga Facing Lawsuit Over Data Breach https://www.infosecurity-magazine.com/news/zynga-facing-lawsuit-over-data/
[21] Hamilton I A 2019 Hackers have stolen a ton of data from IT company whose clients include Oracle, Airbus and Porsche https://www.businessinsider.com/hackers-stole-data-from-citycomp-and-are-blackmailing-the-company-2019-4
[22] TalkTalk case study: reputational risk of cybersecurity attacks 2015 https://www.alva-group.com/blog/the-reputational-risk-of-cyber-attacks-talktalk-case-study/

IOP Conf. Series: Materials Science and Engineering **898** (2020) 012045 doi:10.1088/1757-899X/898/1/012045

Exploring the opportunity for a hybrid methodology in project management: a focus group approach

N Bagiu[1], S Avasilcăi[1], L Alexa[1]

[1] Gheorghe Asachi Technical University of Iasi-Romania, Engineering and Management Department, Blvd. Dimitrie Mangeron, No. 28, 700050, Iasi, Romania

neculai.bagiu@tuiasi.ro

Abstract. Nowadays, project-based work is no longer a challenge for a today company. Optimizing the workforce, better managing the results, delivering in time, quality and budget (the triple constrains [1]) are just a few of the advantages. Still, to re-organize an entire company into a project-based organization requires a lot of efforts and energy across its departments and teams. However, some companies step forward and change the project management methodologies multiple times. That brings another wave of changes, stress and reorganizations combined with a complex organizational phenomenon, which requires far more effort than just to replace the current tools and technologies with new ones [2]. Thus, the main purpose of this paper is to reflect and present the results of the workshop called "Going Agile – the journey", which was held on Project Management Forum 2019, a Conference organized inside of Continental Automotive Romania for all project managers working in the company. Based on the focus group technique there was identified the main struggles associated with the adoption of Agile methodology within Automotive Industry, the similarities with the plan-driven methodologies and willingness to advance the agility in company's daily working life.

1. Introduction

The main concern in adopting Agile methods adoption within software companies has increased due to their promise for higher customer satisfaction, lower defect rates, faster development times and openness to change [1]. The Collabnet and Versionone reported, in their latest State of Agile Report [2], that there are 13 reasons to adopt Agile methodologies (based a yearly survey with professionals from the industry) like: Accelerate software (74% of the respondents), Enhance ability to manage to change priorities (62%) or Increase the productivity (51%) and also there are 13 benefits like Ability to manage to change priorities (69%), Project visibility (65%) or Business/ It alignment (64%) equal with Team morale (64%). Due to the different philosophies and scopes, a 1:1 mapping between Agile methods and practices and the standards is not easy. The specific requirements of the standards e.g. with respect to scheduling, linking information (traceability), independence of quality assurance, reviews, etc. makes it necessary to adapt the existing Agile working methods in order to satisfy these requirements. As a consequence, this may lead to reducing Agility within a company [3]

The most used development model in automotive industry is V-cycle model – that was derived from the waterfall model [4]. This practice is rooted into the very origin of the industry where an increased number of activities have mechanical parts to be produced and assembled [5]. By developing electronic and then software parts in automotive industry, the major players in the industry opened a new chapter in applying the traditional methodology or seeking to find new methods to update the current one. This led to the fact that "the life cycle of products does not respect anymore the profile of the classic curve, this curve turned in the profile of a saw tooth. Before ending the phase of introducing the product on the market, the product is considered already outdated" [6]. The automotive industry is no longer about producing and assembling mechanical components only, but also producing, testing and integrating some other parts, such as sensors, electric motors, software and, more important, algorithms. Therefore, it seems to be a real challenge to select and apply a project management methodology that fit all needs. Another big constrain came from the fact that this industry deals with humans life, the responsibility to implement an efficient break system based on robust algorithm transposed into a software component requires high quality standards in developing, testing and validating the results which do not dependent on time of development, openness to changes or customer satisfaction at that time. The V-cycle model development was recently adopted as the reference model that can be used for ISO26262 [7] for functional safety critical systems [8]. On the other hand, the automotive industry proposes a new paradigm of transportation where every actor – cars, infrastructure, Vulnerable Road Users (VRU) – produce and exchange an enormous quantity of data computed into cloud computing and artificial intelligence clusters to predict the behavioral model, which further will be used in order to decrease the number of accidents and lives lost. This exert an enormous pressure on the industry to open up to innovation, even co-innovation or open innovation, to the fast-delivery methodology, in a reliable manner and customer-oriented products.

2. The context of the research

Continental AG it's a German company that develops intelligent technologies for transporting people and their goods. Founded in 1871 in Hanover [9] as joint stock Company, the main area was manufacturing soft rubber products rubberized fabrics and solid tires for carriages and bicycles. In 1892 Continental was the first German company to manufacture pneumatic tires for bicycles and in 1898 start the production of pneumatic tires for automobiles, in 1943 the patent application for tubeless tires was filled, in 1974 start producing different rubber automobile parts like gaiters, conveyor belt, hydro mounts—special bearing element, etc. in 2007 acquires Siemens VDO and become one of top five suppliers in the automotive industry worldwide. "As a company that is 145 years old history, Continental has lived through a number of changes and has shaped many of them itself. History teaches us that only companies with deeply rooted origins and values can confidently shape their future successfully," declared Dr. Elmar Degenhart [10], chairman of the Continental Executive Board, at the start of the project aimed to reviving the History and Tradition of Continental AG. For Continental AG the seven strategic dimensions are [11]: 1). Value creation; 2). Regional sales balance; 3). Top market position; 4). In the market, for the market; 5). Balanced customer portfolio; 6). Technological balance; 7). Great people culture. In 2019 the corporation registered the following financial numbers in sales of €44.5 billion [12] in five divisions: Chassis & Safety, Interior, Powertrain, Tires, and ContiTech. In 2020 Continental started a new reorganizational project where the Powertrain division was listed as an innovative Continental's spin off company – Vitesco Technologies and the other four divisions were organized into two groups [13] Automotive Technologies, with Autonomus Mobility and Safety and Vehicle Networking and Information business units and Rubber technologies with Tires and ContiTech business units. Continental count 433 location in 59 countries across the world and nine of them in Romania, with more than 241.000 employees in production and also in R&D areas.

The business of the company is mostly project-based structured and with the main concern for its customers. The typical project comprises more than three large mixed teams from minimum two different locations. Therefore, there is a large number of project management professionals structured into different levels (junior/ senior, team leader/ group leader, etc.) and areas of expertise (Software PM, System PM, Tester Manager, Software Project Quality Manager, Integration manager, etc.).

Romania have almost 1000 project managers across all locations and the main methodology implemented in the company's project is waterfall – V cycle - wide used into automotive industry. In 2012 there were some initiative to adopt Agile – SRUM method in some business units as a request from customers, and to minimize the risk of changing the requirements. Starting from there, the initiative grew and going Agile became a big topic in the company. Still there are a lot of business units and projects using V-cycle, that is producing results and is covered by quality process which are well-known in company world-wide. However, the change is slow and for some business areas it's not started yet.

3. Methodology

The Focus group technique emerged as a qualitative data collection approach and a bridging strategy for scientific research and local knowledge [14]. Focus group discussion is perceived to be a "cost-effective" and "promising alternative" in participatory research [15] and it offers a platform for differing paradigms or worldviews [16] [17]. Sociologists and psychologists have used the method since the 1940s (e.g. Merton & Kendall, 1946 [18]; Merton, Fiske & Kendall 1956 [19]).
Focus group discussion consist of four major step process [20]:

3.1 Research design

The traditional Project Management model defines the tasks to be performed along with the desired outcomes of each project phase and assigns the roles to individuals who will perform. This model produces a large amount of documentation, the communication in project is formal, the customer plays an important role in describing the specifications, mostly. In a different way Agile method deals with unpredictability by relaying on people and their creativity rather than on processes [21]. While the standard model focuses on the product, processes and projects Agile relies on the team and the human factor [3]. "A Hybrid approach will increase the end results, improve the delivery of customer requirements and help the organization realize its strategic imperatives" [22].

The main purpose of the research was to collect structured information about the challenges that the project managers face in applying existing project management methodologies, good practices and suggestions for improvements. The structure of research was built on three components common to every project management methodology: roles, tools and processes. Also, it was relevant to understand their dynamics during project phases – four phases in the case of traditional project management approach and three for Agile approach. Completing the structure, if necessary, is another output of this research.

3.2 Data collection

The focus groups sessions were organized in the workshop that was presented during the PM Forum 2019 – a Continental internal conference focused on project management tools, techniques and best practices. The conference brought together almost 100 Project Managers, Team leaders and Managers from all Continental's Romanian offices. Going Agile – the journey workshop had 32 member who participated and were splatted for two hours in 4 small focus groups and presented their results at the end of each iteration of the workshop. The challenge of the workshop was How to craft a path that lead a project/ portfolio/ company from plan-driven methodology (usual known as traditional methodology or waterfall methodology) to a people centric approach (the case of Agile, usually SCRUM but for larger organizations - SAFe)

It was structured in four iterations with different themes/ questions addressed to the focus groups: In iteration one we use a general presentation and the focus was to create a space where all participants could express their feelings about the MP methodologies used, there could be identified the expectations from the focus groups and clarified the used terms. All the experiences and specific cases were welcomed into discussion with the purpose to prepare the audience for the next steps. The participants did recognize the importance of a robust project management methodology – V-cycle – however they did not exclude also the need to be more flexible with client's requests, continuous delivering and integrating the sprint results, better team organization which could be brought by SCRUM methodology. The structured way to capture the inputs was agreed by audience.

For iteration number two the participants were splatted into four groups (2 groups with participants working in waterfall methodology and 2 groups with participants working in Agile environment). Based on overall presentation and their valuable work experience in the area the participants were asked to list the most used tools, processes and roles in their daily work as project managers. It was easy to see the multitude of the roles (11), tools (11) and processes (15) listed by teams analyzed the traditional methodology versus less for those analyzed agile: roles (4), tools (6) and processes (13). Those findings were briefly presented by each team at the end of the iteration.

In iteration number three the participants were asked to analyses and write down on flipchart paper, for each phases of the projects, the three advantages and three disadvantages of using each of methodologies roles/ processes/ tools. This was a time for huge debate in groups to find-out and refine those. Some of the mentioned advantages were: "sprints retrospective have de opportunity to collect lesson learned in very that moment", "in Agile work frame we can see a clear prioritization", "clear and structured quality process" and disadvantages: "to complicate and time consuming development process", "too many meetings", etc. Those findings were briefly presented by each team at the end of the iteration.

For iteration number four the participants were re-organized into two big teams: one team with all participants using Agile and second one with all participants using Waterfall methodology. The challenge was to imagine a brand-new methodology (a hybrid one), which combined the roles/ tools/ processes from both – Agile and Waterfall – and which fitted their current needs in order to deliver robust projects in time, budget and quality. Even the task wasn't complete in given time slot, the teams struggled to combine the best elements of both methodologies in order to create a hybrid one. One proposal used the clarity and structured approach to initiate, define and closure traditional phases in relation with incremental agility of execution phase in project development cycle. The findings were presented by each team at the end of the iteration.

3.3 Analysis and reporting of results

Some of the findings resulted from discussions:

- There is no "absolute" and "infallible" methodology, something that fit to all
- None of the methodologies used (Waterfall or Agile/ SCRUM) perfect fit the need
- Even both provide enough tools to work with, not all these tools are used
- There is a lot of need of clarification regarding knowledge from both areas: waterfall / agile
- Even changing the paradigm it's painful; it's hard to identify what are strong/ week points of each -> to minimize the impact;
- The participants demonstrated a strong knowledge about the topic. They were able to engage deep technical debate at debrief time.
- The debates showed the strong commitment of participants toward the project's goals, teams, quality, etc.
- The participants involvement in entire workshop make a huge contribution in understanding the current situation and prepare a real approach regarding a paradigm's change
- Not the methodology is the problem, but how to make decision in applying this into a project, how to transfer the knowledge and involve the team members into changing process, how this change affect the entire business, are some of the "powerful" questions
- Easy to observe what is missing but hard to craft a new approach that fits the need. It's needed to think twice when you need to change

3.4 Reporting of results

The full report was released and sent to organizers. This was uploaded in the workshop's materials space and distributed to all participants.

4. Conclusions

A better understanding of the new approach, Agile or Hybrid one, can improve the awareness toward project's scope and objective, inside the project team. To reach an acceptable understanding level, is needed more effort in formal training and coaching, in a customized and optimized manner.

For the company it's important to understand the not all projects are similar. The main concern should be manifested especially in analyzing stage and deciding which project needs a different approach and then create space for "special" projects with "special approach/ needs". Thus, by supporting in multiple ways e.g. methodologies, tools, role, those projects will be able to deliver faster, and maybe easier, the expected results.

Another hot topic is standardization of project management methodology. Where standardization didn't add value, we need to consider to blend/ combine/ hybrid the methodology to bring better results. It's important to prepare the organization for these new approaches with knowledge and proper resources even the preparations are never finish or enough.

References

[1] Boehm B Turner R 2004 Balancing Agility and Discipline: Evaluating and Integrating Agile and Plan-Driven Methods *Int Conf on Soft Eng*
[2] Versionone 2019 "The 13th annual state of Agile report" Available: http://www.stateofagile.com/?_ga=2.227818525.1499839566.1588757901-1234154020.1585897582
[3] Müller M Maag B, Siegl M 2018 *How agile development in a regulated automotive environment works* (Frankfurt: KUGLER MAAG CIE GmbH)
[4] I T G Development 2013 *Software Engineering for Embedded Systems, Methods, Practical techniques and Applications* Newnes
[5] Hutanu A Prostean G Badea A 2015 Integrating Critical Chain method with AGILE life cycles in the automotive industry *7th World Conf on Ed Sci* Athens
[6] Prostean G 2001 *Management Prin Proiecte* (Timisoara: Editura Orizonturi Universitare)
[7] I. O. Standardization 2018 "Online browsing platform ISO 26262-6:2018, Available: https://www.iso.org/obp/ui/#iso:std:iso:26262:-6:ed-2:v1:en
[8] Villagra J Acosta L Artuñedo A Blanco R Clavijo M Fernández C Godoy J Haber R, Jiménez F Martínez C Naranjo J E Navarro P J Paúl A Sánchez F 2018 *Intelligent Vehicles, Enabling Technologies and Future Developments* (Butterworth-Heinemann)
[9] Continental AG 2018 "History of Contiental" Available: https://www.continental.com/en/company/history/history-23052
[10] Gress D F 2016 "Continental AG - Our History" Available: https://www.continental.com/en/press/press-releases/-history-and-tradition-8740
[11] Continental AG 2020 "Corporate Strategy" Available: https://www.continental.com/en/company/corporate-strategy/our-seven-strategic-dimensions-57170
[12] Continental AG 2020 "Press Release" Available: https://www.continental.com/en/press/press-releases/fiscal-year-2019-215566
[13] Continental AG 2020 "Corporate Structure" Available: https://www.continental.com/en/company/corporate-structure
[14] A. Cornwall & R. Jewkes 1995 What is participatory research? *Social Science and Medicine,* vol. 14, pp. 1667-1676,.
[15] D. L. Morgan 1995 Focus Groups *Annual review of Sociology* vol. 22, pp. 129-152,.
[16] Guba E G and Lincoln Y S 1994 *Competing paradigms in qualitative research* (CA:Tousand Oaks Sage Publications Inc)
[17] Orr D 1992 *Ecological literacy: Education and the transition to a postmodern world* (Albany, NY: State University of New York Press)
[18] R.K Merton & P.L Kendall 1946 The focused interview *American Journal of Sociology,* vol. 51, pp. 541-557
[19] R.K. Merton, M. Fiske & P.L. Kendall 1956 *Focused Interview: A Manual of Problems and Procedures* 2nd ed New York USA The Free Press, A Division of Macmillian Inc

[20] Morgan D L Krueger R A and King J A 1998 *The focus group kit* Vols. 1–6 (CA: Thousand Oaks, Sage Publications Inc)

[21] Sridhar Nerur, Radhakanta Mahapatra & George Mangalaraj 2005 Challanges of migrating to Agile Methodologies *Communication of the ACM,* vol. 48, no. 5, pp. 73-78

[22] Stanleihg M 2019 "The Hybrid/Agile Project Management Process" Available: https://bia.ca/the-hybrid-agile-project-management-process/

[23] PMI 2004 *A guide to the project management body of knowledge PMBOK Guide (3rd)* (Washington, Newtown Square: Project Management Institute)

[24] S. Sircar, S.P. Nerur and R. Mahapatra 2001 Revolution or Evolution? A comparation of Object-Oriented and structured System Development Methods *MIS Quarterly,* vol. 25, no. 4, pp. 457-471

IOP Conf. Series: Materials Science and Engineering **898** (2020) 012046 doi:10.1088/1757-899X/898/1/012046

Custom vacuum load-lock for the sputtering vacuum chamber, loading/unloading and locking automatisation mechanisms

D.I. Țarcă[1], T. Costea[2], I. Moga[3]

[1,2,3]University of Oradea, Faculty of Management and Technological Engineering, Oradea, Romania

Corresponding address: dan@tarca.ro

Abstract. This paper presents the continuity of my work presented in previous articles regarding my activity as a PhD student at the University of Oradea, Doctoral School of Industrial Engineering. After installing, cleaning, vacuum testing and developing an experimental system for vacuum measurement, we decided that the next best step would be to develop the system with a load-lock further, as the time required for experiments would drastically improve and the need for refurbishing the old positioning mechanism would be eliminated. The paper describes the steps taken, starting with CAD representing the sputtering vacuum chamber for the manufacturer to start their production process, and finalising with automating the whole start/stop and reload sequence for the load-lock with an experimental setup made from readily available equipment in our laboratory, as the initial sequences were manually only and the risk of misusing for non-trained personnel would be high.

1. The work so far

University of Oradea has some experience with vacuum devices and magnetron construction and deposition, but it has been approximately 20 years since experiments were conducted on-site [1], [2]. New emerging technologies and rapid advancements in nanomaterials have sparked the interest of research in this area, and new investments in an advanced materials research infrastructure called SMARTMAT [3] paved the way for new opportunities. Previous achievements of the authors from recent years are presented in [4], [5] and [6].

2. Lock-Load System

After realizing some experiments to determine the quality of the vacuum and the time required to obtain it by automating the measurement process [6], we realized that we could only perform at best 3 depositions in a day, due to the long waiting time until the high-vacuum of at least 10^{-6} mBar was achieved – about 3 to 4 hours. The process of deposition itself would only take up to 10 minutes, then a shutdown of pumps needed to be queued, with a processing time of approximately 30 minutes and eventually returning the whole chamber to atmospheric pressure. We estimated that with a Lock-Load system [7] installed and the vacuum pumps of the Steel chamber permanently running we could realise a sputtering deposition in under an hour, providing that the target on the sputtering gun doesn't need to be changed.

2.1. Preparations for acquiring the system

Building the system ourself would have involved plenty of preparations and the prerequisite of owning the right equipment. As we, at the University of Oradea, are not as well equipped as professionals doing this for a living, and our technology wouldn't have permitted us to build all of the subcomponents necessary, we outsourced this part to an external company. Externalising saved us both time and money.

But all the actions that were done before the actual building of the Lock-Load system were performed at the University of Oradea. First, measurements of the vacuum chamber were made to ensure that the loading mechanism was long enough to push the substrate under the target, and a CAD drawing was made available to the supplier. We identified the coupling mechanism for the Lock-Load onto the chamber – measurements were made only for diameter as the flanges are standard. Next, we had to choose the material, and we finally settled on Aluminium as it is more machinable and easier to find on the market than special alloys of Stainless Steel, from which the chamber is made of [8], [9], [10], [11]. Also, having in mind that very few electrons would escape in the aluminium chamber, causing a possible reaction, this fact was considered negligible from the Load-Lock viewpoint.

We opted for a vacuum solution that is independent of the vacuum mechanism of the main chamber. This raised the price, as individual roughing pump and a vacuum pump, both of small dimensions, needed to be added to the system. We concluded that by doing this, we would add a great deal of versatility to the system. The small pumps could also be used on other experimental vacuum setups that have an educational role. As the option for automating the valves drove the system price higher, and we already had some electro-pneumatic valves in our laboratory, we decided that the automation will be done by ourselves. The full system can be viewed in figure 1

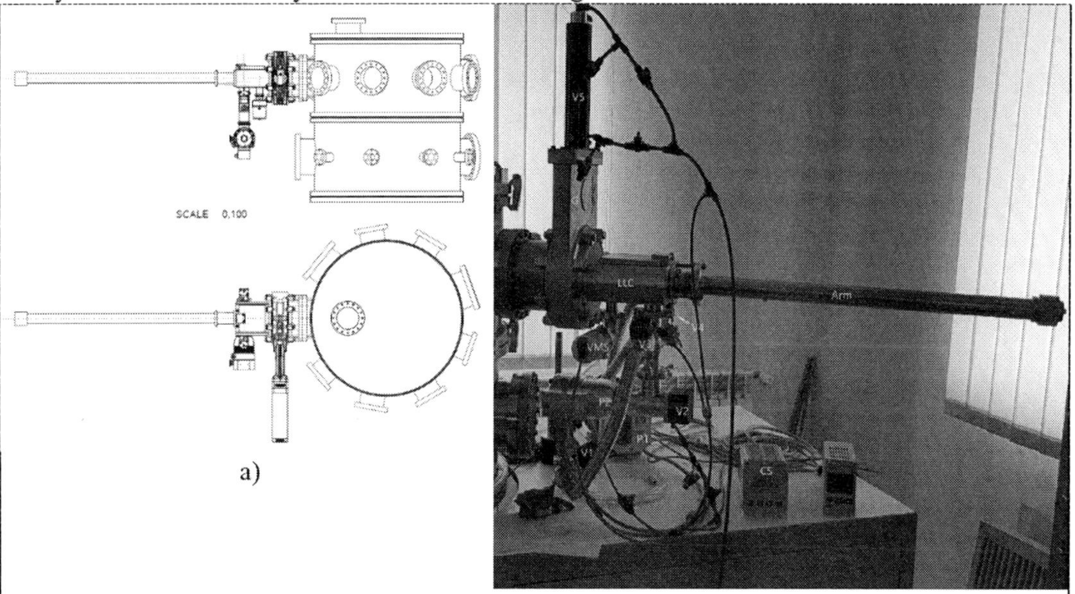

a)

b)

Figure 1. A figure depicting the Load-Lock system. a) CAD representation (left) vs b) actual system (right); V1 – Valve 1; V2 – valve 2; V3 – Valve 3; V4 – Valve 4; V5 – Valve 5; P1 – roughing pump; V2 – turbomolecular pump; CS – control system for turbomolecular pump; VMS – Vacuum measurement system; LLC – Load-Lock Chamber; G – Vacuum Gate; Arm – loading/ unloading mechanism

2.2. Receiving, mounting and testing the system

After receiving the system, we assembled it and ran a short test with the team that delivered it to us. A schematic for the operating procedures is presented in figure 2.

IOP Conf. Series: Materials Science and Engineering **898** (2020) 012046 doi:10.1088/1757-899X/898/1/012046

As it can be seen from figure 2, some steps must be followed in the exact order as in the operating procedures, or else the entire system can be compromised. The High-vacuum pump cannot be started until the whole system has achieved a rough vacuum, as the blades within it will capture big molecules of air that will damage them. Also, the vacuum system of the Load-Lock is more complicated than the one on the Stainless Steel vacuum chamber, on which all the components are mounted in line and are started sequentially, beginning with the rough pump and ending with the turbomolecular pump. The operating principles of both the roughing pump and the vacuum pump are explained in [9] [12] and [13].

The roughing pump has just a mechanical switch, but the turbomolecular pump has a processor and different settings can be changed and viewed, such as power draw, blades speed, current vacuum level etc. This is perfect, as we can connect it to our PXI [6] and get actual readings without the need of an extra Pfeiffer gauge [6].

After mounting the Load-Lock mechanisms with its valves, chambers, pumps and tubes, we needed the control the pneumatic valves. Our first and cheapest solution on the spot at the time we received the system was to manually control them, as we didn't know exactly how the final product will look until they shipped it to us. Further developments are presented below, but unfortunately due to the COVID pandemic, the research has stopped before getting to an actual result in this matter. However, the automatisation process is 95% ready - we only need a controller for the valves and program the sequence presented in figure 2. The pressurised air is and still will be provided by an external pressurised tank of 50L that has an air pump attached and can go up to 8 atm.

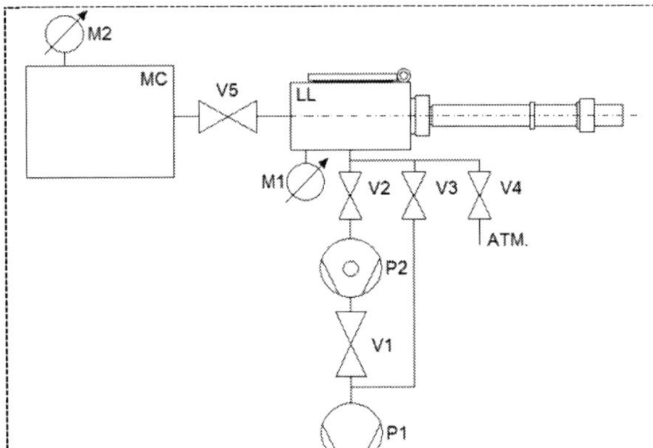

START TURBO:

P1- START
V1 - OPEN
P2 - START = > 1500 Hz (< 0.5A)
(V2, V3, V4, V5 CLOSED)

LL VACUUM:

V1 - CLOSED
V3 - OPEN = >M1 = 5X10-2 mbar
V3 - CLOSED
V1 - OPEN (wait 2sec.)
V2 - OPEN

LOAD / UNLOAD:

M2 - MAIN CHAMBER PRESSURE = LOADLOCK CHAMBER PRESSURE
V5 - OPEN
LOAD TARGET
(SPUTTERING)
UNLOAD TARGET
V5 - CLOSE

RETURN TO ATMOSPHERIC PRESSURE

V2 - CLOSE (V3 ALREADY CLOSED)
V4 - OPEN

Figure 2. Start/Stop process for the Load-Lock

2.3. Working procedures with the Load-Lock

The start of working procedures without the pumps already running have a prerequisite of going through a checklist. If all the answers are yes, then the procedure can begin:

1. Are the pumps connected to the mainline?
2. Is the arm fully retracted from the main chamber?
3. Is the gate valve of the Load-Lock mechanism closed?
4. Are all the pneumatic valves closed?
5. Is the load-lock chamber empty and clean?

Next procedural steps are the following:

1. *Start the rough pump by flipping the switch.
2. *Open the valve that connects to the vacuum pump to achieve primary vacuum in the tubes and the turbomolecular pump.

3. *Start the turbomolecular pump and wait for it to get to operating speed at about 1500 Hz.
4. Put the substrate in the Load-Lock chamber and close the lid.
5. Close the valve that connects the roughing pump to the turbomolecular pump, as we need to evacuate air from the Load-Lock chamber; not closing the valve could result in breaking the turbomolecular pump due to a rapid change in pressure.
6. Open V3 and wait until the pressure inside the Load-Lock chamber is at least $5x10^{-2}$ bar.
7. Close V3 and open V1.
8. Wait 2-3 seconds to get to the first stage of vacuum in case of leaks.
9. Open V2 that connects the turbomolecular pump to the Load-Lock chamber.
10. Wait for the pressure inside the Load-Lock chamber to drop to at least 10^{-5} mBar.
11. Open V5 that actions the gate valve.
12. Load the target by pushing the arm inside.
13. Wait for deposition.
14. Unload substrate from Stainless Steel chamber by retracting the arm.
15. Close V5 – the gate valve.
16. Close V2 as we need to preserve the vacuum inside the pumping and tubes system.
17. Open V4 to get to atmospheric pressure inside the Load-Lock chamber.
18. Open the lid and take out the finished sample.
19. *Stop the turbomolecular pump – and wait for it to slow down to under 500 Hz. If the speed is higher, the blades could break at the change in pressure.
20. *Vent the system by opening V3. If this step is skipped, oil from the roughing pump can penetrate the system. Wait at least 10 seconds for the ventilation process.
21. *Shutdown the roughing pump.
22. *Unplug the system from the power line

To put another sample inside the whole deposition process, we just need to follow the steps from 4-18. The rest of the steps that begin with * are necessary only in the situation that we need to startup or shutdown the pumps.

3. Automation mechanism and process

The automation mechanism presented in Figure 3 consists of 4 electro-pneumatic valves that connect to the pressurised tank and to the valves depicted in Figure 2 – V1, V2, V3, V4 and V5. These are all powered by a 5-20V AC to DC source and controlled by a small programmable board, similar to an Arduino. A step-down is required for the board to function at 5V. Once the automation process is complete, all the processes depicted in figure 2 that do not have an * will be done by the push of two buttons (on/off) and two switches: one for achieving vacuum, and one for returning to atmospheric pressure in the Load-Lock chamber. Powering up and down the two pumps will not be necessary, as they are better left in a permanent running state, due to less friction.

Also, we are considering to implement a failsafe, in which the gate mechanism of the Load-Lock will not be able to close as long as the arm holding the substrate is present in the main chamber. This will prevent severing the arm by mistake. We need to add a Hall sensor, as the arm is magnetically driven in order to obtain a full sealing, and the sensor would detect the magnet.

By now, all that remains to do is to program the board and mount the whole system onto the table that holds the Stainless Steel chamber with the Load-Lock attached. All the other components besides the board are already installed, tested and fully working individually.

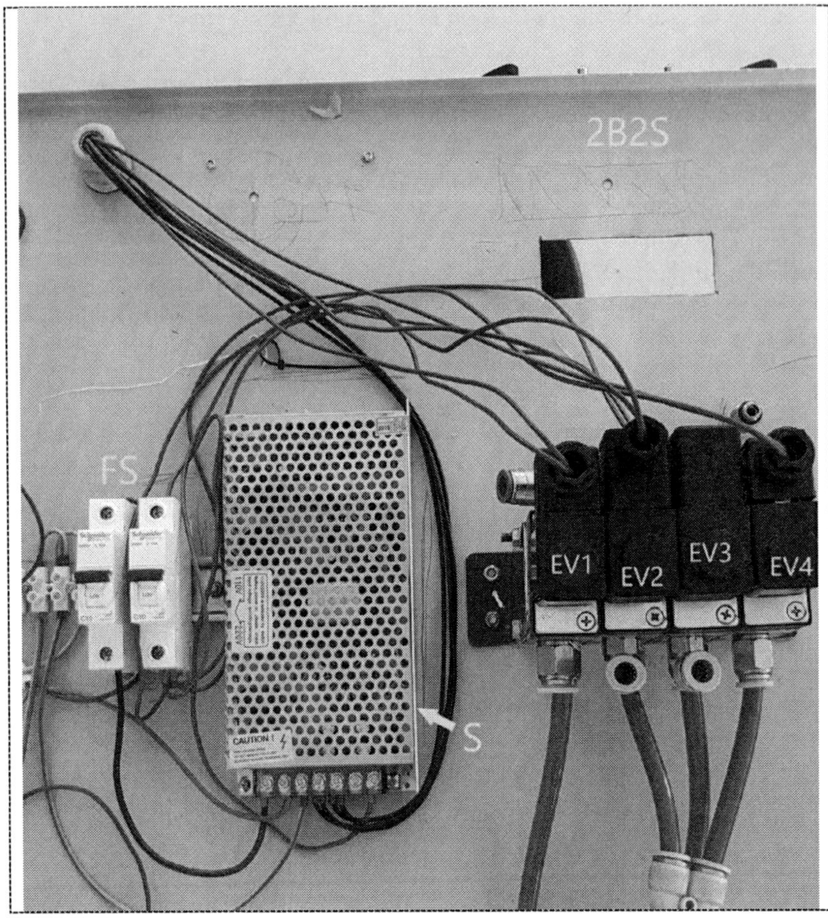

Figure 3. Automatisation system; EV1 – Electro Valve 1; EV2 – Electro Valve 2; EV3 – Electro Valve 3; EV4 – Electro Valve 4; S – AC to DC source; FS – Safety Flip-Switch; 2B2S – 2 buttons and two switches for on/off and start/stop sequences

4. Results and conclusion

Although we did not measure the time necessary for a full substrate change on the PXI [6], we did however make 2-3 measurement by timing the process with a stopwatch, and every time we came under an hour. That means that instead of a maximum of 3 depositions, we can make at least eight depositions per day, effectively tripling our output. This is a great achievement, and the whole experiment helped us understand the industrial approach towards the process, where every minute or second counts. Of course, fine-tuning of the system and further advancements in automation will be considered.

One of our goals is to get the magnetron sputtering deposition system in a state that can be used by other personnel or students from University of Oradea, with minimal training, or a one-sheet instruction page and a check-list ideally. The automation of the Load-Lock mechanism is a step further into reaching this goal, although it can be improved to visually show errors and the state of the valves and sensors.

References

[1] T. Maghiar; P. Ungur; N. Voicu; I. Moga; T. Buidoş, - MAGNETRON – Theory elements, construction, technology (MAGNETRONUL – Elemente de teorie, construcţie, tehnologie), Ed. Univ. din Oradea, ISBN, 973-8083-04-4, 221 pg, 2000;

[2] Ungur P., Maghiar T., Moga I., Vesseleny T. & Mudura P. - Theoretical contribution regarding electron path modfication in cathod – anode interaction space of the magnetron, Proceeding of the 13th International Daaam Symposium, Edit. by B. Katalinic, Published by daaam International 23-26th October,Viena, Austria 2002, ISBN 3-901509-29-1, pp 579-580, 2 pg, 2002

[3] https://erris.gov.ro/Research-Laboratory-for-Adva-1

[4] Moga, I., Ghincu, R., Costea, T., Tarca, D., & Moldovan, O. (2018). Some experiments regarding magnetron sputtering deposition with small capacity devices. *Nonconventional Technologies Review* 22(4), Romanian Association of Nonconventional Technologies, Dec 2018

[5] Dan-Ioan ȚARCĂ, Moldovan Octavian ALIN, Remus GHINCU, Moldovan Ovidiu GHEORGHE, Traian COSTEA, Polojințef Corbu NICOLAE, Installing a high capacity sputtering system at University of Oradea, *Electrical Engineering and mechatronics conference EEMC'19, University of Debrecen, Faculty of Engineering,* May 2019

[6] D I Țarcă, L Csokmai, O Moldovan, R Ghincu and D Potroviță, Experimental data collection system for reading pressure levels in a vacuum environment, *IOP Conf. Ser.: Mater. Sci. Eng.*, 2019, DOI: 10.1088/1757-899X/568/1/012078

[7] https://www.lesker.com/newweb/faqs/question.cfm?id=28

[8] D. M. Hoffman, B. Singh şi I. John H. Thomas, Handbook of Vacuum Science and Technology, Academic Press, 1998.

[9] K. Jousten, Handbook of Vacuum Technology, Second, Completely Revised and Updated Edition, Wiley-VCH, 2016.

[10] R. V. Stuart, Vacuum Technology, Thin Films and Sputtering, New York: Academic Press, 1983.

[11] J. Moore, C. Davis, M. Coplan şi S. Greer, Building Scientific Apparatus, Fourth Edition ed., Cambridge: Cambridge University Press, 2009.

[12] J. F. O'Hanlon, A User's Guide to Vacuum Technology, Third Edition, Hoboken, New Jersey: John Wiley & Sons, 2003.

[13] N. Yoshimura, Vacuum Technology Practice for Scientific Instruments, Berlin: Springer, 2008.

Acknowledgements

The authors would like to thank Prof. Dr. Eng. Habil. Tiberiu Vesselenyi

Partial support for the work was provided by PhD student. Eng. Remus Ghincu.

Partial support for the work was provided by Ş.l..Dr.Eng. Moldovan Ovidiu.

Funding for the present work was provided by the Doctoral School of Industrial Engineering, University of Oradea,

IOP Conf. Series: Materials Science and Engineering **898** (2020) 012047 doi:10.1088/1757-899X/898/1/012047

Quick response manufacturing as a promising alternative manufacturing paradigm

E A Gromova[1]

[1]Peter the Great St. Petersburg Polytechnic University, Politechnicheskaya st., 29, St. Petersburg, 195251, Russia

E-mail: lizaveta-90@yandex.ru

Abstract. The fourth industrial revolution initiates a change in the strategic guidelines of industrial enterprises. The rapid trend towards a wide variety of finished products with a short development and production time has led to a number of problems for enterprises with inventory, overhead and efficiency. The imperative of implementing lean production is becoming irrelevant due to new challenges in the external environment and is giving way to more competitive paradigms. Now the modern production concept should be focused not only on the cost and quality of goods, but also on the speed of response and flexibility. Quick response manufacturing (QRM) is one of them, which is a corporate strategy to reduce time of execution of the order at the industrial enterprise. And it is an extremely new and unfamiliar management paradigm in Russia. The purpose of this study is to analyze quick response manufacturing and to relate it to the current state of the Russian industrial development. Theoretical and practical aspects of this concept are given. The examples of the quick response manufacturing implementation in the Russian industrial sector are presented. Summarizing, QRM is offered as a promising manufacturing paradigm for Russian industry.

1. Introduction
Now great attention is paid to saving time in the course of production and sale of products. This is due to the onset of the fourth industrial revolution – an increase in the degree of unpredictability and market instability, rapid changes in technology and increased competition. With the beginning of the fourth industrial revolution, the business environment is characterized by changes that occur at the highest speed and on a huge scale, as well as described by the systemic nature of the consequences. The rapid trend towards a wide variety of finished products with a short development and production time has led to a number of problems for enterprises with inventory, overhead and efficiency. In this case, attempts to use mass production methods cannot be successful. The idea of "making large batches of products more profitable than small ones", based on achieving "economies of scale" and low unit cost, has lost its relevance. The mass production paradigm does not apply when consumers expect to receive highly customized products produced in a small batch, with an order reference point corresponding to the "custom development" position, and with the provision of additional services and value-adding benefits, such as subsequent product upgrades and reconfigurations, which are considered as important as the product itself. In turn, the imperative of implementing lean production (LP) [1-3] is also becoming irrelevant due to new challenges in the external environment and is giving way to more competitive paradigms – quick response manufacturing (QRM) [4-5] and agile manufacturing (AM) [6-10].

Thus, the purpose of this study is to analyze quick response manufacturing and to relate it to the current state of the Russian industrial development. Noteworthy that Russian sector of economy has traditionally been conservative. The theoretical and practical aspects of the modern production concepts are studied by many researchers. Among them the following scholars are highlighted: T. Ohno (1988), M. Imai (1997), J. Womack and D. Jones (2003), P.T. Kidd (1994), S.L. Goldman, R.N. Nagel and K. Preiss (1991), A. Gunasekaran (1998, 2014, 2017), R. Suri (1998, 2010). Focusing on QRM, R. Suri proposed the concept of quick response manufacturing. It means responding to customer requests so that you can quickly develop and release products that are responsive to those requests. This concept is based on the continuous reduction of the time required to perform all types of activities in the company, while ensuring quality improvement, cost reduction and faster response to changes in the situation. The roots of the concept go to the strategy " time-based competition", which was proposed by G. Stalk and T. M. Hout [11]. This model was concentrated on the use of speed to gain a competitive advantage: a company that uses this strategy provides products or services to consumers faster than competitors.

2. Materials and Methods

Quick response manufacturing is a management concept that aims to radically reduce time costs at all stages of the production cycle and office operations. The external aspect of the concept is a response to the needs of the client, which consists in the rapid development and production of products that take into account all individual characteristics. The internal aspect is to reduce the time of any operation in the context of the whole enterprise. The QRM concept focuses on reducing the overall lead time. The bottom line is the downtime between segments of actual work that take up most of the time. So, the focus of management attention is focused on the critical production path, which covers the time period from the moment when the customer makes an order to the delivery of the first product from this order. Thus, the main idea is to reduce the order lead time due to all the company's operations, both internal and external. Companies that have mastered QRM concepts usually manage to reduce the time for developing new products and the time from receiving an order to shipping finished products by 40-60%, while reducing total costs by 20-30%. It is very important that such impressive results are achieved without major investments, mainly due to the organizational factor-changes in the organization of production and management.

According to the group of authors [12], QRM has a lot in common with the lean production paradigm, such as cross-functional teams, on-demand production, and quality programs. R. Suri [5] suggests that QRM actually strengthens lean production. A. Gunasekaran and Y. Yusuf [13] define QRM as a characteristic of an agile enterprise. Other scientists [14] express the opinion that regardless of the type of product characteristics shown, applying the fundamental principles of LP, based on relentless elimination of production losses to increase value for consumers, will always be an ideal starting point on the path to operational excellence. This also applies to the central principle of QRM-the constant reduction of execution time. The difference between modern concepts is presented graphically (figure 1).

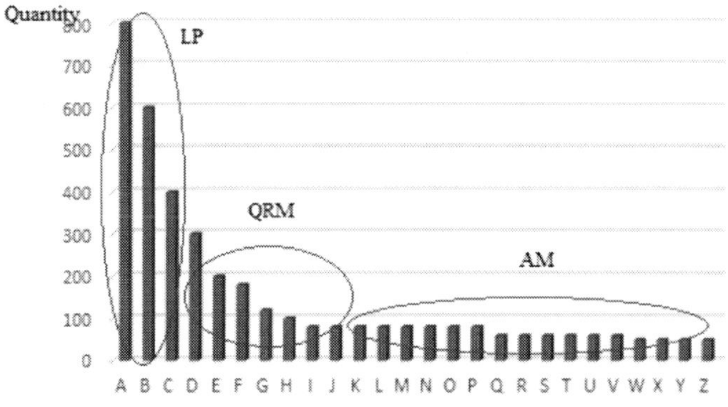

Figure 1. Differences in concepts in the quantity and range of products (developed by the author)

Figure 1 shows that the main difference between LP and AM is that when LP seeks to eliminate both product diversity and process variability (through strict application of standardization), AM systematically pursues the ability to respond by creating and providing flexibility.

3. Results and Discussion

Lean production is the most common modern concept in Russian industrial sector. Quick response manufacturing and agile manufacturing have been implemented sporadically in Russian industrial sector of economy [15-21].

The first Russian QRM Conference was held on April 16-17, 2015 in Perm. The Conference featured presentations by leading developers of the QRM concept and consultants from the USA, Germany, France, Holland, Switzerland and Russia. Participants had the opportunity to get acquainted with the experience of QRM implementation in Russian enterprises, both from the presented reports and during a visit to a company that is successfully implementing this concept.

Public joint stock company "Perm scientific and production instrument-making company" (PJSC" PNPPK") is engaged in the development and production of sensors and systems for navigation, stabilization and orientation of various types of mobile objects. It is one of the leading Russian suppliers of navigation devices and systems. The company has a developed scientific and research base for the production of various types of fibers and fiber-optic components. Together with its subsidiaries, PJSC PNPPK forms the Photonics cluster of fiber-optic technologies.

The company's QRM strategy policy for 2017-2020 has been adopted. The essence of the policy is to achieve the following strategic goal: to reduce the time required to complete customer orders and increase competitive advantage by creating a QRM methodological base and a flexible organizational structure in PJSC PNPPK.

Main policy principles [22]:

- strategic reference point. Creating a strategic reference point in the company-reducing the order lead time;
- comprehensive approach. Applying the principles of the QRM strategy throughout the company: in all production and office divisions, at every workplace;
- improvement. Optimize existing company processes with an emphasis on reducing lead time.
- competence. Education in production and office collectives of multifunctional employees who have the ability to perform related professions and functions, mutual training and mutual assistance in work;
- motivation. Implementation of a comprehensive motivation of the company's employees, which takes into account, along with the main indicators, the guideline for a permanent reduction in order completion time;
- organizational structure. Formation of cellular structural units in the production and office environment aimed at reducing the critical path of production and functioning on the basis of the principles of self-organization, interchangeability and orientation to the overall result;
- planning. Strategic planning of resource utilization and backup capacity, ensuring rapid response to changes;
- management of the relationship. Translation of the principles of the QRM strategy when interacting with partners and contractors of the company.

An unique QRM system is also being implemented at the Public joint-stock company «Chelyabinsk forging and press plant». PJSC «Chelyabinsk forging and press plant» is one of the leading enterprises in the machine-building industry. The main goal of implementing this organizational technology is to respond more quickly and adequately to market changes, and to effectively switch from fulfilling serial orders to producing short batches designed for specific needs of consumers. In the production of dies and molds, so-called cells are created, "sharpened" for the production of certain products. Since November, the cell for the production of round inserts of forming dies and ejectors for forge No. 2 has been operating in full operation. In the near future, there will be six more cells. They will work on the principle of "one window" – when all operations are performed by one team, in one area. The effect of reorganization is obvious. The team speaks "the same language", people solve problems without delay, do everything possible to reduce the number of operations by combining or eliminate unnecessary ones that do not provide an increase in value for the client.

Already, the cycle time for the production of stamps in the shop has been reduced to 15 days. For comparison, a year ago, this figure was 45 days [23]. The goal set for production workers is to reach an average cycle of 7 days. All this, in the end, allows you to reduce the time to develop new products, reduce inventory throughout the production chain. Working within the production cell creates conditions for increasing the level of competence of everyone involved in the process. In particular, machine operators successfully master related professions: toolmaker, slinger, turner, milling machine operator.

According to the head of production of stamps and molds M. Davydkin: "The implementation of the QRM concept is not limited to production halls. Large time losses in the overall cycle from order to delivery of finished products fall on office structures – "efficient producers" of overhead costs. Therefore, special emphasis will also be placed on optimizing office procedures in the near future." Thanks to the acquired flexibility, the output of new products has increased. If in 2015, about 70 new products were developed in the blacksmithing industry, in 2016-200 [23]. These new products helped to cover the decline in the market: the plant makes parts for all mechanical engineering, from the automotive industry to energy engineering, and there in many segments last year, the decline continued. So, the plant has become more responsive to the needs of the market, and as a result, its share has grown. The plant as a whole increased revenue in 2016 by 35%, and production where stamps are used — by 50%. The percentage of deliveries of the plant "just in time" has increased. All the components in the end add up to one big advantage – the plant produces high-quality products in the shortest possible time.

4. Conclusion

In conditions of extremely high competition, which characterizes the current state of business, companies that are able to respond quickly to changes in the external environment, will survive. Many companies strive to implement modern manufacturing paradigms, such as quick response manufacturing (QRM), which is a corporate strategy to reduce time of execution of the order at the industrial enterprise. Thus, the following conclusions can be drawn:

1) the fourth industrial revolution initiates a change in the strategic guidelines of industrial enterprises. Transformation of operational models into new digital models entails an identification of new effective models for the industrial development in new reality. QRM is one of them, which main idea is to reduce the order lead time due to all the company's operations, both internal and external;

2) at this critical time for the Russian industry, the ability of business organizations to quickly respond to customer requests, develop new products in the shortest possible time, and at the same time ensure high quality and reduce costs is of particular importance. In achieving these goals, the unique QRM toolkit, which has been adopted by an increasing number of manufacturing companies in the United States and Western Europe over the past decade, can be very useful. In Russia QRM is an extremely new and unfamiliar management paradigm;

3) successful existing initial examples (PJSC «Perm scientific and production instrument-making company», PJSC «Chelyabinsk forging and press plant») of QRM implementation indicate the possibility of improving the state of the Russian industrial sector of the economy and, in general, the prospects of this concept in Russia.

References

[1] Ohno T 1988 *Toyota Production System*, Productivity Press
[2] Imai M 1997 *Gemba Kaizen: A Commonsense, Low-Cost Approach to Management*, McGraw-Hill
[3] Womack J and Jones D 2003 *Lean Thinking*, New York: Free Press
[4] Suri R 1998 *Quick response manufacturing: a companywide approach to reducing leadtimes*, Productivity Press

[5] Suri R 2010 *It's About Time: The Competitive Advantage of Quick Response Manufacturing*, Productivity Press
[6] Goldman S L, Nagel R N and Preiss K 1995 *Agile competitors and virtual organizations: strategies for enriching the customer*, Van Nostrand Reinhold
[7] Gunasekaran A 1998 Agile Manufacturing: Enablers and an Implementation Framework, *International Journal of Production Research* **36**(5) 1223-1247
[8] Gunasekaran A 2017 *International Journal of Production Research* 1-13
[9] Dubey R and Gunasekaran A 2014 Agile Manufacturing: framework and its empirical validation, *The International Journal of Advanced Manufacturing Technology* **76**(9-12) 2147–2157
[10] Kidd P T 1994 *Agile manufacturing: forging new frontiers*, Addison-Wesley
[11] Stalk G and Hout T M 1990 *Competing against time: how time-based competition is reshaping global markets*, Free Press
[12] Lee W G K, Baines T, Tjahjono B, and Greenough R 2006 Towards a conceptual framework of manufacturing paradigms, *SIMTech Technical Reports*, **7**
[13] Gunasekaran A and Yusuf Y 2002 Agile manufacturing: a taxonomy of strategic and technological imperatives, *International Journal of Production Research*, **40** 1357-1385
[14] Powell D J and Strandhagen J O 2012 21st Century Operational Excellence: Addressing the Similarities and Differences between Lean Production, Agility and QRM, In: Industrial Engineering and Engineering Management (IEEM), 2012 IEEE International Conference, Hong Kong, 449-453
[15] Gromova E A 2019 Digital economy development with an emphasis on automotive industry in Russia, *Espacios* **40**(6) 27
[16] Rudskaya I and Rodionov D 2017 Econometric modeling as a tool for evaluating the performance of regional innovation systems (with regions of the Russian Federation as the example), *Academy of Strategic Management Journal* **16**(2)
[17] Pupentsova S V and Livintsova M G 2018 Qualimetric assessment of investment attractiveness of the real estate property, *Real Estate Management and Valuation* **26**(2) 5-11
[18] Kalinina O and Valebnikova O 2018 Human Capital Management as Innovation Technologies for Municipal Organization, *Advances in Intelligent Systems and Computing* 1315-1322
[19] Rodionov D G, Konnikov E A and Konnikova O A 2018 Approaches to ensuring the sustainability of industrial enterprises of different technological levels, *Journal of Social Sciences Research* 277-282
[20] Glukhov V, Turichin G and Klimova-Korsmik O 2016 Quality management of metal products prepared by high-speed direct laser deposition technology, *Key Engineering Materials* **684** 461-467
[21] Kuladzhi T, Babkin A and Murtazaev S A 2018 Matrix Tool for Efficiency Assessment of Production of Building Materials and Constructions in the Digital Economy, *Advances in Intelligent Systems and Computing* **692**
[22] *Official site of the PJSC "Perm scientific and production instrument-making company"* http://pnppk.ru/ru/home/kachestvo-i-sertifikatsiya/politika-kompanii-v-oblasti-strategii-bystroreagiruyushchego-proizvodstva-qrm (accessed on Apr. 2020)
[23] *Official site of the PJSC "Perm scientific and production instrument-making company"* https://www.chkpz.ru/news/27/el13082/?sphrase_id=23646 (accessed on Apr. 2020)

Acknowledgements

This research work was supported by the Academic Excellence Project 5-100 proposed by Peter the Great St. Petersburg Polytechnic University.

Simulation modelling as a method of risk analysis in real estate valuation

E A Gromova[1] and S V Pupentsova[1]

[1]Peter the Great St. Petersburg Polytechnic University, Politechnicheskaya st., 29, St. Petersburg, 195251, Russia

E-mail: lizaveta-90@yandex.ru

Abstract. The advent of the fourth industrial revolution was marked by the fact that changes became the main characteristics of the new time. It means a high level of uncertainty and instability in the business environment. The real estate market, being a part of the investment market, is changing. The current development of Russia's investment potential leads to an increasing need for professional management of the profitability of such investments. Thus, the main goal of this research is to present non-traditional methods of risk assessment, which should be used in determining the market value of the real estate object at the present time. Simulation modelling is a method of research in which the system under study is replaced by a model that describes the real system with sufficient, with which experiments are conducted in order to obtain information about this system. An example of the application of simulation modelling in the analysis for the purpose of choosing the best and most effective use of a land plot as conditionally free is given. Financing of investment projects is a dynamic process that requires a risk assessment.

1. Introduction

The fourth industrial revolution has an impact on all spheres of activity, and the economic sphere is experiencing the most significant changes. Big changes and a high degree of uncertainty in the external environment are new challenges of our time [1-2]. The real estate market, being a part of the investment market, is the sphere of capital investment in real estate objects as a financial instrument in real form. The current development of Russia's investment potential [3-6] leads to an increasing need for professional management of the profitability of such investments. In addition to evaluating the effectiveness of investment projects, the management functions include:

- qualitative risk analysis (structuring and forecasting of income and expenses for a real estate object as an investment asset, choosing the optimal size of a real estate object when investing in its development, evaluating the rate of return on capital) [7];
- quantitative risk assessments of real estate investments;
- finding ways to reduce risks in investment design [8];
- the development of recommendations about the increase of management efficiency with return on investment in real estate [9].

There are two types of investment in real estate [10]: the purchase of a profitable real estate object and the creation of a new profitable object. An investor almost always first needs to solve a strategic problem: whether to purchase a ready-made object (with or without subsequent reconstruction) or create a new object that meets the modern needs of the market. The final decision is made by the investor based on the analysis performed in order to select the best and most effective use of the object, taking into account the fundamental principles of valuation.

Evaluating the effectiveness of investment in real estate is based on processing a large amount of current input data, and also requires forecasting future cash flows and the value of the rate of return on capital. The characteristics of the investment object (its uniqueness and the lack of reliable market information in the public domain) lead to the fact that it is usually known not the specific values of the quantities used in calculations, but the ranges of their changes. Possible errors in the source data or in the forecast of their changes require the use of methods to account for the impact of such uncertainty on the results obtained.

And so, the final decision is made by the investor at the preparatory stage of the investment project [11], guided by the results of choosing the best and most effective use of the object. This analysis should be accompanied by an assessment of hazard indicators (under conditions of uncertainty) and importance (under conditions of risk). In addition to traditional methods of quantifying possible losses (sensitivity analysis and scenario analysis), it is necessary to use methods of simulation modelling, game theory, quasi metric modelling and hierarchy analysis.

So, the main goal of this study is to present non-traditional methods of risk assessment, which should be used in determining the market value of the real estate object at the present time, which is characterized by a high degree of uncertainty and instability of business environment. An example of the simulation modelling method is done.

2. Materials and Methods

Simulation modelling is a method of research in which the system under study is replaced by a model that describes the real system with sufficient accuracy (the built model describes the processes as they would take place in reality), with which experiments are conducted in order to obtain information about this system. Such a model can be "played" in time, either for a single test or for a given set of tests. The results will be determined by the random nature of the processes. You can use this data to get fairly stable statistics. Experimentation with a model is called imitation (imitation is the comprehension of the essence of a phenomenon, without resorting to experiments on a real object).

The use of the simulation modelling method [12-14] when choosing the best and most effective use of the real estate object allows to:

- simultaneously simulate random changes of several components of the project, taking into account the conditions of correlation;
- automatically generate scenarios from the ranges of possible changes in random variables and selected distribution laws;
- avoid errors when assigning probabilities for each scenario;
- to make a decision, being guided not only by maximizing the effective variable but also by an acceptable (minimal) measure of risk.

3. Results and Discussion

An example of the application of simulation modelling in the analysis for the purpose of choosing the best and most effective use of a land plot as conditionally free is given. The remainder technique is used to determine the value of a land plot.

The distribution of factors (cost of improvements, capitalization coefficient for improvements, net operating income, capitalization coefficient for land) is chosen evenly, since the ranges of input parameters are found for a specific object and it is assumed that in a given interval, the factor can take any value with equal probability. Based on the dependence of the test result on their number, derived by the author in [15], we assume the number of tests equal to 20,000. There is no multicollinearity between factors. The results of the analysis in order to select the best and most effective use of the land plot according to the cost maximization criterion are shown in figure 1 and figure 2.

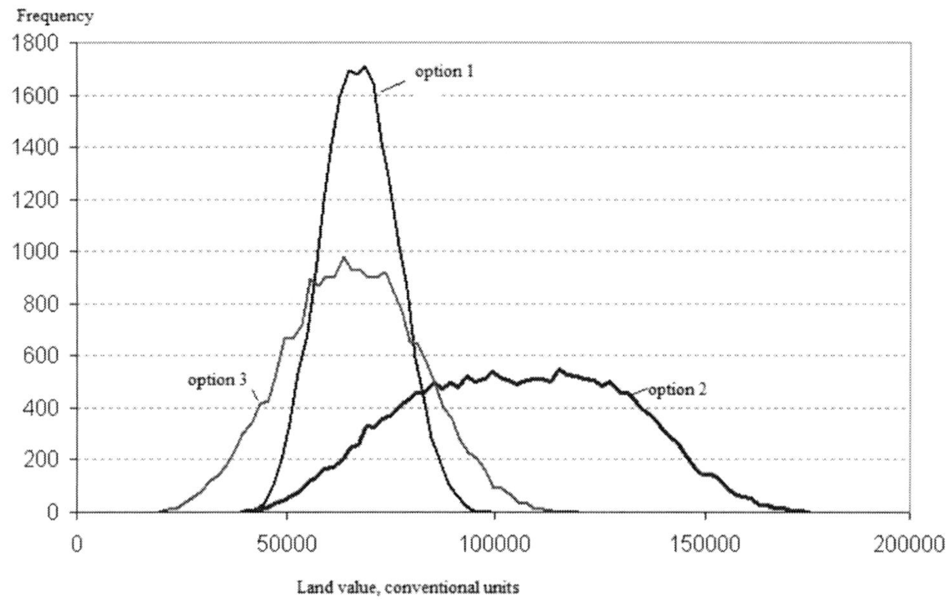

Figure 1. The histogram for the use of land (developed by the authors)

The first option has a wide and flat polygon, which indicates a high risk of uncertainty with a large spread. The polygon of option 1 is shifted to the right along the abscissus axis, i.e. this option corresponds to the maximum cost of land. Options 1 and 3 overlap, so the cost of land is about the same, but option 1 has a smaller spread (the polygon is narrow and drawn up), and therefore less risk than option 2,

Comparing the probability of realizing the value of land in these use cases, you can see that with a probability of 60%, the first of them has a greater value of the land plot in comparison with the third. The second option is 100% likely to have the maximum cost (figure 2).

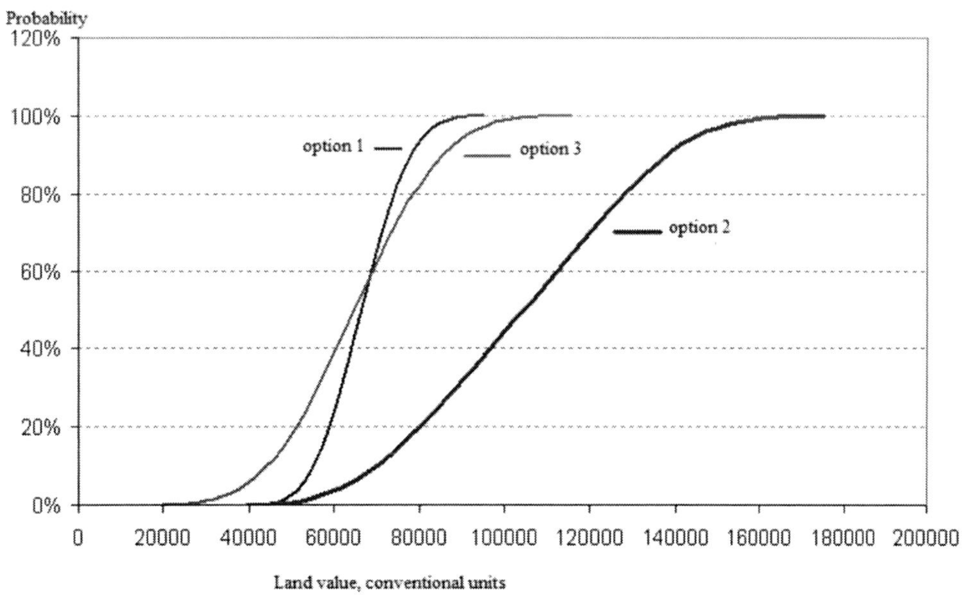

Figure 2. The cumulative probability for the use of land (developed by the authors)

Thus, the analysis confirmed the analyst's decision to choose the second option even at greater risk.

The application of the simulation modelling method will also help to determine the rate of return of the extraction technique and the cost of the object with a profitable approach. In this case, the variable factors will be the rental rate, operating expenses, the cost of reversion, as well as the dynamics of changes in these factors.

When evaluating investments in real estate, the question of the comparability criteria of the compared options in terms of choosing the size of the compared buildings of various functional purposes is not addressed. Meanwhile, it should be borne in mind that in accordance with the principle of added productivity and the above-mentioned principle of increasing and decreasing returns, the value of a land plot will first increase with the growth of the size of the building, and then begin to decrease.

Indeed, for a fixed plot size changes the size of the structure will be mainly due to the increase in the height of the building, which would itself increase the cost of construction per unit area of the latter.

4. Conclusion

In conclusion, it should be noted that a common disadvantage of investment design is the lack of development of the marketing strategy. Insufficient market and competition research leads to an overestimation of projected rental rates and occupancy rates.

In the conditions of uncertainty and instability of business environment provoked by the onset of the fourth industrial revolution, it is proposed to use not only traditional methods of quantitative assessment of possible losses (sensitivity analysis and scenario analysis), but also methods that are not used in standard packages (such as simulation modelling method) at the stage of choosing the best and most effective use of the object at a strategically important stage for the investor.

Financing of investment projects is a dynamic process at any moment, the project implementation conditions may change, which leads to automatic changes in the previously calculated project results. Therefore, much attention should be paid to risk analysis in project performance calculations.

References

[1] Schwab K 2017 *The fourth industrial revolution,* Eksmo, Moscow

[2] Kuladzhi T, Babkin A and Murtazaev S A 2018 Matrix Tool for Efficiency Assessment of Production of Building Materials and Constructions in the Digital Economy, *Advances in Intelligent Systems and Computing* **692**

[3] Rudskaya I and Rodionov D 2017 Econometric modelling as a tool for evaluating the performance of regional innovation systems (with regions of the Russian Federation as the example), *Academy of Strategic Management Journal* **16**(2)

[4] Rodionov D G, Konnikov E A and Konnikova O A 2018 Approaches to ensuring the sustainability of industrial enterprises of different technological levels, *Journal of Social Sciences Research* 277-282

[5] Ilin I, Levina A, Abran A and Iliashenko O 2017 Measurement of Enterprise Architecture (EA) from an IT perspective: Research gaps and measurement avenues, *ACM International Conference Proceeding Series*

[6] Glukhov V V, Ilin I V and Anisiforov A B 2015 Problems of data protection in industrial corporations enterprise architecture, *ACM International Conference Proceeding Series*

[7] Cherepovitsyn A, Metkin D, and Gladilin A 2018 An Algorithm of Management Decision-Making Regarding the Feasibility of Investing in Geological Studies of Forecasted Hydrocarbon Resources, *Resources*, **7**, 47

[8] Nikolova L V, Rodionov D G, and Afanasyeva N V 2017 Impact of globalization on innovation project risks estimation, *European Research Studies Journal,* **20**(2), 396-410

[9] Ozerov E S 2007 *Economic analysis and valuation of real estate*, SPb.:MKS

[10] Tarasevich E I 2000 *Analysis of investments in real estate*, St. Petersburg: MKS

[11] Damodaran A 2002 *Investment Valuation: Tools and Techniques for Determining the Value of Any Asset*, 2nd ed., Oxford, UK: Wiley

[12] Danielsson J, James K R, Valenzuela M, et al. 2016 Model risk of risk models, *Journal of Financial Stability*, **23**, 79-91

[13] Kumar V and Reinartz W 2016 Creating Enduring Customer Value, *Journal of Marketing*, **80**(6)

[14] Chen X, Christensen T, and Tamer E 2018 Monte Carlo Confidence Sets for Identified Sets, *Econometrica,* **86**(6), 1965-2018

[15] Pupentsova S V and Livintsova M G 2018 Qualimetric assessment of investment attractiveness of the real estate property, *Real Estate Management and Valuation* **26**(2) 5-11

Acknowledgements
This research work was supported by the Academic Excellence Project 5-100 proposed by Peter the Great St. Petersburg Polytechnic University.

Designing and validating the questionnaire used to measure the attitude of students towards e-Learning

D Zlatkovic[1], M Ilic[1], N Denic[2] and S Jovkovic[3]
[1]Alfa BK University, Faculty of Mathematics and Computer Sciences, Palmra Toljatija 3, Belgrade, Republic of Serbia
[2]University of Pristina, Faculty of Natural Sciences, Lole Ribara 29, Kosovska Mitrovica, Republic of Serbia
[3]Coledge of Applied Technical Sciences, Aleksandra Medvedeva 20, Nis, Republic of Serbia

dragan.zlatkovic@alfa.edu.rs

Abstract. Over the past two decades, numerous studies have been conducted to analyze the achievements of e-Learning; among other things, user's of information and communication technology (ICT) skills, Internet access and attitude of students towards e-Learning. The majority of studies examine the attitude by establishing the purpose and frequency of computer use, while some measure students' attitudes using a questionnaire without determining the measuring properties of the questionnaire. For this reason, the goal of this paper is to create a questionnaire used to measure the attitudes of students towards e-Learning and to determine its psychometric properties.
Keywords: e-Learning, students, attitude, questionnaire, ICT.

1. Introduction

The implementation of ICT achievements in teaching processes leads to changes at all levels of education, especially at higher education institutions, and requires new standards in teaching [1]. New information is created every day, the amount of ICT knowledge and skills is growing, and this requires "faster" education that is accessible and open, which in turn requires higher student literacy [2]. The basis of computer literacy in *e*-Education implies the use of software support, for example, for writing and text processing, for creating presentations and/or for creating graphic representations and tabular data processing. Furthermore, at the level of undergraduate academic studies (UAS), the description of competences for the majority of occupations strictly defines the necessary ICT skills, and this emphasizes the necessity of constant improvement and improvement of basic professional skills that require ICT [3-6]. Due to changes in the ICT, the teaching process is increasingly relying on multimedia and the Internet, and *e*-Education is mentioned more and more in the context of teaching.

It has been shown that the understanding of the academic community regarding the efficiency of *e*-Learning and the factors that affect the acceptance or non-acceptance of *e*-Learning is insufficient [7, 8]. It has been established that the efficiency of knowledge acquired based on the use of ICT is comparable to the knowledge acquired in traditional teaching [2].

It is a well-known fact that the acceptance of teaching based on *e*-Learning is influenced by motivation, advance ICT skills and the attitudes towards *e*-Learning [7, 9, 10-14].

2. Literature review

Further review of relevant scientific literature yielded the researches and questionnaires used to test the experiences of work with computers and/or to measure the attitude of students and professors towards learning on the Internet. Table 1. shows certain tools used to measure the attitude towards the use of ICT in the learning process of student population [15-18].

Table 1. Certain tools used to measure the attitude of students population towards the use of ICT in teaching process.

Year Country	Authors	Sample	Instrument (Questionnaire)	Number of items	Cronbach Alpha
2006 Austria	Link and Marz [15]	N=1160	Stance and experience of e-learning, online questionnaire	32	a=0.97
2007 India	Mirsha and Panda [16]	N=150	Development and factor analysis of an instrument to measure faculty stance towards e-learning	12	a=0.85
2007 Taiwan	Liaw et al. [17]	N=50 Instructor's	Computer and Internet experience and learner stances towards e-learning questionnaire	25	a=0.95
		N=168 Students		19	a=0.92
2010 United Kingdom	Wilkinson et al. [18]	N=60	Questionnaire for measuring student ICT skills, experience and stances to e-learning	47	a=0.87

Even though the above-mentioned researches have shown high reliability ($\alpha \geq 0.912$), careful examination of the statements has shown that their sense and/or content does not correspond to the situation in our system of higher education, i.e. to the way of work and the habits of our test subjects. All the questionnaires we examined were drafted in English language. The translation of these questionnaires would partially lose the original meaning of the statements, and thus it would be questionable whether any such (translated) questionnaire would indeed measure the attitude of students towards *e*-Learning. The translated questionnaire does not have the same psychometric properties as the original questionnaire. The reliability and validity of the questionnaire translated into Serbian language are not the same as the reliability and validity specified in the original research [3, 19].

3. Methodology

3.1. Research objective

The objective of this research is to design a questionnaire and determine psychometric properties of the questionnaire, such as reliability and validity, and to determine the attitude of students towards *e*-Learning.

The determination of the attitude of students towards *e*-Learning would contribute to a more efficient implementation of *e*-Learning, as well as to the teaching process development strategy in the institutions of higher education [20-22].

3.2. Questionnaire development

After reviewing the relevant scientific literature on *e*-Learning and the questionnaires used to measure the attitude towards *e*-Learning, as well as based on our research done so far [2, 7, 9], a group of over 50 items has been prepared. During the preparation of statements, the rules and recommendations for questionnaire designing have been taken into consideration [9, 14, 18]. A total of eleven items have been taken from the Questionnaire stance and experience of *e*-Learning, online questionnaire [15] and adapted to our research, from the Questionnaire development and factor analysis of an instrument to measure faculty stance towards *e*-Learning [16] a total of nine items have been taken, while five items have been taken from the Questionnaire for measuring student ICT skills, experience and stances to *e*-Learning [18]. For the first version of the questionnaire used to measure the attitude towards e-Learning, 20 items were selected, 10 positive and 10 negative. The positive items describe the awareness, feelings and positive behaviours towards e-Learning. The negative items describe dislike, negative behaviours or awareness towards e-Learning. Both positive and negative items in the questionnaire were given in random order to maintain the attention of the test subjects during the completion of the questionnaire and to reduce untruthful responses to the minimum. The responses to the items were shown on the Likert scale from one to five for positive statements (ranging from "strongly agree" to "strongly disagree") and opposite, from five to one for negative items (ranging from "strongly disagree" to "strongly agree"). The introductory part of the questionnaire includes demographic data about the test subjects: gender, faculty, year of study, place of study, questions concerning the frequency of computer use, questions concerning the purpose of computer use, while the last two questions relate to the familiarity with the concept of *e*-Learning and to attending *e*-Lectures.

3.3. Psychometric properties of the questionnaire

In order to determine the validity of the questionnaire, the exploited factor analysis has been used, specifically Principal Component Analysis, while the Internal-consistency method has been used to determine the reliability, which involved the determination of the Cronbach's Alpha reliability coefficient.

The validity has been determined indirectly by determining the number of factors, while the number of factors has been determined based on the results of the Cattell scree test, through the use of Guttman-Kaiser criterion and by calculating the value of inter-correlation between opposite factors and by calculating the value of inter-correlation between factors and individual statements. An Oblimin rotation of items has been done due to the fact that orthogonal rotation did not result in the sequence of items that would be meaningful. The appearance of the Scree curve and the number of items with the own value of >1 (Guttman-Kaiser criterion for factor reduction was used) have determined the number of factors. An additional reduction of the factors was made when there was a significant correlation between individual factors (for the inter-correlation coefficient >0.6). The critical value of Cronbach's Alpha coefficient of reliability is the value of > 0.70.

According to Guttman-Kaiser's criterion, we only require factors >1. We have five such elements. These five factors explain a total of 70.185% variance. According to the Cattell criterion, we look for a fracture point on the transient diagram. This is point 3, and we keep only the factors above (before point 3), the first two. These two factors explain much more of the variance than the remaining elements.

3.4. Sample of subjects

The analysis covered a total of 121 subjects, of which 51 or 42.1% were male and 70 or 57.9% were female. From the total of 121 subjects, 20 subjects or 16.5% were the students of the first year, 21 or 17.4% were the students of the second year, 32 or 26.4% were the students of the third year and 48 or 39.7% were the students of the fourth year. As many as 93 subjects, i.e. 76.9%, were studying in capital city - Belgrade while 28 subjects or 23.1% came from other cities. The response percentage was 100%.

3.5. Procedure

Prior to the survey, the test subjects were familiarized with the goal of this research, as well as with the questionnaire completion procedure. At the very beginning of the lecture, the students have filled in the questionnaire in very similar conditions. All the subjects have filled in the questionnaire voluntarily and independently, while the time needed for the completion of the questionnaire was around fifteen minutes. The questionnaires were fully filled in by 121 test subjects. The response percentage was 100%.

The research procedure that was carried out during June-July 2019.

4. Interpretation of results

4.1. Validity and Reliability of the Questionnaire

Twenty items on the use of ICT in teaching were subjected to factor analysis (main component analysis).

Firstly, we assessed the suitability of data for factor analysis. By reviewing the correlation matrix, we noticed that there were many coefficients above 0.3. The value of Kaiser-Meyer-Olkin indicator is 0.867 > 0.6. The Bartlett's Test of Sphericity (Table 2.) has reached statistical significance (< 0.05).

Table 2. KMO and Bartlett's test.

Kaiser-Meyer-Olkin Adequacy	Measure of Sampling	0.867
Bartlett's Test of Sphericity	Approx. Chi-Square	1468.441
	df	190
	Sig.	0.000

A two-factor solution explains 52.415% of the variance, where the contribution of the first factor is 36.271%, and of the second factor 16.144%. The rotated solution (Oblimin) has shown that both factors have a lot of large factor loadings and that all the variables yield significant loadings to only one of the factors. There is a moderate positive correlation between these two factors.

A principal component factor analysis has been done, which incorporated a scree curve analysis, a Guttman-Kaiser criterion for the reduction of principal factors, as well as the analysis of inter-correlations between the factors and individual items and analysis of intra-correlation between factors. We have determined two factors: a positive attitude towards *e*-Learning and a negative attitude towards *e*-Learning.

Cronbach's Alpha reliability coefficient for the questionnaire has shown the value of 0.900, which means that the questionnaire is highly reliable (Table 3.). Therefore, a good correlation has been established for all 20 items between individual items and the factors (majority of inter-correlation coefficients amounted to > 0.3), as well as some minor overlapping between two factors. The intra-correlation between these two factors amounted to r=0.305, which describes a good correlation and allows for the measuring of the overall attitude of the students (Table 4.).

Table 3. Reliability Statistics.

Cronbach's Alpha	Cronbach's Alpha Based on Standardized Items	N of Items
0,900	0,896	20

Table 4. Component Correlation Matrix.

Component	1	2
1	1.000	0.305
2	0.305	1.000

4.2. Attitude Towards e-Learning

The overall attitude of students towards *e*-Learning is highly positive and amounts to 78.65±14.35.

The only significant difference was found in the frequency of Internet use. By applying the independent *t-test*, we found that there was a significant difference (p = 0.026) in students' stance towards *e*-Learning comparing students using the Internet weekly (66.00 ± 18.751) in the ratio to students using the Internet daily (79.31 ± 13.876). Students who use the Internet daily have a more positive stance.

IOP Conf. Series: Materials Science and Engineering **898** (2020) 012049 doi:10.1088/1757-899X/898/1/012049

5. Conclusion

The application of this questionnaire to the student population in the Republic of Serbia and the statistic processing of data would allow for a better understanding of attitudes of students towards *e*-Learning. At this moment, there are no rules of profession in relation to the designing of *e*-Courses, there are no recommendations on how much of the lectures should be optimally organized in the traditional way and how much in the form of *e*-Lectures. In education, it is surely important to raise the awareness of the teachers and educational institutions about the use of new forms of teaching, such as *e*-Learning. Now when we know the attitude, we can predict the intention and the behaviour of the students in relation to *e*-Teaching. The results obtained should help with the defining of the changes in the teaching process at universities, and potentially with the designing of guidelines for *e*-Education.

References

[1] Wilkinson A, While A E and Roberts J 2009 Journal *of Advanced Nursing*. **65** 755
[2] Denic N and Zlatkovic D 2017 *Proc. Int. Conf. on Science and Education* (ICONSE 2017 – Antalya) p 30
[3] Nunnally J C and Bernstein L H 1994 *Psychometric theory* (3rd edition). New York: McGraw-Hill.
[4] Misut M and Pribilova K 2015 *Procedia-Social and Behavioral Sciences*. **177** 312
[5] Ezugwu A E, Ofem P O, Rathod P, Agushaka J O and Haruna S 2016 *Procedia-Social and Behavioral Sciences*. **92** 568
[6] Klimova B F 2015 *Procedia-Social and Behavioral Sciences*. **186** 898
[7] Zlatkovic D, Denic N, Petrovic M, Ilic M, Khorami M, Safa A, Wakil, K et al. 2020 *Comput. Appl. Eng. Educ.* (2020) 1. https://doi.org/10.1002/cae.22251
[8] Tomás R, Cano M, Santamarta, J C and Hernándes - Gutiérrez I E 2015 *Procedia-Social and Behavioral Sciences*. 191,1644
[9] Zlatkovic D, Denic N, Petrovic M and Ilic M 2019 *Proc. of 54th International Scientific Conference on Information, Communication and Energy Systems and Technologies (ICEST - Ohrid)* p 194
[10] Festerová M 2015 *Procedia-Social and Behavioral Sciences*. **191** 1275
[11] Khan B H 2005 *Managing e-Learning: Design, Delivery, Implementation, and Evaluation*. Hershey, PA: Information Science Publishing.
[12] Zaharias P and Poylymenakou A 2009 *Inter. J of Human-Computer Interaction*. **25**(11) 75
[13] Crook C 2000 *Motivation, and the ecology of collaborative learning*. In Joiner R, Miell D, Littleton K, Faulkner D (Eds.), Rethinking collaborative learning. London: Free Association Press.
[14] Tseng T, Chiang F and Hsu H 2008 *Computers in Human Behavior*. **24** 940
[15] Link T M and Marz R 2006 *BMC Medical Education*. **6**(34) 1
[16] Mishra S and Panda S 2007 *Asian Journal of Distance Education*. **5**(1) 27
[17] Liaw S, Huang H and Chen G 2007 *Computers and Education*. **49**(7) 1066
[18] Wilkinson J R and While A E 2010 *Computers in Human Behavior*. **26**(6) 1369
[19] Nikolic V, Petkovic D, Denic N and Milovancevic M 2018 Physica *A: Statistical Mechanics and its Applications*. **513** 456
[20] Šumak B, Heričko M and Pušnik M 2011 *Computers in Human Behavior*. **27**(6) 2067
[21] Nikolic V, Kaljevic J, Jovic S, Petkovic D et al. 2018 *Physica A: Statistical Mechanics and its Applications*. **511** 324
[22] Radosavljević M and Anđelković M 2016 *Annals of the University of Oradea Fascicle of Management and Technological Engineering*. **XXV**(XV), 2016-1. p 75

IOP Conf. Series: Materials Science and Engineering **898** (2020) 012050 doi:10.1088/1757-899X/898/1/012050

The influence of innovation onto the logistics process of goods transport by air

N Milošević[1], M Mišić[2] and N Denić[3]

[1] MSc in economics
[2] State University of Novi Pazar - Department of Technical Sciences, Vuka Karadžića bb, Novi
Pazar, Serbia
[3] University of Pristina - Kosovska Mitrovica, Faculty of Sciences, Ive Lole Ribara 29, Kosovska Mitrovica, Serbia

Email: denicnebojsa@gmail.com

Abstract. The logistics of air transport assumes the usage of the airplane and airport warehouse capacities to transport goods from one point to another. Within this paper the process of goods manipulation is explained by using air transport and innovations introduced by various companies have been presented. The modern transport market faces the companies with numerous challenges which these companies wish to address in an adequate manner. By innovating, the companies express their constant tendency to increase the competitive advantage by improving the efficiency and effectiveness of their business processes. The number of innovations is multiple and they exist in every area of air transport. The innovations mostly appear based on the company initiative, but can also be generated by state entities, as well as national and international institutions and associations.

1. Transport

Transport represents the movement of goods or passengers from one destination onto the other in a certain time frame. The transport market supply side consists of transporters which possess the capacities for transport such as trucks, boats and airplanes. The transport market demand side refers to individuals/companies which have the need to transport goods in order to fulfill their business processes. The global market of transport services is developed and all types of transport can be accessed. In some parts of the world, where infrastructure is inadequate, it is more difficult to perform transport activities, but in the 21st century, the number of such countries has significantly been diminished. The contemporary supply of transport capacities within certain types of transport is specialized for different kinds of loads that need to be carried.

The transport market, as is the case with the market of any other product is determined by its basic constitutive elements: the product (the transport service), space, time, supply and demand [1]. The transport service refers to the movement of goods from one place onto the other. Besides the physical manipulation of goods, this service may also include the acquisition of the adequate documentation.

2. Goods air transport

Due to the globalization of economics and the technological advancement, the company connectedness with the global market is very high, which enables these companies to do business activities with other companies around the globe. Therefore, the importance of air transport is increasing and it plays a vital role in the national economic development. The logistics of air transport assumes the usage of an airplane and warehousing services for a quick goods transport from the sending point to the receiving point. That is the fastest way of transport and it offers the advantages of security, speed, geographic and time flexibility.

In order to perform transport and the logistics services by using the air transport and adequate land infrastructure is needed, such as airports, terminals and warehouses. Also, it is needed to develop the infrastructure which would enable a high level of connectedness with the other means of transport, rail tracks and transport access points.

The air cargo can be transported by passenger and cargo planes. In case of using the passenger planes to transport goods, the passengers are seated in the upper part of the airplane, while the cargo is below them, in the space where the baggage is, called the "airplane stomach". The cargo planes are especially designed to transport only cargo. The cargo planes do not have seats but the entire space within them is adjusted to cargo transport.

The service lenders in air transport are a heterogeneous group of operators. They offer various types and levels of logistics services. There are three main categories of operators within the cargo air transport [2][3]:

- Line-haul operators
- Integrated operators, currier services
- Niche operators

Line-haul operators only perform goods transport from airport to airport. They do not negotiate with buyers, but pass over that activity to freight forwarders. These operators can be divided into: cargo, combined and passenger operators.

Cargo operators transport only the cargo within airplanes especially designed for that purpose and their fleet consists of only that type of airplanes.

Combined operators offer the services of cargo transport, both by cargo airplanes as well as passenger airplanes.

Passenger operators only use the space of "airplane stomach" to transport the goods, in case that this space is not used enough.

Integrated operations combine the functions of freight forwarders, air transporters and land transport within one subject, i.e. the management of the complete cargo delivery from the place of origin to the final destination. Most often, these companies have within their ownership the land transport and warehouse capacities and the fleet of cargo airplanes.

Niche operators perform the cargo of specialized equipment, technology and cargo with proportions and characteristics which demand a particular airplane type.

The types of goods transported by air transport are: IT components, devices, easily spoiled goods (food and agricultural produce), weapons, explosives, dangerous goods, construction materials, metal elements and parts, live animals, converters, cars, machines, pesticides, human remains, value shipments (gold, money, noble metals), human organs for transplantation etc [4]. Certain kinds of cargo need to have secured special conditions such as particular warehousing spaces, transport procedure, surveillance and protection measurements. Besides these conditions, it is needed for the employees performing the physical manipulation of the goods to pass through an adequate training.

3. The process of transport

The logistics process of goods transport via air represents a timely set endeavour which assumes the coordination of various subjects [5], transporters, freight forwarders, warehouse agents, customs, goods senders and receivers and the coordination of various activities of loading, unloading, warehousing, customs setting, documentation gathering etc.

The process begins once the goods arrive into the warehouse of the transporting agent who is in charge of goods manipulation [6]. That agent is usually an individual company which has an agreement relationship with the airline transporter, but this job can be performed even by the very transporter, as part of its business, especially within big and frequent airports. It is not such a rare case that one air transporter offers this kind of service to other transporters for a fee. The agent takes care of cargo movement from one airplane to another.

Depending on the type of cargo, the destinations onto which the goods need to be transported and the urgency, the delivery of the goods to the agents must be performed at a certain time before flight departure. That period before take-off is called slot-time.

The entire process of performing air transport by steps is shown within Figure 1.

Figure 1. Scheme of the entire airplane goods transport process.

Within air transport, there is the possibility of cargo being transported to the destination by cargo transfer, i.e. cargo being unloaded from one airplane and reloaded onto the other airplane at some other spot before the final destination. This only refers to airplane transfer, but not the change of the final destination. At that point, the cargo is in transit. It is up to the freight forwarder (who is in communication with the transport company whenever it is necessary) to decide whether to send goods directly or within the process of transit, which depends on the majority of factors such as: price, time (depending in a great deal on the timetable of flights of other airlines), special conditions regarding cargo (security, live animals etc.).

In case of a transfer/transit shipment, the process between flights would correspond to the one shown within Figure 2.

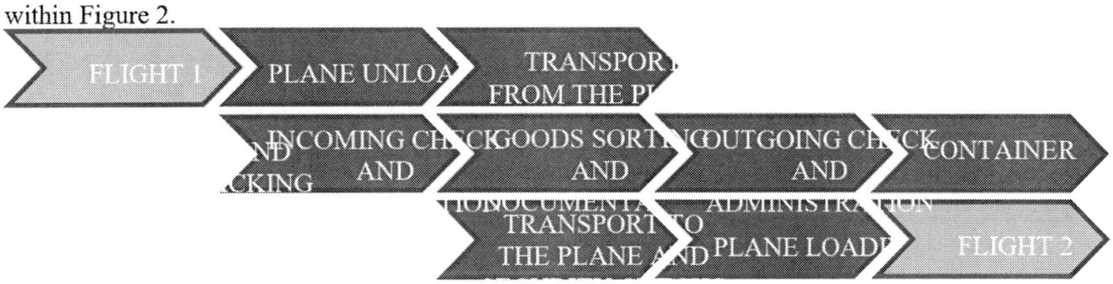

Figure 2. Scheme of the transit shipment process.

In case that goods with high degree of urgency are being transported, and if the flights fit the time frame, the process can be significantly shortened and shown as within Figure 3.

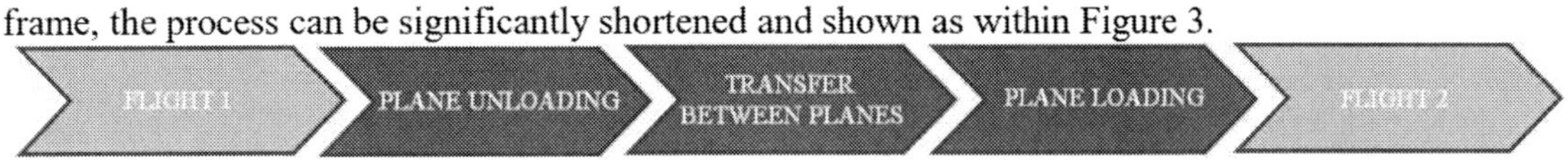

Figure 3. Scheme of the transport process of a very important transit shipment.

The checks before the loading and flight take off are very important for air carriers [7]. A check is performed of physical characteristics of the cargo and the containers, the following documentation, declarations, numbers, marks, security flight checks, etc. The documentation is prepared and temporarily situated and the goods are prepared for loading.

At a certain time before the take-off, the loading shall begin and the accompanying documentation shall be set in an appropriate bag. Once the process of loading is finished, the bag with the documentation is given over to the crew. The doors of the plane are closed and all is ready for take-off.

During the flight, the crew controls the temperature within the trunk. In many airplanes, the temperature and the circulation of air can be set by compartments. When a stable, low temperature is required, containers with cooling equipment are used, together with the isolation equipment, and also ice. In case of transporting bigger live animals (elephants, horses) the airplane can have an animal keeper who will take care of the animals during the flight. The transporters which have this option in offer have special premises where animals spend their time at the airport before the flight.

At a precisely determined time before landing the agent from the country of take-off shall get in touch with the agent from the country of landing and inform him about the details of the flight and delivery. That helps the agent to prepare the goods arrival, to plan the priorities while cargo unloading and to secure an uninterrupted and quick flow of the shipment through the following steps. This notification by the agent from the country of take-off is called the freight forward message. At the destination airport, the cargo and the very flight shall pass through the same procedure once more, only in a reverse order.

Besides physical manipulation, the other important agent functions are [5]:

- To control the total weight and balance of the plane on that side where the cargo is located, to form the loading list and take care of flight security;
- To form a bill of lading for the goods on the plane, for the needs of the air carrier import and export declaration;
- To fill out the form which is called the notification to the plane captain, which contains information for the crew about the possible risks of cargo in case of emergency situations and demanding conditions for the cargo in the trunk;
- To plan and control the reservation, time slots, flow of goods through the warehouse, the flow of goods from the plane and to the plane due to the prevention of hold ups.

4. The innovations within the logistics process

Within the modern transport market, cargo transporters in the air transport industry, especially those coming from the developing markets, are faced with numerous challenges. With the quick growth of the e-sales and production migration to South-Eastern Asia, the rising challenge faced by the modern transport companies is the efficient goods delivery from the aforementioned region to the final destinations. In order to secure business success, the air transporters must efficiently manage the logistics operations. Innovation is imperative in this current and uncertain environment for the survival of an organization [8]. The digitalization of documentation, which shall allow the improved efficiency of documentation processing, can enable the efficient operations management. Investing in information technology is not only an investment in the future but also in the present, where an effective business strategy is focused on the productive use of IT technologies [9]. New technologies enable companies to use information in real time and to easily coordinate business activities. The usage of technology in different operations can secure competitive advantage to business subjects in the way that it shall shorten the time of the cycle duration, enable inter-functional connecting and increase the degree of punctuality and reliability.

Programs, techniques and technologies of the modern IT sector are designed to follow and plan the cargo manipulation and the accompanying actions represent the solution for the modernization of the air transport. The innovations most often appear at the initiative of the companies, but can also appear at the initiative of state organs and national and international institutions and associations.

With the aim to increase the efficiency of logistics operations, to speed up the performance of the entire process of delivery and to enhance innovative thinking, IATA has started the program of air cargo transport transformation called the StB Cargo (Simplifying the Business) [10]. The program consists of six projects: e-freight and e-AWB, ONE record, Interactive Cargo, Smart Facility, ACID (Air Cargo Incidents Database), Cargo Connect.

That aim which is wished to be achieved by using e-freight is the development of the entire process without papers for goods air transport through the regulatory frame, modern electronic deliveries and high level of data quality. To achieve the mentioned goal the prime importance is set on e-AWB and Cargo-XML standard. E-AWB represents an electronic agreement about transport between the sender and the carrier. Cargo-XML standards are used for the electronic communication between air transporters, senders, freight forwarders, agents for cargo manipulation at the airports and custom services. Consequently, three types of paper documents are eliminated: customs, transport and commercial documents of the specialized cargo.

The essence of ONE record is to pass over from the model of message exchange between users onto the model of data sharing. ONE record is not a substitute for e-freight communicating, but is rather developing along its side. Every participant in the process of transport is enabled, with the use of internet technology and cloud hosting, to insert data in the mutual virtual shipment record as well as everyday access to information stored within the data base. For efficient performance measurement, the collection of data and their integration must be simple [11].

The interactive cargo is the project with the aim to develop technology which shall allow interaction of cargo with the data base, which facilitates the participants of the transport process the tracking of state within transport units and locating the cargo in real time.

The first aim of the Smart Facility project is to improve the visibility of the operator service for handling the cargo at the airports and increasing the capacities for performing the basic transport operations. For carrying out the first goal an on-line platform has been developed with publically available information on physical infrastructure, available resources and operator equipment. The second aim of the project is to simplify the procedure of operator capability revision to fulfil the operative standards and the availability of capacities by every air transporter. In order to ensure the safety of air transporters and to diminish the number of revisions and time spent on those activities, a mutually recognized revision has been established, as well as the scheme for the accreditation within the entire branch.

ACID has the aim to collect and secure reports on the incidents concerning cargo, without stating information on the subject and to unite the gathered safety and operative information into a unique data base, where based on these data trend analyses would be performed, the foreseen risk would be diminished and the programs for cargo manipulation improvement would be created.

The aim of Cargo Connect project is to simplify, standardize and modernize the connection with and between the cargo community systems. The absence of relevant and precise information can lead to process inefficiency and the appearance of hidden costs. Therefore, the digital cooperation is important in order to improve the efficiency by creating stronger bonds between air transporters, freight forwarders and other partners.

Besides IATA, companies are also examples of the development and implementation of new solutions. SEW-EURODRIVE is a company which is developing mechanical and software solutions for managing the movement of cargo within airports [12]. The company is familiar with the demands of the airport industry for the air goods transport and is trying to respond to these demands in an adequate manner. The aim is to achieve the highest degree of process automation, which brings to the reduction of costs and shortening the time period needed for the manipulation of goods at the airports. The technological solutions are offered for the entire process of physical manipulation which is performed at the airport, the loading and unloading of airplanes and trucks, the movement of pallets and containers within warehouses and between airplanes and warehouses.

The technology for safety cargo handling assumes the usage of transport rolls, rotation tables, track lines for containers and pallets and software for the control and management of these components. A special feature of the containers is that they demand big contact surfaces. They cannot, as is the case with pallets, be handled by stow sides, but must also be supported in the middle. As a result, todays solutions for warehousing are based on mechanical transport rolls, within which the containers can be moved and placed. Within contemporary solutions, in order to secure such big contact surface, several narrow lines are engaged simultaneously to pull one container.

The majority of airports have limited warehousing capacities so the cargo needs to be lifted and stored high, which results in modern high storage warehouses. The size of the high storage warehouse constantly keeps increasing, so therefore do the demands for the warehousing systems. The solutions which are offered are new vehicles and new systems of automated vehicle management.

To unload the cargo from the truck onto the warehouse and vice versa truck docks are used. Due to the diversity of trucks the height of the truck transfer points must be set using the truck docks. The ramp is self-adjustable and compensates for the height difference in the width of the vehicle. It also has installed protective battens which prevent damages within transport rolls.

Another example is the Web-Cargo Sky platform presented by the company Freightos. It represents a free service which allows the freight forwarders quick access to tariffs and information regarding the availability of capacities of more than 350 airlines and booking in real time [13]. The platform is used a search engine, by entering data on the departure and arrival destinations, data on physical characteristics of the load being transported, and there is a possibility to choose the currency and the desired transporter. The advantage is seen in the simple search of the base platform and the significant diminishing of time needed to acquire information and arrange the transport.

During the last competition organized by IATA in the field of innovation called "2019 IATA Air Cargo Innovation Awards" the winner was the company Unilode Aviation Solutions with the concept of digital transformation [14]. With its partners, Unilode developed a solution for tracking based on Bluetooth. These devices help the users to track the cargo at any moment, regardless of its location. With the aid of this technology data are gathered on temperature, moisture, hits and light, as well the information on current state within the airplane which is important for the users in case of specific cargo transport, such as easily damaged goods, precious goods or other sensitive cargo. The flexible reader network is important for the securing of global visibility and Unilode cooperates with the leading service lenders of transport activities within the branch and with airports in order to fulfil the set aims.

Besides these programs, there is also a notable development of platforms for the simplification and speed increase of performing the customs procedures. The example of such a platform is BE-GATE, developed at the initiative of the Department for customs and excises of Belgium, and supported by the airports in Brussels and Liege [15].

In case of imports, BE-GATE enables the processing of a greater number of notifications concerning the arrival of one such transaction, through the forms approved by the customs service. For deliveries of 22 € and less, this form can also be used as a declaration for placement into circulation. For deliveries greater than 22 € it is needed to present an additional declaration to the PLDA. The results of the selection (packages which must be shown to the customs) shall automatically be sent to the declarant, along with the needed information about the selection.

In case of exports, BE-GATE is used for shipments of value less than 1000 €, and the weight lower than 1000 kg. For greater values and weights it is needed to deliver additional declaration to the PLDA.

The type of goods for which this platform cannot be used is excise goods, goods which require a license, goods with limitations and those goods which have the prescribed measures of control.

5. Conclusions

The growth of importance of air transport is noted on a global level. The investments in infrastructure are greater and more expressed. All stated is followed by a serious information and technological advancement without which modern logistics business cannot be efficient and effective to perform. The development of technique and technology also plays a significant role in securing the uninterrupted performance of activities within the logistics chain. The decreased number of missed flights and their delay achieve savings in time and costs, while the adequate usage leads to the decrease of material losses and damages. By decrease of costs the competitiveness of this type of transport is also achieved.

References

[1] Božić V and Aćimović S 2014 *Marketing Logistika* (Belgrade: Publishing center of the Faculty of Economics, Belgrade University)

[2] Hoel L A, Giuliano G and Meyer M D 2016 *Intermodal Transportation: Moving Freight in a Global Economy* (Washington DC: Eno Transportation Foundation)

[3] Kiso F and Deljanin A 2009 Air freight and logistics services *Promet - Traffic &Transportation* **21** pp 291-298

[4] Drljača M 2017 *Int. Conf. on Traffic Development, Logistics & Sustainability Transport (Opatia)* (Zagreb: Faculty of Transport and Traffic Sciences, University of Zagreb) pp 81-88

[5] Bernala M P, Blascob S V, dé Pellicerc E L and Gonzálezd R S 2012 Optimization of the air cargo supply chain *Journal of Airline and Airport Management* **2** pp 101-123
[6] http://www.transcocargo.com.au/blog/the-process-of-airfreight/, accessed on 08/06/2019
[7] https://www.tc.gc.ca/eng/corporate-services/des-reports-907.htm, accessed on 08/06/2019
[8] Tont D M and Tont M D 2016 An overview of innovation sources in SMEs *Oradea Journal of Business and Economics* **1** pp 58-67
[9] Denić N, Marković S, Spasić B and Milić M 2014 Identification of influential factors of project implementation information systems *Fascicle of Management and Technological Engineering* **23** pp 123-126
[10] https://www.iata.org/whatwedo/cargo/stb/Pages/index.aspx, accessed on 10/06/2019
[11] Đorđević A, Nestić S, Stefanović M, Tadić D, Arsovski S, Doljanica S and Mišić M 2015 *9th Int. Quality Conf.* (Kragujevac: Center for quality, Faculty of Engineering, University of Kragujevac)
[12] http://www.seweurodrive.com/produkt/produkt_sitemap.php, accessed on 11/06/2019
[13] https://www.webcargo.co/2019/06/12/webcargo-skys-launch-with-the-hype-turned-down-to-zero/, accessed on 11/06/2019
[14] https://www.unilode.com/digital-transformation/, accessed on 14/06/2019
[15] https://finance.belgium.be/en/faq/be-gate#q1, accessed on 14/06/2019

IOP Conf. Series: Materials Science and Engineering **898** (2020) 012051 doi:10.1088/1757-899X/898/1/012051

Opportunities, risks and challenges for actual key players of commodities production into the new wave of industry 4.0

L C Simion [1], S Avasilcai [1]

[1]Gheorghe.Asachi Techical University of Iasi,Dept. of Engineering and Management, Iasi, Romania

Abstract:The actual dymanic of the industrial wave created by the concept Industry 4.0 is streaming a serial consequences for the main market consumers which are the direct beneficiaries of comodities or daily consuming goods of food,automotive,gadgets and textile /garments industry.Upon the actual researches available from Aachen University and World bank statistics as Birnabaum report ,if many of the products which belong to food and textile are defined by standard parameters ,so far less dependable of human working factor as direct making of product,the ready made garments production which now is more to be confused with fashion industry,is crossing a period of high challenges to be solved, fighting with the risks to be shrinked until disappearance.The key players which are involved in the whole commodities industry could be defined as the final customers or consumers looking for a permanent variety of items(1),main suppliers of production means(2) and in between we find the producer(3).This one need to handle the pressure of low market prices, frequent economic downturns and motivation of their main resource which are the people either defined as „blue collars" or „white collars".

1 Introduction

The new wave of Industry 4.0 is looking for smart factories producers which are determined by high technology and relations under IoT(Internet of Things) and CRM (Customer relation management)concept which shall be considered as an important opportunity to get an improved quality and short cycle of making.Nevertheless in very few cases,the producers new concerns to get the needed „blue collars" are emphasized. Actually the border between the old „white and blue collars " is very thin and in the future most probably will disappear consequently by the nature of work inteligent parameters of digitalizing&smart factories.The ready garments industry is in a very risky position either will be melted to fashion manufacturing or will not be existing anylonger , trapped to the level of sweat shops of Asian countries doing simple products like the omni-present T-shirts or denim products no matter how „intelligent" fabric have inside .A certain start-up is recorded in Europe,but the road is long and the industry could be saved by integrated solutions with facilities for investments in education and payment systems which is revealing in fact another strategic partner, the government which shall be a vital stake holder of this long term country project.

The means & high tech tools for production have been attaining a high level of digitalization,but they are suitable to an industry of mass production with standardized good.On the other hand,the final client who must consume them,they are living into a market trend which attain the level of mass customization.Here,there is a gap which must be solved,because each of us want somenthing customized,but we don't want to leave the comfort zone as producers,wishing to continue a mass production of goods.

Perhaps the products having a content of standardized patterns in operations(as automotive,food, or technical textiles) are not too much concerned about the human resources involvement who are creating the goods,but a production of garmenting /fashionable goods need to solve as fast as possible the mentioned gap,besides an adapted motivation stimulus concept in order to preserve the cummulated know-how and to maintain the working skills of future generation of employees.The

IOP Conf. Series: Materials Science and Engineering **898** (2020) 012051 doi:10.1088/1757-899X/898/1/012051

statistics of the experts into the field show a slight improvement change of vision but the last figures of Birnbaum[1] report is not so optimistic about solving this challenge.However,an integrated country project must to start activating the political & governments input,besides creating international clusters and also to a lobby movement for industrial field revival.These ideas which have been written before March 2020,should be still considered on this very critical moment of human kind which are facing together globally as worlwide the pandemia of Covid 19.Obviously is a deep crisis not only for humanity but also for economy and sustainable factors , from which we have to learn our lessons and use this misfortunate event to prepare ourselves and extract the opportunities . The best example are the Far East asian countries such as South Coreea which due to previous experiences,succeded to read the signs,become pro-active in research and create alternatives,thus stronger in controlling the health and economy situation. The contribution of Industry 4.0 is different by a comparative analysis with Industry 2.0 and-3.0[2]. Industry 4.0 is neither a new form of technology,nor business ideal,but in fact an approach inspired by new advancements to achieve results that weren't possible 10 years ago. The first industrial revolution saw the landmark the british farmers moving from land work to factory production in the 19th Century. Shall be considered as well that this step was done with a textile weaving machine which was the revolutionary equipment at that time. The second wave spanned the period from the 1850s to World War I from steam energy to steel industrial processing,marking then a new step of revolution with electric power as a source of energy and manufacturing at the size of first mass production.The third industrial revolution was drawing a clear landmarks reffering the nuclear power,developping a break-through era of modern elctronic tehnologies and the culture of total quality production with the first steps of information technology used for globalizing the communication.This period from 1950 to 1970'is marked as well by climbing to new industrial fields as bio-technology and droppring others as poluting extracting industry(i.e.the recession of mining industry in UK of 80's).The actual fourth,wave then the leap owards digitisation. Industry 4.0 introduce the paradigm of IoT (Internet of Things) and the entity of cyber-physical systems ,using the infrastructure of sensors to collect high volume of data which are converted into information and commands able to improve the manufacturing process with fast feed back for further fastdecisions.

Smart plants, which will be the key engine of Industry 4.0, could take the benefit of information provided in real time to revolutionize the processes by both automation and digitisation. It means the machines shall be able of self-optimisation,and even more of self-configuration to achieve the received tasks at superior cost efficiencies and better quality goods or services.The first step of Industry 4.0 is different compare to the previous waves as it began as a concept defined in Germany by an official public report in 2011,known to be as one of the first times that 'Industry 4.0' was defined.A revolutionary idea outlined the commitment of industrial revival on hig-tech basis,whereas the manufacturing operations could be performed fully automated,without need of human intervention.The concept was world wide announced as political statement by Chanceller Angela Merkel,January 2015 at the World Economic Forum in Davos,where she was calling Industrie 4.0[4] the way,quote"to deal quickly with the fusion of the online world and the world of industrial production." The following figures are quite old,which have been increased from 2015, the start-up was going to €200 million (around £146 million,$216 million,or AU$278 million)as German government is investing to encourage research across academia, business and government, and beside Germany,the United State shared the same visionar strategy.USA has developed a non-profit organization under the name of Smart Manufacturing Leadership Coalition (SMLC),gather producers,educational institutions,machines suppliers having the same goal of progress towards Industry.4.0.Here the examples are found more in the private business envioronment,not defined as official strattegical policy.The aim has been defined of creating the smart manufacturing coordinated through an informational network which then create the access to modelling technologies with potential of customization to meet particular needs. One of the crucial chalenge is the need of qualified labor force to develop the services industry and the future need to re-vitalise the industrial sector,which actually is „outsourced" spreaded worlwide.Certain figures report from KPMG has evaluated the component and non-durable(commodities)markets of Industry 4.0 to be worth more than

US$4 trillion by 2020.This shows a result much beyond the initial forecast value of the Internet of Things (IoT) market,which Gartner has calculated to br around US$3.7 trillion by 2020.Business entrepenorial associations were able to persuade the USA goverment involvement to be able to reach important benefit from sharing the Industry 4.0 opportunities,with a recent government report stating the american manufacturing sector shall be around US 445 billion and create around 175,000 jobs.The value of Industry 4.0 comes from improving productivity and removing inefficiency at all levels,with cost reductions from using real-time data.Regarding the strategies into this field,in Romania,unfortunatelly the news are not very promising,so far only several magazines were mentioning about what is happening in the western developped countries of EU,a special master discipline at Technical University Bucharest and the interest of local branches of Siemens,Bosch,FESTO,orVodafone.The last ones have automatically bearing the strategies of headquarter companies,which slowly trigger the movement and start the "engine" of concern related to Industry 4.0 in Romania.In one of his column written in 2016 , the professor Dorel Banabic[2] was drawing the attention that Romania is not yet prepared to face the challenges if the 4-th industrial revolution,which already started since 2011,but is not yet reflected into romanian reality.Unfortunately from the moment of these reviews, not too much progress was recorded and consequently the production of commodities is suffering deeply from this lack of interest either from antrepenorial side or government behalf.According to McKinsey study this "wait and see" attitude poses a high risk and it has to do with a number of implementation barriers faced by the manufacturers with limited progress in Industry 4.0.The business environment as described above could not marching always the working frame for industry of commodities,focused here on textile and garment industry,more limited for invenstments in such development project,risking to certain extend to perish if persist in this non-action status.For instance ,Romania was for more than 30 years one of the most reliable producer into the field of textile and garmenting, recognized as well as one of the most traditional exporter of goods to Europe,USA and worldwide.Later,during the 90's decade,still the garment industry was supporting the export income of the country,well known for the expertise and flexibility into the field.More than that, the medium and high technical education was developed and create a tradition as well,nourished also from the know-how of several antrepeneuors which have the courage and determination to provide production for well-known brands with value added and full product service.The knowledge, experience and expertise gathered along the years could be considered as strengths and competitive advantage among the rest of European countries on top of flexibility and quick response to the market, worthwhile to be continued and integrated into the portfolio of main industrial sectors which could be supported and invested in.Perhaps is not too late to save at least the segment of garmenting/ready to wear products,as the textile industry(the initial link of supply chain)was almost disappearing in Romania.As long as the garmenting has succeeded to survive despite these difficult premises without supplies from Romanian textile enterprises,proved that manufacturers found creative solutions to stand-up and face the difficulties of global economic crises and the lack of financial support/facilities of government.Therefore, the technologies of Industry 4.0 shall be a great opportunity to be identified, applied and customized for this particular segment,in order to increase the productivity and replace the aging labor force and create the interest for the young generation.

2 Researching methods and results.

2.1Preliminary analysis with updated global changes.

The actual paper work was started before the actual drama which we are living globally on the planet, whereas the objective was to draw the attention to the need of re-vitalise and sustainability of romanian textile & garments industry, by integration and adapting into the new era of Industry 4.0, at least to the same level of other European countries.At this very difficult moment recorded on March 2020 of pandemic crisis, the mankind need this industrial field as one of the vital engine, in order to provide the heath care and protections aids long time ignored as masks,protections suits, gloves and other medical items.Before any profit,big international brands holdings as Armani,Louis Vitton or

Balenciaga prove a hign ethical spirit by helping the community,besides the flexibility in practical actions .The old dusty sewing machine/kinitting & weaving loom could be turned into a valuable asset over night.If these important European companies are supported by governments with financial facilities or by the previous results of research & educational investements, in Romania , the textile & garments units are fighting with bear hands,having the benefit of „zero" support,to find resources or logistic guidance no matter if is public institution for homologation of materials or establishing standards of testing the technical/medical products.There is a general mobilisation of antrepenerurs to convert their production from mass or fashion wear into specialized items for medical purpose but is very difficult to bring a consistent assistance,if is missing a clear coordinate strategy at the level of official institutions,testing labs and approved materials to be used.Despite all impediments,still the local antrepeneors are fighting to succeed.By saying this,another argument could be reveal that this industry deserve attention and massive support to be sustained as one the main player for the society.The investment will pay back later ,this field of activity could contribute to the national income,create jobs and stabilize the migration.

2.2 Benchmarks of textile & garmenting industry in the actual environment Industry 4.0
Initially the purpose of the study was to identify the benchmarks of the field to be compared versus actual status of romanian side.The purpose remain the same just the arguments are updated by taking into account the last consequences of pandemic crisis,showing the flexibility of the field to adapt towards the events and the need of this industry into the proximity of each country (outsourcing become suddenly not a good idea).Thus changing the paradigm and the approach of Industry 4.0 of switching by any price to cyberentities and Inthernet of things are sometimes not working in big moments of history .Therefore the pressure of reducing the working cost from operating poor human factor must be transffered to real sources of loss and ineffciency, one of them might be bureaucracy apparatus and army of desk officers living in a virtual reality .Germany's textile industry play a significant role in the digital transformation process of the German industrial sector in general as it is a future key supplier and adopter of digital operations technologies as well as Industry 4.0 solutions.

The industry provides key technologies, especially smart textiles,required for a successful transformation.Smart textile products hold large growth potential.According to market forecasts,the global market for smart textiles will grow from USD 0.1 billion in 2015 to about USD 3 billion by 2026.In order to produce smart textile products in large series or mass production scale and to unlock this potential,the hybrid and typically highly fragmented textile value chains will have to become increasingly interconnected.Germany's textile industry consists of about 700 predominantly small and medium-sized enterprises employing about 63,000 workers in total. Together they generated total revenue of EUR 11.58 billion in 2014.Additionally, there are numerous companies from adjacent sectors handling and processing textile products such as textile machinery, automotive suppliers, and synthetic fiber manufactures.In the past few years, German companies have increasingly focused on manufacturing high-quality and research- intensive technical textiles. As a result, they successfully strengthened their global leadership position while facing severe challenges resulting from increased cost and innovation pressure, ever shorter product and innovation cycles as well as increased customer demands. Today, Germany is the global market leader for technical textiles with annual revenue of about EUR 13 billion [2].Regarding the trend and driver for textile industry in Italy, in September 2016, the Italian Ministry of Economic development presented the *Piano Nazionale Industria 4.0*(the National Industry 4.0Strategy),which provides concrete measures to support enterprises in favor of trigger investments in research and innovation in this area.The strategic decisions to be implement at corporate level,especially when the company intends to be innovative,should be evaluated against the market macro trends.The megatrends act as a long-term development driver that impacts the business,the economy,the society,the cultures and the lifestyle.They originate from the inevitable global changes shaping the world-to-be,defining a set of needs to be answered that cannot be influenced by single entities, groups of companies or even countries.Hence,the business strategy for innovation and development must consider these trend analyses and their involvement in the business. The trend analysis and the drivers of the textile industry,focused on the end user, allowed identifying how the business of this area will be influenced by the textile machinery market trends.In order to satisfy the user needs,the hardest challenge and opportunity in developing new processes and business models in the textile industry is the speed of the production processes ("fast time to market" and "just in time production").The trend and driver analysis allowed identifying the strategic topics of innovation.The European textile and clothing industry is investing in response of the market macro trends. Technology gaps analysis made by ACIMIT with the support of RINA Consulting shows the following results[2].Hereby,specific examples from textile machinery companies that have already implemented a 4.0 approach for each of the three phases of the textile machinery production process are described.The observation between the current situation, represented by the considered sample of companies and the future perspective, represented by the roadmap,allowed to define what are the missing steps that a medium-sized textile machinery company need to implement in order to enable the Industry 4.0 scenario.For each step, the "technology gap" regarding hardware,skills and infrastructures has been evaluated.A feasibility study for the implementation of the identified enabling technologies has been defined for each phase:one star means that the gap is high;five stars represent a narrow gap.It appears that the production and the use and maintenance phases are organized to adopt and to implement the Industry 4.0 technologies.IoT systems and Artificial Intelligence algorithms are some practical examples for remote monitoring and predictive maintenance applications. The design and planning phase, instead, is far from the Industry 4.0 objective. Virtual reality and machine component simulation represent the first steps of innovation in this primary-phase.

2.3Strengths,weaknesses,opportunities,threats(SWOT)analysis

Industry 4.0 is focused on maximizing the synergies and the balance between the three key models of an enterprise:business,organizational and technological.The first two are related to intangible assets;while the third one represents the tool enabling the company to start an innovative process.In order to adopt an Industry 4.0 strategy,the use of new technologies alone is not enough.A cultural change in the company structure is needed by renovating the strategies the enterprise organization and involving people with precise skills.One of the greatest achievement of Industry 4.0 is converting any item into data which by their transforming to information allow real-time decision[5]for business.Furthemore,the next level of outcome is a high level of flexibility in production customization with superior quality,assuring the optimization of productivity for consuming market demanding quick response.

Beside the optimization of production parameters,the level of communication among various levels of teams and management is evolving to share a common goal,involvement and responsibility as innovation become as state of mind,shaping from the traditional pyramid organization model to a flatten trapezoid.From an antrepenorial point of view, applying a managerial tool as a S.W.O.T.analiysis (Strengths,Weaknesses,Opportunities,Threats)will draw out the advantages and limits of the existing business model versus the new one improved by embracing the vision of new productive model of Industry tech 4.0.

Here below, the main features of the benchmarking for textile and garmenting industry are briefly described,taking as basis the examples provided by Germany and Italy(see table 1). Following these examples,a SWOT(strengths,weaknesses,opportunities,threats)is accomplished in order to identify the best strategies to proceed for(diagram 1).The weakness of manufacturers are focused to be solved,by taking the appropiate elements from external business environment such as opportunities and from internal organization strong points.The unfavorable threats from external environment must be scouted permanently to read the sign of economic alert and look for protection measures both for the enterprise and the other entities of the supply chain,especially the actual economic context become extremely volatile,therefore is mandatory for special actions on planning and forecast in the future:

IOP Conf. Series: Materials Science and Engineering **898** (2020) 012051 doi:10.1088/1757-899X/898/1/012051

Table 1. Benchmarking –textile /garmenting industry overall for Germany and Italy

Benchmarking phase 1	Benchmarking phase 2	Benchmarking phase 3
Performance measurement of a component during the use phase in order to improve the component design, throughout the mechanical simulation to identify new geometries and materials.	The use of embedded informatics of different production sections.	The utilization of intelligent management platforms with mobile devices for the monitoring of the work and the maintenance if needed.
The application of virtual reality technologies for machine and components design,combined with simulation software that enable multi-user both local and remote collaboration.	Collaborative robot(cobot) integrated in the industrial network.	Sensors embedded on the machine for data acquisition that permits the machine and product monitoring and the communication with other divisions.
The 3D printing used to create prototype rapidly enables an agile collaboration between different productive sections and allows a new flexible and personalized design	Example of a robot capable to interact with each other in order to optimize the transportation inside the company.	Predictive maintenance to optimize the maintenance and reduce costs.
	Advanced mobile video technologies to facilitate the collaboration and the sharing information between different productive divisions.	Augmented reality applications to support the operator during the maintenance.
		Quality control system combined with gesture recognition technology
Feasability level : 1-5 points (1-big gap/5 –thin gap; 2.5 points	Feasability level : 1-5 points (1-big gap/5 – thin gap; 3 points	Feasability level : 1-5 points (1-big gap/5 – thin gap; 3.5 points

STRENGTHS:

>Speed that can be achieved through the Industry 4.0 plan

>Flexibility of the process and the product introduced by Industry 4.0;

>Domain knowledge:specific expertise and knowledge the company has into;
>Quality:for associated companies products is the starting point to be competitive.

>Quick response/Proximity location machinery production

WEAKNESSES :

>Long chain supply of vertical fields :from yarn to ready made product till shop selling point;

>Training:insufficient academic preparation,a training inside the corporation is needed;

>Standards: the lack of a communication in standard that allows the interoperability between different links of vertical supply chain ;

OPPORTUNITIES:

>Trend and driver of the textile industry: fast fashion, low cost;

>Financial contribution issued by government programms, EU.

>New business marketing models :package of equipments with training and service included;

>new industry transformation trend sponsored by the Industry 4.0 concept

THREATS:

>Time:the market demand time is really FAST so the company organization must be revised in order to supply this trend;
>Lead time:in order to deliver the product quickly to the client, the companies must face off the needed timing to find the required components from the vendors;

>Data security: the hacking is the main issues related to the data acquisition and storage;
>Privacy and data propriety: little expertise, also from the legal point of view

Figure 1. S.W.O.T.Analysis Diagram–textile /garmenting industry overall

2.4 Identifying the position of Romanian textile and garments industry-comparison to benchmarks in SWOT diagram

The analysis will be done following the same methods of evaluation of benchmarks with stars (1-big gap,5-thin gap),only the range of criteriae was enlarged to better emphasize the issues which need immediate solutions ,onto table 2.

Table 2.- Benchmarking analysis for textile/garmenting industry in Romania

BENCHMARK CRITERIAE	TEXTILE UNITS – Romania	GARMENTS UNITS-Romania
Investement programs/Facilities offered by government	No available data	No available data
Feasability level : 1-5 stars(1-big gap/5 –thin gap*	★	★
Education institutions specialized on the trade	-No available data about vocational schools(the old ones were closed); -there is 1 University	-Several high school/vocational schools located in central side of country where are the productions units; -2 Universities
Feasability level : 1-5 stars(1-big gap/5 –thin gap*	★	★★★
Training programs	Internal –developed inside the companies	Internal –developed inside the companies
Feasability level : 1-5 stars(1-big gap/5 –thin gap*	★	★ ★
Local companies building equipments (robots, automates)	No available data	Several representative of german & italian equipments producing spare parts & technical support
Feasability level : 1-5 stars(1-big gap/5 –thin gap*	★	★ ★
Local companies profiled on IT for the industry	No available data	2 units (Gemini and Datas)profiled for CAD and software for real time production control
Feasability level : 1-5 stars(1-big gap/5 –thin gap*	★	★★★
Qualified labor force(middle management, tehniciens, engineers)	Limited only in 2-3 units which are still on production.	Medium number in availabily with high qualification & versatile skills
Feasability level : 1-5 stars(1-big gap/5 –thin gap*	★	★ ★ ★ ★
Qualified labor force (blue collars operating equipment)	Limited only in 2-3 units which are still on production	Medium number in availabily , high qualification & versatile skills
Feasability level : 1-5 stars(1-big gap/5 –thin gap*	★	★ ★ ★ ★
Age of labor force	Around 50 years, aging more as no longer interest /motivation for the profile, due to low payment	Around 40- 50, aging more as no longer interest/motivation for the profile due to low payment
Feasability level : 1-5 stars(1-big gap/5 –thin gap*	★	★ ★
Existing production units investing in future trend of Industry 4.0.	Insignificant ,due to reduced number of production units	Moderate level , due to high prices of equipments and production structure (hectic orders, fast/high fashion)
Feasability level : 1-5 stars(1-big gap/5 –thin gap*	★	★ ★

The above evaluation ,which contain a personal research of data from own experience and practiceof the past 30years,show a not promising image at the first sigth,where the textile field was almost perished along the years and garmenting still existing by certain sporadic foreigngroups investements which was developing the available labor force and also the know-how and existing expertise of technical skilled „white & blue collars".Besides the big gap noticed onto the existing textile domain on all the criteriae,the 2-nd big gap is remarked on the investement programs and facilities offered by government which prove once again the comments made in the previous chapter .The garmenting sector which is functioning better than textile will be more efficient by having a strong and complete supply chain at the level of textile /knitting as source of raw materials which fulfill two lean criteria(proximity for reducing the lead time and spare alternatives for versatlity of customized fashionable production) .Altogether could provide a larger range of production and services for the local community and extending the export trade,enable a larger contribution for national budget.The next diagram(diagram 2) of SWOT analysis focused on Romanian industry of textile/garmenting is aiming to extract the main features which could be solved on short term and most important those element which need long term investment and involvement of all stakeholders (manufacturers, government and schools) .

STRENGTHS: >Reliability: fulifil dead lines in front of customers,sometimes taking over the losses determined by delays of previous links of vertical supply chain >Flexibility: achieved from the experiences with multi-products and complexity of products; >Domain knowledge:re-build the services of full garments, which was reduced in 90's decade to CM lohn. >Quality: obtained by working for premium brands with high standards ; >Quick response to clients:offer positive and constructive solutions for clients demands; >Qualified working force for product development(more than 50 % of existing units doing prototypes and sales collections);	**WEAKNESSES:** >Long chain supply of vertical fields from yarn to ready-made product till shop selling point; >Missing the link of textile local manufacturer >Training:insufficient vocational schools/academic preparation is needed,besides training inside the corporations; >Aging labor force:difficult to recruit fresh employees for generations turnover; >Work nature with most content of human interaction:50 % of time value is depending on sewing which is limited to digitalization; >Low value of product:put permanent pressure on cost with small margins for profit
OPPORTUNITIES: >Alternative products request on the market:besides the usual ready garments, technical/medical/army special products could be on demand; >Financial contribution released by EU programms. >New business marketing models: package of equipments with training and service included; >New industry transformation trend sponsored by the Industry 4.0 concept >IT industry in progress to achieve customized software for product development and production control.	**THREATS:** >Market volatility:high sensitivity to events, seasons,fashion trends changes; >Low level of payment for labor force:influenced by the low prices; >Time: the market demand time is really FAST so the company organization must be revised in order to supply this trend; >Lead time: in order to deliver the product quickly to the client, the companies must face off the needed timing to find the required components from the vendors; >Data security: the hacking is the main issues related to the data acquisition and storage; >Privacy and data propriety: little expertise, also from the legal point of view

Figure 2. SWOT analysis for textile/garmenting industry in Romanian

3 Conclusions and actions .

3.1Conclusions

So far all the achieved data and research results bring us to the main conclusion that there is a potential to be developed in thisa trade ,which was emphasized by the strengths items and on the other hand there are quite many obstacles to pass (the weackness items).For the public which is has just few info about this industry domain,this study look not very different than for the rest of productions fields, suffering more or less from the same matters.In such case,probably a clear direction should be a unified and strong Think -Thank group,quoting as well Dorin Banabic,2016,Magazine of science and sciencemeter,proposing an agenda for Industry 4.0 in Romania,involving the team of technical universities, representatives of business groups and exporters ,IT and oficials which could make lobby to EU Economic Comission in order to get a financial support from the assigned budgets. The main consequence of this approach will be the development of an intelligent manufacturing environment having the capability of communicating and making optimal decisions in an autonomous manner. In order to achieve such an ambitious goal,from the existing info EU could have available more than 1,300 billion EURO in the next 15 years.The major economic and political challenge consists in allowing all industrial domains to take advantage from the digital innovation in products, manufacturing processes and business models.(Dorin Banabic,2016,Magazine of science and sciencemeter).

3.2Potential-actions

A realistic approach in such case is neccesary, in order to see what could be done on short term at the level of manufacturers group and then by a long term at the level of Think-Thank groups and lobby.

3.2.1On-short-term

- manufacturers could make analysis and simmulation plans for production convertible lines(multi products on a versatile line with cellular organization) or short-multi lines (multi products in small lines to generate flexible orders upon request).The opportunity of existing market should be exploited , taking into the possibility of making in parallel smart technical /medical products and value added complexreadywear .

This,,immediate"term is assuming a time horizon of 6 months- 1 year,because feasability study and budget must be configured for the projects involving product development, production layout set-up, needed equipment,personnel training and all the aggregate figures related to profit/loss,depreciation rates for the assets ,sources of financing.

3.2.2On-long-term

-a particular proposal could be a pilot project initiated at the level of partnership between Technical University –DIMA(design industrial and management), a cluster of 1-2 manufacturers in the field & equipment constructers, to create a smart factory model with a product development and set-up with digitalizing the planning,decisions and control of processes and as much as possible the manufacturing. As far as I researched,several examples were done before, in matter of initiative for the complete chain of production activities,one of them for Escada group in Slovenia, in 2013-2014 and another in Germany Aachen with Learning Textile factory 2017,in the next proximity of RWTH Aachen campus [4].So far, not so many haven't have the knowledge , budget and expertise to modernize the complete supply chain either for ready wear or textile .Plenty of examples could be found related to logistic and sales (ware-housing/selling points) or marketing and product development , but having only fragmented activities , this will not be sustainable for the industry for long term.Obviously an assembly approach is needed as previously mentioned but meanwhile costly.That's why association of multi-professional entities are reqeuested to be involved(manufacturers with practical know-how,university which could provide research means &

documenting data bases and also young enthusiastic students involvment,IT division for connections,infrastructure and maintenance,banks or government institutions for financial support).

References

[1] Birnbaum D 2020 *Research and data about Bangladesh garment industry* https://www.linkedin.com/feed/update/urn:li:activity:6618297778269454336/?commentUrn=urn%3Ali%3Aco mment%3A(activity%3A6618297778269454336%2C6618545260580737024)

[2] Online: https://www.techradar.com/news/what-is-industry-40-everything-you-need-to-know

[3] Banabic D 2016 Industry 4.0 started. Is it ready Romania for the challenges of this new revolution?*Magazine of science and scientometric policy,* **vol.5** 194-201

[4] Kusters D, Prass N, Gloy Y-S 2017 *Textile learning factory 4.0-Preparing Germany's textile industry for digital future* 1

[5] Kusters D, Prass N, Gloy Y-S 2017*Executive summary of Industry 4.0:the new challenge for the Italian textile machinery industry* 5-7

[6] Robleck V,Mesko M,Krapez A April-June 2016 *A complex view of Industry 4.0 Sage Open* 4-7

[7] Kuesters D et al 2017 *Procedia Manufacturing* 9 .

IOP Conf. Series: Materials Science and Engineering **898** (2020) 012052 doi:10.1088/1757-899X/898/1/012052

Frictional study of the polyamide/rubber contact materials

M T Lates[1] and C C Gavrila[1]

[1]Product Design, Mechatronics and Environment Department, "Transilvania" University of Brasov, Brasov, Romania

E-mail: latesmt@unitbv.ro

Abstract. The actually researches focused on finding technical solutions with ecologically friendly impact on the environment are include new materials which are characterised by low energy consumptions (loses) during their manufacturing process and their usage. One class of materials characterised by these energy loses conditions is represented by the polyamides. The study regarding the frictional loses of the polyamides being in mobile contacts with steel made spare parts show that they have a good tribological behaviour. The applications domains (mechanical transmissions, tap armatures, food industry, medicine) and the lack of the studies, gives the opportunity of the research in the field of the frictional behaviour of the polyamide/rubber mobile contacts. The paper presents tests regarding the frictional behaviour of the PA46 and PTFE added PA46 polyamide/rubber contacts under different loadings and relative speeds. The tests are performed on a tribometer equipped with a pin-on-disc module. Finally there are highlighted important conclusions regarding the applications and working conditions where the polyamide/rubber couple of materials may be used, in order to get small frictional energy loses.

1. Introduction

Actually solutions regarding the design of spare parts used in wide areas as engineering, medicine, entertainment industry, food and agriculture industry, include materials characterised by low energy consumptions and ecological impact with the environment while they are produced or used. The research of these materials is split, according to the materials typology, in two huge categories: research on metallic materials (steel alloys) and research non-metallic materials (plastics, rubbers, woods etc).

In the field of plastic materials characterised by ecological impact with the environment during their usage and production, the research follow mainly the area of polyamides. These materials, characterised by eco-friendly interactions with the environment during their manufacturing process, are used in combination with metallic made spare elements (in gear or chain drive transmissions, in mechanical transmissions with applications in medicine or food industry where the lubrication is not allowed [1, 2, 3]) or with non-metallic spare elements (polyamides or rubber) in the construction of mechanical transmissions (made from polyamides), in the construction of elastic couplings with rubber made elastic elements (figure 1 [4, 5]) or in polyamide made tap armatures which contain rubber made sealing devices (figure 2 [6]).

One of the criteria which are considered in order to claim that a material is an eco-friendly one, is referring to the energy loses caused by the frictional properties of it. Currently researches are focused on the direction of studying the frictional behaviour of materials; the aim is to find out materials or materials combinations with good frictional properties which assure small energy loses during their usage in combination with other materials.

a)

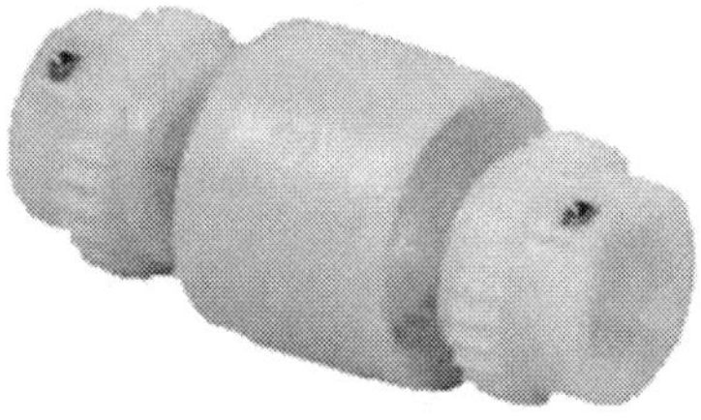

b)

Figure 1. Polyamide made couplings [4, 5].

The tribological researches of the polyamides show that these materials are characterised by small frictions in combination with other polyamides or with steel made elements, reduced wear and small variations of their properties with the increasing of the temperature [1, 2, 3, 7, 8, 9, 10].

Figure 2. Polyamide tap armature [6].

One of the studies regarding the frictional properties of the rubbers is referring on dry conditions for rubber-made elements being in contact with steel parts [11]. Depending on the amount of the carbon black content of the EPDM rubbers there are performed pin-on-plate and roller-on-plate tests (steel-made pin and roller and rubber-made plate) under different test conditions. The conclusion is that the tribological properties of the rubbers are optimised with the increasing of the amount of the carbon black. The friction and wear properties of the rubber elements in contact with surfaces with different

IOP Conf. Series: Materials Science and Engineering **898** (2020) 012052 doi:10.1088/1757-899X/898/1/012052

roughness's are studied in [12]. According to the literature, the rubbers friction has two components: the adhesive type friction and the hysteresis type friction.

The specific literature is weak in studies regarding the frictional behavior of polyamide on rubber type materials; due to the existence in techniques of the mechanical contacts between polyamide and rubber made elements (mechanical couplings, tap armatures etc) it is necessary to make studies regarding the friction of these technical solutions. According to this, the paper presents the frictional behavior of the polyamide/rubber made parts contacts under different loadings and speeds.

2. The tests

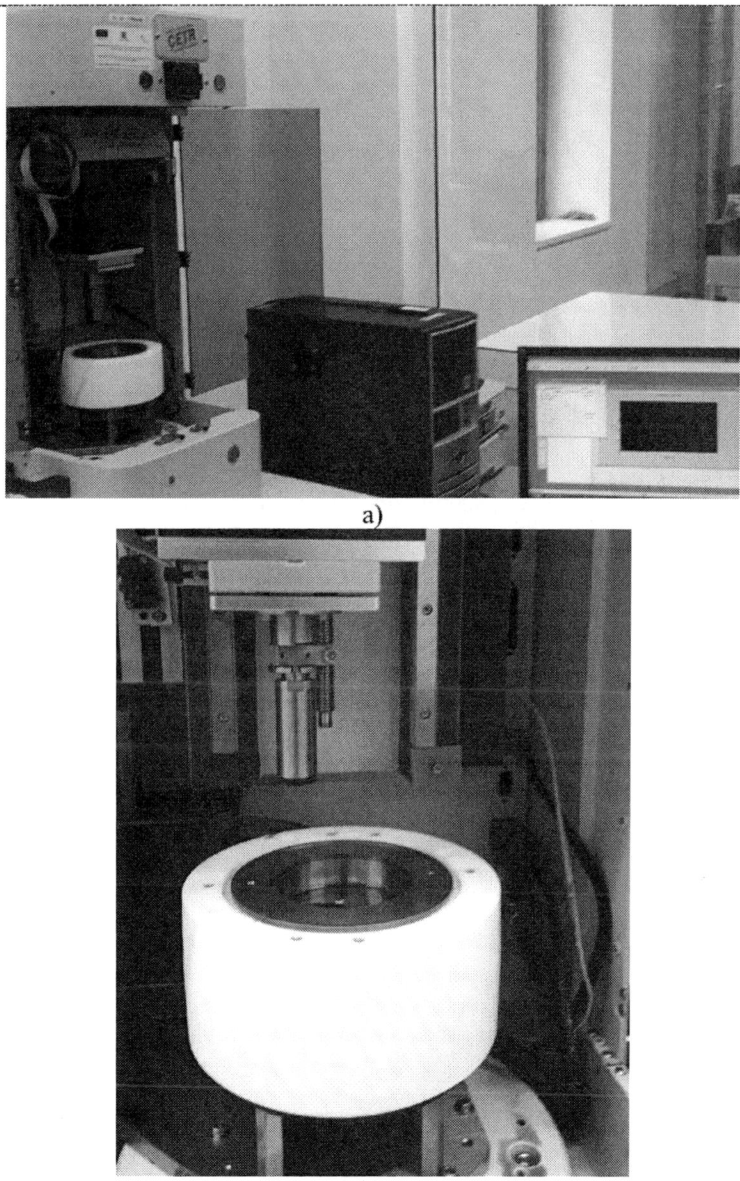

Figure 3. The tribometer with the pin-on-disc module.

The friction tests are performed on an UMT tribometer – figure 3 – equipped with a pin-on disc module which allows tribological tests with variable input parameters as: the normal force, the rotational speed of the disk, the temperature of the oil bath. The variation of the normal force and of the oil bath's temperature may be described continuously, under different variation curves. The values of the rotational speed may be set-up for discrete values.

The tests are performed for two type of materials being in contact: a rubber plate in contact with a P46 polyamide disk and a rubber plate in contact with a PTFE added PA46 polyamide disk (figure 4). The test parameters are represented by: a set of rotational speeds equal with: 500 rpm, 1000 rpm, 2000 rpm and 3000 rpm; the normal equal with:1 N ... 5 N. Each test is performed at the room temperature of 22 °C and has the duration equal with 10 minutes.

a) b)

Figure 4. The polyamide disks in contact with the rubber plate.

3. Results and conclusions

The results present the variation of the friction coefficient with the rotational speed of the disk and with the normal force.

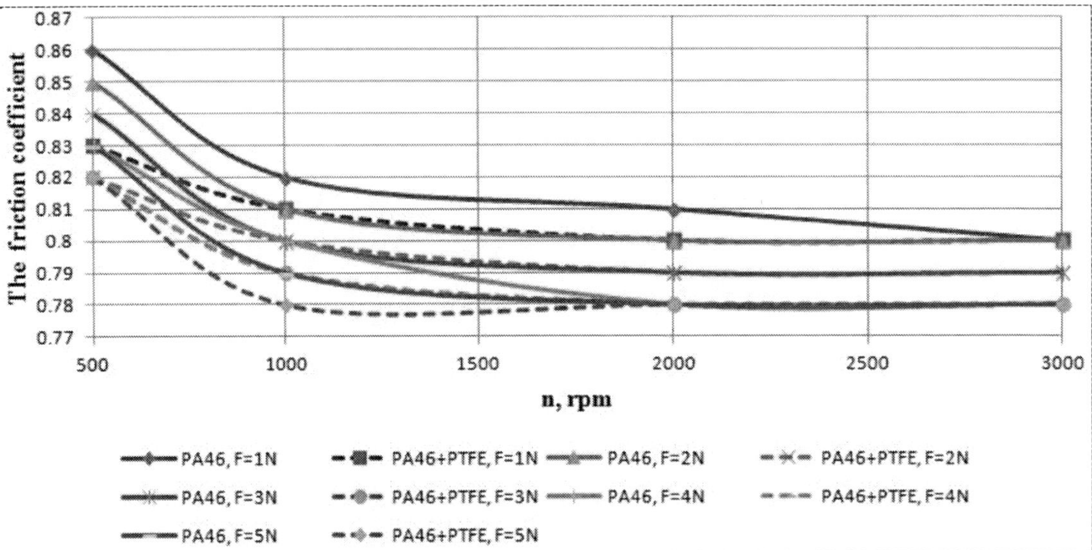

Figure 5. The variation of the friction coefficient with the speed, in the case rubber/PA46 and rubber/PTFE added PA46.

Figure 5 shows the variation of the friction coefficient with the rotational speed for different values of the normal force. The friction coefficient decreases with the increasing of the speed and has smaller values in the case of the contact between the rubber and the PTFE added PA44 polyamide than the contact between the rubber and the PA44 polyamide. After a speed of 2000 rpm, the value of the friction coefficient is stabilised and there are no differences between the rubber/PA46 and the rubber/PTFE added PA46 polyamides.

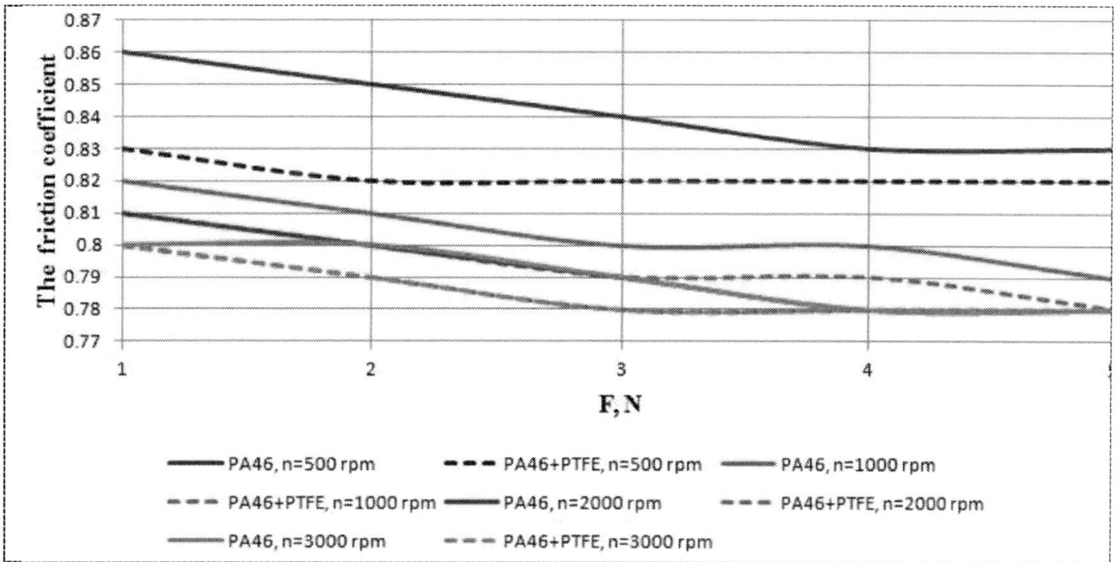

Figure 6. The variation of the friction coefficient with the normal force, in the case rubber/PA46 and rubber/PTFE added PA46.

Figure 6 presents the variation of the friction coefficient with the normal force for different values of the speed. The friction coefficient decreases with the increasing of the normal force and has smaller values in the case of the contact between the rubber and the PTFE added PA44 polyamide than the contact between the rubber and the PA44 polyamide. At high normal forces, the value of the friction coefficient is stabilised and there are small differences between the rubber/PA46 and the rubber/PTFE added PA46 polyamides.

The variation of the friction coefficient at small speeds and small forces can be explained due to the adhesive and hysteresis friction which is characterising the friction in rubber. For small speeds, the adhesive friction has a big influence in the amount of the friction coefficient.

In conclusion, the friction coefficient decreases with the increasing of the speed and of the normal force. In all the cases the friction coefficient has smaller values in the case of the contact between the rubber and the PTFE added PA44 polyamide than the contact between the rubber and the PA44 polyamide; smaller differences are noticed at high speeds and normal forces (where the adhesive friction in the rubber has a smaller influence on the total amount of the friction coefficient). According to these, it is suitable to use PTFE added PA46 polyamides in contact with rubber elements instead of PA46 polyamide at small speeds and normal forces.

References

[1] Hoskins T J, Dearn K D, Chen Y K and Kukureka S N 2014 The wear of PEEK in rolling – sliding contact – Simulation of polymer gear applications *Wear* **309** pp 35-42

[2] Zhaobin C, Tongsheng L, Yuliang Y, Xujun L and Renguo L 2004 Mechanical and tribological properties of PA/PPS blends *Wear* **257** pp 696-707

[3] Kumar S S and Kanagaraj G 2016 Investigation on mechanical and tribological behaviors of PA6 and graphite-reinforced PA6 polymer composites *Arab journal of science engineering* **41** pp 4347-4357

[4] https://www.slideshare.net/rahulgomez/samiflex-elastic-coupling Accessed: 23.03.2020

[5] http://www.acheson.cn/Product/9278533719.html Accessed: 23.03.2020

[6] https://www.industrystock.com/en/company/profile/Praher-Plastics-Austria-GmbH/29973 Accessed: 23.03.2020

[7] Shibata K, Yamaguchi T, Kishi M and Hokkirigawa K 2015 Friction and wear behavior of polyamide 66 composite filled with rice bran ceramics under a wide range of Pv values *Tribology Online* **10 (2)** pp 213-219

[8] Li J and Xia Y C 2010 The friction and wear properties of thermoplastic PA6 composites filled with carbon fiber *Journal of thermoplastic composite materials* **23** pp 337-349

[9] Haixia H, Sirong Y, Mingyu W and Kaixin L 2009 Tribological behavior of polyamide 66-based binary and ternary composites *Polymer engineering and science* **49** pp 2454-2458

[10] Fasahat F, Dastjerdi R, Mojtahedi M R M and Hoseini P 2015 Wear properties of high speed spun multi-component PA6 nanocompositefabrics; abrasion resistance mechanism of nanocomposites *Wear* **322** pp 17-25

[11] Felhos D and Karger-Kocsis J 2008 Tribological testing of peroxide-cured EPDM rubbers with different carbon black contents under dry sliding conditions against steel *Tribology International* **41 (5)** pp 404-415

[12] Persson B N J and Tossatti E 2000 Qualitative theory of rubber friction and wear *Journal of Chemical Physics* **112 (4)** pp 2021-2029

AUTHOR INDEX